Lecture Notes in
Computer Science

Edited by G. Goos and J. Hartmanis



279

Joseph H. Fasel Robert M. Keller (Eds.)

# Graph Reduction

Proceedings of a Workshop
Santa Fé, New Mexico, USA
September 29–October 1, 1986

Springer-Verlag

Berlin Heidelberg New York London Paris Tokyo

**Editors**

Joseph H. Fasel
University of California
Los Alamos National Laboratory
Los Alamos, New Mexico 87545, USA

Robert M. Keller
Quintus Computer Systems, Inc.
1310 Villa Street
Mountain View, California 94041, USA

This volume was edited under the auspices of the U.S. Department of Energy.

CR Subject Classification (1987): F.1.1, F.4.1, C.1, D.1.1, D.3.4

ISBN 3-540-18420-1 Springer-Verlag Berlin Heidelberg New York
ISBN 0-387-18420-1 Springer-Verlag New York Berlin Heidelberg



Printing and binding: Druckhaus Beltz, Hemsbach/Bergstr.
2145/3140-543210

# Preface

On September 29 through October 1, 1986, a workshop on graph reduction was held in Santa Fe, New Mexico, USA, sponsored by the Los Alamos National Laboratory of the University of California and Microelectronics and Computer Technology Corporation (MCC). 66 researchers from the United States, Canada, the United Kingdom, Sweden, Finland, the Netherlands, France, West Germany, and Australia participated in the workshop, which consisted of 28 presentations, combined with much informal discussion. The papers in the present volume are based on the presentations, and because the final versions were prepared after the workshop, they reflect some of the discussions, as well.

## Historical Comment

The term *graph reduction,* referring to an evaluation model for functional languages, was probably first used in the dissertation of Christopher Wadsworth (Oxford, 1971). Similar ideas had earlier been discussed in a thesis at M.I.T. (Suhas S. Patil, M.S. Thesis, *An abstract parallel processing system,* 1967). That numerous investigations of this concept have sprung forth attests to the concept's elegance and seductiveness. Witness the present collection of papers, which represents contributions from a sizeable, but incomplete, subset of the workers in this community.

## Rationale

Some of the benefits of graph reduction can be found in the following traits:

- A mathematically elegant denotational semantics
- Lazy evaluation, which avoids recomputation and makes programming with infinite data structures (such as streams) possible.
- A natural tasking model for fine-to-medium grain parallelism (see *Keller, Lindstrom, and Patil, A loosely-coupled applicative multiprocessing system, Proc. NCC 1979).*

## Brief Introduction to Graph Reduction

The concept of graph reduction can be illustrated by considering a functional expression, such as

$$((a + b) / (c * d)) - ((c * d) / (e + f)),$$

the graphical equivalent of which is shown in Figure 1.

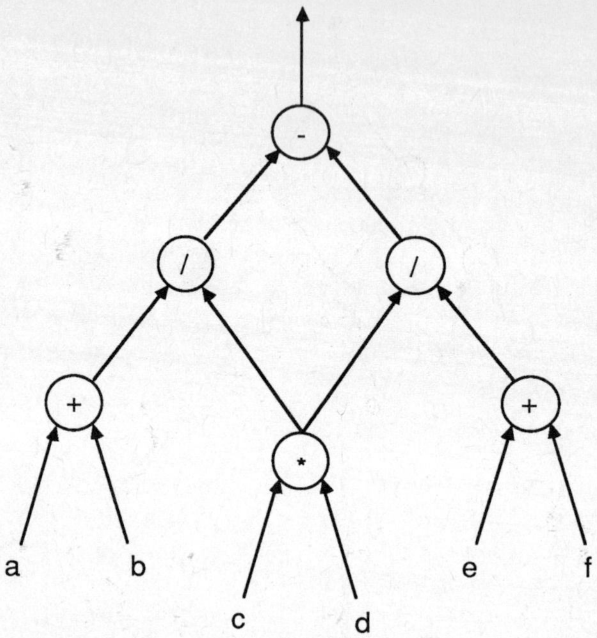


Figure 1

In the figure, the leaves *a, b, c, d, e, f* are assumed to represent values. This graph can be *reduced* by selecting any evaluable operator node; here, the evaluable nodes are those whose inputs are leaves. There are three such operators in the Figure. *Reduction* of such a node is represented by replacing the node with its value, representable symbolically as *(a + b)*, *(c * d)*, and *(e + f)* for each of the three choices in the Figure. Figure 2 shows the result of each of three possible reductions. In practice, only one reduction need be performed at a time, but the *independence* of the three reductions is indicative of the potential *parallelism* in the reduction process. Note also that the subgraph for *(c * d)* is shared, and thus need only be reduced once. This is an advantage of graph reduction, as opposed to string or tree reduction.

Figure 3 begins by showing the result of applying all three reductions, and then continues by applying other reductions, until the result is a single value, again represented symbolically.

In a practical system, we cannot be content with evaluating fixed graphs; we need to be able to build parameterized graphs and reduce them. For example, Figure 4 shows a graph representing the product of the integers from *m* through *n*.

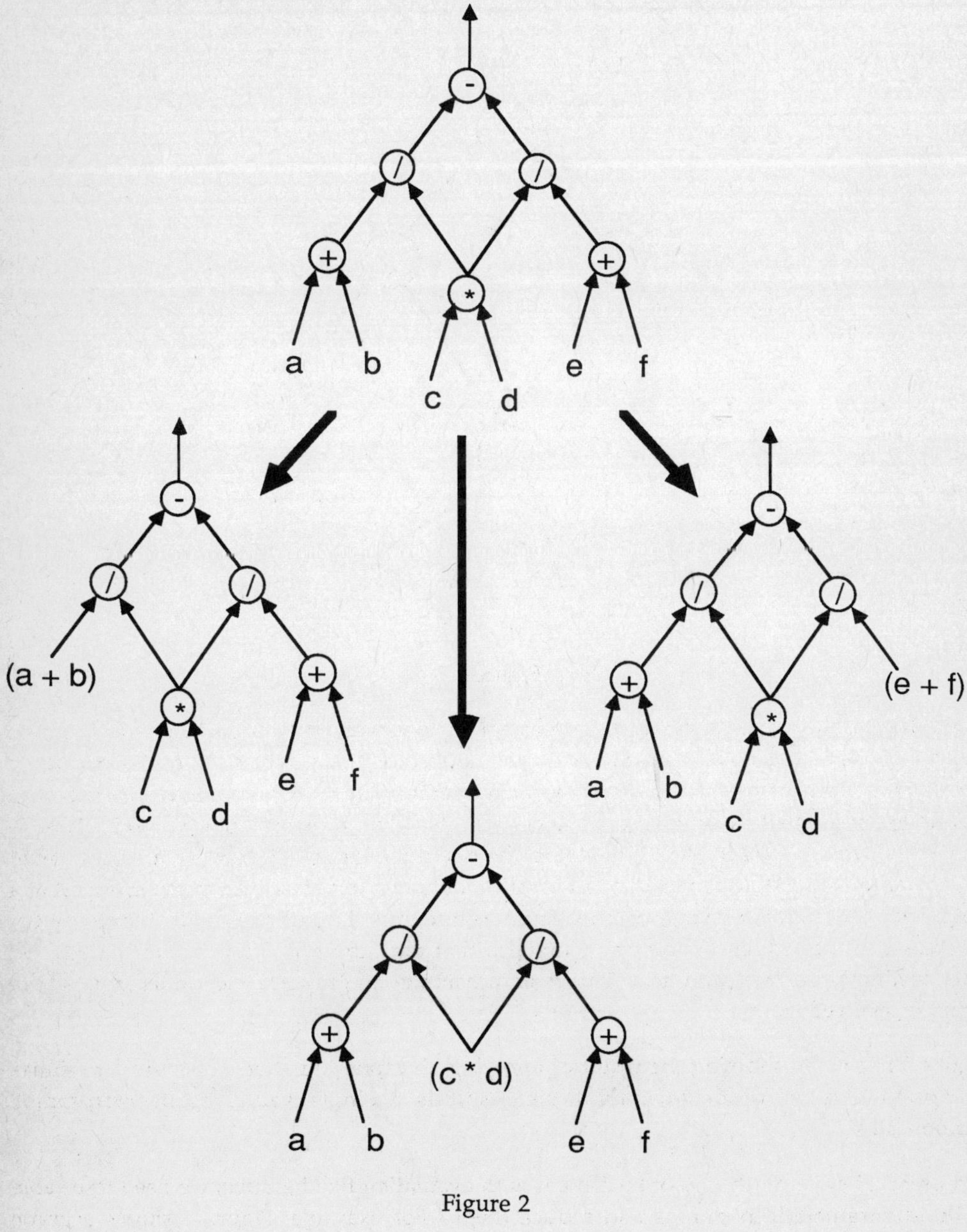


Figure 2

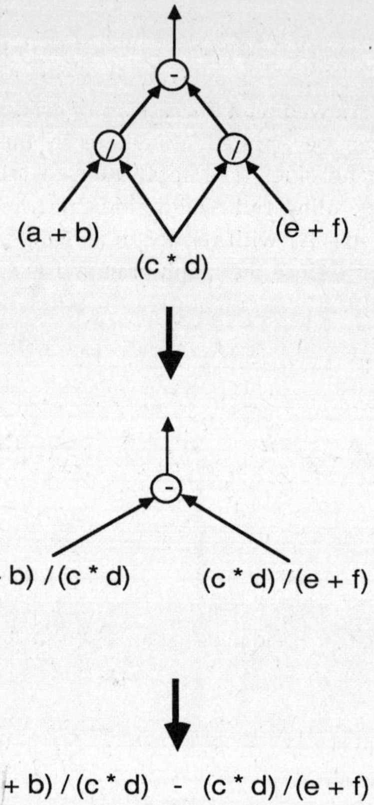


Figure 3

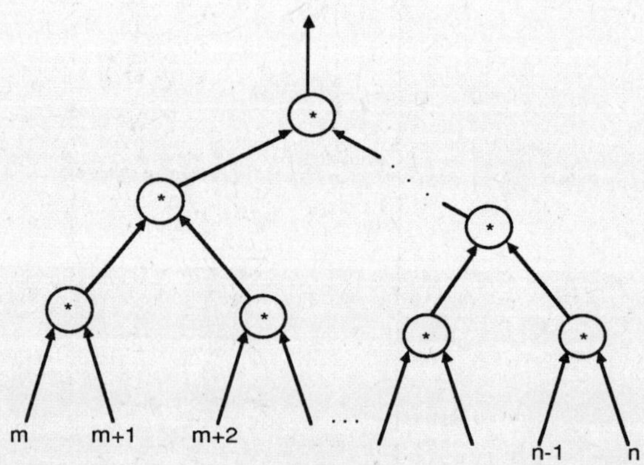


Figure 4

Suppose we wish to construct such a graph for given *m* and *n* and then reduce it. This may be accomplished with a *graph grammar production*, as shown in Figure 5. The idea here is that a node labelled *product* is to be *expanded* into a graph before any reduction. Again, the computation is viewed as a succession of transformations from one graph to another. A production can be applied anywhere in the graph that there is a node corresponding to its left-hand side. The application of this production entails replacing that node with the corresponding right-hand side, splicing in such a way that argument correspondence is preserved. As with reduction, expansion can take place in parallel at several sites within the graph, since the graph grammar is essentially "context-free".

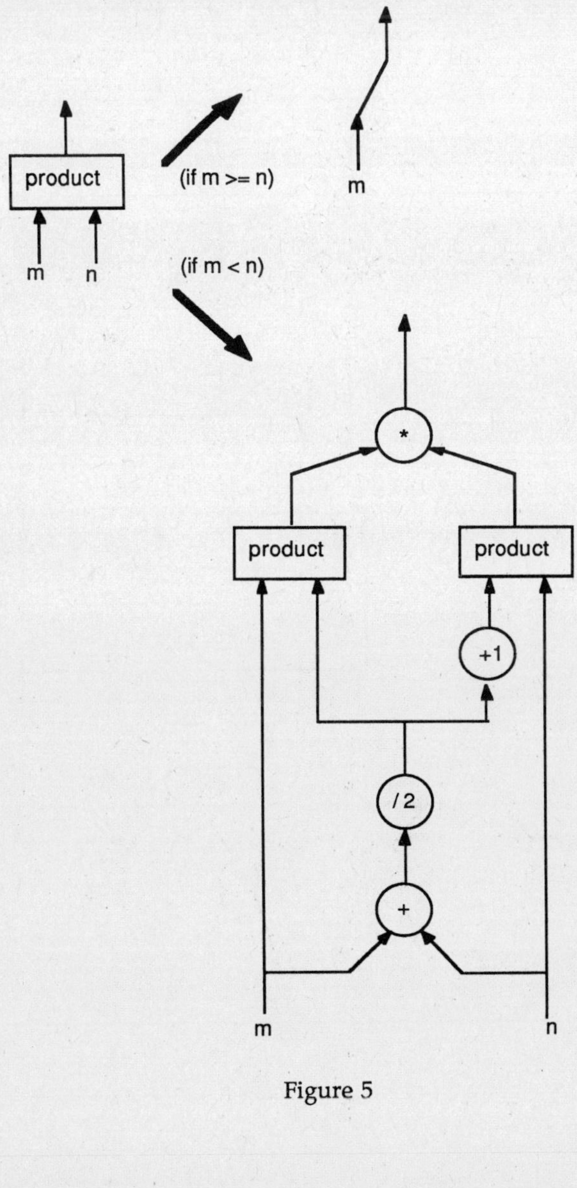


Figure 5

We show in Figure 5 two expansions, conditioned on the relation between *m* and *n*, that must be known before the corresponding right-hand side can be chosen. However, we can achieve the same effect with a single production by introducing a conditional (*if_then_else_* ) node, as shown in Figure 6.

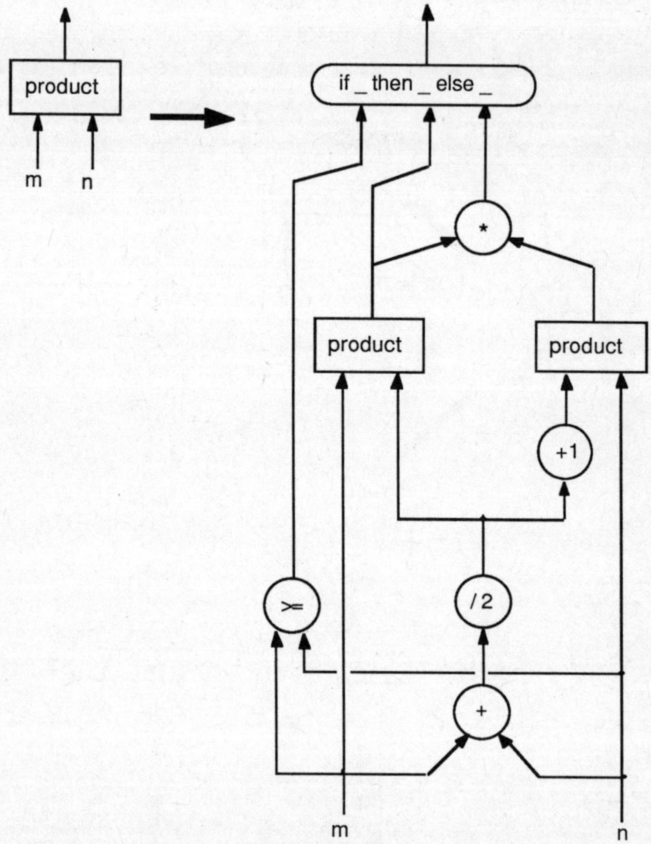


Figure 6

We assume that the conditional node is reduced *lazily*; that is, the *then* and *else* parts are not evaluated before the node is reduced. The principle of *lazy* (or *demand-driven*) evaluation is that no reduction or expansion is performed until its value is needed. This evaluation scheme has the benefit of avoiding redundant computation, and in particular, allows us to use infinite data structures, so long as only finite portions of them are ever demanded. For example, Figure 7 shows an expansion rule for the infinite list (or *stream*) of successive integers beginning with *n*.

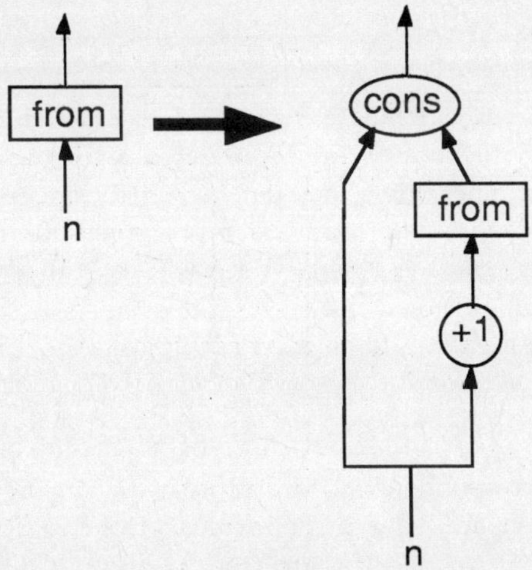


Figure 7

Figure 8 shows the configuration of the stream of integers from 0 after the first two elements have been demanded.

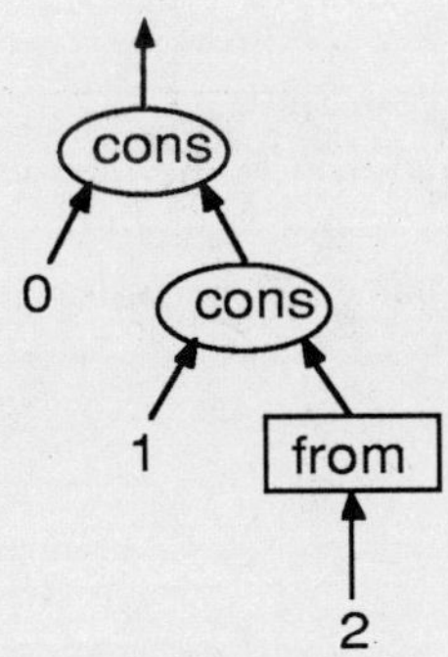


Figure 8

There is, of course, a connection between graph expansion and *beta-reduction* in the lambda calculus. The production itself corresponds to a lambda expression, and the position of the expanded node within the graph corresponds to its arguments, as in the following indication of function application:

(λ(*formals*).*body*) *actuals* --> *body*(with actuals substituted for formals)

Lazy evaluation of graphs corresponds to *normal-order reduction* in the lambda calculus, but need not suffer from the inefficiency associated with the latter, because multiple occurrences of a formal parameter in a lambda abstraction, which result in reevaluation of the corresponding actual parameter, can be represented in the equivalent graph by references to a shared subgraph (the actual), which will be evaluated at most once. It is also noteworthy that the graph representation eliminates the need for the traditional *alpha-reduction*, as the arcs corresponding to bound variable instances in the lambda calculus are anonymous in the graph calculus.

The connection between graph reduction and lambda calculus needs to be completed by mentioning briefly the correspondence to *free variables* in lambda expressions. These are variables that are expected not to get their values from arguments of the immediate lambda binding, but rather to inherit them from outside. This correspondence is best demonstrated by an alternate representation of the production, using an *encapsulated body*. Instead of a named node, which would have been expanded, we use an *apply* node, which expects an encapsulated body as one of its arguments, and which is defined to have a reduction rule resulting in the substitution of the body for the apply node. Figure 9 shows the original production in part (a), the result of beta reduction in part (b), and the equivalent reduction using an *apply* node in part (c).

Now we can use encapsulated bodies to show the equivalent of free variables. These are bodies with arc values entering from *outside* the node, as shown in Figure 10. These values are bound before expansion of the *apply* node, not during. Using this representation, techniques of the lambda calculus can be explained (see *R.M. Keller, Semantics and applications of function graphs, University of Utah 1980*).

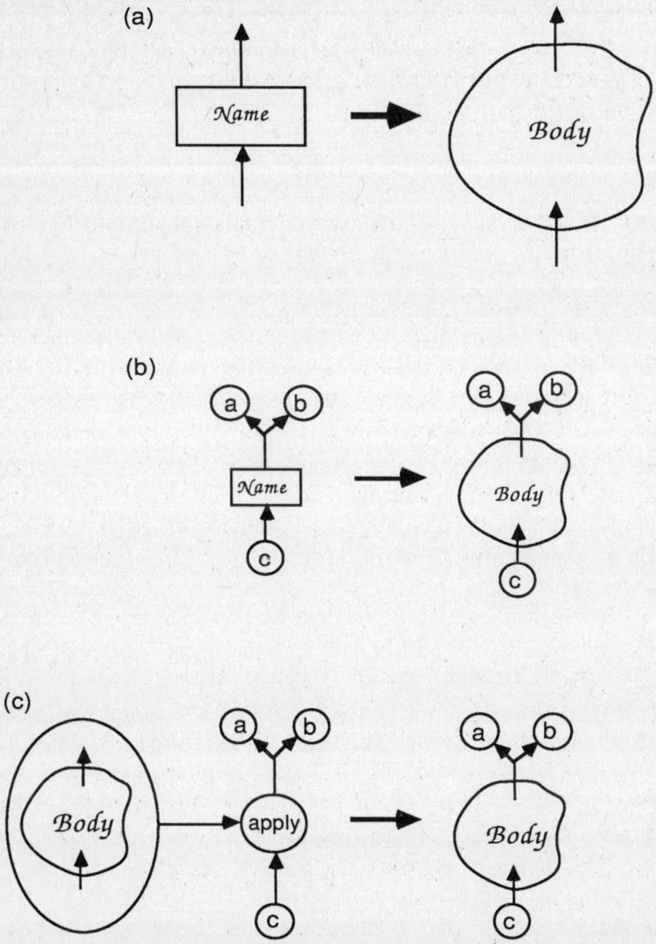


Figure 9

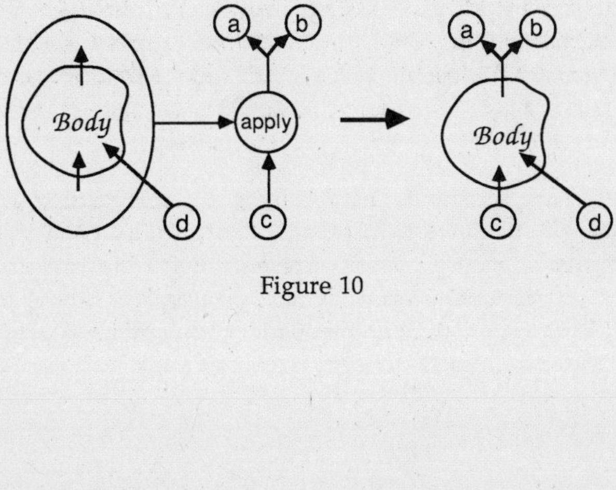


Figure 10

## Open Issues

It is scientifically worthwhile to examine issues such as

- whether graph reduction is the most efficient way to achieve the benefits mentioned in the opening paragraphs (*e.g.* can variants of the SECD model do just as well, or should these variants properly be viewed as instances of graph reduction)
- whether some of the benefits (*e.g.* natural tasking granularity) apply just as well to models that are not purely functional
- whether graph reduction can take on a wider meaning than as an evaluation model for the lambda-calculus, (*e.g.* in a computational symbolic algebra, where the data are *essentially* graphs)
- whether the graph reduction model limits the scope of programming model that can be used (*e.g.* can naturally represent *continuation-based* programming, on which a wide variety of control structures can be represented functionally)

## Contributed Papers

One can find in the papers contributed to the workshop concern for some of the issues mentioned above and related ones. We have grouped the papers into categories, as shown in the Table of Contents. In an area where there is significant interplay of concerns (efficiency, parallelism, language capability, *etc.*), no such categorization can be perfect. Thus, we have merely attempted a coarse grouping to aid ourselves in introducing the topics of concern in each of the papers.

In the category of models for graph reduction, we find papers exposing various views of graph reduction as a means for language implementation. Hans-Georg Oberhauser (*On the correspondence of lambda style reduction and combinator style reduction*) discusses the relationship between two styles of graph reduction, one based on beta reduction in the lambda calculus and the other based on a combinator calculus. Further connections along these lines are exposed by Klaus Berkling (*Head order reduction: A graph reduction scheme for the operational lambda calculus*) and Jon Fairbairn (*A simple abstract machine to execute supercombinators*). Joseph Goguen, Claude Kirchner, and José Meseguer (*Concurrent term rewriting as a model of computation*) discuss the techniques and motivation underlying their Rewrite-Rule Machine project.

An important set of issues surrounds the problems of implementing graph reduction on existing architectures. Benjamin Goldberg and Paul Hudak (*Alfalfa: Distributed graph reduction on a hypercube multiprocessor*) present their implementation work for a multiprocessor with distributed processors and memory. Randy Michelson, Lauren Smith, Elizabeth Williams, and Bonnie Yantis (*Parallel Graph Reduction on a Supercomputer: A Status Report*) present related work oriented toward existing supercomputer implementations. Thomas Johnsson (*Target code generation from G-machine code*) describes a method of functional language compilation using a graph-reduction model.

The graph reduction concept has inspired several architectures specially designed to exploit the parallelism it provides. The papers here reflect work as it has proceeded in three different countries toward such ends: Michel Castan, Guy Durrieu, Bernard Lecussan, Michel Lemaître, Allesandro Contessa, Eric Cousin, and Paulino Ng (*Toward the design of a parallel graph reduction machine: The MaRS project*), P.G. Harrison and M.J. Reeve (*The parallel graph reduction machine, Alice*), and Robert M. Keller, Jon W. Slater, and Kevin T. Likes (*Overview of Rediflow II Development*).

Along with such architectures come issues of how best to allocate physical resources, such as processors and memory. Included in this category are papers by M.C.J.D. van Eekelen and M.J. Plasmeijer (*Specification of reduction strategies in term rewriting systems*), F. Warren Burton (*Controlling reduction partial order in functional parallel programs*), Ashoke Deb (Parallel garbage collection for graph machines), and Ian Watson and Paul Watson (*Graph reduction in a parallel virtual memory environment*).

For proposed architectures, it is important to assess performance before constructing hardware. The papers by Richard B. Kieburtz (*Performance measurement of a G-machine implementation*) and Steven Tighe, Ken Zink, Richard Brice, and William Alexander (*A flexible architectural study methodology*) present results and approaches in this area.

One topic that generated particularly lively discussion at the workshop was that of arrays in functional languages and their implementation by graph reduction. This discussion was inspired by presentations by Paul Hudak on incremental and monolithic array operations and by Arvind on I-structures. This discussion motivated Philip Wadler to devise a new monolithic array operation and to write a paper about it, which he distributed to an electronic mailing list on functional programming. Some further discussion on this subject then followed on the mailing list, and the papers by Paul Hudak (*Arrays, non-determinism, side-effects, and parallelism: A functional perspective*), Philip Wadler (*A new array operation*), and Arvind, Rishiyur S. Nikhil, and Keshav K. Pingali (*I-structures: Data structures for parallel computing*) bear some of the fruit of this discussion.

The final category of papers concerns another area of intense activity with regard to language concepts implementable by graph reduction, that of *logic programming*. These are the papers by Bharat Jayaraman and Gopal Gupta (*Parallel execution of an equational language*), Gary Lindstrom (*Implementing logical variables on a graph reduction architecture*), Uday S. Reddy (*Functional logic languages, Part I*), and John Staples and Peter J. Robinson (*Unification of quantified terms*).

We hope the reader will find these papers useful both as informative sources of current work in the area of graph reduction and as points of departure for future work.

**Acknowledgements**

The workshop was originally suggested by Alfred Hartmann of MCC, and was to be sponsored by Los Alamos, MCC, and the Burroughs Austin Research Center. Unfortunately, not long after planning for the workshop began, BARC was closed.

In addition to the committee members listed below, three others participated in the early phases of organizing the workshop: Henry Tirri, of the University of Helsinki and MCC, and Carl Pixley and Hamilton Richards, then with BARC and now with MCC and the University of Texas at Austin, respectively.

The committee are grateful for assistance and encouragement from several individuals and organizations in Europe. In particular, we thank Ronan Sleep, University of East Anglia; Simon Peyton Jones, University College London; the Alvey Directorate, London, Brian Oakley, Head; and the Division of Software Technology and Advanced Information Processing, Commission of the European Communities, Brussels, John Elmore, Head.

The editors are grateful to Patricia Fasel, Los Alamos National Laboratory, for her generous assistance in the editing and preparation of this volume.

Los Alamos's contribution to the funding of the workshop was provided by the U.S. Department of Energy, Office of Energy Research.

Robert M. Keller
Mountain View, California

Joseph H. Fasel
Los Alamos, New Mexico

June, 1987

**Graph Reduction Workshop Organizing Committee:**

Joseph Fasel, Los Alamos, *General Chairman*
Robert Keller, Quintus Computer Systems, Inc., *Program Chairman*
Randy Michelsen, Los Alamos, *Local Arrangements Chairman*
Wayne Anderson, Los Alamos
Richard Brice, MCC
Scott Danforth, MCC
Alfred Hartmann, MCC
Steven Tighe, MCC
Ken Zink, MCC

# Table of Contents

**Performance Modelling and Simulation**



**Arrays**



**Relationship to Logic Programming**

# On the Correspondence of Lambda Style Reduction and Combinator Style Reduction

Hans-Georg Oberhauser *
FB 10 - Informatik
Universität des Saarlandes
D-6600 Saarbrücken

**Abstract**

A unifying model for lambda style reduction and combinator style reduction is presented. Both reduction types will be described as tree transforming mechanisms thus avoiding as many of the usual implementation details as possible. It will be shown that there is a very strong correspondence between the two reduction mechanisms. This correspondence offers a basis for a fair comparison of the complexities of both mechanisms.

## 1. Introduction

Both lambda style reduction and combinator style reduction mechanisms have been used to implement functional languages [Wis82], [Tur79], [Tur79a], [Hud84], [Sto84], [Cla80], [Hug82]. In the case of lambda style reduction a simple transformation step from a functional program to the corresponding $\lambda$-expression is needed. In the case of combinator reduction an additional translation step from a $\lambda$-expression to the corresponding combinator expression is necessary. In this paper a unifying model for two reduction mechanisms will be described: The well known **eval** function [Hen80], [Wis82] which reduces $\lambda$-expressions using environments and Turners **SKI** combinators [Tur79]. Both

* The author is funded by Deutsche Forschungsgemeinschaft SFB 124 - VLSI Entwurfsmethoden und Parallelität

mechanisms will be regarded as tree transforming mechanisms. Although λ-reduction and combinator reduction seem to be quite different at first glance it will turn out that they are essentially the same. The main difference though lies in the way of "transporting" argument values to the places the corresponding bound variables reside at in an expression. This unifying model was developed to serve as a basis for a fair analytical comparison of the complexities of the two mechanisms.

In the following, most proofs are omitted. They may be found in the full version of this paper. [Obe86]

## 2. λ-trees and c-trees

This section introduces definitions of the tree structures the two mechanisms are based on, namely decorated λ-trees and c-trees (i.e. combinator trees). I choose this representation for λ-expressions, closures and c-expressions (i.e. combinator expressions) in order to keep the corresponding reduction mechanisms as simple as possible and as closely related as possible. Both will be defined by giving tree transformation rules which specify their behavior. Thus the usual implementation details (like reduction stacks) will be avoided and the similarities will become clear.

**Definition** (λ-tree)

A λ-tree is either
- a constant $c$
- a variable $x$
- an application $(fa)$ with function part $f$ and argument part $a$ both of which are λ-trees
- an abstraction $\lambda x.e$ with bound variable $x$ and function body $e$ which is a λ-tree

Figure 2.1 shows pictorial representations of λ-trees in the same order as defined above. Leaves of λ-trees may be constants or variables. Internal nodes are application nodes (represented as circles) or λ-nodes.

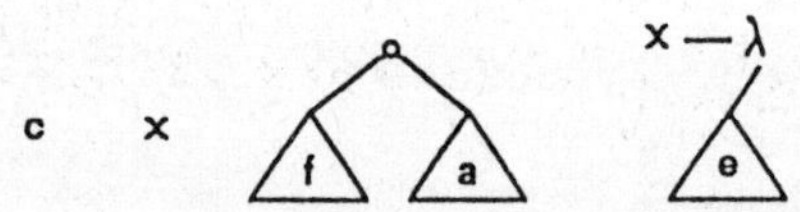


Figure 2.1

The definition of λ-trees given above differs from the usual definition of pure λ-expressions in that constants are allowed as legal λ-trees. This has been done for obvious reasons: I am not interested in λ-calculus (or combinator calculus) as a pure mathematical model but as an implementation mechanism for functional programming languages. The exact choice of constants allowed does not really matter. They are introduced for efficiency reasons only. Pure λ-calculus is as powerful as Turing machines or partially recursive functions: any computable function may be expressed as a pure λ-expression. On the other hand, nobody implementing a functional language would use the low level mechanisms offered by pure λ-calculus to realize addition, subtraction, etc. Thus pure

$\lambda$-expressions are usually enhanced by adding constants like natural numbers, binary operators $(+, -, *, /, cons, lazycons, etc.)$, if-then-else constructs as well as the fixed point operator $Y$. Of course the underlying reduction rules have to be extended to implement the semantics of these constants. The usual $\alpha$- and $\beta$-reduction rules will no longer be enough.

The following notations and conventions will be needed. If $e$ is a $\lambda$-tree then $root(e)$ denotes the root of $e$ and if $p$ is a node in a $\lambda$-tree then $father(p)$ denotes the father of node $p$. Application associates to the left and parentheses will be omitted whenever possible, i.e. $((f\ a)\ b)\ c$ will be denoted as $f\ a\ b\ c$.

**Definition** (bound, free, binds)

Let $x$ be a variable of a $\lambda$-tree $e$ and let $p_1, \ldots, p_k$ be the nodes on the path leading from $x$ to $root(e)$. Let $i$ be minimal, $1 \leq i \leq k$, such that $p_i$ is a $\lambda$-node with bound variable equal to $x$. Then the occurrence of $x$ at $p_1$ is said *to be bound at* $p_i$ or $p_i$ *binds the occurrence of* $x$ *at* $p_1$. If no such $p_i$ exists the occurrence of $x$ at $p_1$ is said *to be free in* $e$. In this case $x$ is a member of $free(e)$, the set of free variables in $e$.

**Example**

Figure 2.2 shows the pictorial representation of $(\lambda x. +\ x\ 1)16$

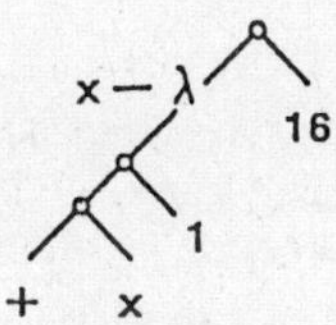


Figure 2.2

Next I am going to enhance $\lambda$-trees by decorating certain nodes with environments calling the resulting trees *closure trees*. The use of this new notion will become obvious in sections 4 and 5 where the reduction mechanisms will be described.

**Definition** (closure, environment)

A closure is a $\lambda$-tree $e$, whose root node is decorated with an environment $env$ and will be denotated as $[e, env]$. An environment is a (possibly empty) list of bindings of variables $x_1, x_2 \ldots, x_n$ to closures $cl_1, cl_2, \ldots, cl_n$ and will be denoted as: $x_1 \rightarrow cl_1 : x_2 \rightarrow cl_2 : \ldots : x_n \rightarrow cl_n$. The empty environment will be denoted as $nil$.

**Definition** (closure tree)

Let $env$ be an environment. Then a closure tree is either
- a closure $[c, env]$, where $c$ is a constant
- a closure $[x, env]$, where $x$ is a variable
- a closure $[f\ a, env]$, where $f$ and $a$ are $\lambda$-trees
- an application $(fa)$, where at least one of $f$ or $a$ is a closure tree and a $\lambda$-tree otherwise
- a closure $[\lambda x.e, env]$, where $e$ is a $\lambda$-tree

Intuitively a $\lambda$-tree will be called a closure tree if some of it's subtrees are not ordinary $\lambda$-trees but closures. A pictorial representation of a closure tree is given in figure 2.3. In addition to the tree edges there is a directed edge (dashed) from node $n$ to the head of the list representing environment *env* if $n$ is decorated with *env*. Again parentheses and the empty environment will be omitted whenever possible. The picture shows the closure tree $+ \; [\; x, \; x \rightarrow [+ \; 1 \; 1, nil] \; : \; y \rightarrow [11, nil] \;] \; 1$. Here node $x$ is decorated with an environment containing bindings to variables $x$ and $y$. Both variables are bound to closures containing empty environments (which are omitted in the picture).

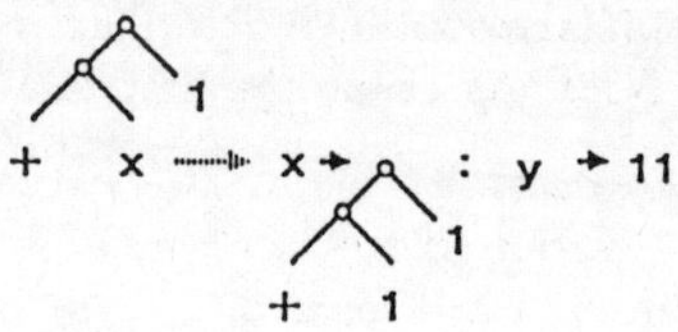


Figure 2.3

**Definition** (c-tree)

A c-tree is either
- a constant $c$
- a combinator
- an application $(fa)$ with function part $f$ and argument part $a$ both of which are c-trees.

As combinators I choose the set defined by Joy [Joy85]: $S$, $S'$, $B$, $B'$, $C$, $C'$, $J$, $J'$, $K$ and $I$. The same set of constants as in the case of $\lambda$-trees is assumed here.

The pictorial representation of the c-tree $(C \; (B \; + \; I) \; 1)$ is given in figure 2.4. In c-trees leaves may be constants or combinators. As internal nodes only application nodes are allowed. As edges only tree edges are allowed.

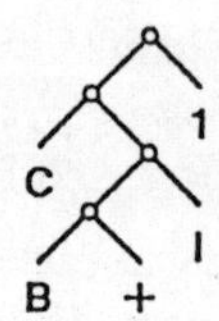


Figure 2.4

## 3. The Translation Algorithm

In this section the translation algorithm which transforms $\lambda$-trees into c-trees will be given. It is almost identical to ABS/J/1 defined in [Joy85] and will be stated for completeness' sake only. The translation algorithm is described using two functions *trans* and *abs*. The purpose of *trans* is to call *abs* at appropriate places such that the abstraction of variables is performed from the inside out. Both *trans* and *abs* are defined inductively on the definition of $\lambda$-trees. The set of free variables of a subtree $e$ is denoted as $free(e)$. In addition to "ordinary" application nodes with two sons the trees which are built up during the translation as intermediate results may contain "tripel" nodes with three sons. A subtree whose root is such a "tripel" node with left son $c$, middle son $f$ and right son $a$ will be denoted as $[c\ f\ a]$ . Here the left son $c$ will always be a tree built up of application nodes and combinator leaves only. At the end of the translation all tripel nodes are "stretched" (by the auxiliary function *stretch*) thus transforming the result into a c-tree. Therefore the translation of an expression $e$ to combinators is performed by the function *translate* such that $translate(e) = stretch(trans(e))$ holds for any $\lambda$-tree $e$. This approach is similar to an idea described in [Ken82]. In the following $x$ is a variable and $e$ and $f$ are trees. As usual, the first rule applicable must be used.

$$\begin{aligned}
trans(a) &= a, \text{ for constants and variables } a\\
trans(fa) &= trans(f)\ trans(a)\\
trans(\lambda x.e) &= abs(x, trans(e))
\end{aligned}$$

$$\begin{aligned}
abs(x,x) &= I &&\\
abs(x,y) &= Ky, && if\ x \neq y\\
abs(x,f\ e) &= [S\ abs(x,f)\ abs(x,e)], && if\ x \in free(f)\ and\ x \in free(e)\\
abs(x,f\ e) &= [B\ f\ abs(x,e)], && if\ x \notin free(f)\ and\ x \in free(e)\\
abs(x,f\ e) &= [C\ abs(x,f)\ e], && if\ x \in free(f)\ and\ x \notin free(e)\\
abs(x,f\ e) &= [J\ f\ e], && if\ x \notin free(f)\ and\ x \notin free(e)\\
abs(x,[c\ f\ e]) &= [(S'\ c)\ abs(x,f)\ abs(x,e)], && if\ x \in free(f)\ and\ x \in free(e)\\
abs(x,[c\ f\ e]) &= [(B'\ c)\ f\ abs(x,e)] && if\ x \notin free(f)\ and\ x \in free(e)\\
abs(x,[c\ f\ e]) &= [(C'\ c)\ abs(x,f)\ e], && if\ x \in free(f)\ and\ x \notin free(e)\\
abs(x,[c\ f\ e]) &= [(J'\ c)\ f\ e], && if\ x \notin free(f)\ and\ x \notin free(e)
\end{aligned}$$

$$\begin{aligned}
stretch(d) &= d, \text{ for constants and combinators } d\\
stretch(fa) &= stretch(f)\ stretch(a)\\
stretch([c\ f\ e]) &= c\ stretch(f)\ stretch(e)
\end{aligned}$$

The usual optimizations based on eta-reduction and it's generalizations are omitted here since they slightly disturb the correspondence to be shown in section 7.

**Example**

In figure 3.1 a $\lambda$-tree and it's translation are shown.

Figure 3.1

I am going to use the following terminology: The call of $trans(\lambda x.e)$ starts the abstraction of variable $x$ out of $trans(e)$. Therefore this new abstraction will be said *to happen because* of the $\lambda$-node that binds $x$ or equally *to be caused* by this $\lambda$-node. Furthermore a call of $abs(x, fa)$ produces a new combinator. Therefore this combinator will be said *to be produced* for $root(fa)$ because of $x$. For the abstraction of more than one variable out of a $\lambda$-tree $e$ the following notation will be used: $abs([x_k, \ldots, x_1], e)$ is standing for $abs(x_k, \ldots, abs(x_1, e) \ldots)$. Of course in the case $k = 0$, $abs([x_k, \ldots, x_1], e)$ $= e$. At the end of this section two easy Lemmata are proved which will be needed later.

**Lemma 3.1**

Let $e$ be a $\lambda$-tree and let $x$ be a variable. Then $x \in free(trans(e))$ iff $x \in free(e)$.

**Lemma 3.2**

Let $e$ be a $\lambda$-tree and let $x$ and $y$ be variables. Then $x \in free(abs(y, trans(e)))$ iff $x \in free(e)$ *and* $x \neq y$.

## 4. The correspondence between $\lambda$-trees and c-trees.

In this section a mapping called $varfree$ will be defined which maps the subtrees of a $\lambda$-tree $a$ to certain subtrees of it's translation $translate(a)$. Before I define the mapping itself three auxiliary definitions are needed. In order to come up with an idea of the structure of a translated $\lambda$-tree $a$ it is necessary to have a precise understanding of the translation process itself. The translation process is controlled by the function $trans$ such that abstractions are performed from the inside out, i.e. innermost $\lambda$-trees are translated first. Since abstractions are caused by $\lambda$-nodes it is necessary to define an ordering on the $\lambda$-nodes of $a$ to reflect the order of abstractions. This is done in the first definition below. Intuitively a $\lambda$-node $p$ binding a variable $x$ should be considered "greater" than a $\lambda$-node $q$ binding a variable $y$ if $q$ is a proper descendant of $p$. In this case the abstraction of $x$ caused by $p$ is performed "later" than the abstraction of $y$ caused by $q$. Additionally for any subtree $e$ one has to specify the set of those $\lambda$-nodes which cause transformations of $e$ during the translation. This is done in the definitions of $useful$, $useless$ and $bounds$ below. Intuitively transformations in a subtree $e$ are performed by abstractions caused by $\lambda$-nodes of two types: First those $\lambda$-nodes binding variables free in $e$. These variables have to be abstracted out of $e$. The second set of $\lambda$-nodes causing transformations of $e$ contains those $\lambda$-nodes which, very informally speaking, sit directly above $root(e)$ and are not separated from $root(e)$ by application nodes and do not bind any variables free in $e$. Although they do not bind any variables free in $e$ they cause transformations of $e$ since any $\lambda$-node causes the generation of at least one combinator. Only an application node is able to block off recursive calls of $abs$. This always happens when a variable $x$ is abstracted out of a tree $e_1e_2$ and $x \notin free(e_1e_2)$. In this case a combinator $J$ is generated at $root(e_1e_2)$ and no further calls to $abs$ are performed. Nevertheless $root(e_1e_2)$ is affected by the abstraction.

**Definition** (node-variable pair, $\ll$)

Let $a$ be a $\lambda$-tree and let $x$ and $y$ be the bound variables of $\lambda$-nodes $p$ and $q$ in $a$ respectively. Then the ordered pair $(p, x)$ is called a node-variable pair. Furthermore $(q, y) \ll (p, x)$ holds iff $q$ is a proper descendant of $p$.

**Definition** (useful, useless)

Let $a$ be a $\lambda$-tree and let $e$ be a subtree of $a$. Let $n_1, n_2, \ldots, n_q$ be the nodes on the path $p$ leading in $a$ from the father of the root of $e$ (i.e. $n_1 = father(root(e))$) to the root of $a$ (i.e. $n_q = root(a)$). Let $m_1, \ldots, m_r$ be the $\lambda$-nodes on $p$ binding variables free in $e$. Then $useful(e)$ is defined as $\{(m_1, x_1), \ldots, (m_r, x_r)\}$, where $x_i$ is the bound variable of $m_i$, for all $i, 1 \leq i \leq r$. Furthermore let $l$ be maximal such that all of $n_1, \ldots n_l$ are $\lambda$-nodes and let $\{o_1, \ldots, o_s\}$ be the subset of $\{n_1, \ldots, n_l\}$ such that $o_i$ does not bind any variable free in $e$, $1 \leq i \leq s$. Then $useless(e)$ is defined as $\{(o_1, y_1), \ldots, (o_s, y_s)\}$, where $y_i$ is the bound variable of $o_i$ for all i, $1 \leq i \leq s$.

**Definition** (bounds)

Let $e$ be a $\lambda$-tree . Then $bounds(e)$ is defined as $useful(e) \cup useless(e)$.

Formally *useful*, *useless*, *bounds* as well as *varfree* defined below should be functions in two arguments. Their definitions reference nodes on the path leading from $father(root(e))$ to the root of the $\lambda$-tree $a$, $e$ is a subtree of. In the following the second parameter $a$ will be omitted assuming that $useful(e)$, $useless(e)$, $bounds(e)$ as well as $varfree(e)$ are interpreted with respect to the correct $\lambda$-tree , i.e to the one corresponding to the functional program to be evaluated. Every $\lambda$-tree under consideration is a subtree of this all encompassing tree, which will be denoted as $a$. Again, note that the values of *bounds*, *useful* and *useless* are static in the sense that they implicitly reference the original $\lambda$-tree $a$ to be reduced. They do not change during the reduction of $a$.

**Example**

Let $e = s\ x\ z$ and let $a = \lambda s.\lambda t.t(\lambda x.\lambda x.\lambda y.\lambda z.e)$ as shown in figure 4.1. Then $free(e) = \{s, x, z\}$, $useful(e) = \{(n_1, z), (n_3, x), (n_7, s)\}$, $useless(e) = \{(n_2, y), (n_4, x)\}$ and $bounds(e) = \{(n_1, z), (n_2, y), (n_3, x), (n_4, x), (n_7, s)\}$. Intuitively any combinator generated at $root(e)$ during the translation of $a$ is caused by $\lambda$-nodes in $bounds(e)$. The subset $useless(e)$ contains all those node-variable pairs in $bounds(e)$ for which combinators $K$, $J$ or $J'$ are generated at $root(e)$. For these elements no combinators are produced at non-root nodes of $e$. Values associated with variables of elements in $useless(e)$ are never needed to reduce $e$. They are blocked off at $root(e)$. The subset $useful(e)$ on the other hand contains all those node-variable pairs in $bounds(e)$ for which combinators are produced not only at $root(e)$ but also at non-root nodes of $e$. Values associated with variables in $useful(e)$ may be needed to reduce $e$. Therefore precautions have to be taken to make them available in $e$. Note that the variables contained in the node-variable pairs of $useful(e)$ are exactly the members of $free(e)$.

**Definition** (varfree)

Let $e$ be a $\lambda$-tree and let $bounds(e) = \{(p_1, x_1), \ldots, (p_k, x_k)\}$ ordered by $\ll$. Then $varfree(e)$, the variable free version of $e$, is defined as:

$$stretch(abs([x_k, \ldots, x_1], trans(e)))$$

The definition of *varfree* above takes into consideration that the variable free version of a subtree $e$ is not entierly specified by $trans(e)$ alone. It also depends on the context of $e$, i.e. those abstractions caused by $\lambda$-nodes in $bounds(e)$. These abstractions are caused by

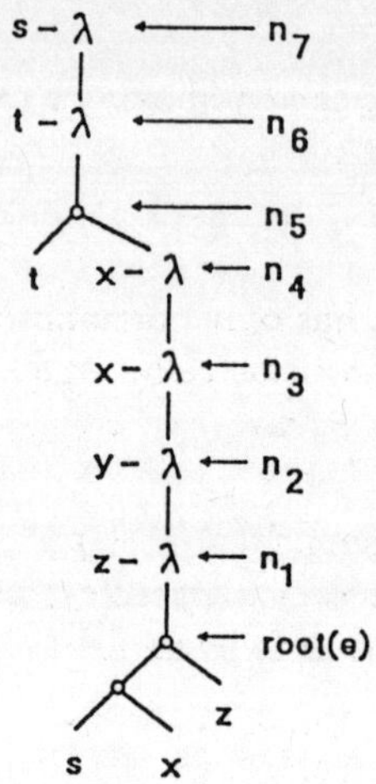


Figure 4.1

proper ancestors of $root(e)$ but in a sense reach down to $e$ thus causing transformations in $e$. Next I am going to prove three Theorems giving a precise understanding of how corresponding subtrees look like. First a couple of Lemmata are needed. They are rather technical and will be needed to prove the Theorems at the end of this section. Essentially they state facts about the structure of the ordered sets *bounds*, *useful* and *useless*.

**Lemma 4.1**

Let $e$ be a $\lambda$-tree. If $father(root(e))$ is not a $\lambda$-node, then $useless(e) = \emptyset$ and $bounds(e) = useful(e)$ hold.

**Lemma 4.2**

Let $e = e_1e_2$ be a $\lambda$-tree, let $(p, x) \in useful(e)$ and let $(p, x) \notin useful(e_1)$ (respectively $useful(e_2)$). Then $x \notin free(e_1)$ (respectively $free(e_2)$).

**Lemma 4.3**

Let $a$ be a $\lambda$-tree, let $e$ be subtree of $a$ and let $(p, x) \in bounds(e)$. If there exists no $(p', x') \in bounds(e)$ with $(p', x') \ll (p, x)$ and $x = x'$ then there exists no $p''$ on the path from $root(a)$ to $father(root(e))$ such that $(p'', x'') \ll (p, x)$ and $x = x''$ hold.

**Lemma 4.4**

Let $a$ be a $\lambda$-tree , let $e$ be subtree of $a$ and let $(p, x) \in bounds(e)$. Then $(p, x) \in useful(e)$ iff $x \in free(e)$ and there exists no $(p', x') \in bounds(e)$ such that $(p', x') \ll (p, x)$ and $x = x'$ hold.

**Lemma 4.5**

Let $a$ be a $\lambda$-tree, let $e$ be subtree of $a$ and let $(p, x) \in bounds(e)$. Then $(p, x) \in useless(e)$ iff $x \notin free(e)$ or there exists $(p', x') \in bounds(e)$ such that $(p', x') \ll (p, x)$ and $x = x'$ hold.

**Notation**

In the following $D_i$ stands for one of the combinators $S$, $B$, $C$ or $J$ and $D'_i$ stands for one of the combinators $S'$, $B'$, $C'$ or $J'$.

**Definition** (pure c-tree)

A pure c-tree is a c-tree $t$ such that t has only combinators as leaves and $t$ is of the form $D'_n(D'_{n-1}(\ldots(D'_2 D_1)\ldots))$. $D_i$ or $D'_i$ will be called the i-th combinator in $t$.

**Theorem 4.1**

Let $e = e_1 e_2$ be a $\lambda$-tree and let $bounds(e) = \{(p_1, x_1), \ldots, (p_k, x_k)\}$ ordered by $\ll$, $k \geq 1$. Then $varfree(e) = c\ varfree(e_1)\ varfree(e_2)$ and $c$ is a pure c-tree, such that the i-th combinator $D_i$ in $c$ is defined as follows:

$$D_i = \begin{cases} S'\ (or\ S,\ for\ i = 1), & \text{if } (p_i, x_i) \in useful(e_1) \text{ and } (p_i, x_i) \in useful(e_2), \\ B'\ (or\ B,\ for\ i = 1), & \text{if } (p_i, x_i) \notin useful(e_1) \text{ and } (p_i, x_i) \in useful(e_2), \\ C'\ (or\ C,\ for\ i = 1), & \text{if } (p_i, x_i) \in useful(e_1) \text{ and } (p_i, x_i) \notin useful(e_2), \\ J'\ (or\ J,\ for\ i = 1), & \text{if } (p_i, x_i) \notin useful(e_1) \text{ and } (p_i, x_i) \notin useful(e_2). \end{cases}$$

$c$ will be called the pure c-tree produced for $root(e)$ (See figure 4.2).

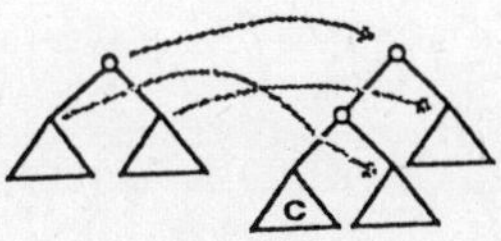


Figure 4.2

**Proof:**

Let $bounds(e) = \{(p_1, x_1), \ldots, (p_k, x_k)\}$ ordered by $\ll$. Then by definition of $varfree$ and $trans$, $varfree(e) = stretch(abs([x_k, \ldots, x_1], trans(e_1)trans(e_2)))$. I am going to prove the following invariant by induction on the number $l$ of variables abstracted:
After abstracting $x_1, \ldots, x_l$, $l \leq k$, in this order out of $trans(e_1)trans(e_2)$ one arrives at a tree of the form $[c\ e'_1 e_2]'$ where

$e'_1 = abs([x_{i_s}, \ldots, x_{i_1}], trans(e_1))$ and

$e'_2 = abs([x_{j_t}, \ldots, x_{j_1}], trans(e_2))$ and

$\{(p_{i_1}, x_{i_1}), \ldots, (p_{i_s}, x_{i_s})\} = useful(e_1) \cap \{(p_1, x_1), \ldots, (p_l, x_l)\}$ consists of the first $s$, $s \leq l$, elements of $useful(e_1)$ and

$\{(p_{j_1}, x_{j_1}), \ldots, (p_{j_t}, x_{j_t})\} = useful(e_2) \cap \{(p_1, x_1), \ldots, (p_l, x_l)\}$ consists of the first $t$, $t \leq l$, elements of $useful(e_2)$ and

$c$ is a pure c-tree and all the claims stated in Theorem 4.1 about the i-th combinator in $c$ hold for all $i$, $1 \leq i \leq l$.

$l = 1$ : Assume $(p_1, x_1) \in useful(e)$.
If $(p_1, x_1) \in useful(e_1)$ and $(p_1, x_1) \in useful(e_2)$ then $x_1 \in free(e_1)$ and $x_1 \in free(e_2)$ and therefore $x_1 \in free(trans(e_1))$ and $x_1 \in free(trans(e_2))$ by Lemma 3.1. Thus $abs(x_1, trans(e_1)trans(e_2))$ yields $[S\ abs(x_1, trans(e_1))abs(x_1, trans(e_2)))]$ in this case.
If $(p_1, x_1) \notin useful(e_1)$ and $(p_1, x_1) \in useful(e_2)$ then $x_1 \notin free(e_1)$ by Lemma 4.2 and $x_1 \in free(e_2)$. Therefore $x_1 \notin free(trans(e_1))$ and $x_1 \in free(trans(e_2))$ by Lemma 3.1. Thus $abs(x_1, trans(e_1)trans(e_2))$ yields $[B\ abs(x_1, trans(e_1))abs(x_1, trans(e_2))]$ in this case.
By a symmetrical argument one gets that $(p_1, x_1) \in useful(e_1)$ and $(p_1, x_1) \notin useful(e_2)$ implies $[C\ abs(x_1, trans(e_1))trans(e_2)]$ as result.
Now assume $(p_1, x_1) \notin useful(e)$.
This is equivalent to $(p_1, x_1) \notin useful(e_1)$ and $(p_1, x_1) \notin useful(e_2)$. Furthermore it is equivalent to $(p_1, x_1) \in useless(e)$. By Lemma 4.5 $x_1 \notin free(e)$ and therefore by Lemma 3.1 $x_1 \notin free(trans(e_1))$ and $x_1 \notin free(trans(e_2))$. Thus the result of the abstraction is $[J\ trans(e_1)trans(e_2)]$ in this case. Therefore the claim holds in all cases for $l = 1$.

$l > 1$: Now for the induction step assume that the claim holds for $abs([x_l, \ldots, x_1], trans(e_1)\ trans(e_2)) =: [c'e_1''e_2''], 1 \leq l < k$ and consider $abs(x_{l+1}, c'e_1''e_2'')$.
Assume $(p_{l+1}, x_{l+1}) \in useful(e)$.
If $(p_{l+1}, x_{l+1}) \in useful(e_1)$ and $(p_{l+1}, x_{l+1}) \in useful(e_2)$ then $x_{l+1} \in free(e_1)$ and $x_{l+1} \in free(e_2)$ and therefore $x_{l+1} \in free(e_1'')$ and $x_{l+1} \in free(e_2'')$ by Lemma 4.4 and 3.2. In this case the result is $[(S'c')abs(x_{l+1}, e_1'')abs(x_{l+1}, e_2'')]$.
Like for $l = 1$ in the other cases one gets $[(B'c')e_1''abs(x_{l+1}, e_2'')]$, if $(p_{l+1}, x_{l+1}) \notin useful(e_1)$ and $(p_{l+1}, x_{l+1}) \in useful(e_2)$ and $[(C'c')abs(x_{l+1}, e_1'')e_2'']$, if $(p_{l+1}, x_{l+1}) \in useful(e_1)$ and $(p_{l+1}, x_{l+1}) \notin useful(e_2)$.
Now assume $(p_{l+1}, x_{l+1}) \notin useful(e)$.
This is equivalent to $(p_{l+1}, x_{l+1}) \notin useful(e_1)$ and $(p_{l+1}, x_{l+1}) \notin useful(e_2)$. Furthermore it is equivalent to $(p_{l+1}, x_{l+1}) \in useless(e)$. If $(p_{l+1}, x_{l+1})$ is not the first pair with variable $x_{l+1}$ in $bounds(e)$ then $x_{l+1} \notin free(e_1'')$ and $x_{l+1} \notin free(e_2'')$ by Lemma 3.2. If on the other hand $(p_{l+1}, x_{l+1})$ is the first pair in $bounds(e)$ with variable $x_{l+1}$ then $x_{l+1} \notin free(e_1)$ and $x_{l+1} \notin free(e_2)$ by Lemma 4.5. Therefore $x_{l+1} \notin free(e_1'')$ and $x_{l+1} \notin free(e_2'')$ by Lemma 3.2. Thus in both cases the result is $[(J'c)e_1''e_2'']$ and the claim holds for $i + 1$.
Thus the invariant holds for all $l$, $1 \leq l \leq k$.

Since $useful(e_1) \subseteq bounds(e)$ and $useful(e_2) \subseteq bounds(e)$ the variables abstracted out of $trans(e_1)$ and $trans(e_2)$ are exactly the ones contained in the node-variable pairs of $useful(e_1)$ and $useful(e_2)$ respectively. Since $useless(e_1) = useless(e_2) = \emptyset$ by Lemma 4.1 and since in this case $bounds(e_1) = useful(e_1)$ and $bounds(e_2) = useful(e_2)$ hold the proof of the Theorem follows immediately by the definition of *varfree*.

Theorem 4.1 intuitively says that any subtree $e_1e_2$ of a $\lambda$-tree $e$ is transformed into a subtree of the form $c\ e_1'e_2'$ during the translation of $e$. $e_1'$ and $e_2'$ are the variable free versions of $e_1$ and $e_2$ respectively and $c$ is a pure c-tree consisting of combinators generated for $root(e_1e_2)$. This c-tree consists of combinators of two "types": The $S$, $B$

and $C$ combinators as well as their primed versions were generated at $root(e_1e_2)$ because of variables free in $e_1e_2$. During the reduction these combinators control the transport of argument values into the variable free version of the $\lambda$-tree $e_1e_2$. These arguments are needed to reduce $e_1e_2$. The $J$ and $J'$ combinators on the other hand were generated for variables whose values are not needed to reduce $e_1e_2$. During the reduction these values will be "thrown away" at the root of $e_1e_2$. These ideas will be formalized in section 7. The following Theorem takes care of the special case $bounds(e) = \emptyset$. In this case no pure c-tree is generated at $root(e_1e_2)$. Intuitively this should be clear since during the reduction of $a$ no values of free variables have to be transported into $e_1e_2$ (since $useful(e_1e_2) = \emptyset$) and since no superfluous arguments have to be thrown away at $root(e_1e_2)$ (since $useless(e_1e_2) = \emptyset$). Thus it is only necessary to transform $e_1$ and $e_2$ to their variable free versions to obtain the variable free version of $e_1e_2$. No combinators are needed at $root(e_1e_2)$.

**Theorem 4.2**

Let $e = e_1e_2$ be a $\lambda$-tree with $bounds(e) = \emptyset$. Then $varfree(e) = varfree(e_1)varfree(e_2)$.

Theorem 4.3 finally states that the variable free version of a function $f = \lambda x.e$ is equal to the variable free version of it's function body $e$. This follows immediately from the definition of $varfree$.

**Theorem 4.3**

Let $f = \lambda x.e$ be a $\lambda$-tree. Then $varfree(f) = varfree(e)$ holds.

**Example**

In figure 4.3 a $\lambda$-tree and it's translation are shown. The mapping $varfree$ is indicated by arrows in the obvious way.

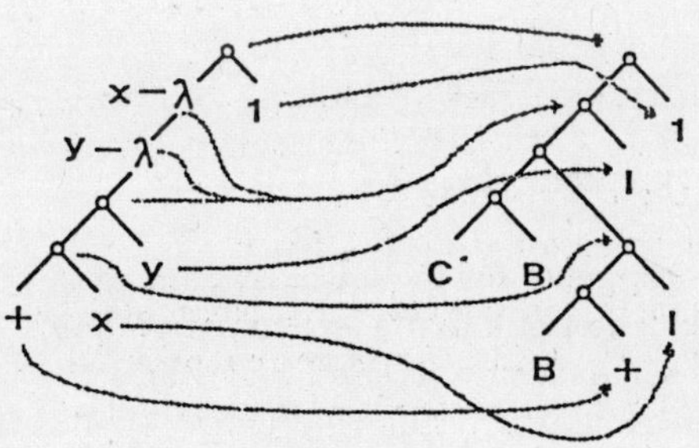


Figure 4.3

Now I am ready to define the two reduction mechanisms which allow us to reduce $\lambda$-trees and c-trees to their normal forms, i.e. to execute functional programs. The two mechanisms will be called $\lambda$-red and c-red respectively. As reduction strategy normal order will be assumed in both cases.

## 5. The $\lambda$-Reducer "$\lambda$-red"

The $\lambda$-reducer $\lambda$-red will be controlled by a system of tree transformation rules which essentially behaves like the well known eval function (but incorporates the reduction stack which is needed by eval as an additional feature to store closures). The $\lambda$-reducer may be regarded as a tree transforming machine : It moves over a closure tree and keeps transforming it (according to the reduction rules to be defined) until the normal form is reached (if it exists). In order to pin down the position of $\lambda$-red in the tree, the subtree $e$ at whose root node $\lambda$-red is currently standing will be denoted as $\langle e \rangle$. A closure tree $e$ and the node in $e$ $\lambda$-red is currently standing at will be called a $\lambda$-configuration. To reduce a $\lambda$-tree $e$ (which is assumed to have no free variables) the reducer $\lambda$-red starts in $\lambda$-configuration $[\langle e \rangle, nil]$ as shown in figure 5.1. The root node of $e$ is marked to show the current position of $\lambda$-red. The reduction rules of $\lambda$-red will have the following form: $a \triangleright b$, where $a$ and $b$ are $\lambda$-configurations. The meaning of this sample rule is: replace $\lambda$-configuration $a$ by $\lambda$-configuration $b$. Thus in any configuration $\lambda$-red tries to match a subtree it is currently standing in with the left hand side of one of the reduction rules. If such a rule is found the matched subtree is transformed according to this rule yielding the next $\lambda$-configuration. If no reduction rule is applicable at the node the $\lambda$-reducer is currently standing at, the reduction process stops. I now give the tree transformation rules for $\lambda$-red. Again the first rule applicable must be used.

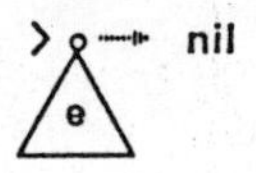


Figure 5.1

$\lambda$-1: $[\langle x \rangle, env]$ $\triangleright$ $[\langle e \rangle, env']$,
where $x \rightarrow [e, env']$ is the leftmost binding of $x$ in $env$

$\lambda$-2: $[\langle \lambda x.e \rangle, env]\ [b, env']$ $\triangleright$ $[\langle e \rangle, x \rightarrow [b, env'] \ : \ env]$

$\lambda$-3: $[\langle fb \rangle, env]$ $\triangleright$ $[\langle f \rangle, env]\ [b, env]$

$\lambda$-4: $[\langle c \rangle, env]$ $\triangleright$ c, for constant $c$

Rule $\lambda$-1 searches the environment for the leftmost binding of variable $x$ to some closure $[e, env']$ and starts the reduction of $e$ in it's environment $env'$. Rule $\lambda$-2 performs a new binding of variable $x$ to some closure $[b, env']$ and starts the reduction of the function body $e$ in an updated environment. Rule $\lambda$-3 simply builds a closure containing the argument $b$ of a function application $fb$ and the environment $env$ and then starts the reduction of the function part $f$ in the environment $env$. Rule $\lambda$-4 finally reduces a constant $c$ to itself. After performing this reduction, $\lambda$-red stops yielding the result $c$.
A $\lambda$-reducer controlled by these rules will reduce a $\lambda$-expression not always completely to it's normal form, but to it's so called "head normal form" if it exists. This form is reached, when a constant occurs on a function position or a predefined function and too few argument are supplied. In this case the reducer would have to reduce possible argument expressions of this constant to their normal forms to arrive at the normal form

of the whole expression. This case is not dealt with here and the c-reducer will have the same property.

The rules defined above may be extended in a very straightforward way to cope with operators like $+$, *lazycons*, $if - then - else$, $Y$, etc. Of course the resulting rules will be a little more complicated, since they may contain recursive calls to λ-red. The following rules should be inserted in the list of rules λ-1 to λ-4 somewhere before rule λ-4 to make sure that they are tried first.

λ-3.1: $[\langle + \rangle, env]\ [e_1, env]\ [e_2,\ env] \ \triangleright\ r_1 + r_2$
$$where\ [\langle e_1 \rangle, env]\ \triangleright \ldots \triangleright\ r_1 \text{ and } [\langle e_2 \rangle, env]\ \triangleright \ldots \triangleright\ r_2$$

λ-3.2: $[\langle lazycons \rangle, env]\ [e_1, env]\ [e_2,\ env] \ \triangleright\ lazycons\ [e_1, env]\ [e_2, env]$

λ-3.3: $[\langle if \rangle, env]\ [e_1, env]\ [e_2, env]\ [e_3, env] \ \triangleright\ [\langle e_i \rangle, env]$

$$where\ i = \begin{cases} 1, & \text{if } [\langle e_1 \rangle, env]\ \triangleright \ldots \triangleright\ true; \\ 2, & \text{if } [\langle e_1 \rangle, env]\ \triangleright \ldots \triangleright\ false. \end{cases}$$

λ-3.4: $[\langle Y \rangle, env]\ [f, env] \ \triangleright\ [\langle f \rangle,\ env]\ ([Y,\ env]\ [f, env])$

In rule λ-3.1 two recursive calls to λ-red are made to reduce the two argument trees. The results $r_1$ and $r_2$ are then added. Similar rules can be defined for $-$, $*$, $/$, *cons* and other strict binary operators. In the case of *lazycons* in rule λ-3.2 no recursive call is necessary. The reduction simply stops at this point. In rule λ-3.3 the boolean guard is reduced first after a recursive call to λ-red. Then the *true* or *false* part of the conditional is reduced depending on the value of the guard. How these recursive calls are implemented does not matter here. The same phenomenon will show up in the combinator reducer.

**Example**

Figure 5.2 shows a sequence of configurations the λ-reducer is stepping through when reducing the λ-expression $(\lambda x. +\ x\ 1)\ 16$ in an empty environment. The position of λ-red in the tree is marked by $>$.

The following Lemma gives a precise understanding of what the environment contains at any time during the λ-reduction process. It will be needed in section 7 to show the correspondence between c-red and λ-red.

**Lemma 5.1**

Let $a$ be a λ-tree without free variables and let $e$ be a subtree of $a$. Let $n_1, \ldots, n_p$ be the λ-nodes on the path in $a$ leading from $father(root(e))$ to the $root(a)$. Let $x_1, \ldots, x_p$ be the bound variables of of $n_1, \ldots, n_p$ respectively and let $(n_1, x_1) \ll \ldots \ll (n_p, x_p)$. Let λ-red start in λ-configuration $[\langle a \rangle, nil]$. Then for any closure $[e, env]$ built during the reduction the environment $env$ has the form $x_1 \rightarrow cl_1 : \ldots : x_p \rightarrow cl_p$, where $cl_1, \ldots, cl_p$ are closures.

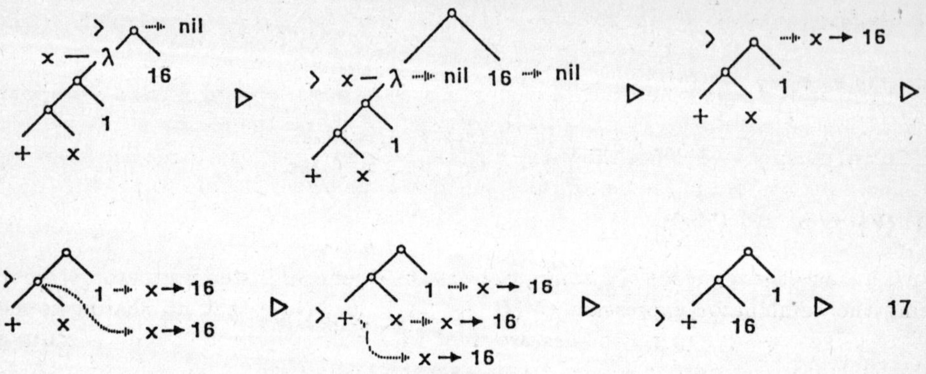

Figure 5.2

## 6. The Combinator Reducer "c-red"

In analogy to $\lambda$-red the combinator reducer c-red too will be regarded as a tree transforming machine controlled by a system of reduction rules. The same notational conventions apply. Here are the rules for c-red. Again the first rule applicable must be used.

c-1 : $\langle f\ a\rangle \quad \triangleright \quad \langle f\rangle\ a$
c-2 : $\langle I\rangle\ x \quad \triangleright \quad \langle x\rangle$
c-3 : $\langle S\rangle\ a\ b\ c \quad \triangleright \quad \langle a\ c\ (b\ c)\rangle$
c-4 : $\langle B\rangle\ a\ b\ c \quad \triangleright \quad \langle a\ (b\ c)\rangle$
c-5 : $\langle C\rangle\ a\ b\ c \quad \triangleright \quad \langle a\ c\ b\rangle$
c-6 : $\langle J\rangle\ a\ b\ c \quad \triangleright \quad \langle a\ b\rangle$
c-7 : $\langle S'\rangle\ a\ b\ c\ d \quad \triangleright \quad \langle a\ (b\ d)\ (c\ d)\rangle$
c-8 : $\langle B'\rangle\ a\ b\ c\ d \quad \triangleright \quad \langle a\ b\ (c\ d)\rangle$
c-9 : $\langle C'\rangle\ a\ b\ c\ d \quad \triangleright \quad \langle a\ (b\ d)\ c\rangle$
c-10: $\langle J'\rangle\ a\ b\ c\ d \quad \triangleright \quad \langle a\ b\ c\rangle$
c-11: $\langle K\rangle\ a\ b \quad \triangleright \quad \langle a\rangle$
c-12: $\langle c\rangle \quad \triangleright \quad c$, for constant $c$

For the translation algorithm and the reduction rules given above the law of abstraction holds:

$$abs(x, e)\ x \triangleright \ldots \triangleright e$$

I am going to use the notation $a \rightarrow b$ if a c-configuration $b$ is obtained out of a c-configuration $a$ by a possibly empty series of reductions according to rule c-1 followed by a single reduction according to a rule other than c-1. As well as in the case of $\lambda$-red this system of reduction rules may be extended in exactly the same manner to cope with predefined functions, i.e. operators. Again operators must be able to make recursive calls to c-red and then perform operations according to their semantics. How this problem is solved does not matter here. The important point is, that the same solution may be adopted in both reducers.

c-11.1: $\langle +\rangle\ e_1\ e_2 \quad \triangleright \quad r_1 + r_2$
$\qquad$ *where* $\langle e_1\rangle \triangleright \ldots \triangleright r_1$ and $\langle e_2\rangle \triangleright \ldots \triangleright r_2$

c-11.2: $\langle lazycons\rangle\ e_1\ e_2 \ \triangleright\ lazycons\ e_1\ e_2$

c-11.3: $\langle if\rangle\ e_1\ e_2\ \dot{e}_3 \ \triangleright\ \langle e_i\rangle$

$$where\ i = \begin{cases} 1, & \text{if } \langle e_1\rangle \triangleright \ldots \triangleright true; \\ 2, & \text{if } \langle e_1\rangle \triangleright \ldots \triangleright false. \end{cases}$$

c-11.4: $\langle Y\rangle\ f \ \triangleright\ \langle f\rangle\ (Y\ f)$

Figure 6.1 shows the sequence of c-configurations the c-reducer is stepping through when reducing the combinator expression $C\ (B\ +\ I)\ 1\ 16$. Note that no sharing is assumed in c-red as well as no recipes are used in λ-red. Of course c-red is started in c-configuration $\langle e\rangle$ where e is the combinator expression to be reduced. (A c-configuration is a c-tree $t$ and a node of $t$ c-red is currently standing at).

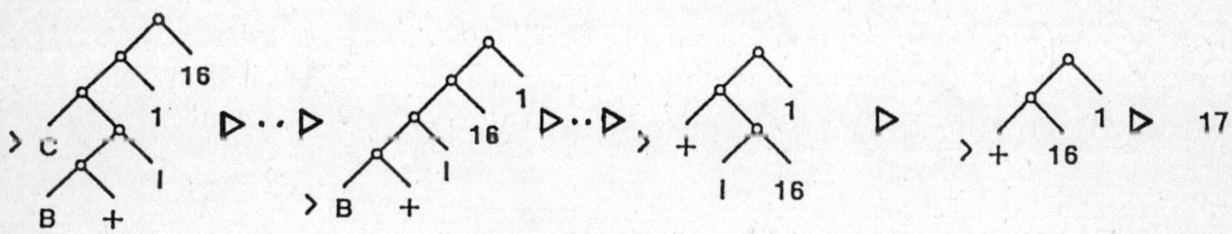


Figure 6.1

## 7. The Correspondence between λ-red and c-red

In this section the correspondence between λ-red and c-red will be shown. Both reducers will be seen to perform essentially the same operations modulo their different abilities to handle the transport of argument bindings. I will first give an intuitive understanding of the correspondence by means of an example and then prove it formally.

Let λ-red start in a λ-configuration $[\langle a\rangle,\ nil]$. During the reduction process it arrives at λ-configurations of the form: $[\langle e\rangle,\ env]\ cl_1 \ldots cl_m$, where $[e,\ env]$ and $cl_1, \ldots, cl_m$ are closures. Let us have a closer look at the closure $[e,\ env]$. The environment $env$ is of the form:

$$x_1 \rightarrow [e_1,\ env_1] : \ldots : x_n \rightarrow [e_n,\ env_n]$$

By Lemma 5.1 the variables $x_1 \ldots x_n$ are bound to λ-nodes $p_1, \ldots, p_n$ on a path that leads from the father of $(root(e))$ to $root(a)$. Furthermore the free variables of $e$ have to be a subset of the variables bound in $env$ (this is the whole idea of a closure). How could the corresponding subtree the combinator reducer is working at look like? It somehow has to incorporate the variable free form of $e$ and certain parts of the environment $env$. This seems to be a reasonable assumption since c-red too somehow has to be able to access values variables are bound to. The following example may illustrate this.

### Example

This example is intended to give an intuitive idea of the correspondence between λ-red and c-red. In the left column a series of snapshots of λ-configurations λ-red will be stepping through is given. In the right column the corresponding c-configurations are shown. The expression to be reduced is $(\lambda f. + (f2)11)((\lambda x.\lambda y. + xy)1)$.

In figure 7.1 both reducers are shown in their start configurations. The circled subtrees on the right are the variable free versions of the corresponding subtrees on the left. After one application of rule λ-3 and λ-2 λ-red is in the configuration shown in figure 7.2. The combinator reducer has not done anything since there is nothing in the world of combinator reduction that corresponds to binding. The right combinator subtree in figure 7.2 may be interpreted as corresponding to the environment-part in the λ-tree in figure 7.2 as well as corresponding to the right subtree on the left of figure 7.1.

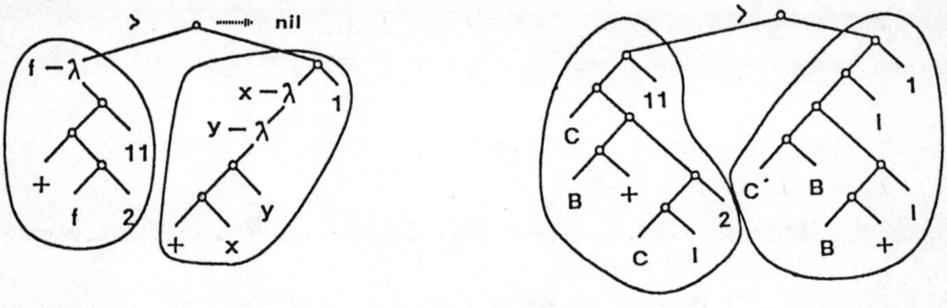


Figure 7.1

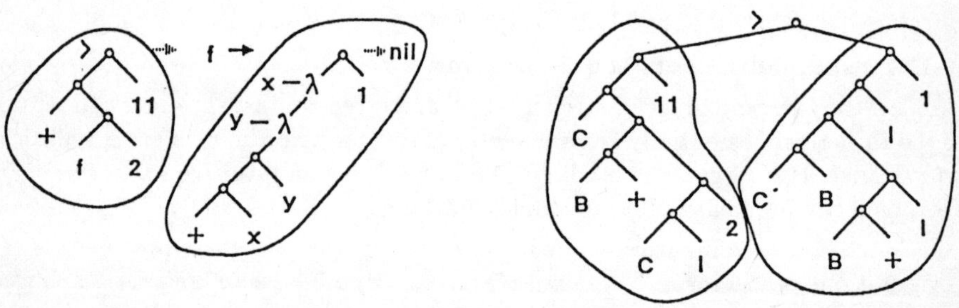


Figure 7.2

Now λ-red starts transporting the environment containing the binding of $f$ towards the leaf $f$ and arrives at the λ-configuration depicted in figure 7.3. (environments will only be shown at those subtrees they are really needed in.) The combinator reducer c-red does something similar by reducing according rule c-5 and arrives at the c-configuration depicted on the right in figure 7.3. Again corresponding tree parts are encircled (e.g. $B + (CI2)$ is the variable free version of $+(f2)$).

After handing the environment to $f2$ + and one recursive call λ-red arrives at the λ-configuration shown in figure 7.4. c-red has made one reduction according to rule c-4.

Next λ-red moves the environment to the leaf node $f$ and c-red moves the subtree corresponding to the environment by reducing according to rule c-5. This situation is shown in figure 7.5.

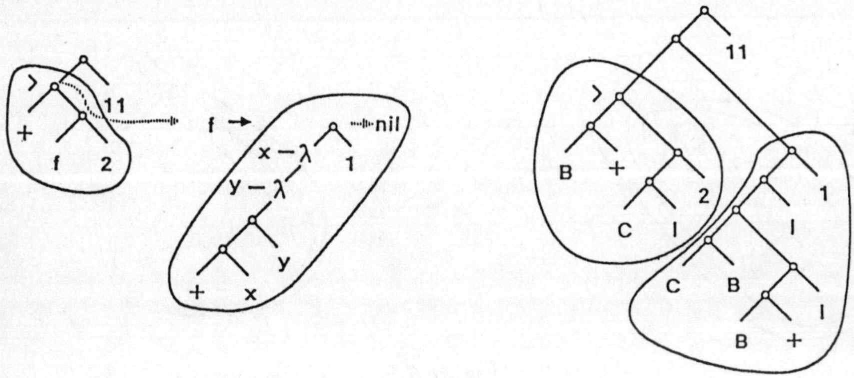


Figure 7.3

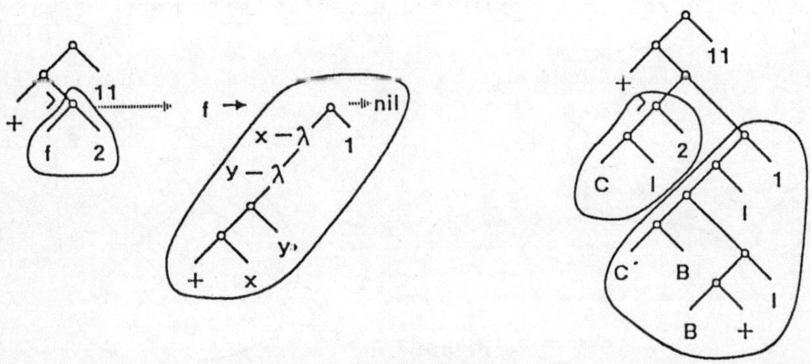


Figure 7.4

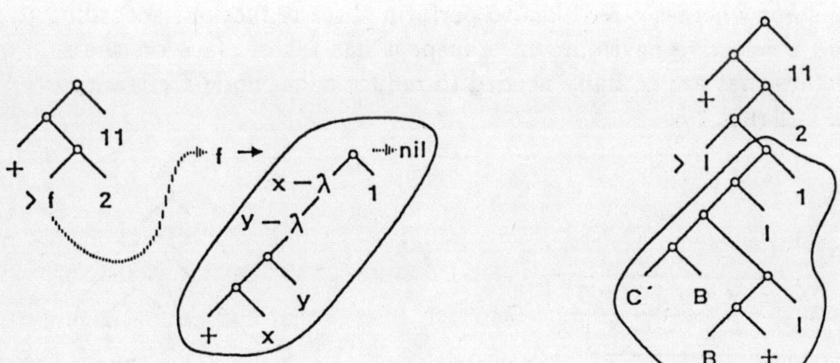


Figure 7.5

The move corresponding to $\lambda$-red's environment lookup is a reduction according rule c-2 on the combinator side yielding the following situation shown in figure 7.6.

Again the bindings performed next by $\lambda$-red lead to no move of c-red. Subtrees 1 and 2

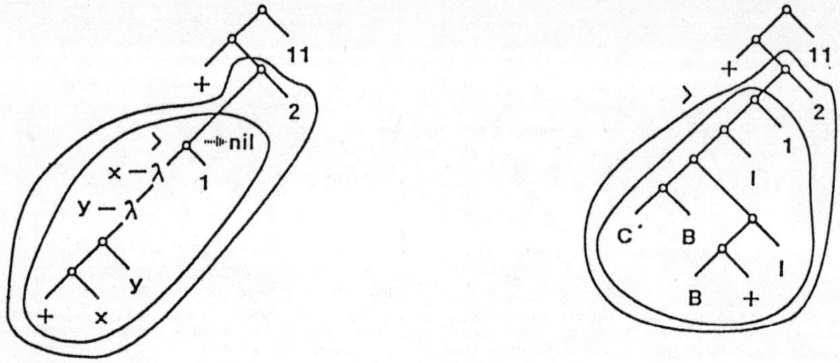

Figure 7.6

on c-red's side may be interpreted as corresponding to the arguments 1 and 2 on $\lambda$-red's side as well as belonging to the newly built environment (figure 7.7). Note the order of the variables bound in *env* and the order of arguments in the corresponding c-tree !

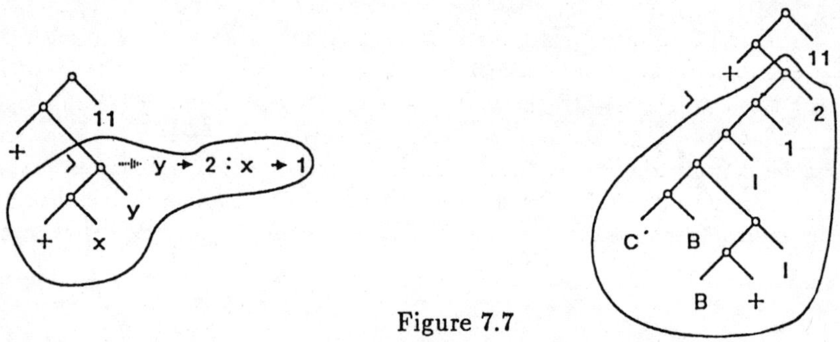

Figure 7.7

Next $\lambda$-red moves the environment containing bindings of $x$ and $y$ to the leaves $x$, $y$ and + in two steps whereas c-red has to perform three reductions according to rule c-9 and c-4. Here a selective environment-transport has taken place on the side of c-red. Only those values that are actually needed to reduce a leaf node are transported to that leaf as shown in figure 7.8.

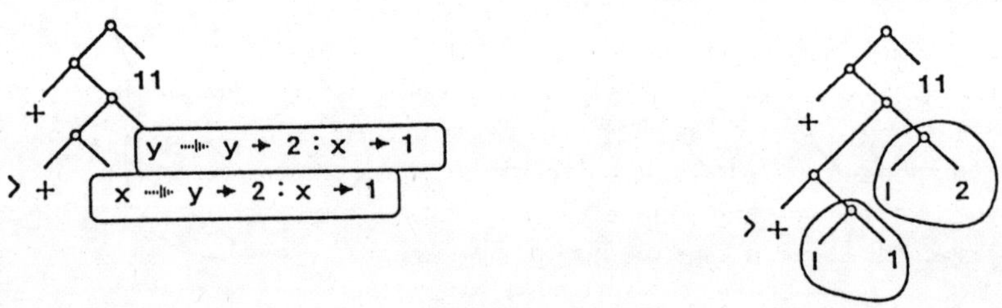

Figure 7.8

The last picture shows the configuration both reducers are in after having made two environment lookups and two reductions according rule c-2 respectively (figure 7.9).

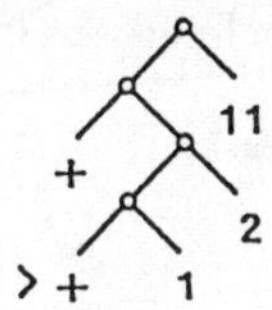


Figure 7.9

Intuitively the following has happened: Both reducers transported argument trees to the corresponding leaf positions in the trees. The $\lambda$-reducer had to make explicit bindings and transported it's environment "in one piece" but had to search the environment at variable leaf nodes to find the right trees. The combinator reducer had no bindings to make but transported it's argument trees separately and selectively through the tree. For one reduction step of $\lambda$-red according to rule $\lambda$-3 a series of combinator reductions was necessary on c-red's side. The purpose of this series of reductions at the root of a subtree $e$ is twofold: First, all those argument trees which were needed to reduce $e$ were transported selectively to subtrees $e_1$ and $e_2$ of $e$ by reducing according to the reduction rules of combinators $S$, $B$, $C$ or their primed versions. On $\lambda$-red's side these argument trees were bound in *env* to variables contained in node-variable pairs in $useful(e)$. Second, all those argument trees which were not needed in $e$ were blocked off at $root(e)$ by reducing according to combinators $K$, $J$ and $J'$. On $\lambda$-red's side these argument trees were bound in *env* to variables contained in node-variable pairs in $useless(e)$. This selective transport of arguments on c-red's side ensures that exactly the right arguments arrive at variable leaves making environment lookup trivial. For an example of the second purpose see figure 7.10

Figure 7.10

In the following these observations will be formalized: I first give a definition which essentially is an extension of *varfree* to closure trees. Then I prove a couple of Lemmata which will be needed to show the main Theorem of this paper. It states that there is a strong correspondence between $\lambda$-reduction and combinator reduction. It will be proved, that the two mechanisms behave almost identical modulo their different abilities to transport argument trees to variable leaves.

**Definition** (corr)

Let $a$ be a $\lambda$-tree and let $e$ be a subtree of $a$. Let $p_1, \ldots, p_n$ be the $\lambda$-nodes on the path from $father(root(e))$ to $root(a)$ with bound variables $x_1, \ldots, x_n$ respectively. Let $[e,\ env]$ be a closure and let $env = x_1 \rightarrow [e_1,\ env_1] : \ldots : x_n \rightarrow [e_n,\ env_n]$. Let $bounds(e) = \{(p_{i_1}, x_{i_1}), \ldots, (p_{i_k}, x_{i_k})\}$ ordered by $\ll$. Then the c-tree corresponding to $[e,\ env]$ is called $corr([e,\ env])$ and is defined recursively as:

$$corr([e,\ env]) = \begin{cases} varfree(e)\ corr([e_{i_k}, env_{i_k}]) \ldots corr([e_{i_1},\ env_{i_1}]) & \text{, if } env \neq nil; \\ varfree(e) & \text{, otherwise.} \end{cases}$$

**Example**

Consider the closure $[e, env]$ on the left of figure 7.3 and let $p$ be the $\lambda$-node binding $f$. Now $e = +(f2)$, $env = f \rightarrow [e', nil]$ where $e' = (\lambda x.\lambda y.+xy)1$. Furthermore $varfree(e) = stretch(abs(f, trans(e))) = B + (CI2)$ and $varfree(e') = stretch(trans(e')) = C'B(B + I)I1$. Therefore one gets: $corr([e, env]) = varfree(e)corr([e', env']) = varfree(e)varfree(e')$ since $bounds(e) = \{(p, f)\}$ and $env' = nil$. $varfree(e)varfree(e')$ is shown on the right of figure 7.3 as the application of the two circled subtrees.

Lemma 7.1 proves a connection between a binding of a variable $x$ in an environment $env$ of a closure $[e, env]$ and the node-variable pair $(p, x)$ in $bounds(e)$. Intuitively it states that whenever $x_i \rightarrow [e_i, env_i]$ is the leftmost binding of $x$ to some closure in $env$, i.e. if it is the first binding of $x$ found by $\lambda$-red when searching the environment $env$, then $(p, x)$ is in $useful(e)$. This fact will be used to show that c-red reduces the variable free version of $x$ applied to suitable arguments to the correct c-tree, namely the one corresponding to $[e_i, env_i]$.

**Lemma 7.1**

Let $a$, $e$, $bounds$ and $env$ be defined as in the definition of $corr$ above. Let $x \in free(e)$ and let $x_i \rightarrow [e_i, env_i]$ be the leftmost binding of $x$ in $env$ and let $x_i = x$ be the bound variable of $\lambda$-node $p_i$. Then $(p_i, x_i) \in useful(e) \subseteq bounds(e)$.

Lemmata 7.2, 7.3 and 7.4 state how the variable free versions of a constant $c$, a variable $x$ and an application $e_1e_2$ reduce when applied to a suitable number of arguments. These three lemmata form the core of the main theorem in this paper.

**Lemma 7.2**

Let $c$ be a constant and let $bounds(c) = \{(p_1, x_1), \ldots, (p_k, x_k)\}$ ordered by $\ll$. Then $\langle varfree(c)\ a_k \ldots a_1\rangle$ reduces to $c$.

**Lemma 7.3**

Let $x$ be variable and let $bounds(x) = \{(p_1, x_1), \ldots, (p_k, x_k)\}$ ordered by $\ll$. Then $\langle varfree(x)a_k \ldots a_1\rangle$ reduces to $\langle a_i\rangle$, where $i, 1 \leq i \leq k$, is minimal with $x = x_i$.

**Lemma 7.4**

Let $e = e_1e_2$ be a $\lambda$-tree , let $bounds(e) = \{(p_1, x_1), \ldots, (p_k, x_k)\}, k \geq 1$, $bounds(e_1) = \{(p_{i_1}, x_{i_1}), \ldots, (p_{i_m}, x_{i_m})\}$ and $bounds(e_2) = \{(p_{j_1}, x_{j_1}), \ldots, (p_{j_n}, x_{j_n})\}$ all ordered by $\ll$. Then $\langle varfree(e)a_k \ldots a_1\rangle$ reduces to $\langle(varfree(e_1)a_{i_m} \ldots a_{i_1})(varfree(e_2)a_{j_n} \ldots a_{j_1})\rangle$.

Lemma 7.4 gives a precise understanding of how the transport of argument trees works. As mentioned in section 4 for any application $e_1e_2$ two "types" of combinators are generated:

If $(p, x) \in useful(e_1e_2)$ then a combinator $S$, $B$ or $C$ (or a primed version thereof) was generated depending on the membership of $(p, x)$ in $useful(e_1)$ and $useful(e_2)$. During the reduction process these combinators ship the corresponding argument trees selectively to the subtrees $e_1$ and $e_2$.
If however $(p, x) \in useless(e_1e_2)$ holds then the argument tree associated with $x$ is never needed to reduce $e_1e_2$. The combinator $J$ or $J'$ generated at $root(e_1e_2)$ for $x$ gets rid of this useless argument tree. In a sense the root of $e_1e_2$ gives rise to the first opportunity to throw away useless argument trees by reducing according to the reduction rules associated with $J$ and $J'$. There is nothing in the world of combinators which corresponds to the notion of binding in the world of $\lambda$-reduction. The $\lambda$-reducer would bind the argument tree to $x$ but would never look it up in the environment. This is not possible for the combinator reducer. It explicitly has to throw away an argument which is not needed. This is exactly what happens at $root(e_1e_2)$. Now I am ready to state the main theorem of this paper.

**Theorem 7.1**

Let $a$ be a $\lambda$-tree without free variables. Let $\lambda$-red started in $\lambda$-configuration $[\langle a\rangle, nil]$ arrive at $\lambda$-configuration $[\langle e\rangle, env]\ cl_1 \ldots cl_m$ after a certain amount of $\lambda$-reduction steps. Then c-red started in c-configuration $\langle translate(a)\rangle$ builds the combinator tree $corr([e, env])\ \ corr(cl_1)\ \ldots\ corr(cl_m)$ after a certain amount of combinator reduction steps and stands either at the root of $corr([e, env])$ or at some node on the left backbone of $corr([e, env])$.

**Proof:**

I first show that the claim holds for the start configurations. The start configuration of $\lambda$-red is $[\langle a\rangle, nil]$. The start configuration of c-red is $\langle translate(a)\rangle$. It suffices to show that $translate(a) = corr([a, nil])$, since $m = 0$ and $env = nil$. Clearly $useless(a)$ is empty by definition of $useless$. Furthermore $free(a)$ is empty by assumption. Therefore $bounds(a)$ is empty and $translate(a) = stretch(trans(a)) = varfree(a) = corr([a, nil])$ hold by definition of $varfree$ and $corr$. Thus the claim holds for the start configurations.
For the induction step let $\lambda$-red be in $\lambda$-configuration $[\langle e\rangle, env]cl_1 \ldots cl_m$. Furthermore let $env = x_1 \rightarrow [e_1, env_1] : \ldots : x_n \rightarrow [e_n, env_n]$ and $bounds(e) = \{(p_{i_1}, x_{i_1}), \ldots, (p_{i_k}, x_{i_k})\}$ ordered by $\ll$.

case 1: $\lambda$-red reduces according to rule $\lambda$-4 ($e = c$). In this case $\lambda$-red yields the result:

$$c\ cl_1 \ldots cl_m$$

and stops. By the induction hypothesis the combinator reducer c-red has built the c-tree :

$$corr([c, env])corr(cl_1) \ldots corr(cl_m)$$

and is currently standing at the root of $corr([c, env])$ or somewhere on it's left backbone. After a couple of reduction steps according to rule c-1 it arrives at the leftmost leaf of $corr([c, env])$. Since

$$corr([c, env]) = varfree(c)corr([e_{i_k}, env_{i_k}])\ldots corr([e_{i_1}, env_{i_1}])$$

it now reduces $corr([c, env])$ to $c$ by Lemma 7.2 and stops yielding the result:

$$c\ corr(cl_1)\ldots corr(cl_1).$$

case 2: $\lambda$-red reduces according to rule $\lambda$-1 ($e = x$). Here the next $\lambda$-configuration is:

$$[\langle e_l\rangle, env_l]cl_1 \ldots cl_m$$

where $x_l \to [e_l, env_l]$ is the leftmost binding of $x$ in $env$ (i.e. $x = x_l$ holds). By the induction hypotheses the combinator reducer c-red has built the c-tree :

$$corr([x, env])corr(cl_1)\ldots corr(cl_m)$$

and is currently standing at the root of $corr([x, env])$ or somewhere on it's left backbone. Now

$$corr([x, env]) = varfree(x)corr([e_{i_k}, env_{i_k}])\ldots corr([e_{i_1}, env_{i_1}])$$

Let $p_l$ be the $\lambda$-node binding $x_l$. By Lemma 7.1 $(p_l, x_l) \in useful(e) \subseteq bounds(e)$. By Lemma 4.4 $(p_l, x_l)$ is minimal in $bounds(e)$ with $x_l = x$. Thus again after a couple of reduction steps according to rule c-1 c-red reduces $corr([x, env])$ to $corr([e_l, env_l])$ by Lemma 7.3 and yields the c-configuration:

$$\langle corr([e_l, env_l])\rangle corr(cl_1)\ldots corr(cl_m)$$

case 3: $\lambda$-red reduces according to rule $\lambda$-3 ($e = fb$). Here the next $\lambda$-configuration is:

$$[\langle f\rangle, env][b, env]cl_1 \ldots cl_m.$$

By the induction hypotheses the combinator reducer c-red has built the c-tree :

$$corr([fb, env])corr(cl_1)\ldots corr(cl_m)$$

and is currently standing at the root of $corr([fb, env])$ or somewhere on it's left backbone. Now

$$corr([fb, env]) = varfree(fb)corr([e_{i_k}, env_{i_k}])\ldots corr([e_{i_1}, env_{i_1}])$$

Again after a couple of reduction steps according to rule c-1 the combinator reducer reduces $corr([fb, env])$ to $corr([f, env])corr([b, env])$ yielding the c-configuration :

$$\langle corr([f, env])corr([b, env])\rangle corr(cl_1)\ldots corr(cl_m)$$

This follows immediately from Lemma 7.4 and the definition of $corr$. After another reduction step according to rule c-1 c-red stands at the root of $corr([f, env])$ and the claim holds.

In the case that $bounds(fb) = \emptyset$, $corr([fb, env]) = varfree(fb) = varfree(f)$ $varfree(b)$ by Theorem 4.2. In this case c-red does not have to do anything but at most one reduction according to rule c-1 to arrive at the corresponding c-configuration since $bounds(fb) = \emptyset$ implies $bounds(f) = bounds(b) = \emptyset$ and therefore $varfree(f)varfree(b) = corr([f, env])corr([b, env])$ holds.

case 4: $\lambda$-red reduces according to rule $\lambda$-2 ($e = \lambda x.b$). Here the next $\lambda$-configuration is:

$$[\langle b\rangle, x \rightarrow cl_1 : env]cl_2 \ldots cl_m,$$

if $m \geq 1$. (If $m = 0$ then $\lambda$-red stops and yields $[\lambda x.b, env]$ as result.) By the induction hypotheses the combinator reducer c-red has built the c-tree :

$$corr([\lambda x.b, env])corr(cl_1) \ldots corr(cl_m)$$

and is currently standing at the root of $corr([\lambda x.b, env])$ or somewhere on it's left backbone. Now

$$\begin{aligned}
&varfree(\lambda x.b) = \\
&abs([x_{i_k}, \ldots x_{i_1}], trans(\lambda x.b)) = \\
&abs([x_{i_k}, \ldots x_{i_1}, x], trans(b)) = \\
&varfree(b)
\end{aligned}$$

since $bounds(b) = \{(p, x), (p_{i_1}, x_{i_1}), \ldots, (p_{i_k}, x_{i_k})\}$ ordered by $\ll$, where p is the $\lambda$-node whose bound variable is $x$. (i.e. $p = father(root(b))$) by Theorem 4.3. Therefore

$$\begin{aligned}
&corr([\lambda x.b, env])\ cl_1 \ldots cl_m = \\
&varfree(\lambda x.b)\ corr([x_{i_k}, env_{i_k}]) \ldots corr([x_{i_1}, env_{i_1}])\ corr(cl_1) \ldots corr(cl_m) = \\
&varfree(b)\ corr([x_{i_k}, env_{i_k}]) \ldots corr([x_{i_1}, env_{i_1}])\ corr(cl_1) \ldots corr(cl_m) = \\
&corr([b, x \rightarrow cl_1 : env])\ corr(cl_2) \ldots corr(cl_m)
\end{aligned}$$

by definition of *corr*. In this case c-red did not have to do anything and is now standing on the left backbone of $corr([b, x \rightarrow cl_1 : env])$.

case 5: $\lambda$-red reduces according to one of the rule $\lambda$-3.i, $1 \leq i \leq 4$, i.e. it reduces a subtree of the form $[op, env][a_1, env] \ldots [a_n, env]$. By the induction hypotheses the combinator reducer has built the corresponding subtree which is of the form $op\ corr([a_1, env]) \ldots corr([a_n, env])$ since $bounds(op) = \emptyset$. Therefore c-red too makes recursive calls to itself in order to reduce the argument trees standing on parameter positions the operator *op* is strict in. The $\lambda$-reducer will do exactly the same.

This finishes the proof of Theorem 7.1.

Remark to case 4 of the proof above:
As shown $\lambda$-red stops in the case $m = 0$ and yields the result $[\lambda x.b, env]$. The combinator

reducer has built the c-tree $varfree(b)corr([x_{i_k}, env_{i_k}]) \ldots corr([x_{i_1}, env_{i_1}])$. For $root(b)$ exactly $k+1$ combinators have been produced since the cardinality of $bounds(b)$ is $k+1$. Therefore c-red may reduce according to the reduction rules associated with the first $k$ combinators produced for $root(b)$. The first combinator produced is the last one to be reduced. There is no reduction possible according to the reduction rule associated with this combinator, since the preceding $k$ combinator reductions have "used up" $corr([x_{i_k}, env_{i_k}]) \ldots corr([x_{i_1}, env_{i_1}])$. Thus the combinator reducer may proceed a little further than $\lambda$-red in this case.

## 8. Relation with other Work

The problem of comparing $\lambda$-style and combinator style reduction mechanisms was first considered by Simon Peyton Jones [Jon82] on an experimental basis. The next approach was made by T.Ida and A.Konagaya [Ida84]. Both approaches are not quite satisfactory. The former since to my opinion it is not fair to compare the two mechanisms on an experimental basis: Too many implementation details could clutter up the results. Furthermore the two mechanisms profit by entirely different optimizations, e.g. there is nothing on the $\lambda$-side which corresponds to the benefits the combinator reducer enjoys since it is implemented using graph reduction. Recipes as described in [Hen80] only partially solve this problem. (However the substituting $\lambda$-reducer described in [Wad71] is a solution.) To come up with a fair analytical comparison the two mechanisms have to be boiled down to the very basic mechanisms they are built on. The unifying model described in this paper already proved to be a promising way to go in this direction. The second approach mentioned above does not take into consideration the tradeoff between the relatively high argument transportation costs the combinator reducer is suffering from and the trivial environment-lookup it has to perform (reduction according to I).

## 9. Future Work

A new graph reducer has been developed based on the comparisons made in this paper [Obe86a]. The new reducer is in a sense hybrid between c-red and $\lambda$-red: It employs an environment-like structure to handle the transport of argument bindings and uses control information similar to combinators to guide the transport of arguments through the graph. Thus it avoids the quadratic growth in code size of c-red but maintains full lazyness. The new reducer may be regarded as implementing Wadsworth's lambda-style graph reducer. [Wad71]

## 10. Acknowledgements

Thanks are due to R.Wilhelm, D. Maurer, F.Müller and A. Lucks for reading earlier versions of this paper and for many interesting discussions.

## 11. Literature

Cla80 T.J.W. Clarke, P.J.S. Gladstone, C.D. MacLean and A.C. Norman, "SKIM - The S, K, I Reduction Machine", *Procedings of the 1980 ACM Lisp Conference*, p. 128 - 135, Aug. 1980.

Hen80 P. Henderson, "Functional Programming: Application and Implementation", *Prentice-Hall International Series in Computer Science*, 1980.

Hud84 P. Hudak and D. Kranz, "A Combinator-based Compiler for a Functional Language", *Eleventh annual ACM Symposium on Principles of Programming Languages*, January 1984.

Hug82 R. J. M. Hughes, "Super-Combinators ", *Conference Record of the 1982 ACM Symposium on LISP and Funtional Programming1-10*, 1982.

Ida84 T. Ida and A. Konagaya, "Comparison of closure reduction and combinator reduction schemes", *unpublished*, 1984.

Jon82 Simon Peyton Jones, "An Investigation on the Relative Efficiencies of Combinators and Lambda Expressions", *Conference Record od the 1982 ACM Symposium on Lisp and Functional Programming*, p. 150-158, Aug. 1982.

Joy85 M.S. Joy and V.J. Rayward-Smith, F.W. Burton, "Efficient Combinator Code", *Computer Lang. Vol. 10, No. 3/4*, 1985.

Ken82 J.R. Kennaway, "The Complexity of a Translation of λ-calculus to Combinators", *Report CS/82/23/E*, p. University of East Anglia, 1982.

Obe86a H.G. Oberhauser, "A Fully Lazy Lambda Style Graph Reducer", SFB 124 -C1, 06/1986, Universität des Saarlandes, West Germany, 1986.

Obe86 H.G. Oberhauser, "On the Correspondence of Lambda Style Reduction and Combinator Style Reduction", SFB 124 -C1, 02/1986, Universität des Saarlandes, West Germany, 1986.

Sto84 W.R. Stoye, T.J.W. Clarke and A.C. Norman, "Some Practical Methods for Rapid Combinator Reduction", *1984 ACM Symposium on Lisp and Functional Programming*, p. 159-166, 1984.

Tur79 D.A. Turner, "A new implementation technique for applicative languages", *Software – Practice and Experience*, 9, p. 31-49, Sept. 1979.

Tur79a D.A. Turner, "Another algorithm for bracket abstraction", *The Journal of Symbolic Logic*, 44, June 1979.

Wad71 C. P. Wadsworth, "Semantics and Pragmatics of the Lambda-Calculus.", *Ph.D Thesis*, Oxford University, 1971.

Wis82 D.S. Wise, "Interpreters for Functional Languages", *Functional Programming and it's Applications, an advanced course. Cambridge University Press*, 1982.

# Head Order Reduction: A Graph Reduction Scheme for the Operational Lambda Calculus

Klaus Berkling
CASE Center
Syracuse University

**Abstract**

This paper introduces a new reduction order for the lambda calculus, called *head order*. It lends itself to a graph reduction implementation. The head normal form of a lambda expression corresponds closely to K. J. Greene's lazy normal form of his LNF-calculus, that is, the head order normal form will have the same variable (respectively, constant) as head and the same number of arguments as its normal form. In the context of future implementations, the head normal form is also called the head normal form skeleton (respectively, outline). This skeleton is produced first. Then the normal forms of the arguments are obtained and inserted into this skeleton concurrently, since they do not interact anymore. Head order reduction automatically exposes concurrent processes.

The procedure for generating a head normal form is described in terms of beta-reduction in-the-large, eta-extension in-the-large, and identity-reduction in-the-large. These in-the-large operations represent a new concept. They take sequences of bindings and/or applications as initial form, and produce a result in-the-large. They comprise a multitude of single reductions (*e.g.*, beta, eta) occurring in a specific pattern.

A geometric, visual correspondence is described which pictures the process of obtaining a head normal form in terms of smoothing out a zigzag line (representing a lambda expression). The claim is that this image gives insight into the reduction process of very large lambda expressions. A formal definition is given to the concept of *environment*, which corresponds to a special zigzag line. An environment lookup corresponds to a selector function applied to a sequence of arguments.

Special attention is given to the problem of sharing. It is not generally possible, as opposed to the case of combinators, since an argument may contain free variables. They might get bound erroneously when the argument is substituted. Head order reduction has the advantage of normal order: no argument is reduced unless actually needed. How often an argument is copied depends on the occurrence of free variables and the number of different contexts it is substituted in. The implementation difficulties to make this information available on the run are touched upon.

Finally, some preliminary simulation results are reported. A set of lambda expressions was abstracted by K. J. Greene's abstraction algorithm. Then all combinators are expressed in terms of lambda expressions. Normal forms of the abstracted lambda expressions are compared to the normal forms of the original expressions. They are all respectively equal up to eta-conversion. This and a comparison of results obtained by the GMD string reduction machine give a fairly good indication of the correctness of the method.

## Introduction

The recent surge of interest in functional and relational programming and their implementations has also revived interest in the lambda calculus and its operational characteristics. Although there exist many abstract machines to describe its operational characteristics, none of them is really satisfactory. They are either not complete [Aiello 1981], lack certain properties (Church-Rosser, reduction semantics), or, last but not least, lead to implementations which are simply too inefficient to become a serious competitor to the von Neumann type computer. Notorious difficulties arise from free variables (relatively free with respect to some expression), which are not compatible with a straightforward stack implementation. Systems for functional programming may forbid certain evaluation sequences and avoid the free variable in this way, but on the other hand, certain desirable capabilities with respect to higher order functions are lost. The difficulties with free variables are so large that serious attempts to make combinators operational have been undertaken successfully. A system based on combinators typically has a set of combinators as an instruction set implemented in hardware or firmware. However, the programmer does not think in terms of combinators, and compilation and decompilation phases have to be introduced between soft user programs and hard combinator instructions. Experience has shown that user programs translate into incomprehensible and, at best linear growing [Turner 1979], very large combinator expressions. The object code is always larger than the source code, because, no matter what the underlying architecture is, it is always necessary to express complicated operations recursively in terms of simpler operations. Combinators, in spite of all their nice properties, (Church-Rosser, reduction semantics, sharing, laziness, variable-freeness), are not totally convincing. The execution of a combinator is actually more expensive in terms of memory cycles than the execution of a von Neumann type instruction, but very little is accomplished. The size of combinator expressions requires large memories which in turn require large addresses, which require in turn wide data paths. All this is, of course, technically possible, but neither cheap nor simple [Clarke 1980, Scheevel 1986]. The price paid to overcome the free variable problem seems too high. The idea of supercombinators [Hughes 1982] addresses this apparent disparity by reverting to conventional machine language and a compilation technique which allows free variables at the user level, but binds all free variables at the next surrounding group of bindings [Johnsson 1983].

One might give the following reasoning why combinators are not the final answer. The historical line of development notwithstanding, combinators may be considered as abbreviations for closed lambda expressions. Closed lambda expressions do not contain free variables. An enumeration of closed lambda expressions ordered by size contains the well-known combinators among others which do not have a name (see section 5).

What is not the case, however, is that lambda expressions higher up in this enumeration (corresponding to user programs) can be assembled from a fixed set of earlier lambda expressions in the enumeration in a simple straightforward way, for example, by dividing up the large expression into subtrees, because many variables can occur relatively free, that is, between the variable occurrence and its corresponding binding instance are many other bindings interspersed.

When it comes to producing the normal form of a given lambda expression, we can start from either the original lambda expression or its translation into a combinator expression. There are now three paths open to obtain a normal form:

1) Using beta-reduction on the original expression.
2) Using combinator-reduction on the translated version.
3) Using beta-reduction on the translated version, but after all combinator names have been replaced by their corresponding lambda expressions.

The advantage of combinator-reduction in this comparison is that the execution of a combinator does not require the full-fledged lambda calculus. It is enough to rearrange a few pointers without ever going to the trouble of dealing with variable names at all.

Suppose, however, one had at hand a fast graph reduction implementation of the lambda calculus. The ensuing pointer rearrangements would be very similar, if not identical, and so would the number of memory cycles used. As a consequence, the space-time complexities of paths two and three would not be substantially different.

The open question is how the space-time complexity of path one relates to the others under the assumption that there is a fast graph reduction implementation of the lambda calculus that can cope correctly and completely with the problem of free variables. The result given in this paper is that there is such a fast graph reduction implementation of the lambda calculus, and it is so fast that path one has the lowest space-time complexity of the three.

## 1. Reduction Order

Much research has been devoted to the order of reduction, or evaluation, and the consequences for the termination properties of some chosen specific order. Call by value, call by name, call by reference, normal order, applicative order, lazy evaluation, fully lazy evaluation, and its derivatives are the terms used in this context. Functional notation is often used to describe a particular order. For example, call by value can be denoted by the formula

(eval combination) = (apply (eval rator) (eval rand)),

and call by name by the formula

(eval combination) = (apply (eval rator) rand).

*Apply* and *eval* are not essentially different; evaluation recursively implies application and conversely. Thus, in general, we may write (informally)

(eval combination) = (eval-apply (eval rator) (eval rand)).

The recurrence of *eval* on the top level is necessary since every *apply* may yield new operator-operand combinations. Taking the functional notation seriously, this corresponds to the occurrence of arbitrarily many *eval* or *red* metaoperators at every node of a program expression. Indeed, an implementation requires scanning and rescanning the expression to be reduced until no more reducible subexpressions are left in it. Evaluation or reduction cannot be described as a function on a tree in terms of applying the same metaoperator recursively on its subtrees and only on its subtrees. This simple recursive call does not provide for the multiple occurrences of eval at every node needed to rescan for newly created redexes

It is desirable, however, to organize the reduction of an expression in such a way that parts already done do not have to be revisited over and over again. This is important to make concurrent processing easier, but it is also important to get away from the revisiting implementation philosophy for the following reason: To eradicate an expression from all combinations (redexes) altogether might not be the goal when considering the problem in a larger context. It is reasonable to expect that the expression (3 + 4) reduces to 7, but only when we implicitly assume that the goal is to coerce it to a number. For the equation (3 + 4) = (a + b) the goal might be to establish a coercion of a to 3 and b to 4 (or a to 4 and b to 3) in a larger context.

A hint in the direction of exposing done parts is given by the concept of a *head normal form* [Wadsworth 1971]. The emphasis on the head is the interesting concept. For the following it is advantageous to consider lambda expressions represented by Prefix Polish notation such that applications are denoted explicitly by an *ap* constructor abbreviated to **p**. An arbitrary lambda expression can then be considered to consist of:

1) a sequence of *n* bindings, possibly zero, followed by
2) a sequence of *m* **p**'s, possibly zero, followed by
3) the *head*, and
4) followed by *m* expressions of the form just described.

What is the *head?* This expression cannot begin with a **p**, because it would not be the head and *m* would be larger by at least one. Thus it can only be a variable or an expression starting with a lambda, but otherwise continue as an expression of the form just described above. For a more formal description of this we refer to Barendregt [Barendregt 1981, p.168 ]:

$$\lambda x_1 \cdots\cdots \lambda x_n\, p_1 \cdots\cdots p_m\, \{x_i \mid \lambda y \cdots\cdots\}\, e_m \cdots\cdots e_1 \;; \qquad n,m \geq 0$$

The notation for variables is another concern. While character strings are appropriate at the user interface, the operational representation has to be absolutely unique and safe, that is the possibility of confusion and collision of variable names cannot be tolerated. The only system fulfilling these requirements are binding indices [Berkling 1982, De Bruijn 1972]. They depend *only* on the structure of the lambda expressions and are completely implementation independent, but nevertheless lend themselves to a specific implementation technique. One could also think about pointers already at this level, because pointers are unique via the memory locations to which they point. But this entails to take into account all the properties of a random access memory and the implementation is then location dependent. (On the other hand, the view of a pointer as the address of a memory location is somewhat narrower than the concept in general usage dictates.) In neither case did we accomplish an explicit and direct physical connection among the variable occurrences belonging together. In both cases, additional operational procedures are necessary to realize the connection. Binding indices are counting binding levels and have to be changed when bindings disappear as a consequence of beta-reductions. These necessary changes are often held against binding indices, but it is also necessary to check for possible collisions and to do changes if needed in other implementations. Wadsworth's [Wadsworth 1971] suggestion of backpointers from variable occurrences to binding instances and then to arguments is of very limited value only. Only if *all* variables are replaced by pointers, and only if *all* pointers ultimately point to constants—only then can an output expression be reconstructed from the pointer structure. Here's why: If there are any free variables left in the structure, the information necessary to alter these variables to avoid collisions has been removed from the structure. Wadsworth's method may be useful in the context of evaluation where constants are mapped into constants. We are, however, interested in an operational, pure, unrestricted lambda calculus where lambda expressions reduce or expand to lambda expressions.

The operational representation of a lambda expression in *head form* (This is somewhat misleading, since all lambda expressions are in that form; head form is a particular way of looking at it.) using binding indices is:

$$H \Rightarrow \{\, \lambda^{n}\ p^{m}\ H\ A^{m} \mid \# \,\} \;; \qquad n > 0\,,\, m \geq 0 \qquad (1)$$

$$A \Rightarrow \{\, \lambda^{n'}\ p^{m'}\ H\ A^{m'} \mid \# \,\} \;; \qquad n' \geq 0\,,\, m' \geq 0$$

The head *H*, as well the argument *A*, may be a number # with the following restriction: The number occurring at the end of the path from the root of an lambda expression must be smaller than the number of bindings encountered on this path. That means we only consider closed lambda expressions. The conversion algorithm to indices from character string representations of variables leaves these so called "free variables", which are free with respect to the the expression considered, unchanged. They are operationally treated as names for some unknown, but assumed closed lambda expressions. Thus, the term "free variables" is not appropriate in this context.

If one attaches the notion *lambda expressions in-the-large* to the productions in (1), there is also the corresponding notion of *beta-reduction in-the-large*, which extends the notion of beta-reduction in the same way as the simple production rule "a well-formed formula is either a variable, a combination, or a lambda expression" are extended by the productions in (1). In the following, we will develop a graphical representation to support the insight in beta-reduction in-the-large and its ensuing implementation.

## 2. Betasequences and Apsequences

There are *betas-aps* and *aps-betas*. The meaning of this terminology becomes immediately clear if we consider the production of a lambda expression from (1):

$$\lambda^* \, p^* \, \lambda^* \, p^* \, \lambda^* \, p^* \cdots\cdots\cdots p^* \, \# \cdots\cdots \tag{2}$$

We call $\lambda^*$ a *betas* and $\mathbf{p}^*$ an *aps,* where each * may be replaced by an arbitrary integer. If we represent a betas by a line at 45 degrees from upper left to lower right, and an aps by a line at 45 degrees from upper right to lower left, then a terse representation of a lambda expression looks like the zigzag line in Figure 1. We have disregarded all A's (arguments, operands) pending to the right of this zigzag line. These A's are also zigzag lines.

An aps-betas is very different from a betas-aps, as shown by Figures 2 and 3. Disregarding the redexes in the A's at first, a non-degenerate aps-betas contains at least one beta-redex explicitly, as opposed to a betas-aps. The "corner" of the aps-betas is a beta-redex, which when reduced exposes the next one. Thus beta-reduction proceeds, redex by redex, until either the λ's or the **p**'s are exhausted. Considering the whole zigzag line of Figure 4, we can now define beta-reduction in-the-large. The cut C-C obviously takes out a maximal possible aps-betas. The numbers 1 to 18 denote only subtrees and are not binding indices! This identification is necessary to explain the equivalence preserving transformation shown in Figure 4. The cut C-C has been moved down the tree to the farthest possible position, namely before the bindings 17 and 18, but this is not the only place where the cutout has to be inserted; it also has to go before all subtrees from 11 to 16, which we denoted by underlining, except for one explicit occurrence serving as an example.

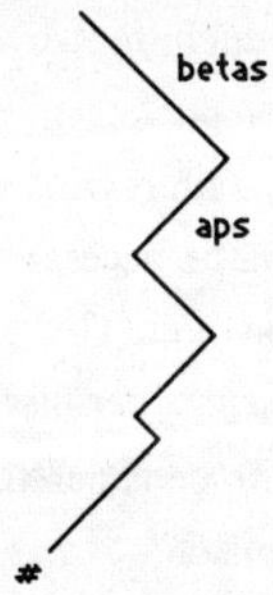


Figure 1. A Zigzagline

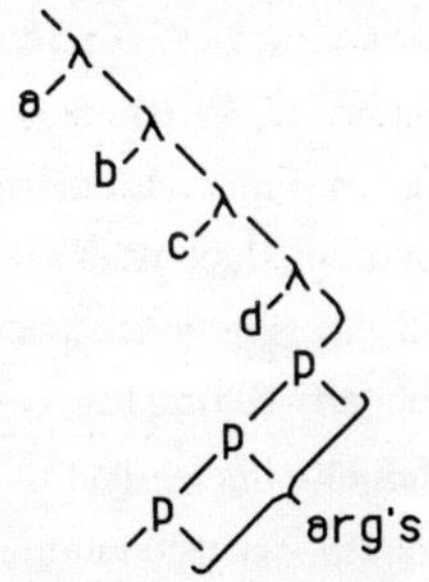


Figure 2. Enlarged betas-aps

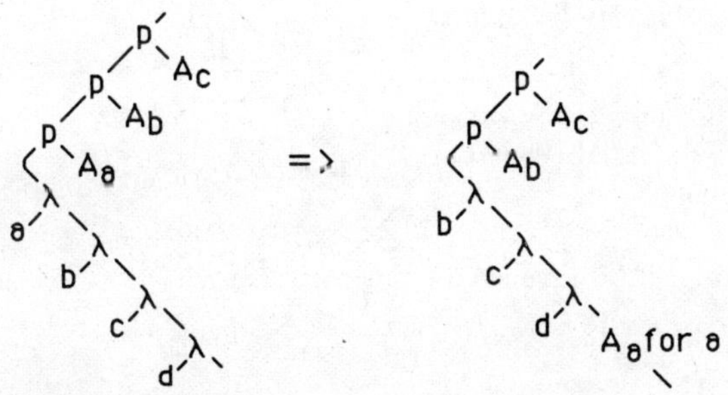


Figure 3. Enlarged aps-betas and Beta-reduction

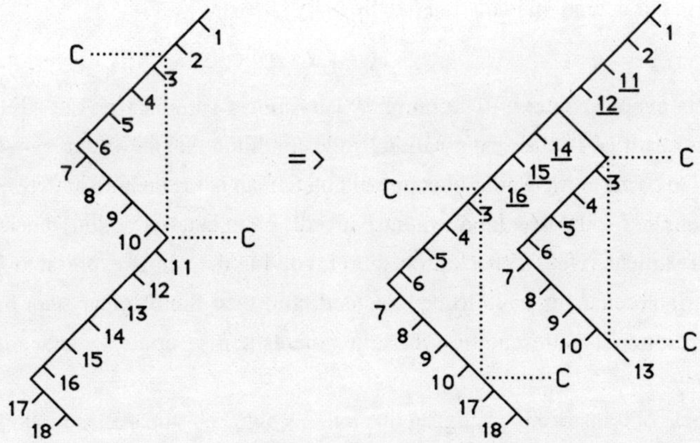


Figure 4. Beta-reduction in-the-large

In the Figure 4, the λ's are exhausted before the **p**'s. The case in which the **p**'s are exhausted before the λ's is shown in Figure 5. At first sight, one would be tempted to take the cut C-C to execute the transformation performed in Figure 4, but this leads to complications with respect to bindings 10 and 11. To maintain the binding structure, far reaching adjustments would be necessary if cut C-C is moved over bindings 10 and 11. A more elegant way employs an extension of cut C-C to cut D-D. To make this possible, the betas-aps 10' 11' 11* 10* has to be inserted. This is a permissible operation based on eta-extension. Namely, in classical notation, M and λx.(M x) are considered extensionally equivalent under the eta rule. Eta-extension in-the-large is multiple eta-extension:

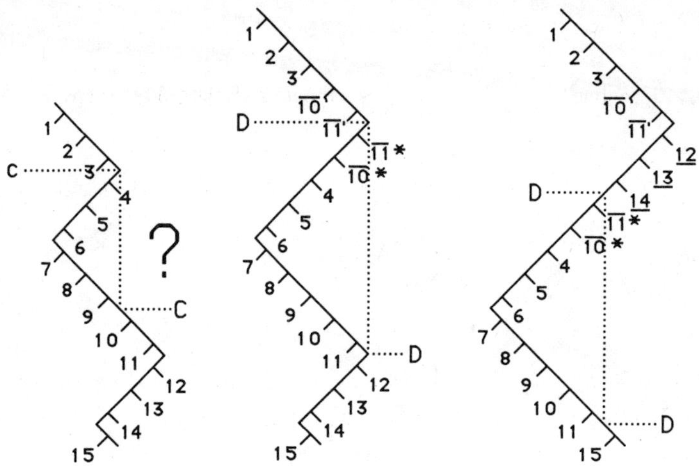

Figure 5. Beta-reduction in-the-large with preceding Eta-extension in-the-large

M and λxλy.((M y) x) are considered extensionally equivalent under the eta rule in-the-large. This corresponds to a betas-aps in our operational notation like this:

⟩λλ pp⟨ "2" ⟩0 1⟨

where the angular brackets denote context. The number 2 in quotes summarizes the adjustments necessary. Eta-extension is qualified in classical notation by the condition that the variables used in the extension are not permitted to occur in M. In our operational notation an adjustment to all free variables in M is necessary to compensate for the new bindings introduced. As in beta-reduction, this compares favorably to the classical treatment. Here a check procedure involving the whole expression M has to be initiated, if needed, variable changes have to be executed, and then the original scan has to be resumed for M. In our operational representation the adjustments can be done while scanning M, which has to happen anyway.

A more involved example of transforming a zigzag line into a single betas-aps-betas is shown in the 6 figures in Appendix A. The order of reduction is such that this procedure is repeated for all arguments. All betas-aps-betas's terminate in a variable the corresponding binding instance of which is in the aps-betas. A aps-betas, however, with a variable at the lower right end has a special meaning. We will discuss this in the next section.

## 3. The Abstract Concept of Environment

We have shown in the last section that beta-reduction in-the-large combined with eta-extension in-the-large transforms an arbitrary zigzag line into a single betas-aps-betas-#-args. In particular, this string contains a balanced string aps-betas ending in a variable. What is a betas-#? In classical notation, for example, we have $\lambda x \lambda y.x \equiv K \equiv I^m{}_n$; m=1, n=2. The parameterized identity combinator is due to Kleene [Kleene 1952] and Abdali [Abdali 1976]. In general a selector or identity combinator is of the form

$$\lambda x_1 \lambda x_2 \cdots\cdots \lambda x_n \,.\, x_m \equiv I^m{}_n \qquad\qquad 1 \le m \le n.$$

If such an selector is applied to the appropriate number of arguments *n*, it reduces to—or selects—the $m^{th}$ argument:

$$(\lambda x_1 \lambda x_2 \cdots\cdots \lambda x_n \,.\, x_m\,)\; a_1\, a_2 \cdots\cdots a_n \;\Rightarrow\; a_m$$

Since beta-reduction in-the-large as described in the last section always collects all bindings because of the use of eta-extension in-the-large, we always get a selector which is in our operational notation of the form

$$\lambda^n\, i \qquad\qquad 0 \le i \le n\text{-}1.$$

The identity-reduction in-the-large,

$$p^n\, \lambda^n\, i\; a_{n-1} \cdots a_0 \;\Rightarrow\; a_i\,,$$

is equivalent to a look-up in an environment. Obviously, this environment is best implemented as an array with index i accessing the appropriate location containing a pointer to the argument. The argument is either a variable or an expression. If it is an expression, we obtain again an arbitrary zigzag line and the procedure of applying beta-reduction in-the-large, eta-extension in-the-large, and identity-reduction in-the-large is repeated. If it is a variable, however, we obtain a single betas-aps at the top level terminating in a variable. This result has an important property: The top level betas-aps line is not altered by later transformations, no matter what reduction sequences take place in the arguments. Moreover, these arguments do not influence each other from now on. Thus, we have the opportunity to perform the reductions in these arguments concurrently. However, these arguments are not the original expressions, they are prefixed by aps-betas's (equivalent to environments) originating from beta-reductions and eta-extensions in-the-large. Since any implementation would take advantage of the representation of aps-betas's as arrays with index-access, and since these arrays are partially shared, this sharing impairs the full utilization of concurrency. But, on the other hand, full concurrency would require excessive copying of environments.

Nevertheless, every argument can be recursively reduced by the algorithm described above with the slight modification that parts of the zigzag line representing the expression are represented as arrays.

## 4. Head Order Reduction

An appropriate name for the reduction order developed in the previous sections is head order. It is closely related to *head normal form* [Wadsworth 1971], and *lazy normal form* [Greene 1985]. In a combinator context the head is a combinator symbol which leads to reductions in the context of preceding application nodes just as the aps-betas does. In the case of combinators the head is considered done if reduced to a constant. The whole expression is now considered in lazy normal form. In the case of the lambda calculus the head is considered *done* if it is converted to a bound variable (not associated to an argument), and the whole expression is now in head normal form.

However, the order of subsequent reductions has not yet been specified for the order developed here. The state of the expression after finishing the head is a betas-aps-#, a sequence of bindings followed by a sequence of applications ending in a bound variable. The obvious idea with respect to concurrent reductions is to reduce all arguments in the aps concurrently. This is possible directly for combinators. It is also possible for the lambda calculus if a copy of the appropriate environment is attached to the argument. As a matter of fact, the arguments are already augmented by the aps-betas's as a consequence of beta-reduction, eta-expansion, and identity-reduction in-the-large. Only part of the argument, namely the prefixed aps-betas, is conveniently represented as a pointer into one shared environment. Any attempt to implement sequentially or concurrently such a system would try to preserve this sharing property. Since environments expand and shrink, a naive sharing would lead to corruption if several argument processes share a single environment. The answer to this problem is to get proper control over the shrinking or cutting back of the environment. Conventionally, this is done when returning from a function. But, both terms, "returning" and "function", refer to implementation features, not concepts of the lambda calculus. A general lambda expression with its seemingly arbitrary nesting of applications and bindings does not exhibit discernible "functions". Conventionally, a function has a sequence of formal parameters and a body wherein these formal parameters actually occur. Generally, however, all parameters in the body might be relatively free with respect to this body. Here the terms function and parameter have lost their conventional meaning. Thus, the "functional languages" are restricted to certain patterns—functions—of bindings and applications and fall somewhat short of resembling the lambda calculus. Consequently, we cannot tie the cutting back of the environment in the general lambda calculus to returns from functions since there are no discernible objects of that sort in it.

The time to cut back the environment is determined as follows: The reduction process leads to a state in which environments are "pushed down" to variables. The consequent selection process, corresponding to a look-up, reduces the aps-betas to one of the arguments in the aps. Unfortunately, the environment representation of an aps-betas is shared, and environments do not and cannot disappear in the manner aps-betas's do. The method to cut back the environment suggested for a sequential implementation is based on two concepts: a) to collect all arguments with the same environment in one package, and b) to cut back after all these processes in this collection are done.

This method prohibits the destruction of environment entries before they are used. Considering again Figure A1 (appendix), we can see that groups of arguments (not necessarily adjacent) with the same environment may occur, provided the lambda expression in question is sufficiently large. (Most of the problems we are trying to solve here do not become apparent in small examples.) These groups obviously can be reduced concurrently while sharing a common trunk environment in a read-only fashion, but are otherwise completely decoupled, which is a desirable property with respect to concurrent processing.

From the discussion above and from looking at Figure A1, one might get the impression that the environment remains contiguous under all circumstances. Unfortunately, this is not the case. "Contiguous" in this context means that all variable look-ups are indexed accesses into an array. This corresponds to the statement that all environment aps-betas's are completely contained in all succeeding aps-betas's. This statement is false, because a head variable may be associated to an argument to be reduced in an earlier environment. The intervening environment entries, however, cannot be overwritten by reductions at the head, since there are arguments to be processed in the aps. Instead, the environment generated while reducing at the head has to be stacked on top of the current one with a provision to bridge the intervening environment entries. Thus, an environment representation in terms of an array instead of individual aps-betas's has *gaps*, which means that indexing is not straightforward. Each gap leads to an out-of-range test of the index value with ensuing corrective action corresponding to bridging the gap.

Head order reduction compares to call by value (applicative order) and call by name (normal order) as follows: The reduction of an expression in operand position is postponed until its appropriate environment is reached while cutting back the environment array. This is later in sequence than when encountering it, and later than when variables associated to it are encountered. This has the advantage of normal order: an argument is not reduced if no variable occurrence is refering to it. This also has the advantage of call by value of reducing it once and then sharing it, provided the argument is a closed expression (which remains closed even if reduced), which is the same as saying the argument expression maps constants into constants. This is the case for all ordinary programming languages which do not allow functions (procedures) containing free variables as arguments.

In contrast to the above, expressions appearing in head positions—either immediate or via a variable binding—are reduced when encountered at the head. The head position is an operator position. An expression brought into an operator position can never be shared since this expression and parts of the head context may participate in further reductions in a manner not forseeable at this time.

Head order also has the termination properties of normal order since an argument is always reduced in its appropriate environment (definition point) as well as in its appropriate context (variable occurrence). The delay in reducing arguments at the latest possible point in sequence has some other advantages. The theoretically postulated lambda definability of all programming language constructs is somewhat academic if tried with conventional reduction orders or evaluation schemes. Mishaps like reducing both branches of an IF (with the possibility of ensuing non-termination) or the isolated reduction of an Y operator, which is a non-terminating process, do occur and spoil the stability of such systems. Head order reduction removes all these difficulties, since work is done only if absolutely necessary, and at the last possible point in sequence. Lastly, we repeat the advantage of having a display of all processes which could be done concurrently under the condition of sharing information maximally.

We conclude this section with the remark that the term *environment* in this paper is always used in the context of some implementation that is based on the abstract structure of the aps-betas's.

## 5. To Share or not to Share

In the previous section we have discussed the problems of sharing and the advantages which may be derived from it. The possibility to share can be recognized easily for obviously closed lambda expressions, while the benefits are diminished by the necessity to test for closedness at run time.

What are the "obviously closed lambda expressions"? Combinators are equivalent to closed lambda expressions, and we will treat combinators as names for closed lambda expressions. The pragmatic essence of names is the sharing property. The name of one object may have multiple occurrences, all of which refer to the same object. Considerable savings are thus realized in real life and in computing because copies of the object do not have to be fabricated, or if copying is not possible, the object does not have to be moved.

In an implementation of a system based on the ideas set forth in this paper, multiple combinator name occurrences would share the single representation of the lambda expression which is named by its combinator symbol. This is similar to the von Neumann type computer, where multiple occurrences of the same instruction code share a single instruction execution sequence realized in hardware or firmware. There is, however, a significant difference. Only combinator name occurrences in operand positions can be shared; a combinator name in operator position must always be expanded and a copy of the named lambda expression generated, but this copy process is intrinsically an execution process as described in the previous section. In this sense the system is also a combinator reduction system, which reduces combinator expressions to lambda expressions. By employing the same abstraction algorithm at the back end which mere combinator reduction systems use at the front end, one may obtain combinator expressions, too.

Table 1 is an enumeration of closed lambda expressions ordered according to the number of bindings (size of the topmost betas), and then roughly according to the size of the expression (size of the top most aps). Attached to each lambda expression is the generally agreed upon name, in some cases several names exist. The smallest one is the identity combinator, I = (-1 . 0). The coding of the canonical representation is based on binding indices and the Lisp list notation: A *cons* node corresponds to an application, except where the *car* part is a negative integer, which encodes the number of bindings. Some expressions are given also in the usual notation.

There are some parameterized combinators in a notation due to Abdali [Abdali 1976]. This notation can be naturally extended to encompass the Church numerals [Church 1941]. A parameterized combinator corresponds to a whole set of expressions with a specific formation law. There is an obvious pattern for each such parameterized combinator; the class of expressions so formed can be characterized as *rational* [Morris 1980], since a representative of a particular pattern can be denoted by an index or a set of indices. A parameterized combinator is closely related to the concept of type, which is obvious for the Church numerals.

| | | |
|---|---|---|
| I | (-1 . 0) | λx.x |
| TL | (-1 0 -2 . 0) | λx.(x λyλz.z) |
| HD | (-1 0 -2 . 1) | |
| Y | (-1 (-1 0 . 0) -1 1 0 . 0) | |
| LISTP | (-1 (0 . $I_{33}$) . $I_{22}$) | |
| NOT | (-1 (0 . $I_{22}$) . $I_{21}$) | |
| PREDCh | (-1 (0 -1 (((0 . I) . $I_{22}$) . $Nt_0$) SUC . 0) . $I_{43}$) | |
| NULL | (-1 (((0 . $I_{22}$) . $I_{21}$) . $I_{33}$) . $I_{21}$) | |
| $I_{21}$ | (-2 . 1) | λxλy.x |
| $I_{22}$ | (-2 . 0) | λxλy.y |
| FALSE $Nt_0$ | (-2 . 0) | |
| TRUE K | (-2 . 1) | |
| EXPCh R | (-2 0 . 1) | λxλy.(y x) |
| $Nt_n$ | (-2 1 1 1 1 ····· 1 . 0) n 1's | |
| W | (-2 (1 . 0 ) . 0) | λxλy.((x y) y) |
| AND | (-2 (1 . 0 ) . $I_{22}$) | |
| OR | (-2 (1 . $I_{21}$) . 1) | |
| $I_{31}$ | (-3 . 2) | |
| $I_{32}$ | (-3 . 1) | |
| $I_{33}$ | (-3 . 0) | |
| SUC | (-3 1 (2 . 1) . 0) | |
| MULTCh | (-3 2 1 . 0) | |
| B | (-3 2 1 . 0) | λfλgλx.(f (g x)) |
| CONS | (-3 (0 . 2) . 1) | |
| C | (-3 (2 . 0) . 1) | λfλgλx((f x) g) |
| S | (-3 (2 . 0) 1 . 0) | λfλgλx((f x)(g x)) |
| IF | (-3 (2 . 1) . 0) | |
| NBB | (-4 3 2 1 . 0) | |
| PLUSCh | (-4 (3 . 1)(2 . 1) . 0) | |
| BB | (-4 (3 . 2) 1 . 0) | |
| CC | (-4 (3 2 . 0) . 1) | |
| SS | (-4 (3 2 . 0) 1 . 0) | |

Table 1. List of Combinator Equivalents

Let us now consider the possibility of sharing lambda expressions. Since closedness cannot be decided by looking at the root of the expression, either one has to copy as often as a beta-reduction process prescribes it, or one has to test for closedness. The latter is clearly undesirable because of its recursive nature, which comes to bear in particular for non-trivial expressions. The point in time to make the decision to share or not to share is after the reduction of the expression in question, since the expression will be reduced at least once. Can we produce the information of whether the result expression is closed or not as a side effect of the reduction process? The answer is "yes". The necessary information can be obtained from observing the behaviour of environment lookups.

The combination of head order reduction and graph representation makes sharing the normal state of affairs. Since environments are shared, a simple replacement of the unreduced expression by the reduced one in the environment produces sharing. Unfortunately, this leads to mistakes, because of the well known effect of other bindings' capturing variables that are free in the shared expression. Thus, once a relatively free variable is discovered in a subexpression, one has to copy it.

Because of head order reduction and the graph representation, only the respective topmost head normal form skeleton is copied while its arguments are still shared. The decision to share or not to share is therefore repeated for every head normal form skeleton.

To each head normal form skeleton belongs a distinctive piece of the environment. A variable refering to an entry beyond this piece is obviously free relative to the respective skeleton. Because of the substitution from the environment, the skeleton may become closed or not, depending on what has been substituted. If an expression has been substituted, nothing can be decided now.

However, all variables point to entries in the environment. Bindings which are not associated with applications find their way into the environment via eta-extensions. The value parts of such entries are marked correspondingly and contain information to correct the variable indices to be substituted.

If a variable is brought in from beyond the environment piece belonging to the respective head normal form skeleton, the latter is not closed and sharing may not be possible. This is why: Different occurrences of the same skeleton in question may get different variable indices, which clearly requires copying.

An implementation would monitor this import of variables and make the information available when the decision to share or not to share has to be made. Since copying takes place only if necessary, an optimal shared structure is obtained.

## 6. Lambda Expressions and Combinator Expressions

Here are some correspondences between lambda expressions and combinator expressions: Considering a selector (betas-#), for example, λa λb λc λd λe λf .d, which is equivalent to (-6 . 2) we can abstract it to the form:

( K K K (B . K) (B . K) . I)

It takes 11 beta-reductions to reduce the expanded expression back to (-6 . 2). The general correspondence is

$$(\text{-n} \; . \; \text{j}) \equiv (\underbrace{\text{K K K} \; \cdots \; \text{K}}_{\text{n-1-j times}} \; \underbrace{(\text{B . K})(\text{B . K})(\text{B . K}) \cdots (\text{B . K})}_{\text{j times}} \; . \; \text{I})$$

In the next example, four variables are abstracted from a combination under the assumption that all variables occur in the operator part as well as in the operand part. This results in the following structure:

(( S (B . S) (B B . S) (B B B . S) . ⟨operator part⟩) . ⟨operand part⟩)

An expansion and 26 following beta-reductions reduce it to

(-6 ((((5 . 3) . 2) . 1) . 0) (((4 . 3) . 2) . 1) . 0).

This could be in turn considered as a parameterized combinator $SN_j$ with the reduction rule for j=4 (using left-association):

$$SN_4 \; f \; g \; x_1 \; x_2 \; x_3 \; x_4 \Rightarrow (f \; x_1 \; x_2 \; x_3 \; x_4 \; (g \; x_1 \; x_2 \; x_3 \; x_4))$$

The general formation law is easily seen, yielding, for example,

$$SN_1 \equiv S \qquad SN_1 \; f \; g \; x \Rightarrow f \; x \; (g \; x).$$

Church numerals have this correspondence:

$$Nt_n \equiv (\text{-2} \; \underbrace{1 \; 1 \; 1 \; 1 \cdots 1}_{\text{n times}} \; . \; 0)$$

$$\equiv (\underbrace{(S \; . \; B) \; (S \; . \; B) \; (S \; . \; B) \cdots (S \; . \; B)}_{\text{n times}} \; K \; . \; I$$

Some correspondences may become awkward. While the Church numeral successor combinator has a simple lambda expression equivalent (see table 1), the predessor combinator as a lambda expression does not look very elegant. The following version shown in Figure 6 is due to M. Martelli:

$$(\text{-1} \; (0 \; \text{-1} \; (((0 \; . \; I) \; . \; I_{22}) \; . \; Nt_0) \; SUC \; . \; 0) \; . \; I_{43})$$

$$(\text{-1} \; p \; p \; 0 \; (\text{-1} \; p \; p \; p \; p \; 0 \; I \; I_{22} \; Nt_0 \; (p \; SUC \; 0)) \; I_{43})$$

It is so complicated that its structure is not easily seen in a linear form no matter what syntax we choose. Compare Figure 6 for a tree-structured representation. A Church numeral is transformed twice following this pattern:

$$(((\text{-2} \; 1 \; 1 \; 1 \cdots 1 \; . \; 0) \; . \; A) \; . \; B) \Rightarrow (A \; A \; A \; A \cdots A \; . \; B)$$

Let us first consider the outermost transformation. Here we have

$$A = (\text{-1} \; p \; p \; p \; p \; 0 \; I \; I_{22} \; Nt_0 \; (p \; SUC \; 0))$$

$$B = (\text{-4} \; . \; 1) = I_{43}.$$

A applied to B yields $Nt_0$ by selection. Hence we need only to consider what A applied to Church numerals reduces to. Considering then the innermost transformation we have

$$A = I$$

$$B = FALSE = I_{22}$$

The innermost transformation therefore transforms all Church numerals, including zero, to FALSE (= $I_{22}$) which subsequently selects SUC . <Church numeral>. Since (A . B) is zero and not one, the predecessor transformation is accomplished.

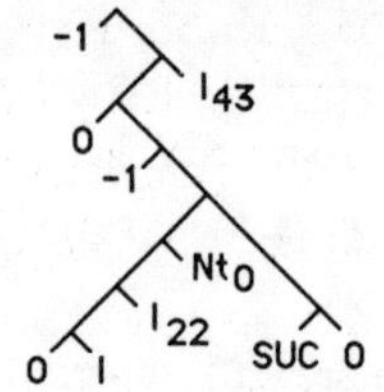


Figure 6. Predecessor for Church Numerals

In order to demonstrate the correctness of the predecessor operator for Church numerals we reasoned about lambda expressions using metavariables A and B by using the equations

$$(A\ .\ B) = Nt_0$$

$$(A\ .\ Nt_j) = (SUC\ .\ Nt_j)$$

The seemingly reasonable assignment A = SUC yields the equation

$$(SUC\ .\ B) = Nt_0$$

In other words, we expect B to be a lambda expression representation of -1. For the predecessor given above A ≠ SUC. Does the equation have a solution for B and thus constitute a different lambda expression representation or not?

When asking questions of that sort we discover that equality of lambda expressions is not lambda definable. This leads to a new set of concepts and ideas which establish the connection to logic programming.

## 7. Summary

The head-order reduction scheme developed in this paper has significant advantages over conventional call-by-value or call-by-name sequencing. The main point is that the skeleton of the head normal form remains unaltered once obtained. All arguments are recursively reduced to head normal form, but the sequence among these arguments is not determined. However, if parts of the expressions are represented as environments, the sequencing is restricted due to the sharing of environments. In either case is it possible to generate the result in preorder sequence. Since the environment uses memory space in an orderly manner, too, no garbage collection is needed. The head-order scheme also exhibits a formal correspondence between environment structure and expression structure.

The implementation of the lambda calculus outlined in this paper is complete and correct. Variables may occur free or bound without any restriction. This means in particular, that an expression cannot be considered as a composition of functions. Namely, a function is an expression with a distinguished set of formal parameters, all of which actually occur in its body. Occasionally, a free parameter may also occur in the body. But this is considered an exception because of the difficulties of implementing it correctly. However, in the general case of a lambda expression, there are no discernible objects like functions. To explain this, let us consider a subtree that starts with a sequence of lambda bindings. We call such a subtree a function if no subtree of it starts with other bindings; that is, it contains only constants and variables as leaves. Another prerequisite for calling it a function is that most of these variables are bound in the leading sequence of lambda bindings. Let us contrast the function case with a situation where most of the variables are not bound in the leading sequence of lambda bindings and the subtree contains many other subtrees that start with lambda bindings. Such a subtree is certainly well-formed in terms of the lambda calculus, but is not a function in the usual sense.

Head-order reduction combines the termination properties of normal order reduction with the efficiency of applicative order, since arguments are reduced only if required in the result structure, and arguments are reduced only once if they can be shared.

All closed lambda expressions (*i.e.*, not containing variables relatively free in this expression) may be named and thus become combinators. In this sense, we also have described a combinator reduction scheme. Any combinator expression (obtained by an abstraction algorithm) can be expanded into a bigger lambda expression by replacing each combinator symbol by its lambda expression equivalent. Then the big lambda expression is reduced to head normal form. This is, of course, a costly detour. The direct reduction of the original expression, which is much smaller than its corresponding combinator expression, would have used a much smaller number of beta-reductions. Measurements show factors up to 300. The abstraction process is considered compilation and not taken into account.

The implementation of this head-order scheme based on graph reduction has been checked against the string reduction machine [Berkling 1975] for about 50 expressions, some consisting of up to 300 symbols. We consider it safe to claim correctness because of the vastly different methods of obtaining the result. This, of course, does not preempt the necessity of a formal proof.

## 8. Bibliography

[Abdali 1976]
Abdali, S.K., "An Abstraction Algorithm for Combinatory Logic", *The Journal of Symbolic Logic 41,* 1, March 1976, pp. 222-224.

[Aiello 1981]
Aiello, L. and Prini, G., "An Efficient Interpreter for the Lambda-Calculus", *Journal of Computer and System Sciences 23,* pp. 383-424 (1981).

[Barendregt 1981]
Barendregt, H.P., *The Lambda Calculus: Its Syntax and Semantics,* North-Holland Publishing Co., Amsterdam, Netherlands, 1981 and 1984.

[Berkling 1975]
Berkling, K.J., "Reduction Languages for Reduction Machines", *Proceedings of the 2nd International Symposium on Computer Architecture,* Houston, January 1975, pp. 133-140.

[Berkling 1982]
Berkling, K.J., and Fehr, E., "A Consistent Extension of the Lambda Calculus as a Base for Functional Languages", *Information and Control 55,* 1-3, pp. 89-101.

[Church 1941]
Church, A., *The Calculi of Lambda-Conversion,* 1941, Princeton University Press.

[Clarke 1980]
Clarke, T.J.W., Gladstone, P.J.S., Maclean, C.D., and Norman, A.C., "SKIM - S, K, I, Reduction Machine", Proceedings 1980 Lisp Conference, Stanford, California, August 1980.

[DeBruijn 1972]
De Bruijn, N.G., "Lambda-calculus notation with nameless dummies: A tool for automatic formula manipulation with application to the Church-Rosser theorem", *Indag. Math. 34,* pp. 381-392.

[Hughes 1982]
Hughes, R.J.M., "SUPER-COMBINATORS: A New Implementation Method for Applicative Languages", *1982 ACM Symposium on Lisp and Functional Programming, Pittsburgh, Pennsylvania,* August 1982.

[Greene 1985]
Greene, K.J., "A Fully Lazy Higher Order Purely Functional Language with Reduction Semantics", CASE Center Technical Report No. 8503, Syracuse University.

[Johnsson 1983]
Johnsson, T., "The G-Machine: An Abstract Machine for Graph Reduction", Chalmers University of Technolgy, Goteborg, Sweden, August 1983.

[Kleene 1952]
Kleene, S.C., *Introduction to Metamathematics,* Van Nostrand, New York, 1952.

[Morris 1980]
Morris, F.L., "Computing Cyclic Liststructure", School of Information and Computer Science, Report 2-80, Syracuse University, June 1980.

[Scheevel 1986]
Scheevel, M., "NORMA: A Graph Reduction Processor", *1986 ACM Conference on LISP and Functional Programming,* August 1986.

[Turner 1979]
Turner, D.A., "A New Implementation Technique for Applicative Languages", *Software—Practice and Experience 9,* September 1979, pp. 31-49.

[Wadsworth 1971]
Wadsworth, C.P., *Semantics and Pragmatics of the Lambda Calculus,* Ph.D. Thesis, Oxford University, September 1971.

**Appendix A**

We show in 6 figures the steps executed towards a head normal form of a randomly chosen lambda expression. The state before is displayed on the left side; the state after, on the right.

In Figure A1 an eta-extension takes place, and the triangle *A* is now the one to be pushed down. In Figure A2 triangle *A* has been moved over the application nodes of *a;* the triangle *A* is duly noted at the rand expressions, which are not explicitly represented. To maximize the next triangle an eta-extension is introduced in Figure A3 yielding a new triangle *B*. Again, in Figure A4, the triangle *B* is pushed down over the application nodes *b* with triangle *B* recorded at the rand expressions. The process has now produced triangle *C*, which does not require an eta-extension before being pushed down over application nodes g as shown in Figure A5. The last figure in this sequence, Figure A6, shows the formation of triangle *D*, which requires a single eta-extension. Because of the limited space, further actions are not shown; they would depend on the expression extending downward. To be consistent with the eta-extension in Figure A6, this expression would have to start with application nodes. The eta-extensions mentioned above are actually eta-extensions in-the-large.

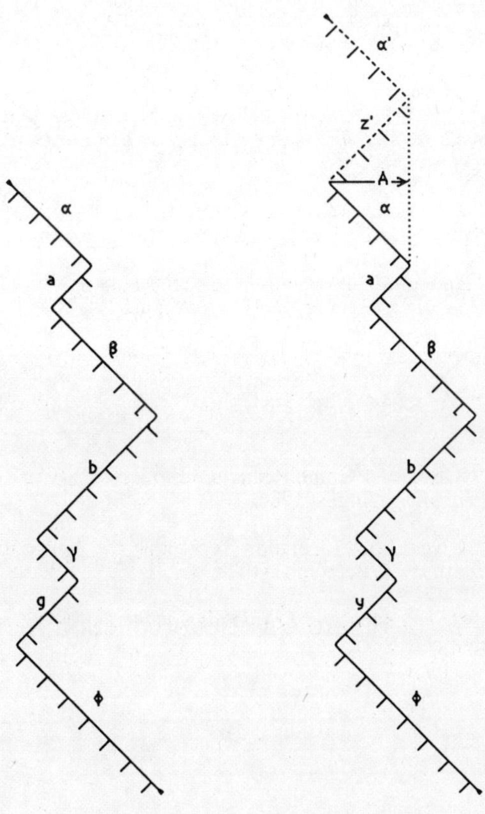


Figure A1. Eta-extension z'–α'

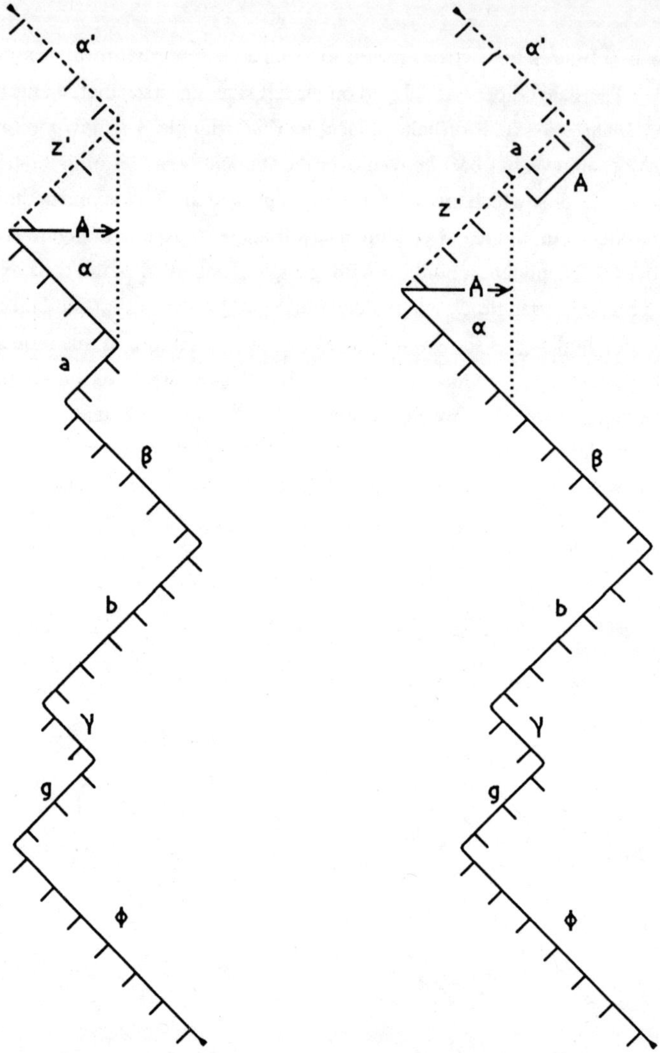


Figure A2. Beta-reduction A.

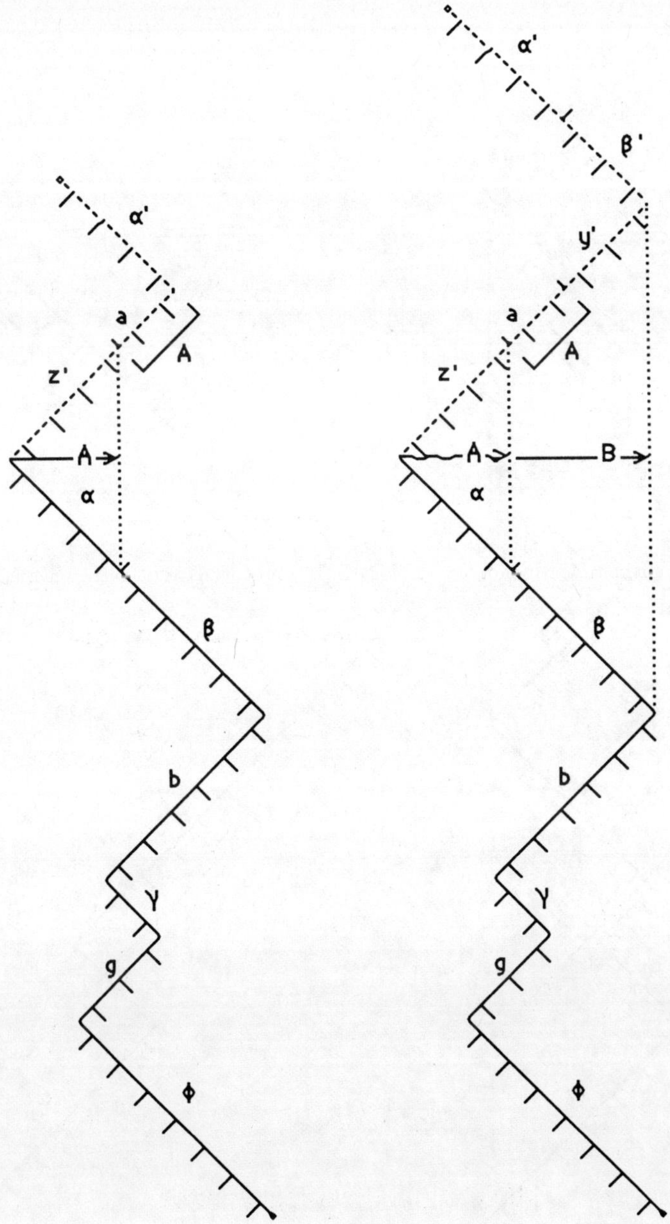


Figure A3. Eta-extension y'–β'

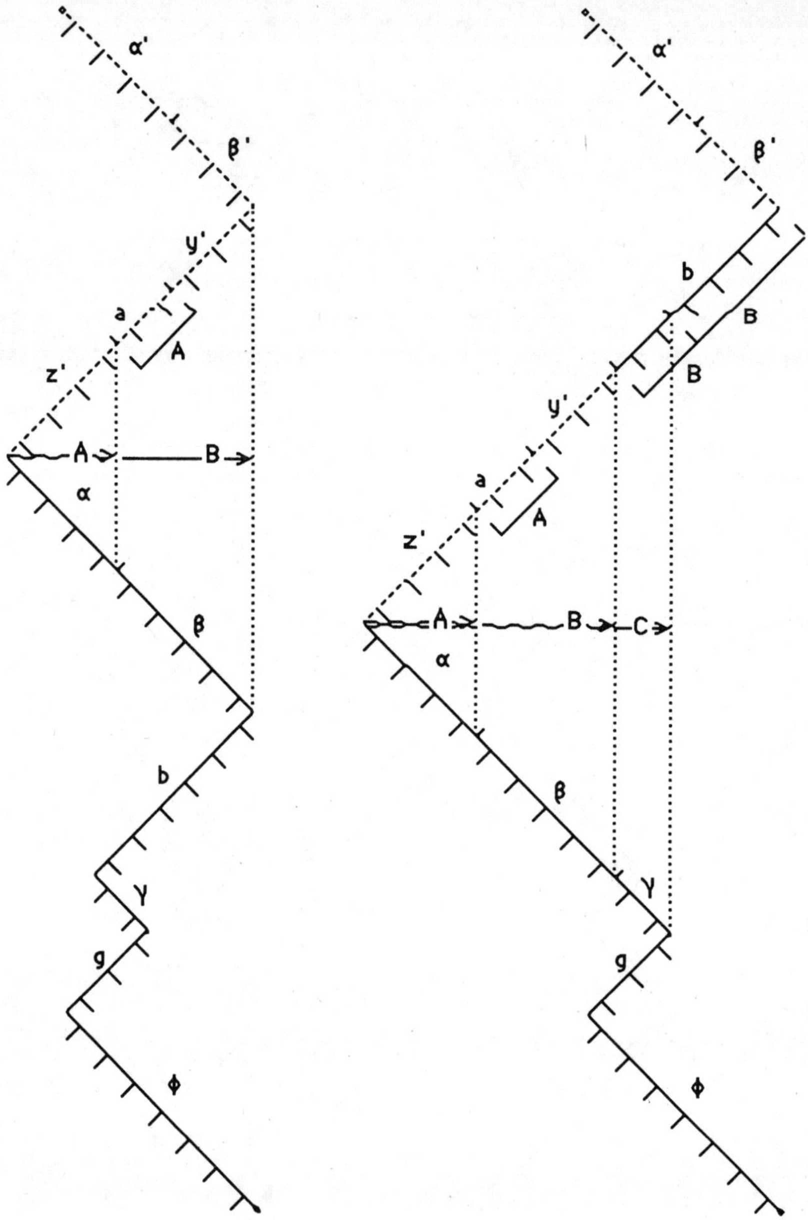


Figure A4. Beta-reduction B.

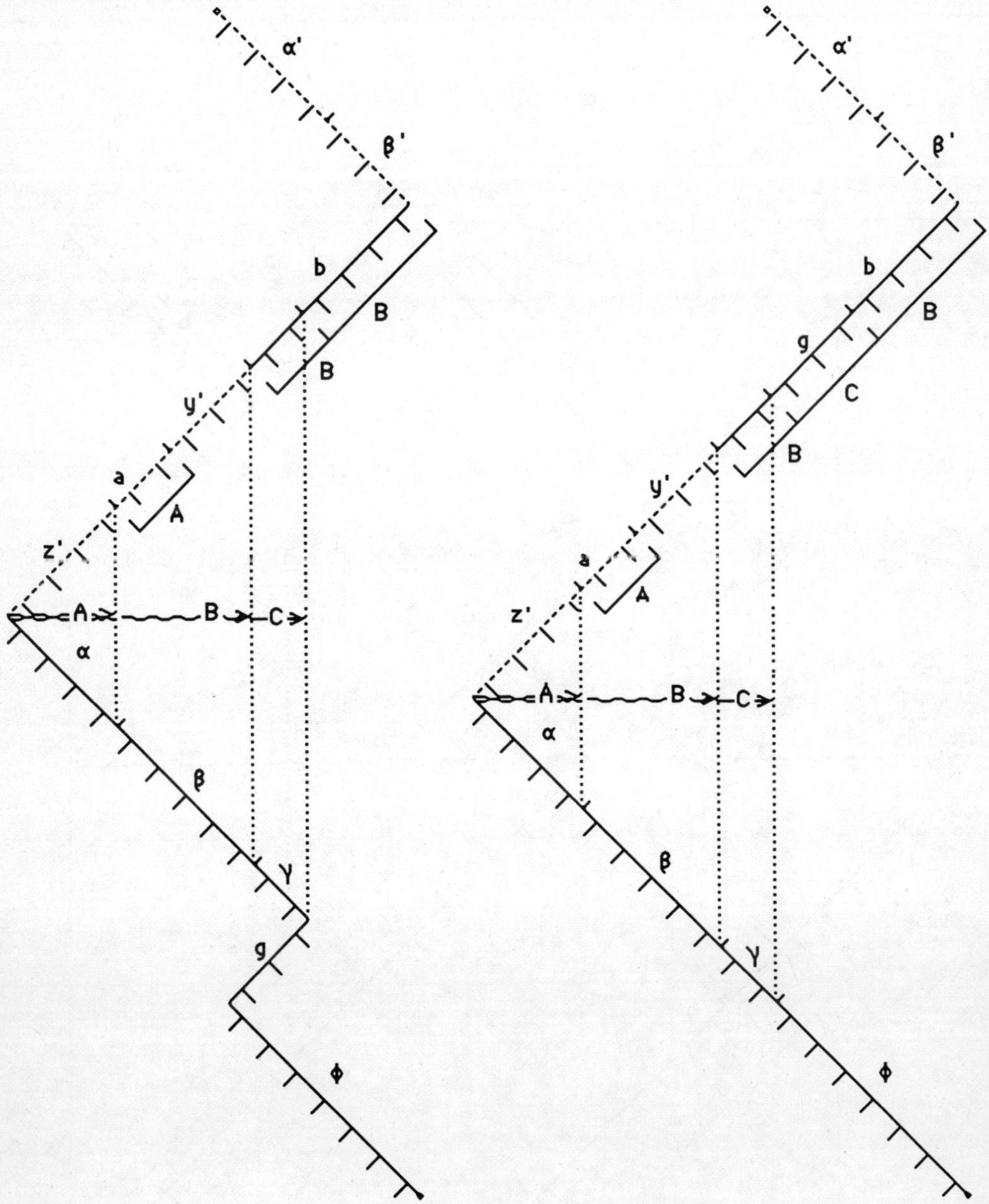


Figure A5. Beta-reduction C.

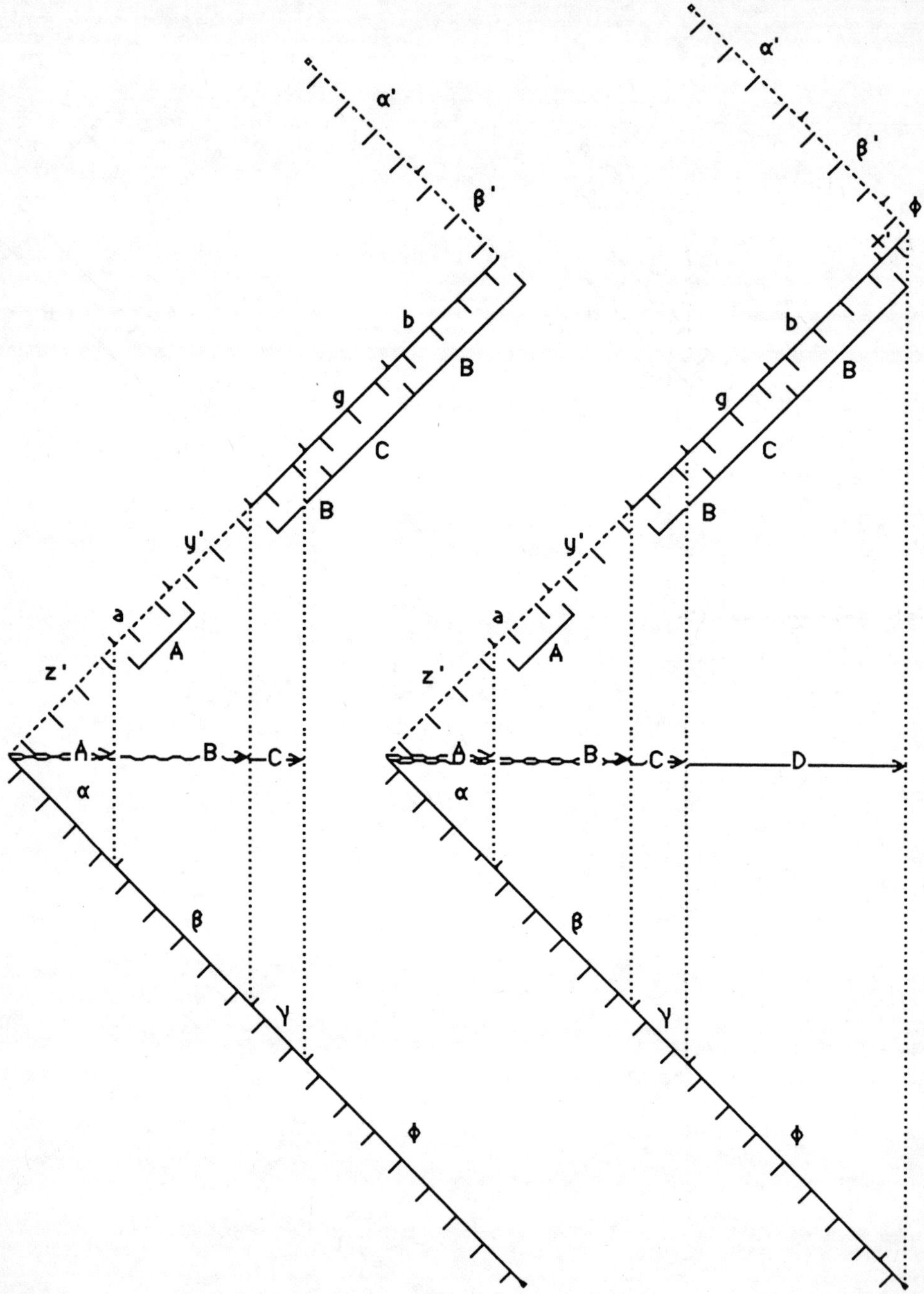


Figure A6. Eta-extension x'–ϕ'.

# A Simple Abstract Machine to Execute Supercombinators

Jon Fairbairn
Cambridge University Computer Laboratory,
Corn Exchange Street,
Cambridge CB2 3QG,
United Kingdom.
Telephone +44 223 334649, Telex 81240 (CAMSPL-G)
Mail address: jf@UK.AC.Cam.CL

**Abstract**

This paper gives a brief description of an abstract machine into which supercombinators may be compiled.

The description comprises a presentation of the instruction set of the machine, followed by the operational semantics of the normal order version, the algorithm to convert combinators to instructions and an outline of how the machine may be altered to perform lazy evaluation.

## 1. Introduction

The present compiler for Ponder [Fairbairn 83, Fairbairn 85, Tillotson 85] uses an abstract machine [Fairbairn & Wray 86] that is similar in appearance to the G-machine [Johnsson 83]. The machine described in this paper arises out of the observation that it is difficult to improve the performance of a functional language implemented in this style past the present level. The reason for this is that part of the core of the earlier abstract machines is essentially interpretive, and code executed in this manner runs significantly more slowly than directly executed code.

The techniques of strictness analysis and compile time sharing detection help reduce the amount of interpretive code, but neither of these can produce perfect code in all cases. The information derived by strictness analysis definitely helps improve performance, but higher order strictness information gives little or no improvement [Fairbairn & Wray 86]. While we expect that the combined use of higher order and non-flat domain strictness analysis [Hughes 85b, Kieburtz & Napierala 85, Wadler 85] would detect many more strict contexts, it is difficult to take advantage of this information at run-time. Similarly it is almost impossible to use sharing detection to any advantage at run time.

The new machine comes from an attempt to restructure things in such a way that these techniques of analysis can be used more effectively. The result shares properties with the SECD machine [Landin 64], the architecture used at Yale [Hudak & Goldberg 85, Hudak & Kranz 84] and frame based reduction engines. The experiments with the new machine are at such an early stage that it is difficult to tell whether it does allow analysis to be used more effectively. We feel that the new machine is at least of intellectual interest.

The work has been carried out in conjunction with Stuart Wray, who provided most of the impetus and ideas for the normal order version of the machine.

## 2. The Architecture

The machine comprises an argument stack, a frame heap and a code stream. Each object is represented as a pair consisting of a frame pointer and a code pointer. The argument stack holds arguments to functions as they are being built up, and the frame heap holds frames of arguments for combinators. The code stream comprises labelled sequences of instructions.

There are registers PC to point to the current code, CurrentFrame to address the current frame and ArgP to hold the stack. You can think of the state of the machine as representing a function applied to some arguments; the function is represented by PC and CurrentFrame and the arguments are on the argument stack in order.

There are just three instructions.

**Take** $n$

Takes $n$ items off the argument stack and puts them in a new frame, adjusting CurrentFrame to point to this new frame. Take represents the beginning of a combinator.

**Push <item>**

Pushes an <item> onto the argument stack. <item> can be **arg** $n$ — the $n^{\text{th}}$ argument in the current frame, Combinator C — the code sequence for the combinator C, or Label L — a label within the current combinator (in this last case, the frame part of the item is taken to be the current frame).

**Enter <item>**

Item is as in Push (except that it will never be <label>), and the effect is to load CurrentFrame and PC with the item.

Formally the machine can be represented as a tuple <PC, ArgP, CurrentFrame, Frames>. Here code pointers are represented by the sequence of instructions to which they point. While frames could also be represented in this way, it will be better to regard them as indexing a structure. We will write $F$ for frame heaps, $f$ for indexes of frames in heaps and $F[f \mapsto (a_1, \ldots, a_n)]$ for a heap extending $F$ where $f$ indexes $(a_1, \ldots, a_n)$. Each $a_i$ is of course of the form $<c, f>$.

The operation of the machine is described by the following rewrite rules:

$$<[\texttt{Take}\ n; I], (a_1, \ldots, a_n, A), f_0, F> \Rightarrow <I, A, f, F[f \mapsto (a_1, \ldots, a_n)]>$$
$$<[\texttt{Push arg}\ n; I], A, f, F[f \mapsto (\ldots, a_n, \ldots)]> \Rightarrow <I, (a_n, A), f, F[f \mapsto (\ldots, a_n, \ldots)]>$$
$$<[\texttt{Push label}\ l; I], A, f, F> \Rightarrow <I, ((l, f), A), f, F>$$
$$<[\texttt{Push combinator}\ c; I], A, f, F> \Rightarrow <I, ((c, 0), A), f, F>$$
$$<[\texttt{Enter arg}\ n], A, f, F[f \mapsto (\ldots, <c_n, f_n>, \ldots)]> \Rightarrow <c_n, A, f_n, F[f \mapsto (\ldots, <c_n, f_n>, \ldots)]>$$
$$<[\texttt{Enter combinator}\ c], A, f, F> \Rightarrow <c, A, 0, F>$$

## 3. Converting combinators to code

We assume that a combinator is of the form $\lambda a_1, \ldots a_n.\mathit{expression}$, and that a expression is either an *atom* (i.e. an $a_i$ or a combinator name) or an application of one expression to another.

In the traditional style, $C[\![e]\!]$ generates code for the expression $e$:

$$C[\![\lambda a_1, \ldots a_n.\mathit{body}]\!] \Rightarrow [\texttt{Take}\ n; C[\![\mathit{body}]\!]]$$
$$C[\![\mathit{atom}\ e_1\ \ldots\ e_n]\!] \Rightarrow [P[\![e_n]\!]; \ldots; P[\![e_1]\!]; E[\![\mathit{atom}]\!]]$$
$$P[\![a_n]\!] \Rightarrow \texttt{Push arg}\ n$$
$$P[\![\mathit{combinator}]\!] \Rightarrow \texttt{Push combinator}\ [C[\![\mathit{combinator}]\!]]$$
$$P[\![e]\!] \Rightarrow \texttt{Push label}\ [C[\![e]\!]]$$
$$E[\![a_n]\!] \Rightarrow [\texttt{Enter arg}\ n]$$
$$E[\![\mathit{combinator}]\!] \Rightarrow [\texttt{Enter combinator}\ [C[\![\mathit{combinator}]\!]]]$$
$$E[\![e]\!] \Rightarrow \texttt{Enter label}\ [C[\![e]\!]]$$

## 4. Making it Lazy

In a supercombinator evaluator the only way that sharing can be introduced is via an argument to a supercombinator. Thus the only overwriting needed it to update an argument in a frame with a reduced version of itself. Such overwriting will always be with an expression that is not a redex *ie* it will be a combinator applied to too few arguments. To make the machine lazy requires a mechanism whereby representations of such applications can be created when they are needed. A further alteration is that in order to take advantage of the updating of arguments that may be shared we can no longer just push the argument onto the stack; instead we must push the address of an Enter instruction for this argument. This means that the argument in the frame is not read until it is about to be entered; subsequent evaluations of the same argument can then take advantage of the updated version.

This involves adding a new register `ShareLimit` to the machine. `ShareLimit` points to a place on the argument stack at which the current application is shared. At this place on the stack is a pointer to the argument that needs to be updated and a new value for `ShareLimit`. Whenever a Take instruction is executed there must be a check to see if there are enough arguments on the stack above `ShareLimit`. If there are, we can just proceed as normal; if there are not, we know that the combinator that we are entering is being shared applied to too few arguments. In this second case we must make a frame out of the arguments above `ShareLimit` on the stack and overwrite the argument addressed at `ShareLimit` with a pointer to a piece of code that pushes the arguments in this new frame back onto the stack and goes back to the Take instruction.

If the abstract machine is implemented in hardware, the tests on `ShareLimit` can take place concurrently with the rest of the Take instruction, making updates effectively interrupts, so there is no overhead when evaluating something that is not shared.

## References

[Fairbairn 83]:
Jon Fairbairn,
*Ponder and its Type System,*
University of Cambridge Computer Laboratory Technical Report No. 31, 1983.

[Fairbairn 85]:
Jon Fairbairn,
*Design and Implementation of a Simple Typed Language Based on the Lambda-Calculus,*
University of Cambridge Computer Laboratory Technical Report No. 75, May 1985.

[Fairbairn & Wray 86]:
Jon Fairbairn & Stuart Wray,
*Code generation techniques for functional languages,*
1986 ACM Conference on Lisp and Functional Programming (proceedings) pp 95–104

[Hughes 85b]:
John Hughes,
*Strictness Detection in Non-Flat Domains,*
Institutionen för Informationsbehandling, Chalmers Tekniska Högskola, Göteborg, Draft, August 1985

[Hudak & Goldberg 85]:
Paul Hudak & Benjamin Goldberg,
*Serial Combinators: "Optimal" Grains of Parallelism,*
Yale University Department of Computer Science 1985

[Hudak & Kranz 84]:
Paul Hudak & David Kranz,
*A combinator based compiler for a functional language,*
11th ACM Symposium on Principles of Programming Languages ACM Jan 1984, pp 121–132

[Johnsson 83]:
Thomas Johnsson,
*The G-Machine: An Abstract Machine for Graph Reduction,*
Proceedings of SERC Declarative Programming Workshop at UCL, April 1983

[Kieburtz & Napierala 85]:
Richard B. Kieburtz & Maria Napierala,
*A studied laziness—strictness analysis with stuctured data types,*
Oregon Graduate Centre, Extended Abstract, July 1985.

[Tillotson 85]:
Mark Tillotson,
*Introduction to the Functional Programming Language "Ponder",*
University of Cambridge Computer Laboratory Technical Report No. 65, May 1985.

[Wadler 85]:
Phil Wadler,
*Strictness Analysis on Non-Flat Domains,*
Programming Research Group, Oxford University, November 1985.

# Concurrent Term Rewriting as a Model of Computation*

**Joseph Goguen, Claude Kirchner† and José Meseguer**
**SRI International, Menlo Park CA 94025**

**Abstract:** A new model of computation, **concurrent term rewriting**, is proposed as a bridge between a class of easily programmed ultra high level languages and advanced massively concurrent architectures. At the highest level of abstraction, this model views computation as replacing selected subterms by others, at multiple sites concurrently. In this view, concurrent term rewriting provides a standard of correctness, and the choice between using trees or graphs to represent terms is a matter of convenience and efficiency. After introducing the basic concepts and properties of concurrent term rewriting, we discuss some basic implementation issues. A second, more concrete model of computation, called **partitioned concurrent term rewriting**, takes account of the fact that a (possibly very large) term may be partitioned into fragments that reside on different processors, with each processor concurrently rewriting its own fragment. A number of implementation and optimization issues are also discussed, including overlapping rewrites, rule ordering, compilation, and flow analysis. **Concurrent E-strategies** are introduced as a flexible control mechanism to optimize performance and facilitate systems programming tasks. All mathematical definitions are gathered in one appendix, while another describes the OBJ language used in examples.

## 1 Introduction

The role of a model of computation is to provide a simple bridge between the software and the hardware levels of description. For example, the traditional von Neumann model essentially says that a computation is a sequence of fetch/compute/store cycles. Not only is this model inherently sequential, but it also involves heavy fetch/store traffic between the processor and memory elements. Of course, there are many well-known techniques to reduce that traffic; but the multiplicity of such techniques only reinforces the central point, that the processor/memory link is a major bottleneck (the so-called "von Neuman bottleneck" [2]).

This paper introduces a new model of computation, **concurrent term rewriting**, to bridge the gap between our ultra high level programming languages featuring implicit concurrency, and our massively parallel architecture, the Rewrite Rule Machine (RRM). This model is inherently concurrent, and has no analog of the von Neumann bottleneck. In practice, the gap between software and hardware is large enough so that models at several levels of abstraction are desirable.

---

*Supported by Office of Naval Research Contracts N00014-85-C-0417 and N00014-86-C-0450.

†CNRS (Centre de Recherche en Informatique de Nancy) France; research performed while on leave at SRI International.

This paper presents our two most abstract models: concurrent term rewriting and partitioned concurrent term rewriting.

The following subsections of this introduction briefly discuss the overall goals of the RRM project, the software issues that our models address, and the Rewrite Rule Machine. Section 2 studies the most abstract level of concurrent rewriting, while Section 4 treats partitioned term rewriting. Some implementation issues are discussed in Section 3, including tradeoffs between the tree and graph representations for terms. Section 5 discusses some compilation techniques, and Section 6 discusses strategies to control the order of evaluation. Precise mathematical definitions are given in Appendix A, while Appendix B summarizes OBJ, the functional programming language used in some examples.

We thank Mr. Timothy Winkler for his valuable assistance with this paper, both with its LaTeX arcana and with its technical content. We also thank the other members of the RRM and OBJ Projects with whom we have had extensive discussions on these ideas, particularly Drs. Sany Leinwand and Ellen Munthe-Kaas and Prof. Ugo Montanari.

## 1.1 Goals of the RRM Project

The purpose of the RRM project is to design, construct and evaluate a general purpose, massively parallel computer architecture to execute ultra high level declarative languages very efficiently. Promising application areas include:

1. powerful program development environments, including intelligent editors, compilers, libraries of reusable software, rapid prototyping tools, specification languages, program verifiers, debuggers, test case generators, etc.;

2. artificial intelligence applications, including robot planning, natural language processing, high level vision, expert systems, etc.; and

3. hardware simulation.

However, the machine is general purpose and can also do concurrent numerical computation.

The RRM consists of a large number of processors, each a custom VLSI chip for processing structured data independently and very efficiently (see Section 1.3). Our ultra high level languages (see Section 1.2.2 and Appendix B) combine object-oriented, functional, and logic programming with powerful type systems that handle subtypes, generics and exceptions.

## 1.2 Software Issues

From the software viewpoint, term rewriting has the following two advantages: computations can be described independently of execution order; and there are abundant opportunities for concurrent execution.

### 1.2.1 Generality and Expressiveness

We assume familiarity with the tree representation of terms, and we freely identify terms and trees in the following discussion; this is harmless except when discussing how to represent terms at the hardware level, where graphs are another alternative to be considered.

There are many applications in which the data are naturally tree-structured (in the sense that there is a natural hierarchy, with a "root" node at the top and with one or more branches from each node that is not a tip node), and for which term rewriting is a highly efficient and natural mode of execution. Some examples of naturally occurring tree structure are:

- Menus, such as occur in interactive graphics,
- Expressions, such as $(A + B)(A^2 + 3)$ and, more generally, programs,
- Natural language syntax, and many other structures that occur in natural or artificial languages, such as plans and explanations, and
- Most generally, any abstract data type.

In fact, tree structure is a fully general data structure, since any (semi-)computable function can be seen as working on tree-structured data by rewriting subtrees into other subtrees. **Term rewriting** (or replacement or reduction) refers to the replacement of one subterm by another, whenever a template is matched. **A rewrite rule** consists of two such templates, one for the subterm to be replaced, and another that determines what it is to be replaced by.

In particular, such processes as selecting a particular item from a menu, evaluating an arithmetic expression, verifying a program, assigning a meaning to a sentence, editing a program, constructing a plan, and compiling or interpreting a program, can all be conveniently described by term rewriting. The popularity of attribute grammars attests to the ubiquity of tree structure. Moreover, the feasibility of writing nontrivial programs with rewrite rules has been shown by experience with numerous functional programming languages, including OBJ, Hope, Miranda, FP, and the work of Hoffman and O'Donnell. In particular, several different language interpreters have been written in OBJ, including one for OBJ.

### 1.2.2 Programmability

We feel that *programmability* is one of the most critical issues blocking progress in parallel computation: it does little good (except for very homogeneous problems) to provide very many processors if the programmer has to explicitly assign processes to processors. Our concurrent term rewriting model enables efficient execution of our ultra high level languages on our massively parallel RRM without any explicit concurrency control at the language level, because concurrency is implicit in the rewrite rules themselves. Indeed, our simulations of concurrent term rewriting [24] show that naive OBJ programs can capture most of the concurrency that is available in problems; moreover, these programs can be modified relatively easily to achieve optimum performance, e.g., by replacing list structures by balanced binary trees.

The RRM Project takes OBJ2 [6] as its basic language. OBJ2 is a very advanced functional language based on term rewriting, having a uniquely powerful generic module facility and type system. This basis has been extended to include logic programming in Eqlog [11] and to include object-oriented programming in FOOPS [12].

There are some cases where explicit scheduling of the evaluation process is highly desirable, and Section 6 shows how our **concurrent E-strategies** (E is for evaluation) accomodate these in a natural, declarative style. E-strategies free us from subservience to lazy, eager, or any other fixed evaluation strategy, and our simulation results show that E-strategies can be very effective for avoiding explosive growth of required space, while still preserving time efficiency. Section 6.1 shows how E-strategies can also be used to support systems programming functions, such as scheduling, synchronization, and input/output, as well as real-time programming.

## 1.3 The Rewrite Rule Machine

The Rewrite Rule Machine (RRM) project aims to produce a highly parallel computer for enormously efficient term rewriting. A number of novel architectural design ideas have been developed to meet this goal, including the following:

- **Rewriting occurs directly in active memory**, rather than in a processor that must access separate memory; this eliminates the von Neumann bottleneck;

- **Multi-Grain Parallelism**, whereby the finest-grained tightly-coupled chip level does concurrent term rewriting in SIMD[1] mode, while the three coarser, progressively more loosely coupled, levels run in MIMD mode. The following sketches our current design for a medium size RRM:

  1. *Cells*, which are very small processors, arranged in a regular pattern, holding the individual nodes of a data term; each cell can hold up to 8 nodes, so the physical representation of the term may involve overlapping nodes.
  2. *Chips*, containing 144 cells plus a controller and some interface circuitry; the controller broadcasts a stream of micro-instructions which are obeyed in lock-step (SIMD) mode by all the cells on a chip.
  3. *Boards*, containing about 512 chips, including 128 custom processors and some backup memory (which will also aid with fault tolerance), operating in MIMD mode; cells on the same board will share a common name space (i.e., mapping of tokens into small integers). (We are considering various connection topologies at this level, including a hexagonally tessellated torus.)
  4. A *rack* of 16 boards, plus a general communication network, and a conventional minicomputer to store rewrite rule sets and manage global load balancing, constitute a total machine, in which term rewriting can occur many places concurrently, in both SIMD and MIMD modes.

  This design is intended to minimize communication costs, without requiring any exotic switches, special packaging techniques, or unusually high speeds.

- **Declarative-style concurrency control**, in which such systems programming tasks as scheduling and input/output are viewed as term rewriting (see Section 5).

- **Local communication** predominates, since terms involve only local connectivity.

Our preliminary simulation results [24] are very exciting, indicating both that there is enormous parallelism available in typical OBJ programs, and that SIMD rewriting at the chip level is a *big win*, since many instances of the same rule tend to occur together, so that the resulting economy of design should permit a speed which much more than compensates for the need to perform rewrites one rule at a time. Furthermore, our explorations of compiler technology for the RRM suggest that many optimizations should be possible (see Section 5).

A preliminary performance estimate, assuming MOSIS fabrication facilities and conventional VME boards, gives an astonishing raw computational power of 600,000 million rewrites per second, for a reasonable 20 MHz clock speed. Of course, raw computational power is a totally misleading (though unfortunately very common) measure, but our initial simulation results encourage us to expect a realistic peak performance of about 30,000 million rewrites/second (for the medium size RRM described above, with 2,048 rewrite chips), which is roughly equivalent to 1 million MIPS (i.e., 1,000,000 million instructions/second) on a conventional processor! (These very preliminary estimates are based on the architectural assumptions sketched above, and represent a significant advance over those in [10]; see [21] for more detail.)

[1] SIMD stands for Single Instruction, Multiple Data, while MIMD stands for Multiple Instruction, Multiple Data. These provide a rather crude classification of architectures.

## 1.4 Compiling

- Given the above information about the software and hardware components of the RRM Project, it follows that a compiler must transform sets of rewrite rules into sequences of micro-instructions to be broadcast by each chip's controller to all its cells, for execution in SIMD mode. Of course, there is a bit more to it than this. Since the languages that we are developing for the RRM are somewhat higher level than just sets of rewrite rules, their compilers must have front ends that transform source code into rewrite rule sets. There should also be a back end for optimization; this paper discusses several techniques that should be useful in that regard (see Sections 3, 4, and 5). Notice that different micro-instruction sequences can be unfolding on different chips, since in general they may contain different sorts of data and thus require different rule sets.

There are two very particular tasks for compilers that follow from the need to reconcile arbitrary expressiveness at the language level with limited resources at the chip level: translate all $n$-ary operations, for $n > 2$, into binary operations; and translate all rules with lefthand side depth greater than (say) 3 into two or more rules with depth no greater than 3. Both these transformations allow us to handle relatively rare situations without the need for additional hardware, and both are simply accomplished by introducing extra operation symbols.

An RRM compiler should be *incremental*[2], because it will most likely be used interactively: sometimes new modules will be defined, presenting new rules sets to be compiled, while other times new expressions will be given for evaluation.

# 2 Concurrent Term Rewriting

This section is an informal introduction to concurrent term rewriting. The formal definitions that are essential for a rigorous mathematical development are given in Appendix A, along with some basic properties.

## 2.1 Main Concepts

In the concurrent term rewriting model of computation, data are **terms**, constructed from a given set of **function symbols** (also called **operations**), including constants, and **programs** are sets of **equations** that are interpreted as left-to-right rewrite rules. The left- and right-hand sides of an equation are both terms constructed from **variables** as well as function symbols. A variable may be instantiated with any term of appropriate sort, and a set of instantiations for variables is called a **substitution**.

**Term rewriting** (or **reduction**) has two phases: first, **matching**, which is finding a substitution, called a **match**, that yields a subterm of the given term when applied to the lefthand side of the rule; and then **replacing** that subterm by the corresponding substitution instance of the righthand side of the rule. This process should proceed until there are no rules that apply, in which case the term is said to be **reduced**, or **normalized**, or in **normal form**.

For example, the lefthand side of the rewrite rule

```
(*) fibo(N) -> fibo(N - 1) + fibo(N - 2)
```

(where `fibo` is the function symbol for Fibonacci numbers) matches to `fibo(6)` in the term `(fibo(6) + fibo(5)) + 0` (which can be also represented by the tree in Figure 1), after instantiating the variable `N` to the constant `6`. The subterm where a match occurs is called a **redex**,

[2]This means that old code does not have to be recompiled when new code is added; rather, only the new code is processed, and the result is added to the old compiled code.

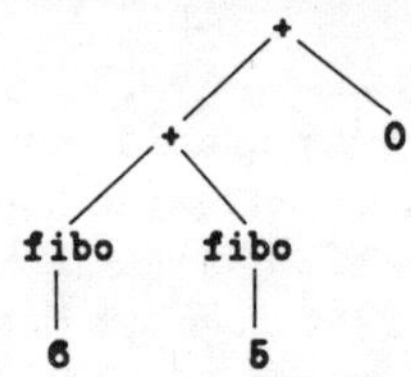


Figure 1: The Tree of an Expression

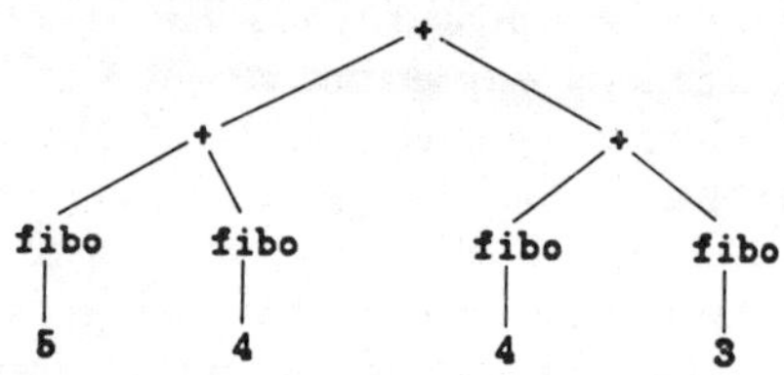


Figure 2: Result of a Concurrent Computation

which is `fibo(6)` in this example. The redex subterm is then replaced by the corresponding substitution instance of the righthand side, and in this example, the original term is rewritten to

```
((fibo(6-1) + fibo(6-2)) + fibo(5)) + 0 .
```

Rewriting by applying rule at exactly one location at a time is called **sequential term rewriting**. If the rewrite rule (*) had been applied to `fibo(5)` instead, one step of sequential rewriting would have yielded the term

```
(fibo(6) + (fibo(5-1) + fibo(5-2))) + 0 .
```

Notice that the rule (*) could also have been applied to *both* `fibo(6)` and `fibo(5)`, yielding

```
((fibo(6-1) + fibo(6-2)) + (fibo(5-1) + fibo(5-2))) + 0 .
```

This is **concurrent term rewriting**, where rules may be applied at several places simultaneously, and each match of each rule is rewritten, constituting a single step of concurrent computation. For example, two steps of concurrent rewriting can transform the term in Figure 1 into the term in Figure 2: first, concurrently apply the rule (*) at two sites, and then concurrently apply a rule for subtracting one with the rule `N + 0 -> 0`.

This simple example shows that term rewriting is by its very nature concurrent. We emphasize that no explicit concurrency constructs are required at the language level to achieve or to describe concurrency. We see this as a major advantage. It is also worth remarking that with concurrent term rewriting, the number of steps required to compute `fib(n)` using (*), plus `fibo(0)=0` and `fibo(1)=1`, is linear in `n`, whereas it is exponential for sequential rewriting.

It seems worthwhile to distinguish a third kind of rewriting, intermediate between sequential and concurrenct rewriting, which could be called **data concurrent rewriting**: here, several instances of the *same* rule are applied at once. Data concurrent rewriting is a special case of general concurrent rewriting that is especially appropriate for hardware, since it involves SIMD execution; it is also what we expect to implement in the Rewrite Rule Machine.

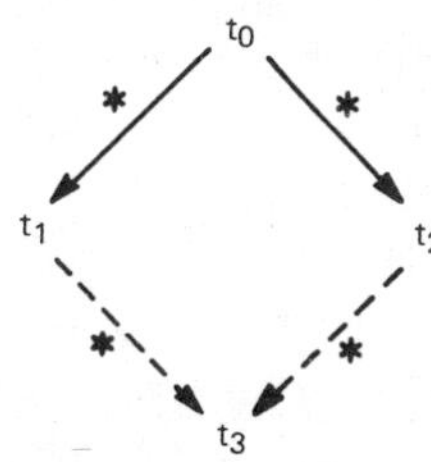


Figure 3: The Church-Rosser Property

A set of rules is called **terminating** if every possible way to concurrently rewrite a term eventually yields a normal form (although in general, more than one normal form may be reached). Some equations are nonterminating and should not be used as rewrite rules; for example, the commutativity rule `X + Y -> Y + X` could lead to an infinite chain of rewrites,

```
fibo(6) + fibo(5) -> fibo(5) + fibo(6) -> fibo(6) + fibo(5) -> ... .
```

Recall that a **normal form** is a term that is **reduced**, i.e., to which no rewrite rule applies. Both the sequential and the concurrent rewriting models of computation take normal forms as the **results** of computing; that is, a computation proceeds by applying rewrite rules, until a normal form is reached, which is then considered to be the final answer.

For functional computation, one expects that the final result does not depend on the particular chain of rewritings that led to it, i.e., one expects all normal forms to be equal. This property is guaranteed if the rules satisfy the **Church-Rosser property**, illustrated by Figure 3, where the starred arrows indicate 0, 1, or more steps of (possibly concurrent) rewriting. The Church-Rosser property says that any two rewrite sequences starting from the same term (e.g., along the two solid arrows) can always be reconciled by further rewriting (e.g., along the two broken arrows). For a set of rewrite rules that is both Church-Rosser and terminating, every term has a *unique* normal form, i.e., every computation terminates with a unique answer; this result holds for both sequential and concurrent rewriting (Appendix A gives a more precise discussion).

However, in nondeterministic computation, one expects that different final results can be reached, and therefore one cannot expect the Church-Rosser property to hold.

There are two basic data representations for finite terms: trees, and directed acyclic graphs, abbreviated **dags**. We assume that the tree representation is already familiar. A dag is a directed graph in which there are no cycles. The dag representation permits identical subterms to share the same representation, and thus is more space efficient. Sometimes we may use the word "tree" when we really mean "term;" this should be harmless except when discussing how to represent terms.

The notion of **occurrence** is useful for describing the locations of particular nodes in a tree. Basically, an occurrence is a sequence of positive numbers, describing a *path* from the root to some other node. Thus, the root node has occurrence $\varepsilon$, the empty sequence, and for example, 2.1 denotes the path from the root which first takes the second-to-left edge, and then takes the leftmost edge; see Figure 4.

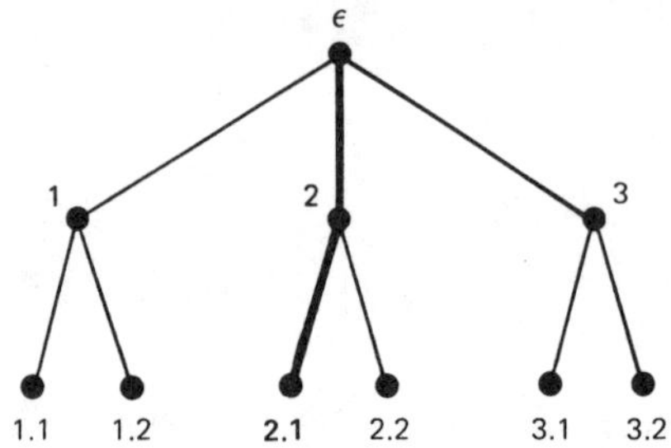


**Figure 4: A Tree with Some Occurrences**

### 2.1.1 Sorts and Semantics

This paper largely ignores type structure issues. However, OBJ is a strongly typed language, in the sense that every function symbol, variable, and term has a unique sort; therefore, the operational semantics of OBJ must be based on "many-sorted" rewriting. Actually, OBJ goes further than this, by allowing *inclusions* between sorts (for example, the natural numbers are a **subsort** of the rational numbers) and operation **overloading** (for example, + for both the naturals and the rationals); therefore OBJ is an **order-sorted** language. Terms no longer have a unique sort, but under certain weak assumptions, they have a unique "most descriptive" sort [13].

Order-sorted term rewriting can be implemented as if sorts did not exist, because many-sorted term rewriting can be reduced to unsorted term rewriting [22], and order-sorted rewriting can be reduced to many-sorted rewriting [9]. However, order-sorted rewriting can be implemented more efficiently if some sort information is checked during matching.

Although this paper is primarily focussed on operational rather than on mathematical semantics, we should mention an important result that gives a denotational semantics for term rewriting models of computation. A set $R$ of rewrite rules using a given "signature" $\Sigma$ of function symbols, can always be viewed as a set of $\Sigma$-equations. This allows us to consider the class of all $\Sigma$-algebras that satisfy $R$. Now this class always contains **initial algebras**[3], which in a sense are the "most prototypical" models of $R$. The theorem states that the set of all normal forms of a terminating and Church-Rosser term rewriting system $R$ constitute an algebra that is initial among all the algebras satisfying $R$.

The basic idea of **initial algebra semantics** is to take an initial algebra (it turns out nicely that any two are isomorphic) as the *denotation* of a set of equations. So the theorem says that term rewriting is a *correct* implementation; since the denotational semantics of OBJ is initial algebra semantics, the theorem says that concurrent term rewriting correctly implements OBJ. (This theorem is proved in [7] for sequential term rewriting, and extends easily to concurrent term rewriting.) The connection with initial algebra semantics also carries over to the order-sorted case.

One can gain a better intuitive understanding of what it means to be an initial $\Sigma$-algebra satisfying $R$ by considering the following two conditions, which together are equivalent to initiality:

- **no junk:** every element of an initial algebra is denoted by some $\Sigma$-term;
- **no confusion:** two $\Sigma$-terms denote the same element of an initial algebra iff they are *provably equal*, using the equations in $R$ as axioms.

The rules of deduction for many-sorted algebra are given in [23] along with a proof of the above characterization; rules of deduction for order-sorted algebra are given in [13].

[3]More precisely, an initial algebra in a class $C$ is an algebra that has a unique homomorphism to each member of $C$.

Figure 5: Overlapping and Non-overlapping Redexes

## 2.2 Linearity and Overlapping

If the lefthand (resp., righthand) side of a rule has only one instance of each of its variables, then the rule is called **left-** (resp., **right-**) **linear**. For example, (*) above is left-linear. We will see that left-linear rules can be easier to match than non-left-linear rules, because it is not necessary to check for equality of subterms.

A possible problem for concurrent rewriting occurs when the lefthand sides of two rules match a term in such a way that their instances partially overlap; then we say that the rules are **overlapping**. When a rule overlaps with itself, it is called **self-overlapping**. For example, the associativity rule

```
(X + Y) + Z -> X + (Y + Z)
```

overlaps with itself in matching the expression `((5 + 7) + 9) + 7`, where both the entire expression and the subterm `((5 + 7) + 9)` are redexes. Overlapping redexes pose a problem for the well-definedness of concurrent rewriting, as illustrated in Figure 5, since it may be less clear which nodes of the rewritten subterms should be part of the final result. Section 3.3.4 discusses a technique for tolerating most such clashes.

## 2.3 A Binary Tree Program

Figure 6 shows a simple OBJ program for sorted lists of integers represented as binary trees with a function for inserting a new integer. (Appendix B explains the OBJ features used here.) Leaves are represented as integers in curly brackets, e.g., `{ 7 }`, and any tree larger than a single leaf is of the form `[ T N T' ]`, with a left subtree `T`, a "content" integer `N`, and a right subtree `T'`. Since concurrent rewriting allows computation to happen everywhere, we can overlap the insertion of new elements. Indeed, it is not hard to check that for a tree of depth $d$, inserting a list of length $l$ can be done in less than $d + 4l + 3$ steps of concurrent rewriting, including evaluations of the usual conditional `if_then_else_fi`, and assuming that equality and inequality tests on integers take just one step. Figure 7 shows some snapshots during the concurrent rewriting of a small example. Executing this code in SIMD mode would only take about 3 times as long, which is a very encouraging result.

# 3 Some Implementation Issues

This section discusses some of the issues that arise in trying to implement concurrent term rewriting efficiently. The discussion tries to achieve independence from details of the RRM design where

```
obj BINARY-TREE is
    protecting INT .
    sorts Tree .
    op {_} : Int -> Tree .
    op [___] : Tree Int Tree -> Tree .
    op insert : Int Tree -> Tree .
    vars T T' : Tree .
    vars N M : Int .
    eq : insert(N,{ M }) = if N == M then { M }
                                    else if N < M then [ { N } M { M } ]
                                             else [ { M } N { N } ] fi fi .
    eq : insert(N,[ T M T' ]) = if N == M then [ T M T' ]
                                          else if N < M then [ insert(N,T) M T' ]
                                                   else [ T M insert(N,T') ] fi fi .
endo
```

Figure 6: Binary Tree Code

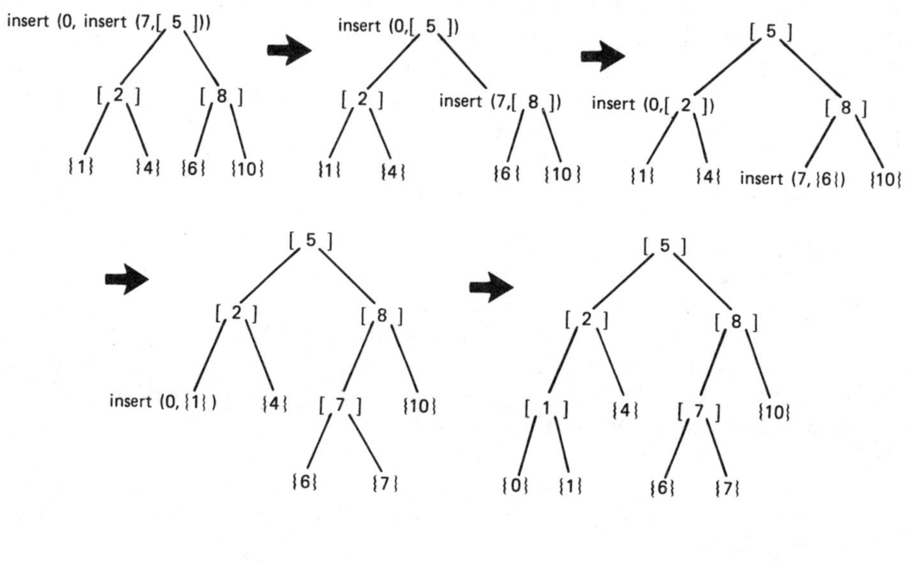


Figure 7: Some Reductions of a Binary Tree

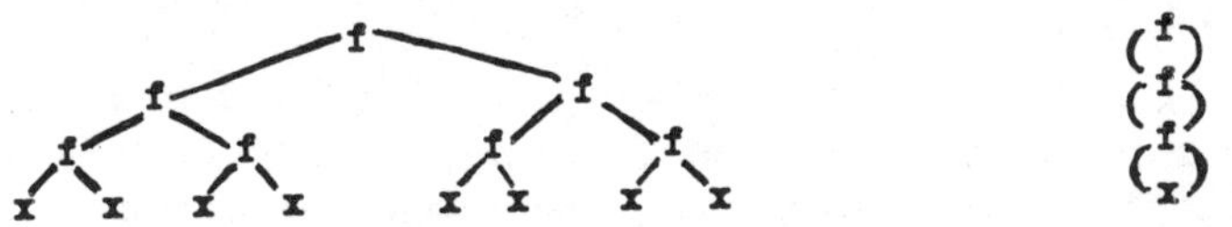

Figure 8: Two Representations of a Term

that is possible, but also uses the RRM as a reference point, to show that certain ideas really may be practical.

## 3.1 Trees and Dags

An important issue is whether to represent terms as directed acyclic graphs (**dags**), in order to permit the sharing of identitical subterms, or as trees, which prohibit sharing, but require copying. The tree/dag issue arises at each level of modelling, with different tradeoffs at each level. Ultimately, simulation will determine which levels should use dags. Here, we discuss dags and strict trees for the concurrent term rewriting level. The tree/dag problem is not new, and is closely related to work exploiting graph representations in dataflow and other architectures [20,19].

A dag is a directed graph in which there are no cycles. The dag representation permits identical subterms to share the same physical representation, and thus can be more space efficient. We have already given several examples of the tree representation. Notice that a tree is a particular kind of dag, having a unique **root** node (i.e., a unique node that is not the target of any edge) and such that every other node is the target of exactly one edge.

Any dag with nodes labelled by function symbols and having a unique root from which all other nodes are reachable, can be transformed into a tree by **unfolding**, which means successively splitting the nodes that are targets of multiple edges; completely unfolding a dag yields a tree. We can view all the dags that unfold into the same tree as different representations of the same term. For example, the term

$$f(f(f(x,x),f(x,x)),f(f(x,x),f(x,x)))$$

has, among many others, the two representations shown in Figure 8.

Obviously the dag representation is more space efficient than the tree representation. In fact, there is always a unique dag having the minimum number of nodes, called the **fully shared dag**. For example, in Figure 8, the fully shared dag on the right requires only 4 nodes, while the tree occupies $2^4 - 1$. Although the space needed by the tree representation can be exponentially greater than that needed by the dag representation, most terms are not regular enough to gain exponential space from full sharing. Simulation results [24] on the space efficiency of the dag representation show that trees generally require at most a few times as many nodes for the examples considered.

## 3.2 Matching

Matching a lefthand side may require checking the equality of subterms and/or multiple read access to a node. The subsections below discuss these issues.

### 3.2.1 Subterm Equality

In order to match a non-left-linear rule such as $x + x \rightarrow x$, it is necessary to know that the two subterms matching the variable $x$ are equal. This can be expensive for trees, but it is trivial if we are sure that the data structure is a fully shared dag. However, this advantage is more than offset by the inefficiency of implementing the fully shared dag representation, which would require checking to see whether every new subdag might possibly also occur somewhere else. On the other hand, the more daggy the representation is, the faster equality can be checked, and any successful dag equality test can be used to increase sharing, by sharing the subdags that test equal. Also, note that if cells have a flag indicating whether or not the substructure that they head is known to be reduced, equality tests can sometimes be greatly speeded up by waiting to test equality until both substructures have been reduced; for example, if the reduced forms are atoms, this may be worth doing.

An interesting possibility is to allow programmers to annotate variables in the lefthand sides of rules with some special symbol (for example, ! as in X!) to indicate that this variable can only match reduced subterms.

### 3.2.2 Multiple Read Access to a Node

Simultaneous multiple read access at a node is never necessary in the tree representation, since there is always a unique path to each node. Even for dags, most cases that might seem to need simultaneous read access at a single node during matching actually do not, under the assumption that matching is accomplished by a micro-instruction stream broadcast from a chip's controller (i.e., in SIMD mode). For example, matching the rule $f(h(x), h(g(y))) \rightarrow q(x, y)$ to the dag 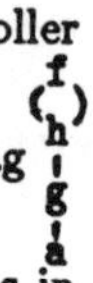 requires two reads of the node labelled h; but they will occur at different micro-instructions in the rule's stream, and therefore cannot conflict.

Assuming SIMD control, the only way that simultaneous multiple read access can occur at a node is if it participates in the same role in two different instances of the lefthand side of the same rule. For example, in applying the rule $f(g(x)) \rightarrow r(x)$ to the dag fragment 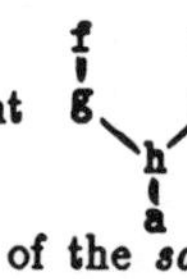, the cell containing h must be accessed twice, once by each of the two instances of the *same* micro-instruction from the controller. This will require some special hardware to serialize the reading, or else to inhibit one of the rewrites. Notice that this problem can never arise with a tree representation. One compiler technique discussed below (see Section 5.2) is "multiplexing" several rules into a single micro-instruction sequence in order to reduce the total number of micro-instructions broadcast. If rule multiplexing is possible on a large enough scale, multiple read access to nodes might arise rather often. (Section 6.1 discusses a related problem that can arise in systems programming applications.)

A similar situation could arise in testing subdag equality, but here we get a bonus: if a multiple simultaneous read access occurs, this means that the test succeeds without further effort, since the subsubdags actually have the same location!

## 3.3 Replacement

Among significant issues for replacement are: copying, direct modification of node labels, garbage collection, and handling overlapping rules.

### 3.3.1 Copying

Some rules are not right-linear, for example, the distributive rule

$$x * (y + z) \rightarrow (x * y) + (x * z).$$

For dags, duplicating a variable in rewriting only requires copying a pointer, but for trees, duplicated subtrees must be copied, and they could be very large. Simulation results [24] show that for most of the examples considered, only a small percent of the tree needs to be copied, between 2 and 9 percent; however, one sorting program reached 25 percent, which is problematic. This argues for allowing at least some sharing.

### 3.3.2 Direct Modification of Node Labels

Directly modifying node labels is always correct for trees, but not for dags. For example, we can rewrite the tree 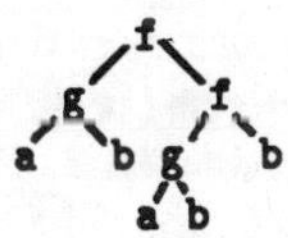 with the rule 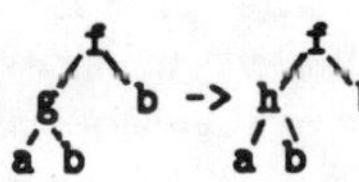 just by modifying the node at occurrence 2.1, to obtain 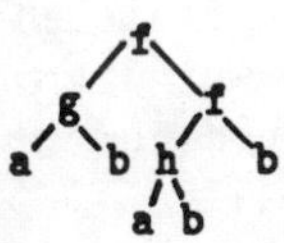, whereas the dag representation 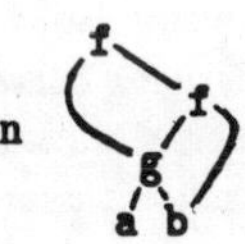 requires copying **g**, with result 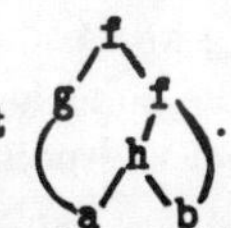. However, even for dags, it is always correct to overwrite the label at a non-shared node or at the top of a redex.

### 3.3.3 Garbage Collection

After rewriting, some substructures may be orphaned, in the sense that they are no longer referenced; in this case, the storage that they occupy should be freed, i.e., they should be **garbage collected**. This may happen, for example, when applying a rule with strictly fewer variables on its righthand side than its left. It seems likely that relatively cheap hardware support can be provided for garbage collection: each cell can have its own reference count register, and a cell is considered **free** if its reference count is zero; physically adjacent cells can have one bit links to signal when they are freed; this signal would decrement the local reference count of child cells[4]. Such a scheme would operate at least as fast as the micro-instruction clock, concurrently with micro-instruction execution. In the event that logically linked cells are physically distant, it will be necessary to use a slower communication channel; but it should not be slower than the channels that must anyway be used for rewriting or reconfiguring physically distant cells.

---

[4]Garbage collection is simpler for trees, since reference counts are not needed.

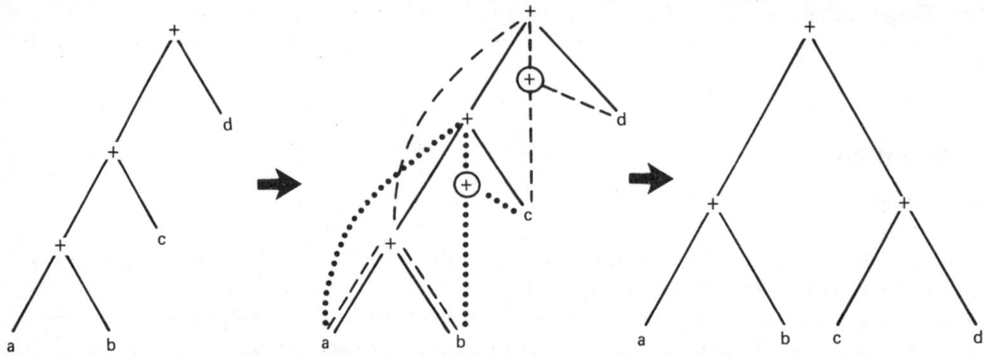


Figure 9: Concurrent Dag Rewriting with the Associativity Rule

### 3.3.4 Overlapping Rules

It is a remarkable fact that garbage collection allows simultaneous rewriting with overlapping rules. For example, in Figure 9 the lefthand side of the associative rule $((x+y)+z) \rightarrow (x+(y+z))$ matches at the root of the tree in the right part of the figure; the result of rewriting at that occurrence is shown with dashed lines in the central part of the figure. The lefthand side of the associative rule also matches at occurrence 1 of the tree in the left part of the figure, and the result of rewriting there is shown with dotted lines in the central part. Whether these two rewrites are performed simultaneously or sequentially, two new nodes (shown with circles around their + labels) are added, one of which turns out to be garbage, since it is no longer reachable from the root; one of the original nodes, at occurrence 1, also becomes garbage. After collecting these nodes, we get the tree shown in the right part of Figure 9. A key point is to directly *overwrite* the information in the root node of the redex (including its label and pointers to its children), instead of creating a new node containing that information.

It is very helpful to collect garbage right after rewriting, since otherwise further rewrites may do strange and useless things to the garbage. If garbage collection is fairly efficient, this will be a very nice technique.

This technique is fully general, except that applying two *different* rules at the *same* redex must be forbidden. However, SIMD operation guarantees that this cannot happen; in fact, it also guarantees that only a self-overlapping rule, like associativity, will need concurrent rewriting. Of course, rule multiplexing (see Section 5.2) could change this. Finally, notice that if overlapping rewriting starts with a tree, then it always ends with a tree, although dags can arise as intermediate forms.

## 3.4 Summary and Discussion

Using fully shared dags seems hopelessly inefficient, because it requires searching a possibly enormous data structure for identical, possibly very small, subdags. However, some sharing seems inevitable, because the tree representation may require copying very large substructures; it is also undesirable to test large substructures for equality, but this should be needed less often. Thus, the best representation for terms seems to be dags, but without enforced full sharing. The disadvantages are that sometimes a few extra nodes may have to be copied (see Section 3.3.1), that garbage collection may be more complex, and that some opportunities for sharing will be missed, thus giving rise to the inefficiencies associated with trees; these seem relatively minor.

The disadvantage of missed sharing can be minimized in various ways. One useful technique is to have the compiler detect duplication on the righthand side of a rule, and represent all instantiations of the rule with sharing. For example, if the righthand side of a rule is `f(k(y,h(x)),h(k(y,h(x))))`, it should be represented by the dag 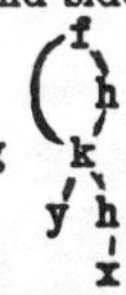 .

A second technique is to **memoise** a function, in order to avoid recomputing its values (which are the reduced forms of the constant terms having that function at the root) by placing those values in a table where they can just be looked up. In this way, all instances of a given function call implicitly share the same table entry. This can be accomplished at the language level by annotating the function with the keyword `memo` when it is declared. For example,

```
op fibo :  Nat -> Nat [memo]
```

will ensure that values of `fibo` are only computed once; this converts sequential computation by the naive rule (where `>=` is "greater than or equal")

```
fibo(N) -> fibo(N - 1) + fibo(N - 2) if N >= 2
```

from exponential time to linear, and for concurrent term rewriting, it converts the space from exponential to linear. There are many variants, including memoizing "phrases" (the instances of some term with variables), local memoization (which only applies to calls that occur in some particular rule or set of rules), and lazy memoization [17] (which is useful for the lazy evaluation of list structures). Memoization has been very useful for sequential OBJ, but its value for concurrent programming largely depends on hardware considerations. For example, it seems likely that memo tables cannot be stored inside the basic processor chip, so that slow access will make memoizations not worthwhile except for very complex computations.

# 4 Partitioned Concurrent Term Rewriting

Since each RRM processor has limited capacity, data exceeding a certain size must be distributed across several processors. However, this is actually a blessing! Concurrent SIMD rewriting can achieve only limited parallelism on a large inhomogeneous term involving very different function symbols in different parts, because match attempts for a given rule can only succeed within part of the data. But, if the data is partitioned so that each processor has a relatively homogeneous fragment, and if different processors apply rules completely independently (MIMD mode across processors), then matches will succeed much more often.

Sections 4.1 and 4.2 below address two data partitioning issues: when and where to partition data; and how to minimize communication and reconfiguration costs when rewriting across processors.

## 4.1 Criteria for Data Partitioning

Data partitioning is needed when initializing a computation, and also when a given processor gets too full[5]. If there are not enough processors, then we must send some data into storage. At partition time, static information (based on the program) is always available, and there may also be dynamic information (based on the history of the computation). When partitioning is necessary, at least three kinds of data are elegible for dispersal:

[5]In the current RRM design, reconfiguration thrashing may occur above a certain occupancy level, as a processor strives to ensure that logically adjacent cells remain physically adjacent during rewriting.

1. "foreign" subterms, which contain function symbols unrelated to the rest of those in the processor, and therefore require different rules, thus slowing down the SIMD reduction process;

2. subterms that will be needed little or not at all, for example, an already fully reduced part of the final answer that will not be needed later; and

3. subterms that will later become very large and swamp the processor.

An important objective is to minimize communication costs. The following factors seem relevant:

- *Activity levels* of subterms could be determined periodically during a computation, and then used to decide which subterms to ship elsewhere (see [18]). Subterms with very low activity level could be sent to memory, while subterms with very high activity level could be sent to other processors.

  A valuable clue about a subterm is whether it is reduced. RRM cells should have a flag indicating if the subterm headed by that cell is (known to be) reduced.

- *Flow analysis information* might help determine boundaries between different "homogeneous components" of the data, within which entirely different rules are relevant. Function symbols that are closely (flow) related should be clustered together, with each cluster constituting a "homogeneous component." If a node marks the boundary to a different component, then it and the relevant rules could be sent to a new processor. However, such sudden transitions often may not exist, and more experience with the technique described in Section 5.4 will be needed to assess its value.

  We also note that users also partition the set of rewrite rules through their use of modules. However, this partition represents the conceptual structure of the program, and the extent to which it is relevant to data partitioning is not clear.

- *Growth rates* of subterms might sometimes be estimated from the equations for their function symbols (plus equations for the other function symbols that they use); then high growth subterms could be sent to other processors, especially if they are already large. Sometimes it might even be possible to estimate the maximum space needed for reducing a given subterm. However, since we cannot expect a compiler to put complexity theorists out of business, this approach is of very limited value. It might seem more practical for programmers to estimate space growth rate and/or maximum required space; however, this will be almost as hard for them as for a compiler.

## 4.2 Rewriting Across Processors

Some interprocessor communication is unavoidable, because matching may involve subterms resident in different processors. But our objective must be to minimize interprocessor communication, because it is enormously more expensive than intraprocessor communication. Therefore we consider not only how to handle matching attempts across a boundary between processors, but also how to do replacement and reconfiguration after a match succeeds.

There are at least three approaches to cross-processor matching:

1. Ship subterms from child processors to the parent processor requesting the match;

2. Ask each child processor to do part of the match; and

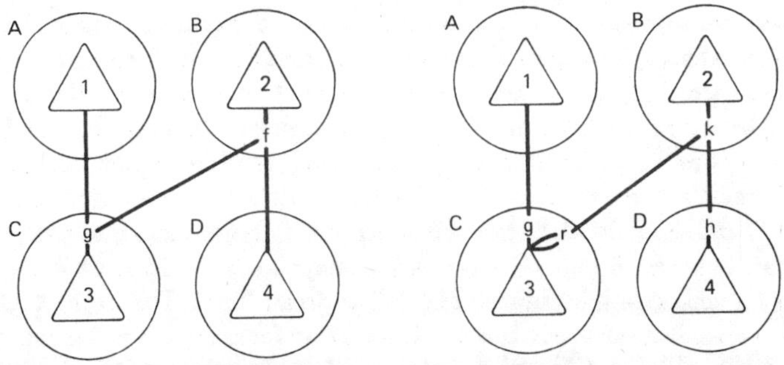


**Figure 10: Reconfiguring a Dag after Rewriting**

3. Duplicate part of the term in both processors, up to the maximum depth of lefthand sides, then match in one of the processors, shipping the result to the other.

Let us consider left-linear rules first. The first alternative requires sending up subterms even if the match fails, whereas the second alternative only requires sending down the rule number and a micro-instruction pointer. The third alternative requires no communication for matching, but may require considerable overhead in maintaining consistency between the duplicated parts.

For a non-left-linear rule in the first alternative, there is no bound on the possible size of what may have to be sent up. For the second alternative, there is no communication problem as long as the subterms being tested for equality lie in the same processor; but if they lie in different processors, again there is no bound on the communication that may be needed. The third alternative just doesn't work for non-left-linear rules.

In the first alternative, the material that was sent upward could be replaced in either processor, but since only the parent processor has all the data, it is simpler to do it there, rather than sending instructions and bindings downward for replacing parts of the match. In the second alternative, replacement is distributed by definition, and moreover, bindings may have to be sent both upward and downward, or even "crossward," between processors that are not directly connected (the likelihood and cost of this will depend on board level topology). Nonetheless, since matches fail more often than they succeed (our simulation experiments [24] show about 5% of nodes are involved in matches during SIMD computation), the second alternative remains more attractive than the first, which requires sending all this information to the parent, whether or not the match succeeds. The third alternative needs communication to keep the processors involved consistent, in case of a successful match; moreover, it is even possible that two different processors will both succeed in matching on the duplicated subterm (either with different rules or with the same rule). Here, it will probably pay to send things downward, since only the rule number, micro-instruction pointer, and bindings need be sent, and in general we expect rewriting to be more frequent in the child processors. The fact that matches are relatively rare, means that multiple matches in the duplicated subterm will even more rare, and therefore the synchronization between duplicated subterms need not be tremendously efficient; however, just freezing the relevant part of one or both processors until they agree again may be too expensive. In summary, it seems that the third alternative will be best only if we canfind a sufficiently efficient way to assure consistency of duplicated subterms; otherwise, the second approach should be used; for non-left-linear rules, the second approach should be used in any case. Perhaps the technique for handling overlapping rewrites in Section 3.3.4 could be helpful in assuring consistency. Finally, note that there may be some hybrid of the second and third alternatives.

We now consider an example of reconfiguration after a successful match, shown in Figure

10. Here processors are represented by circles and named by capital letters A,B,C,D, while dag fragments are represented by triangles numbered 1,2,3,4. The rule $f(g(x),y) \rightarrow k(r(x,x),h(y))$ matches across processors B,C,D. Note that processors A and B both refer to subterm 3 in processor C. Thus, to correctly replace the match by its corresponding righthand side, the function symbol $g$ must be kept as a link from A to C, even though the rule in question deletes $g$. Notice also that the righthand side $k(r(x,x),h(y))$ has been partitioned across processors B, C and D, with $k$ in B, $r$ in C and $h$ in D. Although this is only one choice among several, placing $r$ in C needs only one link between B and C, whereas placing $r$ in B would need 2; a major goal for reconfiguration is to minimize the number of interprocessor links. For trees, it always helps to send function symbols downward, but this need not be so for dags, since a given node can have more references to it than it has descendent nodes, in which case there will be fewer interprocessor links if it is sent upward. In practice, this will probably not be very common, but it is possible to check at run time and then do the best thing, assuming that cells contain both a reference count and know their descendents[6].

# 5 Compiler Techniques

By its nature, compiling involves some considerations that are relatively architecture specific; in this case, we use the RRM. One compiler technique discussed below is "multiplexing" several rules into a single micro-instruction sequence in order to reduce the total number of micro-instructions broadcast. A second technique is to optimize the order in which rules are applied. A third technique is flow analysis.

## 5.1 Flags and Reduction

RRM cells contain a number of flags that can be set[7] in response to instructions broadcast from their controller. For example, the controller could ask all cells that contain the function symbol + to set `Flag1`.

Some flags will have fixed interpretations. Perhaps the most important of these is the Reduced Flag, which tells whether or not the subterm headed by a given cell is reduced. This flag is useful for optimizing non-left-linear rules, and is also needed for some of the E-strategies described in Section 6. Determining whether a subterm is reduced is not hard if we know which function symbols are constructors[8], since a subterm consisting entirely of constructors is necessarily reduced. The required communication can be done with (a relatively small amount of) special hardware, running concurrently with micro-instruction execution, just as with reference counts for garbage collection (see Section 3.3.3).

Even without knowing constructors, we can determine whether a subterm is reduced by checking that every potentially applicable rule has been tried and has failed. This may not be as expensive as you first think. Because the compiler knows exactly which rules could potentially apply to a node headed by a given function symbol (in practice, there are not very many — we have rarely seen more than 5), it can arrange to set a flag in a cell whenever a rule that might apply to it in fact fails; when all these flags are up and all child cells are reduced, the cell's Reduced Flag can be set. Checking constructors is certainly more efficient, but also note that the

[6] Probably an $n$-ary function symbol will be encoded as $n-1$ binary cells at the hardware level.

[7] The values of flags are "up" and "down," also called "on" and "off," and to "set" a flag is to turn it "on."

[8] We intend to augment OBJ with optional constructor declarations.

two methods can be combined. Another possibility is to keep track of whether a rule is reduced with respect to various subsets of the total rule set.

Two additional flags are discussed in Section 6; they are used to inhibit rewriting.

## 5.2 Multiplexing Rules

It should be possible to get greater performance from our single instruction stream chip level design by "multiplexing" instructions in various ways. For a simple example, suppose that three operations obey the same rule, say the associative rule. Then we can match and replace for the associative law for all three operations using just one instruction sequence, instead of three, with some simple tricks, e.g., assigning internal codes to these three operations that have some subset of bits in common, and moreover are the only operations having exactly that bit pattern in their codes. Then a cell needs only check those bits to know whether it has one of these three operations. Whether it is a good idea to multiplex overlapping rules depends on the cost of garbage collection, which we may use to handle overlapping rewrites (see Section 3.3.3); our current guess is that this technique will pay off. (In any case, the problem must be handled for self-overlapping rules.)

Augustsson [1], from the G-machine group, has studied multiplexing for pattern matching, and our group has also worked on merging patterns for OBJ3. However, the RRM should compile not just pattern matching, but also replacement, into cell level micro-instructions. Work of Huet and Levy [15] and Hoffman and O'Donnell [14] is also related.

## 5.3 Ordering Rules

Since some rules are much more expensive to apply than others, we should try to reduce both the number of applications and the redex sizes of the expensive rules; the best way to do this is to first apply rules that tend to simplify terms. Among the factors affecting rule cost are the following:

1. Size of the left- and right-hand sides;
2. Linearity of the lefthand side;
3. Whether a rule is self-overlapping; and
4. The nature of the underlying data structure.

The tradeoffs among these factors are complex. We use the rational numbers to illustrate (OBJ code is given in Figure 11; Section B.2 explains the additional features of OBJ used here).

It is best to first apply cheap rules that greatly simplify terms. The most promising rule is a left-linear rule with its righthand side much smaller than its lefthand side, particularly if it has fewer variables in its righthand side than its lefthand side. For example, the rule `R * 0 = 0` is wonderful from this viewpoint, since it just deletes a subterm without having to examine it at all!

Although rule 5 in Figure 11 seems to have the same character, the situation is actually more complex, because we need to know that the denominator of the fraction is non-zero before we can apply the rule. This is easy, we just check the sort before applying the rule. However, this also seems to imply that the rule will not be useful for large expressions, since we are (usually) not sure that a subterm is really non-zero until it has been reduced. The possibility of **retracts** (see Section B.3) means that this is not really so. However, retracts impose some extra overhead, both from inserting retracts above each "/" sign whose denominator is non-atomic at parse time, and

```
obj RAT is
    protecting INT .
    sorts NzRat Rat .
    subsorts Int < Rat .
    subsorts NzInt < NzRat < Rat .
    op _/_ : Rat NzRat -> Rat .
    op _/_ : NzRat NzRat -> NzRat .
    op -_ : Rat -> Rat .
    op -_ : NzRat -> NzRat .
    op _+_ : Rat Rat -> Rat [comm] .
    op _*_ : Rat Rat -> Rat [comm] .
    op _*_ : NzRat NzRat -> NzRat [comm] .
    vars R S : Rat .
    vars R' S' : NzRat .
    vars I J : NzInt .
    eq : R' / R' = 1 .                                          -- 0
    eq : R / (R' / S') = (R * S') / R' .                        -- 1
    eq : (R / R') / S' = R / (R' * S') .                        -- 2
    ceq : J / I = quot(J,gcd(J,I)) / quot(I,gcd(J,I))
                      if gcd(J,I) =/= 1 .                       -- 3
    eq : R / 1 = R .                                            -- 4
    eq : 0 / R' = 0 .                                           -- 5
    eq : R / (- R') = (- R) / R' .                              -- 6
    eq : - (R / R') = (- R) / R' .                              -- 7
    eq : R + (S / R') = ((R * R') + S) / R' .                   -- 8
    eq : R * (S / R') = (R * S) / R' .                          -- 9
endo
```

Figure 11: Rational Numbers Code

---

also from removing them with a special rule when the denominator finally becomes an atomic non-zero rational. It is not clear that this overhead is worth the trouble in this particular case, but it is clear that retracts are a powerful and reasonably cheap general solution to the problem of a type system that seems, at first glance, overly restrictive.

Rule 4 also simplifies terms; although the simplification is less dramatic, it does not require knowing that its variable is non-zero; in fact, rule 4 is the *only* rule in this example that does not need to check the sort of a denominator.

In general, non-left-linear rules may require inspecting an unbounded region to test equality, whereas left-linear rules need only inspect a region bounded by the size of the rule's lefthand side. For example, the rule `X + X = X` has the power to halve the complexity of large terms, but matching in such a case may require an expensive equality check. For this reason, matching really should not be attempted for non-left-linear rules unless the subterms matched to occurrences of multi-occurrence variables are already reduced, or unless we somehow know that terms will be small and/or have a lot of sharing. Of course, whether these terms are (currently known to be) reduced is easily seen from their flags. The only non-left-linear rule in Figure 11 is rule 0.

Although rule 0 seems attractive because it can halve the complexity of a term, it is also non-left-linear, and therefore should only be applied to already reduced subterms. Thus, paradoxically, this rule is very cheap to apply, but does little to simplify terms.

Self-overlapping rules are probably not very expensive, assuming that they are handled as indicated in Section 3.3.4 using garbage collection. Only rules 1 and 2 are self-overlapping in this example.

Rules 3 through 9 are ll both left-linear and non-self-overlapping.

While the rules in Figure 11 are an elegant Church-Rosser and terminating order-sorted specification of the rational numbers, it is not clear that they are in the most efficient form for rewriting. For example, it takes 4 steps of sequential rewriting to rewrite an addition of the form `(A / B)+(C / D)` to the reduced form `((C * B)+(A * D))/(B * D)`, requiring more than twice as many micro-instructions as would the single rule

```
(A / B)+(C / D) -> ((C * B)+(A * D))/(B * D).
```

Perhaps it would be better to add some derived rules of this kind to those in Figure 11. On the other hand, they are more specialized rules and would apply less often; also, it would take more time to run through this larger rule set. It would be an interesting and surprising result if the rule set in Figure 11 were more efficient for concurrent term rewriting than the more traditional bottom-up alternatives.

## 5.4 Flow Analysis

There are at least three potential applications for the flow analysis of rules, although two of these seem rather speculative at this time:

1. To help determine what rule sets should be sent to processors from the central rule depository;

2. To aid with data partitioning; and

3. To aid with compiling code for a particular term, either initially or at runtime.

The idea is to group together rules having the same function symbol on top of their lefthand side, and then relate these rule sets by seeing what other rules use their function symbols. The following notions are therefore relevant:

**Definition 1** *A function symbol f* **flows weakly** *into a function symbol g iff there is a rule with f at the top of its lefthand side such that g occurs somewhere in its righthand side; if, in addition, g does not occur in the lefthand side of the rule, then f* **flows strongly** *into g.* □

Notice that strong flow is a subrelation of weak flow. Figure 12 shows the flow graph for the equations in rational numbers of Figure 11; here, broken lines indicate weak flow and unbroken lines indicate strong flow. Note that flow graphs can cut across different OBJ modules; for example, `gcd` and `quot` are integer functions imported from the `INT` module.

The principal application of flow analysis is to help determine what rules should be sent to processors when they get new data. It is important to send as few rules as possible, not only because of the limited storage capacity of processors, but also because time can be wasted in trying to apply rules that are not actually relevant to the subterm in question. Of course, all rules for function symbols that actually occur in the data should be sent. Definition 1 suggests criteria for deciding what other rules are most closely related to these rules, and therefore could also

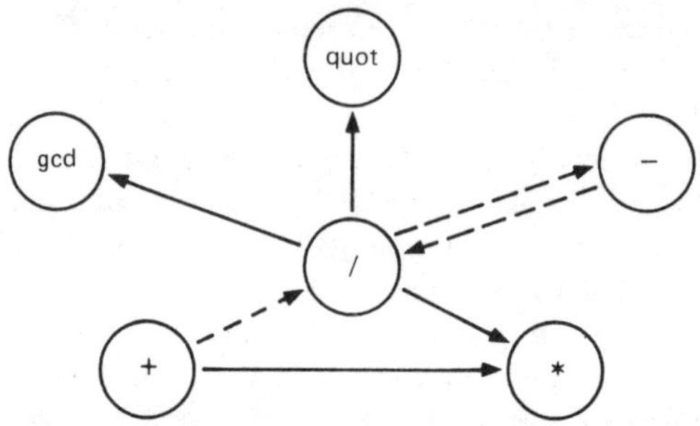


Figure 12: Flow Graph for the Rational Numbers

potentially be sent. Flow analysis can identify some rules that could never apply to a particular term. For example, consider the term

```
(3 / 6) / (7 * (16 / 12)).
```

The flow graph of Figure 12 shows that only the rules 0,1,2,3,4,9 and the rules for **gcd** and **quot** could possibly be needed.

Our experience with OBJ2 shows that there is considerable "locality" to rule sets, in the sense that once a given rule applies, the rules most closely related to it are the ones most likely apply in the future. This is what makes the notion of cache viable for conventional programming, and it is what flow analysis formalizes for rewrite rules.

A possible application to data partitioning was mentioned in Section 4.1; the idea is that clustering in the flow graph might reveal critical function symbols marking transitions between function clusters. Thus, flow analysis might be helpful in allocating subterms to processors when initially setting up a computation having a large enough initial term. It seems unlikely that flow analysis could be useful for the dynamic reallocation of subterms, because of the very limited computational resources available on a chip.

The third application seems more problematic, since it requires knowing the term at compile time, whereas we expect to use of the RRM in a cycle of incremental compilations and term executions; in general, it would be too expensive to recompile everything for every new term. However, flow analysis should be helpful in deciding what already compiled code to down-load to processors.

## 5.5 Miscellaneous

It seems a good idea to let the compiler decompose rules with large lefthand sides into two or more rules, by introducing some new function symbols; then we can bound the depth of lefthand sides (4 seems about right). The rules obtained in this way should be executed as a sequence, since if the first rule fails, we can just skip the rest. This would require communication from cells to their controller; an easy solution is just to AND together all cells' responses; then a "false" signal will appear on the single line that actually goes into the controller iff no cell can execute a rule.

Such a simple communication network can also be used for other optimizations, for example, testing whether a processor contains any instances of some function symbol, so that if there are none, the controller need not bother with rules involving that function on their lefthand side; moreover, this prohibition can continue until there is a match of a rule that introduces the function symbol from its righthand side (or until the symbol is imported from another processor).

Two further remarks:

- It may be more efficient to implement associativity with a special representation that gives the effect of rewriting modulo the associative equation (see Section B.4), rather than implement it with a rewrite rule; and
- Equality checks should be used to increase sharing; even a partially successful equality check can be used in this way (see Section 3.2.1) .

# 6 Evaluation Strategies

For most computations, the order of evaluation should not affect the final result. If the Church-Rosser property (Figure 3) holds for a set of rewrite rules, then starting from a given term, different initial choices of which rules to apply can always be continued by further rewriting to yield the same result. It is a pleasant fact (stated more formally in Appendix A) that this property holds for concurrent rewriting iff it holds for sequential rewriting. Indeed, the Church-Rosser property ensures that concurrent term rewriting correctly implements initial algebra semantics, because it allows rewrite rules to be applied in any order. But even for concurrent rewriting, some way of controlling evaluation order is very desirable, for the following reasons:

- **Space Efficiency.** Some orders of rewriting a given term may involve much greater intermediate term sizes than others.
- **Termination.** Some orders of rewriting a given term may fail to terminate, while others terminate with a normal form.
- **Concurrent Programming.** Explicit concurrency control may be needed for applications such as scheduling, synchronization and input/output.

Application in this last category are very important for actually using a machine like the RRM, and frequently involve behavior that is not Church-Rosser (see Section 6.1).

OBJ2 [6] introduced **E-strategies** (E for evaluation) to improve the efficiency of sequential rewriting. For example, given a three-argument function symbol $f$ with strategy `[1 2]` and an expression $f(t_1, t_2, t_3)$ to be reduced, the first argument $t_1$ (indicated by the number 1 in [1 2]) must be reduced first, say to $t'_1$, before reducing the second argument $t_2$ to $t'_2$ (indicated by the number 2); then we must rewrite at the top of $f(t'_1, t'_2, t_3)$; we might or might not eventually go on to reduce $t_3$; for example, it might be deleted by rewriting at the top. Thus, E-strategies free OBJ2 from blind subservience to lazy, eager, or any other fixed order of evaluation; they are enormously useful for saving time, space, and termination.

In sequential term rewriting, redexes must be tried one at a time, and the problem is to find the best rule and redex to try next; it is therefore little extra trouble to *prevent* evaluation within a given subterm. But since concurrent term rewriting tries all instances of a given rule at once (within a given processor), it takes some special effort to prevent evaluation within a designated subterm. Although the sequential E-strategies of OBJ2 are not appropriate for concurrent rewriting, they can be generalized. For example, we might want to evaluate all three major subterms of $f(t_1, t_2, t_3)$ concurrently, delaying evaluation at the root until at least the first and second subterms are reduced. This suggests that concurrent evaluation strategies should use *sets* rather than *sequences* of numbers. Thus, we might write {1 2} to indicate that the first and second subterms should both be reduced before rewriting at the top, while still allowing reduction within $t_3$. More formally now,

**Definition 2** *Let $f : s_1 s_2 \ldots s_n \to s$ be a function symbol of arity n. Then*

1. *A* **basic concurrent E-strategy** *for f is a subset $I = \{i_1, \ldots, i_k\}$ of $\{1, \ldots, n\}$, with the operational interpretation that arguments in I must be fully reduced* before *f is evaluated (i.e., before a rule is permitted to apply at the root of the tree headed by an occurrence of f).*
2. *A* **prioritized concurrent E-strategy** *for f is a sequence $I_1, I_2, \ldots, I_k$ of non-empty disjoint subsets of $\{1,\ldots,n\}$, with the following operational interpretation:*
   - *concurrently evaluate all arguments of f, giving first priority to arguments in $I_1$, second priority to arguments in $I_2$, ..., and last priority to any arguments in $I_k$; but*
   - do not *evaluate f at the top until all arguments in $I_1, \ldots, I_k$ are reduced.*
3. *A* **sequentialised concurrent E-strategy** *for f is a sequence $P_1 \mid P_2 \ldots \mid P_k \parallel I_{k+1}$ of disjoint prioritized E-strategies $P_1, \ldots, P_k$, with an optional disjoint final subset $I_{k+1}$, having the following operational interpretation:*
   - *first reduce all arguments in $P_1$ (according to priority)* without *reducing any arguments in $P_i$ for $i > 1$ or in $I_{k+1}$;*
   - *then reduce all arguments in $P_2$ (according to priority)* without *reducing any arguments in $P_i$ for $i > 2$ or in $I_{k+1}$; ...*
   - *then reduce all arguments in $P_k$ (according to priority) still* without *reducing any arguments in $I_{k+1}$;*
   - do not *evaluate f (i.e., at the root) until all arguments in $P_1, \ldots, P_k$ are reduced.*

□

Notice that the nodes in $I_{k+1}$ *never* get executed.

We can somewhat extend this notation by allowing subsets $I$ of $\{0, \ldots, n\}$ instead of just $\{1, \ldots, n\}$, where "execution at 0" is interpreted to mean applying a rule at the *root* of the tree. Notice that once a rule has been applied at the root, it is not meaningful to continue to follow the strategy — one should shift to the strategy of the *new* root (even if it is the same function symbol, it is thought of as a new copy ot it). In particular, for sequentialized E-strategies, it is not meaningful to have $0 \in P_i$ unless $i = k$; i.e., unless $P_i$ is the last set of arguments.

Here are some examples of basic strategies. In concurrent OBJ, a basic E-strategy $\{i_1, \ldots, i_k\}$ may be written just $[i_1 \ldots i_k]$, without any set brackets, or commas. First, the usual conditional should have strategy the $\{1\}$, so its OBJ declaration would be

```
if_then_else_fi : Bool S S -> S [1]
```

which requires reducing the Boolean argument before trying rules at the top, while addition might look like

```
_+_ : Int Int -> Int [1 2]
```

which calls for fully evaluating both arguments before trying to add them, and finally the insert function from the program in Figure 6 might be

```
insert : Tree Int Tree -> Tree
```

indicating the *default* strategy $\emptyset$, which just says that rewriting is allowed at the top as well as in any subtree.

To implement a basic strategy $I$, all we need is a Rewrite Inhibit Flag in the root cell, with access to the Reduced Flags of all its child nodes; the Inhibit Flag is initially set, and is removed when the Reduced Flags of all child nodes in $I$ have been set (the Reduced Flag is discussed in Section 5.1).

The operational interpretation of prioritized E-strategies is that the controller should first focus on reducing child nodes in $I_1$ until they are all reduced, then focus on those in $I_2$ until they are all reduced, .... Notice that this does not *prevent* reduction anywhere in the term, except at the very top node, $f$. We can imagine a refinement of prioritized E-strategies in which the subsets $I_i$ need not be disjoint. This would allow expressing more refined priorities, that a child in both $I_i$ and $I_{i+1}$ need not be reduced before shifting priority from $I_i$ to $I_{i+1}$; however, it is not clear how useful this refinement would be.

Prioritized strategies are more difficult to implement than basic E-strategies. The intention that the controller's choices of which rule(s) to try next should reflect priority implies that different rules may get emphasized as different subterms become reduced. It can be clear enough what the controller should do in some rather extreme cases, for example, if it is necessary to put some subterms into storage in order to have room for reducing others; but in many other cases, a prioritized strategy may be asking for overly subtle distinctions. The middle ground, containing cases for which a compiler can fairly easily generate fairly efficient code, is not yet well explored, but the expressive capability of this feature does seem attractive. (Some examples are given in the next subsection.)

Basic and prioritized E-strategies involve *optimistic* evaluation, in the sense that they allow evaluating arguments that may be not needed, in subterms that may entirely disappear later. There is no harm in this when rewriting is used to compute values, but it may be undesirable when rewriting is used for synchronization, scheduling, etc. For this reason, we introduce sequentialized E-strategies, which go further by actually *inhibiting* subterm evaluation. Notice that the optional final subset for a sequentialized E-strategy will completely inhibit evaluating some subterms, at least until after the top has been evaluated. (Examples are deferred until the next subsection.)

The OBJ2 sequential E-strategy `(1 2)` corresponds to the sequentialized concurrent E-strategy $\{1\} \mid \{2\}$; more generally, any sequential OBJ E-strategy corresponds to a sequentialized E-strategy of disjoint singleton sets. Notice that the empty basic E-strategy, $\emptyset$, and the empty prioritized and sequential E-strategies, $\epsilon$, are the same "default" strategy, which does not restrict evaluation in any way; $\emptyset$ as a sequentialized E-strategy is also the default. Notice also that an argument that is not mentioned in any subset is never inhibited, and also has no effect on when the root is evaluated; in this sense, it is a *background* process, living on whatever it can get from the rules of more activate processes, and not effecting the order of evaluation in any way.

Let us now consider how the RRM could handle the rewrite inhibits required by sequentialized E-strategies. Each cell will need two flags, "Rewrite Inhibit" and "Subterm Inhibited." The first flag allows rewrites to apply only if it is off. The second flag has three values, to indicate whether the Rewrite Inhibit flags in the subterms below are all on, all off, or some of each. The instructions issued by the local controller for rewriting must include commands that keep all these flags up to date. The expense of doing this is in direct proportion to the depth of the tree that must be first inhibited, and then later uninhibited. For the intended system programming applications, this depth should not be very great; moreover, the necessary signals can be propagated very quickly, since they are just one or two bits, and require no elaborate logic. It might also be reasonable to set a maximum depth for which inhibition is effective, i.e., to allow rewriting below that depth. Although some extra hardware is needed for communication and flag storage, it seems well worth the extra power that is gained for system programming tasks, as indicated in the subsection below; however, it is not clear that sequentialized E-strategies would be very useful for optimizing the performance of ordinary programs.

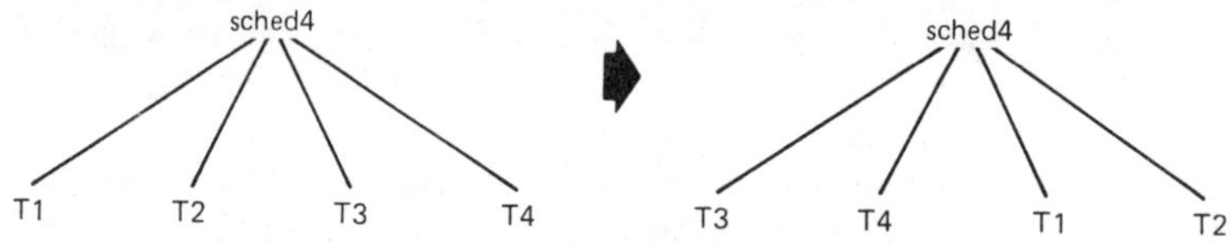


**Figure 13: Rule for a Scheduler**

## 6.1 Systems Programming

For a Church-Rosser rewrite rule set, E-strategies are just *declarative* annotations, in the sense that they cannot cause incorrect answers[9]; although they can cause failure to halt, they can also *avoid* failure to halt, and they can save time and space. Moreover, a compiler can often generate E-strategies that optimize performance; feasability of this has been shown for sequential E-strategies in sequential OBJ2 [6]. But for non-Church-Rosser rewrite rule sets, E-strategies can strongly affect the results of computation; in our view, that is their *purpose*, in support of systems programming tasks, such as scheduling and synchronization. We feel that even here, they retain a declarative character, especially in comparison with the more elaborate mechanisms found in languages like Concurrent Prolog. Moreover, E-strategies seem particularly appropriate for the RRM, whose only mode of computation is term rewriting.

We first consider a simple scheduler that concurrently works on all four of its arguments (we will call them *tasks*), giving priority to the first two until they are reduced; then it gives priority to the third and fourth. All four must be reduced before rules are tried at the top.

```
sched1 :  T T T T -> Sched1 [{1 2},{3 4}]
```

Here T is a sort for some class of tasks, and Sched1 is a sort dedicated to this particular scheduler; again, our notation allows omitting the commas called for in Definition 2. On the other hand, the scheduler

```
sched2 :  T T T T -> Sched2 [{1 2} | {3 4}]
```

first works concurrently on the first two tasks, while doing nothing on the third and fourth, until the first two are finished; then it works on the third and fourth; again, all four subterms must be reduced before trying rules at the top. Next, the scheduler

```
sched3 :  T T T T -> Sched3 [{1} {3 4}]
```

first gives priority to the first task, and then to the third and fourth; when these three are reduced, it allows reductions at the top. The second task is a background task.

For a more complex example, the following scheduler concurrently works on one task, with a second lower priority task also working in the background, and two more tasks inhibited; but when the first task is finished, it switches to working on its third task, with the fourth in the background, using the rule shown in Figure 13; when the third is finished, it switches back to the first task, with the second in the background. Thus, exactly two tasks are enabled and two tasks are inhibited in each of the two states.

```
sched4 :  T T T T -> Sched4 [{1} || {3 4}]
```

This is a sequentialized E-strategy that makes strong use of the final inhibit set. Of course, the rule in Figure 13 leads to non-terminating behavior; but that is just what we wanted.

Figure 14 shows a rule for an output queue, along with an example queue. Here the nodes

[9] This is not quite true, because the final inhibit set in a sequentialized E-strategy can actually prevent reduction.

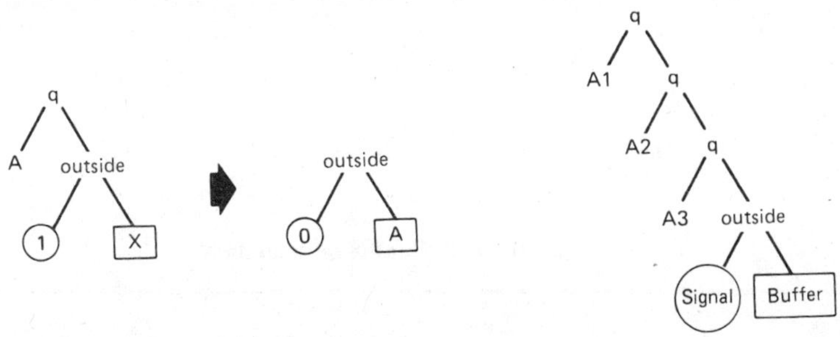


**Figure 14: An Output Queue**

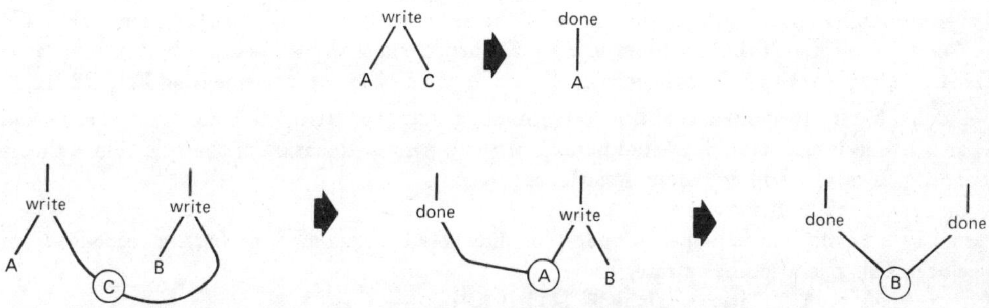


**Figure 15: Writing into a Cell**

shown with a circle and a square are very special cells in an RRM chip that actually connect to the outside world. The cell with a circle contains a one bit Signal that indicates whether or not the square Buffer cell is ready to receive a new token; when the Signal cell contains a "1" it is ready. The rule in Figure 14 places the next item from the queue into the Buffer cell, when the Signal cell indicates readiness, and then resets the Signal to 0. The outside world is responsible for setting the Signal cell back to 1 when it is ready. X and A are variables, and the function symbol `outside` marks the boundary with the external Signal and Buffer cells.

A feature of this example that may be shocking at first, is that a subterm can change from being reduced to being non-reduced. In particular, the rule in Figure 14 does not apply if the Signal is 0, and so the subtree is reduced; but once the outside world resets the Signal to 1, the subtree is reducible. This implies that Reduced Flags should not be allowed to inhibit future rewriting at an `outside` function symbol.

We can also handle concurrent writing into a cell without semaphors or other software mechanisms to lock and release resources, using low-level mechanisms that are already part of concurrent rewriting. Consider the rule for writing into a cell shown in Figure 15. Here, A, B, C are value variables, and `write` and `done` are function symbols. What is tricky here is that the cell shared (originally containing C), and shown here circled) must *not* be copied, despite the fact that our

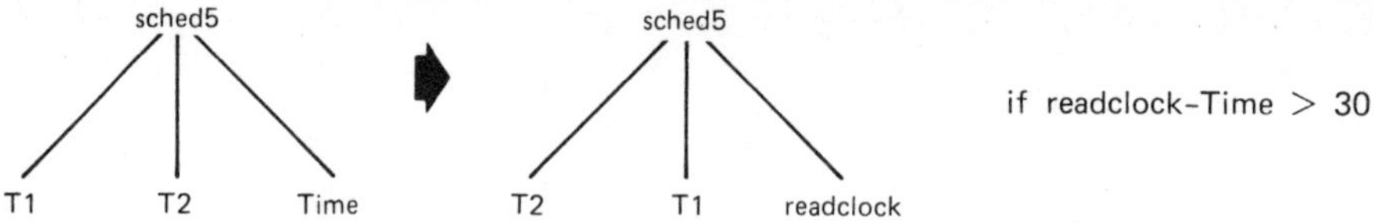


**Figure 16: A Real-time Scheduler**

standard operational semantics for dag rewriting would in fact *require* copying (see Section 3.3.1); but here, correctness demands that the cell remain shared. The solution is for the hardware to sequentialize these two writes into the same cell, since they arise from applying the same rule, as previously discussed for reads in Section 3.2.2. This implies that we may need still another flag, to inhibit copying of such cells during rewriting. Notice that the result is non-deterministic.

We are really dealing with *objects* here, in the sense of "object-oriented programming," and the Copy Inhibit flag is the key to ensuring that they are actually treated as objects, that is, as shared updatable entities. These considerations provide a basis for implementing FOOPS [12].

Finally, Figure 16 shows a real-time programming example. Here, there are two tasks, T1 and `T2`, and a timestamp `Time`, deposited here from the previous execution of the rule. Since this is a conditional rule, it will not fire unless the condition

```
readclock - Time > 30
```

is true; here `readclock` is a special operation that gets the current time (say in seconds) from the clock. This operation has strategy

```
sched5 :  T T Time -> Sched5 [{1} | {2}]
```

and the effect of the whole thing is to work on the two tasks alternately for 30 seconds each, with the other task frozen.

This world of systems programming, which may be strongly non-Church-Rosser as well as non-terminating, with its many flags and strategies, where terms may spontaneously turn from reduced to reducible, may seem like Alice's Wonderland to the theorist, and he might well long for a return to the simplicity of initial algebra semantics. However, the problems posed by the need to connect with the real world, including input/output and real-time programming, are not going to go away. The contribution of this subsection is to show that these problems can be handled within our concurrent term rewriting model of computation. We feel that this approach is powerful and excitingly simple, in comparison with the features found in most other languages (e.g., metacall in Parlog); this is probably due to that fact that we reluctantly worked our way outward from an especially clean mathematical semantics. It now becomes a challenging research problem to find a similar relatively clean algebraic denotational semantics for the systems programming features described above, preferably somehow based on initial algebra semantics. The semantics of FOOPS, based on abstract machines, is one approach; another is to consider power algebras of infinite tree algebras.

Of course, we can also expect the systems programming ideas to undergo further development, perhaps becoming even simpler and more declarative. We wish to emphasize that these relatively exotic techniques are not needed for *computation*, and in particular, programs written in OBJ really do have a nice initial algebra semantics; thus, our denotational programming language is not being used for systems programming, but rather our operational concurrent term rewriting model of computation.

# A Appendix: Formalism

Although there have been some theoretical studies of dag rewriting, including [3] and [5], as far as we know, no previous study has considered it as a model of computation for advanced computer architectures, or has dealt with concurrent dag rewriting. On the other hand, there is a very rich literature on sequential term rewriting on which to build such a study, including [16], [15] and [4]. Our notation generally follows [23] and/or [16], and these papers should be consulted for more detail on sequential rewriting.

## A.1 Definitions

Given a set $X$ of variable symbols, and a set $\Sigma$ of function symbols, each with **arity** a natural number $\alpha(\sigma)$, the **free $\Sigma$-algebra** over $X$, denoted $T_\Sigma(X)$, has as its elements the **$\Sigma$-terms** (over $X$), where a $\Sigma$-term $t$ (over $X$) is a partial function from the free monoid $P^*$ on the positive natural numbers $P$ to $\Sigma \cup X$, defined on a finite subset of $P^*$, called its **domain** and denoted $O(t)$. The elements of $O(t)$ are called the **occurrences** of $t$, and for $m \in O(t)$, $t(m)$ is called the **label** of $m$ in $t$. $O(t)$ must include:

1. $\varepsilon$, the empty occurrence for the root or top of $t$; and
2. the elements $m.i$ for $i = 1, ..., n$, denoting the $n$ child nodes of $m$, when occurrence $m$ in term $t$ has label $t(m) = f$ of arity $\alpha(f) = n$.

The occurrences of $t$ are partially ordered by $t < t'$ if $t$ is a prefix of $t'$.

Let $\mathrm{Var}(t)$ denote the set of variables of $t$, let $t|_m$ denote the subterm of $t$ at occurrence $m$ (for $m \in O(t)$), and let $t_{[m \leftarrow t']}$ denote the term obtained by replacing $t|_m$ by $t'$ in $t$. A term is **linear** iff for any $x \in \mathrm{Var}(t)$, $x$ has only one occurrence in $t$. The set of nonvariable occurrences in a term is denoted $\bar{O}(t)$. Many-sorted terms can be defined similarly [23].

A **substitution** $\sigma$ is an endomorphism of $T_\Sigma(X)$ with a finite domain, i.e., such that $\sigma(x) \neq x$ for only a finite set of variables $x \in X$. A substitution $\sigma$ may be denoted $(x_1 \leftarrow t_1, \ldots, x_n \leftarrow t_n)$. The composition of substitutions $\sigma$ and $\rho$ is denoted $\sigma.\rho$. Let $\sigma|_Y$ denote the restriction of substitution $\sigma$ to a subset $Y$ of $X$. If $F$ is a set of substitutions then $F|_Y = \{\sigma|_Y \mid \sigma \in F\}$ is the set of elements of $F$ restricted to $Y$.

An **axiom** is a pair $(t, t')$ of terms, denoted $t = t'$. A **rewrite rule** is a pair $(t, t')$ of terms, denoted $t \rightarrow t'$, such that $\mathrm{Var}(t')$ is a subset of $\mathrm{Var}(t)$. A set of rewrite rules is called a **Term Rewriting System**. A rewrite rule $t \rightarrow t'$ is called **left-linear** (resp., **right-linear**) iff $t$ (resp., $t'$) is a linear term, i.e., no variable occurs more than once in $t$ (resp., $t'$).

The **rewriting relation** $\rightarrow^R$ of a term rewriting system $R$ is defined as follows: $t \rightarrow^R t'$ iff there exists an occurrence $m$ in $\bar{O}(t)$, a rule $l \rightarrow r$ in $R$ and a substitution $\sigma$ such that $t|_m = \sigma(l)$ and $t' = t_{[m \leftarrow \sigma(r)]}$. $t|_m$ is called a **redex** in $t$ at occurrence $m$ under the rewrite rule $l \rightarrow r$. For $t$ fixed, a redex may be represented by the 3-tuple $(m, l, r)$. For example, for the rule $x * e \rightarrow x$, $(1 * 2) * e$ is a redex in the term $2 * (2 * ((1 * 2) * e))$ at occurrence 2.2. We write $t \rightarrow^R_{[m,\sigma,l \rightarrow r]} t'$ if we want to make explicit the occurrence, the substitution, and the rule involved in a rewrite.

A term $t$ which is not a single variable, **overlaps** a term $t'$ at occurrence $m \in \bar{O}(t')$ iff there exists a substitution $\theta$ such that $\theta(t) = \theta(t'|_m)$. For example the term $x * e$ overlaps at occurrence 1 the term $(2 * y) + 3$. A set $R$ of rules is called **non-overlapping** iff for any pair of rules

$l \to r$, $l' \to r'$ in $R$, $l$ and $l'$ do not overlap each other at any (nonvariable) occurrence. Otherwise the set is said to be **overlapping.** A rule $l \to r$ is called **self-overlapping** iff $l$ overlaps itself at an occurrence different from $\varepsilon$; associativity is a typical self-overlapping rule.

For $t$ a term and $R$ a term rewriting system, let $R(t)$ denote the set of all the redexes in $t$ under $R$, i.e.,

$$(u, l, r) \in R(t) \Leftrightarrow l \to r \in R \text{ and } \exists\sigma \text{ such that } t_{|u} = \sigma(l).$$

**Definition 3** *A subset $W$ of $R(t)$ is said to be* **non-overlapping** *iff for any redexes $(u,l,r)$ and $(u', l', r')$ in $W$,*

- *$u$ and $u'$ are incomparable, or*
- *$u < u'$ and $\exists v \in O_{Var}(l)$ s.t. $u.v \leq u'$ where $O_{Var}(l)$ is the set of variable occurrences in $l$,*

$$O_{Var}(l) = \{m \mid m \in O(l) \text{ and } l(m) \in X\}.$$

- *If $l$ is non-left-linear for the variable $x$ and if there exists $w \in O_x(l)$, the set of all occurrences of $x$ in $l$, such that $u' = u.w.v$ for some $v$, then $\{(u.w'.v, l', r') \mid w' \in O_x(l)\} \subseteq W$.*

*Let $\Delta(t)$ be the set of all non-overlapping subsets of redexes in $t$.* □

This definition is illustrated by Figure 5 in the main text; the need for the third "completeness" condition came as a bit of a surprise. We are now ready for

**Definition 4 Concurrent rewriting,** *with relation denoted $\to^{R\|}$, is defined by*

$$t \to_W^{R\|} t' \Leftrightarrow \begin{cases} W = \{(u_i, l_i, r_i) \mid 1 \leq i \leq n\} \in \Delta(t) \\ \text{and} \\ i < j \Rightarrow u_i \not< u_j \\ \text{and} \\ t \to^R_{[u_1, l_1, r_1]} t_1 \to \ldots \to^R_{[u_n, l_n, r_n]} t'. \end{cases}$$

□

The second condition says that we get the result of applying the set of redexes in $W$ with a bottom-up sequential strategy, while the last condition requires that the redexes are non-overlapping.

Definition 4 assumes that all rewrites involved occur in parallel, starting and finishing within the same interval. Although this assumption is entirely false in practice, concurrent rewriting in the sense of Definition 4 can still be used to *simulate* true concurrent rewriting, in which different rewrites start and finish at different times, to any desired degree of accuracy. The method involves considering each rewrite in the sense of Definition 4 to take less time and to have less concurrency. In particular, we may let the time for a rewrite in the sense of Definition 4 correspond to RRM micro-instruction cycle time, and then freeze the cells involved in a match until enough cycles have passed to complete the rewrite; this permits rewrites to take different lengths of time and also to overlap in time.

**Definition 5 (Locally) maximal concurrent rewriting,** *with relation $\to^{R\sharp}$, is defined by*

$$t \to_W^{R\sharp} t' \Leftrightarrow \begin{cases} t \to_W^{R\|} t' \\ \text{and} \\ W \text{ maximal in } (\Delta(t), \subseteq). \end{cases}$$

**Numerically maximal concurrent rewriting,** *with relation* $\rightarrow^{R\sharp^{num}}$, *is defined by*

$$t \rightarrow_W^{R\sharp^{num}} t' \Leftrightarrow \begin{cases} t \rightarrow_W^{R\sharp} t' \\ \text{and} \\ |W| \text{ maximal in } \Delta(t) \text{ for } \leq . \end{cases}$$

□

Notice that $\rightarrow^{R\sharp^{num}} \subseteq \rightarrow^{R\sharp} \subseteq \rightarrow^{R\|}$. Maximal concurrent rewritings are concurrent rewritings with a set of non-overlapping redexes that cannot be further extended. Numerically maximal concurrent rewritings are maximal concurrent rewritings that have the maximum possible number of non-overlapping redexes. Some simulation results for maximal concurrent term rewriting are given in [24]. The concepts of Definition 5 are largely of theoretical interest, because they capture the greatest amount of concurrency that could possibly be achieved in one step. It is not practical to implement maximal concurrent rewriting, and even if it were, it would sometimes be undesirable; the RRM uses mixed grain SIMD/MIMD rewriting.

## A.2 Properties

**Definition 6** *Let* $\rightarrow$ *be a relation on the set of terms, and let* $\xrightarrow{*}$ *and* $\xleftrightarrow{*}$ *be respectively its reflexive transitive and symmetric reflexive transitive closures. Then* $\rightarrow$ *is* **terminating** *iff there is no sequence of the form*

$$t_0 \rightarrow t_1 \dots t_n \rightarrow t_{n+1} \cdots ,$$

*and* $\rightarrow$ *is* **Church-Rosser** *iff*

$$(\forall t_1, t_2)\ t_1 \xleftrightarrow{*} t_2 \Rightarrow \exists t' \text{ such that } t_1 \xrightarrow{*} t' \text{ and } t_2 \xrightarrow{*} t' .$$

*Then we say that a set R of rewrite rules is* **sequentially terminating** *(resp.,* **sequentially Church-Rosser***) iff the sequential rewriting relation* $\rightarrow^R$ *is terminating (resp., Church-Rosser), and that R is* **concurrently terminating** *(resp.,* **concurrently Church-Rosser***) iff the concurrent term rewriting relation* $\rightarrow^{R\|}$ *is terminating (resp., Church-Rosser).* □

**Proposition 7** *Let R be a term rewriting system. Then*

- *if sequential rewriting with R is terminating, so is concurrent rewriting, and in particular, so is maximal concurrent rewriting;*
- *if sequential rewriting with R is Church-Rosser, so is concurrent rewriting, and in particular, so is maximal concurrent rewriting.*

□

**Corollary 8** *If a term rewriting system R is sequentially terminating and Church-Rosser, then it is also concurrently terminating and Church-Rosser, and in particular, maximal concurrent rewriting is terminating and Church-Rosser.* □

This result insures the correctness of concurrent term rewriting as an implementation of initial algebra semantics. Results of the same form are not as robust as one might think, as shown by Example 10 below; but first, we need

**Definition 9** *A term rewriting system is* **weak terminating** *iff every term has a normal form.*
□

Note that terms can have more than one normal form in a weak terminating term rewriting system.

**Example 10** *If $R$ is given by*

$$\begin{array}{lll} f(x,x) & \to & a \\ g(x) & \to & f(x,g(x)) \\ b & \to & g(b) \\ f(x,g(x)) & \to & b \end{array}$$

*then $R$ is weak terminating, since every term has $a$ as a normal form; but the concurrent rewriting relation $\to^{R\sharp}$ associated to $R$ is* not *weak terminating, since $b$ has no normal form, e.g.,*

$$b \to^{R\sharp} g(b) \to^{R\sharp} f(g(b), g(g(b))) \to^{R\sharp} f(f(g(b), g(g(b))), \ldots) \to^{R\sharp} \ldots .$$

*Notice that the first rule is non-left-linear, and also that $R$* is *Church-Rosser, but is* not *interreduced (i.e., some righthand sides can be reduced by applying rules to them).* □

This example shows that maximal concurrent rewriting can be dangerous for systems that are Church-Rosser and only weak terminating. This is somewhat interesting because many important examples are Church-Rosser and only weak terminating. However, as noted above, maximal concurrent rewriting cannot really be implemented efficiently, and anyway, we could always use a concurrent E-strategy to block maximal concurrent rewriting and ensure termination.

# B Appendix: OBJ

This appendix describes the OBJ2 ultra high level language [6]. OBJ is *declarative*, in the sense that its programs try to say what the problem is, rather than how it should be solved. This allows the compiler and operating system much more leeway in detecting and using whatever concurrency is available. By contrast, today's most popular languages (including parallel languages) are *imperative*, in the sense that their programs consist of commands that not only say exactly what to do, but also exactly when and where to do it. This not only limits compile-time and run-time optimization, but also greatly limits portability among different architectures, and makes programs harder to read, write and debug. Also, OBJ programs are generally much smaller than imperative programs.

OBJ has a number of features not shared by other functional programming languages with which we are familiar (such as Miranda and FP) including the following:

1. **User-definable abstract data types**, not limited to constructors like most other functional languages;

2. **Generic modules**, which not only give a powerful modularization facility, going far beyond mere functional composition, but also give the power of higher order programming without its difficulties in understandability and verification (see Section B.5);

3. **Subsorts**, which support exception handling and overloading in an elegant way (see Section B.3);

4. **Pattern matching modulo equations**, including the associative, commutative, idempotent, and identity laws, which greatly increases the power of matching, and hence the expressive power of the language (see Section B.4);

```
obj BITS is
    protecting INT .
    sorts Bit Bits
    ops 0 1 : -> Bit .
    op nil : -> Bits .
    op _._ : Bit Bits -> Bits .
    op length_ : Bits -> Int .
    var B : Bit
    var S : Bits
    eq : length nil = 0 .
    eq : length B . S = inc length S .
endo
```

Figure 17: String of Bits Code

5. **E-strategies**, which, as explained in Section 6, avoid enslavement to any fixed evaluation strategy, such as eager or lazy, and thus allow greater efficiency, in both time and space; E-strategies also support system programming tasks;

6. **Module expressions**, which allow constructing (sub)systems from previously defined modules (see Section B.5);

7. **Theories and views**, which permit users to specify module interfaces and also to state non-executable properties of modules (see Section B.5); and

8. **Very simple denotational semantics** given by the initial algebra interpretation of the equations in a program.

The last point explains why OBJ is declarative: its statements can be read as equations stating properties that one wants the solution to have, i.e., as describing the *problem* to be solved. OBJ is particularly suitable for the RRM because its operational semantics is term rewriting. OBJ has perhaps the most powerful generic module mechanism available in any current programming language, and is "wide spectrum" in the sense that it integrates coding, specification, design, and verification into a single framework. Let us now further discuss some of these features; the reader who wants more detail than we can give here should see [6].

## B.1 A Simple Example

The simple program for strings of bits in Figure 17 illustrates some basic features of OBJ2. OBJ2's most basic entity is the **object**, a module encapsulating executable code. The keywords `obj ... endo` delimit the text of an object. Immediately after the initial keyword `obj` comes the object's name, in this case `BITS`; then comes a declaration indicating that the built-in object `NAT` is imported. This is followed by declarations for the new sorts of data, `Bit` and `Bits`, for the constants `0`, `1` and `nil`, and for the operations `_._` and `length_`, each with information about the distribution and sorts of arguments and the sort of the result; underbar characters "`_`" are used to indicate argument places for mixfix operations; thus `_._` is infix, and `length_` is prefix. Finally, variables of sorts `Bit` and `Bits` are declared, and the two equations constituting the body of the object are given; `inc` is the increment (i.e., add 1) function on naturals, imported from `NAT`.

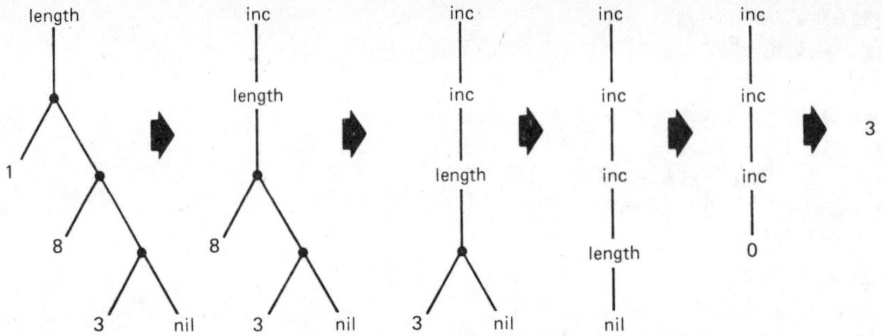


Figure 18: Reduction of a Bits Expression

Terms are constructed from constants and function symbols, for example, `3 . (7 . nil)` and `length 1 . (8 . (3 . nil))`; the latter evaluates to 3, applying the two equations as left-to-right rewrite rules in Figure 18. (This example does not exploit concurrency, but could easily be changed to do so, e.g., with balanced binary trees.)

## B.2 Some Further Features

The keyword `protecting` indicates that whatever is defined by the imported module is not affected by anything within the module to which it is imported. This means that

1. no new data items of sort `Nat` are introduced by `BITS`; and

2. no old data items of sort `Nat` are identified by the equations in `BITS`.

In the terminology of Section 2.1.1, the first condition means that there is "no junk" and the second means that there is "no confusion." OBJ also allows two weaker import declarations: **`extending`**, which only guarantees that there is no confusion; and **`using`**, which makes no guarantees at all.

OBJ2 provides two kinds of conditional. The first is the usual mixfix conditional function, `if_then_else_fi`, defined by the equations

```
    if true then X else Y = X .
    if false then X else Y = Y .
```

whenever `X` and `Y` match to subterms of the same sort (plus the first argument must be of sort `Bool`). This polymorphic conditional is used in both of the equations in the `BINARY-TREE` example of Figure 6; moreover, these equations actually embed one instance of this conditional within another. The second kind of conditional is the **conditional equation**, indicated by the keyword `ceq` as in equation 3 of Figure 11; the operational semantics says that the rule is applied only if the condition, which follows the keyword `if`, evaluates to `true`; the denotational semantics says that the equality holds if the condition is true.

The binary tree example of Figure 6 also contains the equality function `==`, which, like the mixfix conditional, is polymorphic, in the sense that `X == Y` is defined whenever `X` and `Y` have the same sort; moreover, it evaluates to `true` iff `X` and `Y` evaluate to the same reduced form, and otherwise evaluates to **false**. The rational numbers example of Figure 11 contains the disequality function symbol =/=, which is just the negation of ==.

## B.3 Subsorts and Polymorphism

One sort of data is often contained in another; for example, the natural numbers are contained in the integers, which are contained in the rationals; thus, the sort Nat is a **subsort** of Int, and Int is a subsort of Rat, written

```
Nat < Int < Rat.
```

Moreover, an operation may be restricted to subsorts of its arity and its value sort, and still be "the same" operation. For example, each addition operation

```
_+_ : Rat Rat -> Rat,
_+_ : Int Int -> Int,
_+_ : Nat Nat -> Nat
```

is a restriction of the preceding one; such operations are said to be **overloaded**; this provides another simple, but powerful, form of polymorphism[10].

OBJ2's very flexible subsort mechanism also provides the following, all within the framework of an initial algebra semantics:

- **Multiple inheritance** in the sense of object-oriented programming permits one sort to be a subsort of two (or more) others, each having various defined operations; then all these operations are inherited by the subsort. For example, we might have both

  ```
  RegisteredVehicle < Vehicle
  RegisteredVehicle < TaxedObject,
  ```

  with say a speed operation on Vehicle and a tax-amount operation on TaxedObject; then both of these operations are *inherited* by items of sort RegisteredVehicle.

- The familiar difficulties for abstract data types with "partial" operations (such as tail for lists and push for bounded stacks) disappear by viewing the operations as total on the right subsorts.

- **Errors** can be treated in several styles, without need for special syntactic or semantic "error handling" mechanisms.

See [13] for the details of order-sorted algebra, which provides the formal semantic foundation for all these features.

Another important feature is **retracts**, which give the "benefit of the doubt" to expressions that strictly speaking do not type check, but might after evaluation. Assuming that the factorial function _! is only defined for natural numbers, consider the expression (- 6 / - 2)!; it does not parse as given, because the factorial needs a natural number but has been given a rational. However, the naturals are a subsort of the rationals, so evaluating the rational expression could lower its sort to Nat, and indeed, does so in this case. OBJ inserts a formal retract function symbol r.Rat.Nat which lowers the rational expression to sort Nat, i.e., the original expression is replaced by the expression r.Rat.Nat(- 6 / - 2)!. The semantics of this retract is given by the equation

```
r.Rat.Nat(N) = N .
```

---

[10]Parameterized objects provide still another kind of polymophism, since a generic object's operations are available to all its instances.

```
obj LIST-OF-NAT is
    protecting NAT .
    sorts List .
    subsorts Nat < List .
    op __ : List List -> List [assoc id: nil] .
endo
```

Figure 19: List of Natural Numbers Code

where N is a variable of sort Nat. Thus, the retract only disappears if the expression to which it has been applied rewrites to one of the right sort; otherwise the retract remains provides an informative error message.

## B.4 Rewriting Modulo Equations

OBJ2 also supports rewriting modulo various equations, including associativity, commutativity, identity and idempotence. For example, in the rational numbers of Figure 11, both addition and multiplication are declared commutative by giving them the attribute comm. In particular, this means that the rule

```
R * (S / R') = (R * S) / R'
```

will match, modulo commutativity of multiplication, the expression (19 / 5) * 7, with R=7, S=19, R'=5.

Similarly, the LIST-OF-NAT module in Figure 19 has an associative list concatenation operation with nil as its identity, i.e., declared with attribute [assoc id: nil], meaning that it is associative and has identity nil. This is about as simple a way of getting lists as you can imagine; also, it has exactly the right syntax,in that the user can write (1 2 3) instead of (1 . 2) . 3, and will also see the unparenthesized form printed; and finally, it provides string matching that is associative with identity; e.g., in the SORTING module of Figure 21, the rule

```
ceq : sort(L E L' E' L'') = sort(L E' L' E L'') if E' < E .
```

matches the term sort((4 7) 3) modulo associativity and identity, with substitution L=4, E=7, E'=3, L'=L''=nil.

## B.5 Modules, Theories and Views

OBJ2 has three basic kinds of entity:

- **Objects** declare new types (our notion of type includes both the operations and the sorts of data), and give equations that define the operations. These equations are executable code when interpreted as rewrite rules.
- **Theories** define the requirements for interfaces; they also declare sorts, operations, and equations, but these equations need not be executable as rewrite rules.
- **Views** express bindings of actuals to requirements at module interfaces; they are used to put modules together to form larger program units, and also to declare that certain properties are true of a module.

```
th POSET is
   protecting BOOL .
   sort Elt .
   op _<_ : Elt Elt -> Bool .
   vars E E' E'' : Elt .
   eq : E < E = false .
   ceq : E < E'' = true if (E < E' and E' < E'') .
endth
```

Figure 20: Theory of Partially Order Sets

```
obj SORTING[X :: POSET] is
    protecting LIST[X] .
    op sorted : List -> Bool .
    op sort : List -> List .
    vars E E' : Elt .
    var L L' L'' : List .
    eq : sorted(nil) = true .
    eq : sorted(E) = true .
    eq : sorted(E E' L) = (E == E' or E < E') and sorted(E' L) .
    ceq : sort(L E L' E' L'') = sort(L E' L' E L'') if E' < E .
    ceq : sort(L) = L if sorted(L) .
endo
```

Figure 21: Sorting Module

Combining the first two of these gives the class of **modules**. Modules can import other modules, and thus use the capabilities that they provide.

An important aspect of OBJ2 is its systematic use of generic (parameterized) modules. Encapsulating related code makes it more reusable, and generics are even more reusable, since they can be "tuned" for a variety of applications by choosing different parameter values; moreover, debugging, maintenance, readability and portability are all enhanced. The interface declarations of OBJ2 generics are not purely syntactic, like Ada's; instead, they may contain *semantic* requirements that actual modules must satisfy before they can be meaningfully substituted. This can prevent many subtle bugs.

For example, the interface of a generic `SORTING` module should not just impose the syntactic requirement that only a function of type `X X -> Bool` can be given as the actual, but it should insist that that function be a partial ordering, i.e., that it satisfy the requirement theory `POSET` given in Figure 20. Then we can define the generic `SORTING` module as shown in Figure 21, with the interface requirement that any actual parameter `X` must satisfy the theory `POSET`. Note that `SORTING` imports another generic object, `LIST`, for lists of elements (this module is the obvious modification of the code in Figure 19).

```
view INT-DESC of INT as POSET is
     sort Elt to Int .
     vars X Y : Elt .
     op : X < Y to : Y < X .
endv
```

Figure 22: A View of the Integers as a Poset

### B.5.1 Instantiation and Views

To **instantiate** a parameterized object means to provide actual objects satisfying each of its requirement theories[11]. In OBJ2, the actual objects are provided through **views**, which *bind* required sorts and operations to those actually provided (i.e., a view maps the sort and operation symbols in the formal requirement theory to those in the actual object[12]) in such a way that all axioms of the requirement theory are satisfied. For example, the view in Figure 22 views `INT` with descending order as a poset; in the line

```
    op :  X < Y to :  Y < X ,
```

the first `<` is from `POSET`, while the second `<` is from `INT`. Instantiating a parameterized object with a view is indicated by replacing the parameter part with the view name. Thus, instantiating `SORTING[P :: POSET]` to sort integers in descending order is indicated by `SORTING[INT-DESC]`; it is also allowed to put an explicit view inside square brackets in this way.

Note that there can be more than one view from a theory to an actual; for example, another instantiation of `SORTING` with `INT` uses a view that maps `<` in `POSET` to the (irreflexive) divisibility relation in `INT`. Often there is a `default view` that "does the obvious thing," for example, binding `<` to an operation with the same name in the actual (default views only exist when there is a unique "obvious" mapping; see [8] for more detail).

A very unusual feature of OBJ2 is its support of **parameterised programming** [8], which allows modifying and combining modules to form larger systems, with the interfaces between program units described by theories. Thus, (a form of) program transformation is available within the language itself. In OBJ2, these commands appear in **module expressions**, which describe (and create) complex combinations of modules. This capability goes beyond that provided by the Module Interconnection Languages (MILs) that sometimes appear as a separate level above a programming language, since in general MILs only describe program structure, but cannot be used to create it.

For example, given a generic module `ITER` that iterates an associative binary operation `_*_` and has the theory `MONOID` of monoids as its requirement theory, and given a generic module `POLY` that forms polynomials on a given ring of coefficients and a given set of variables and has two requirement theories, `RING` for coefficients and `TRIV` (the trivial theory) for variables, we can extend polynomials on the rational numbers (with identifiers used as variables) by an operation `poly-pi` that multiplies arbitrary lists of polynomials, with the module expression

```
    ITER[POLY[RAT,ID]] * (iter to poly-pi)
```

where a default view has been used to bind the function symbol `_*_` in the `MONOID` requirement theory to the multiplication function having the same syntax in `RAT`, and where `* (iter to`

[11] Parameterized objects can also be seen as "object generators."

[12] More generally, a view can map an operation in the formal to an expression in the actual, as illustrated in just a moment.

**poly-pi)** indicates that the iteration over lists has been renamed to **poly-pi**.

In the same way, we can extend the rational numbers with operations **rat-sigma** and **rat-pi** that (respectively) add and multiply arbitrary lists of rationals, with the two module expressions below:

```
ITER[RAT+] * (iter to rat-sigma)
ITER[RAT*] * (iter to rat-pi)
```

where RAT+ and RAT* are views of RAT as an additive and as a multiplicative monoid respectively.

The kind of module composition that is supported by module expressions has several advantages:

1. This composition is more powerful than the purely functional composition of functional programming, in that a single module instantiation can perform many different function compositions at once. For example, a complex arithmetic module CPXA that takes real arithmetic modules as actual parameters could be instantiated with

   - single precision reals, CPXA[SP-REAL],
   - double precision reals, CPXA[DP-REAL], or
   - multiple precision reals, CPXA[MP-REAL],

   each involving the substitution of dozens of functions into dozens of other functions; this would be *much* more effort with just functional composition available.

2. The logic remains first order, so that understanding and verifying code is much simpler than with higher order logic.

3. Semantic declarations are allowed at module interfaces (given by requirement theories).

4. Strong type checking will catch many errors.

5. Besides instantiation, module expressions also allow renaming of module parts (as above) and "adding" modules (i.e., making the functions in several available at once).

# References

[1] Lennart Augustsson. *Compiling Pattern-Matching.* Technical Report 25, University of Göteborg, Programming Methodology Group, September 1986.

[2] John Backus. Can programming be liberated from the von neumann style? *Communications of the Association for Computing Machinery*, 21(8):613–641, 1978.

[3] M. Bauderon and Bruno Courcelle. *Graph expressions and graph rewritings.* Technical Report, Université de Bordeaux 1, 1985.

[4] Gérard Boudol. Computational semantics of term rewriting systems. In Maurice Nivat and John Reynolds, editors, *Algebraic Methods in Semantics*, pages 169–236, Cambridge University Press, 1985.

[5] Cynthia Dwork, Paris Kanellakis, and Larry Stockmeyer. *Parallel Algorithms for Term Matching.* Technical Report, MIT, 1986.

[6] Kokichi Futatsugi, Joseph Goguen, Jean-Pierre Jouannaud, and José Meseguer. Principles of OBJ2. In Brian K. Reid, editor, *Proceedings of 12th ACM Symposium on Principles of Programming Languages*, pages 52–66, Association for Computing Machinery, 1985.

[7] Joseph Goguen. How to prove algebraic inductive hypotheses without induction: with applications to the correctness of data type representations. In W. Bibel and Robert Kowalski, editors, *Proceedings, Fifth Conference on Automated Deduction*, pages 356–373, Springer-Verlag, 1980. Lecture Notes in Computer Science, Volume 87.

[8] Joseph Goguen. Parameterized programming. *Transactions on Software Engineering*, SE-10(5):528–543, September 1984.

[9] Joseph Goguen, Jean-Pierre Jouannaud, and José Meseguer. Operational semantics of order-sorted algebra. In W. Brauer, editor, *Proceedings, 1985 International Conference on Automata, Languages and Programming*, Springer-Verlag, 1985. Lecture Notes in Computer Science, Volume 194.

[10] Joseph Goguen, Claude Kirchner, Sany Leinwand, José Meseguer, and Timothy Winkler. Progress report on the rewrite rule machine. *IEEE Computer Architecture Technical Committee Newsletter*, 7–21, March 1986.

[11] Joseph Goguen and José Meseguer. Eqlog: equality, types, and generic modules for logic programming. In Douglas DeGroot and Gary Lindstrom, editors, *Functional and Logic Programming*, pages 295–363, Prentice-Hall, 1986. An earlier version appears in *Journal of Logic Programming*, Volume 1, Number 2, pages 179-210, September 1984.

[12] Joseph Goguen and José Meseguer. Extensions and foundations for object-oriented programming. In Bruce Shriver and Peter Wegner, editors, *Research Directions in Object-Oriented Programming*, MIT Press, to appear 1987. Preliminary version in *SIGPLAN Notices*, Volume 21, Number 10, pages 153-162, October 1986.

[13] Joseph Goguen and José Meseguer. *Order-Sorted Algebra I: Partial and Overloaded Operations, Errors and Inheritance.* Technical Report To appear, SRI International, Computer Science Lab, 1987. Given as lecture at Seminar on Types, Carnegie-Mellon University, June 1983.

[14] Christoph M. Hoffman and Michael O'Donnell. Programming with equations. *Transactions on Programming Languages and Systems*, 1(4):83–112, 1982.

[15] Gérard Huet and Jean-Jacques Levy. *Computations in Non-ambiguous Linear Term Rewriting Systems.* Technical Report, INRIA Laboria, 1979.

[16] Gérard Huet and Derek Oppen. Equations and rewrite rules: a survey. In Ron Book, editor, *Formal Language Theory: Perspectives and Open Problems*, Academic Press, 1980.

[17] John Hughes. *Lazy Memo-Functions.* Technical Report 21, University of Göteborg, Programming Methodology Group, September 1985.

[18] Bharadwaj Jayaraman and Robert Keller. Resource control in a demand-driven data-flow model. *IEEE Conference on Parallel Processing*, 118–127, August 1980.

[19] Robert Keller, F. C. H. Lin, and J. Tanaka. Rediflow multiprocessing. *IEEE Compcon*, 410–417, February 1984.

[20] Robert Keller, Gary Lindstrom, and S. Patil. A loosely-coupled applicative multi-processing system. *AFIPS Conference Proceedings*, 613–622, June 1979.

[21] Sany Leinwand and Joseph Goguen. Architectural options for the rewrite rule machine. In Steven and Svetlana Kartashev, editors, *Proceedings, Second International Supercomputing Conference*, Van Nostrand, 1987. To appear.

[22] José Meseguer and Joseph Goguen. *Deduction with Many-Sorted Rewrite Rules*. Technical Report CSLI-85-42, Center for the Study of Language and Information, Stanford University, December 1985. To appear in *Theoretical Computer Science*.

[23] José Meseguer and Joseph Goguen. Initiality, induction and computability. In Maurice Nivat and John Reynolds, editors, *Algebraic Methods in Semantics*, pages 459–541, Cambridge University Press, 1985.

[24] Timothy Winkler, Sany Leinwand, and Joseph Goguen. Simulation of concurrent term rewriting. In Steven and Svetlana Kartashev, editors, *Proceedings, Second International Supercomputing Conference*, Van Nostrand, 1987. To appear.

# Alfalfa: Distributed Graph Reduction on a Hypercube Multiprocessor *

Benjamin Goldberg
Paul Hudak

Yale University
Department of Computer Science

November 1986

**Abstract**

Alfalfa is implementation of a functional language on the Intel iPSC multiprocessor. It is based on a heterogeneous abstract machine model consisting of both graph reduction and stack oriented execution. Alfalfa consists of two major components, a compiler and a run-time system. The source language, Alfl, contains no constructs that allow the programmer to specify parallelism or synchronization and thus it is the task of the compiler to detect the exploitable parallelism in a program. The run-time system supports dynamic scheduling, interprocessor communication, and storage management. A number of statistics gathered during execution are presented.

## 1 Introduction

It has been suggested that Functional Languages provide a natural methodology for programming parallel computers. However, until recently parallel architectures have not been available to test this hypothesis. In addition, functional programing implementations on sequential machines have tended to be rather inefficient and there has been concern that such inefficiency may nullify any possible gains when extended to the parallel case.

Alfalfa is a research project at Yale University that attempts to address these issues via an implementation of a functional programming language on the Intel iPSC multiprocessor. There are two major components to Alfalfa, a compiler and a run-time system. Unlike

*This research was supported in part by the US Government under DOE grant FG02-86ER25012

other compilers for functional languages, the Alfalfa compiler performs a large number of optimizations specifically aimed at making the execution of functional programs efficient on multiprocessors. The Alfalfa Run-time System provides much of the mechanism for the distributed execution and performs such tasks as dynamic scheduling and storage management. It also performs the important task of gathering statistics about program execution that will be necessary for evaluation and modification of the algorithms used by both the Alfalfa compiler and the Alfalfa Run-time System.

## 2 The Intel iPSC

The Intel iPSC (for Personal SuperComputer) is a MIMD multiprocessor that can be configured with up to 128 Intel 80286 microprocessors. Each processor has its own memory and there is no shared memory. The processors are linked via a hypercube network, that is each processor sits at a vertex of an N dimensional hypercube where N is an integer between 0 and 7.

In an N dimensional hypercube, there $2^N$ processors. Each processor has N neighboring processors with which it can communicate directly. All communication between processors in the iPSC occurs via the passsing of messages. The operating system provides the user with only a few communication primitives such as send, blocking receive, and non-blocking receive[3].

On each processor board, there are N Ethernet chips, each of which is responsible for point to point communication with one of the processor's neighbors. The longest distance a message must travel within the iPSC is through N links (or "hops"). Message routing is performed by the operating system and is transparent to the programmer.

Unfortunately, the current version of the iPSC suffers from a very large communication overhead, both in terms of message latency and processor drain.

## 3 The Source Language

The functional programs that get executed on the iPSC by Alfalfa are written in Alfl, a functional language designed at Yale[6]. It is similiar in many ways to other existing functional languages such as SASL[15], ML[13], and HOPE[1]. It is weakly typed, however, and forces Alfalfa to perform run-time type checking.

Perhaps the most important aspect of Alfl as it applies to this paper is the fact that it contains no explicit constructs for expressing parallelism. Thus the programmer is dependent on the Alfalfa compiler and run-time system to detect and exploit the inherent parallelism in his program[1]

## 3.1 The abstract machine

### 3.1.1 Motivation

One of the motivations for designing a system like Alfalfa is to be able to execute functional programs on "conventional" multiprocessors, architectures that are available now and in the near future. The processors that constitute these multiprocessors have been designed to execute sequential programs using a stack and stack oriented instructions. Although it may be argued that we would be better off designing processors specifically targeted for graph oriented evaluation, such as in [2,5,12], we believe that we can build a system that can exploit the strengths of the current machines. At the same time, we strive to avoid, as much as possible, the weakness of current multiprocessors, namely the cost of interprocessor communication.

The overall design philosophy in the Alfalfa system can be summarized as "Separate the issues of parallelism, graph reduction, and stack based execution". Graph Reduction is a powerful evaluation strategy for supporting lazy evaluation and higher order functions, but involves unnecessary overhead if stack based execution would suffice. The same can be said for distributed evaluation. If there is parallelism worth exploiting, then a multiprocessor can be a powerful computing engine. However, if we try to distribute work that is inherently sequential, then performance will be degraded. In other words, we want to accomplish as much as possible with as little mechanism as possible.

### 3.1.2 Serial Combinators

In previous papers[10,9], we have discussed transforming the source program into a new set of functions, called serial combinators, whose bodies are sequential and can be executed on a single processor without loss of parallelism. In addition, they are the largest individual functions exhibiting this property. It may also be the case that a sequence of calls to serial combinators may exhibit no parallelism. In section 3.1.3 we discuss how such a sequence is detected and automatically executed single processor.

In section 4.3.2 we describe the mechanism for executing serial combintors. It is worth mentioning here, however, that the code generated for each serial combinator is native code for the multiprocessor. It is not a sequence of instructions for an abstract machine, e.g. a "graph reduction engine", that needs to be interpreted at run time. If the execution of a serial combinator requires some synchronization or some other mechanism not provided

[1] There is a dialect of Alfl, called Par-ALFL, which contains constructs for expressing parallelism.

by the hardware, then the compiler simply generates a call to a system routine provided by the Alfalfa run-time system.

### 3.1.3 Heterogeneous Evaluation Model

As stated above, we would like to minimize the overhead involved in execution by utilizing the mechanisms to support parallelism and lazy evaluation only when required. In doing so Alfalfa becomes a hybrid of three evaluation models which have varying degrees of power and accompanying overhead. The three models are:

1. Parallel Graph Reduction

2. Sequential Graph Reduction

3. Sequential Stack-based evaluation

and are listed in decreasing order of both power and overhead cost. In this section we describe the three machine models and how a functional program may be transformed to utilize them.

In the distributed execution of functional languages, graph reduction serves two basic purposes, heap (i.e. graph space) allocation of closures and a return mechanism for parallel function applications. Heap allocation of closures is necessary to support lazy evaluation via the construction of unevaluated expressions and shared objects. Closures are also constructed to represent higher order functions that are used as first class objects in a program.

In addition to closures arising out of lazy evaluation and higher order functions, closures are allocated in the heap to serve as activation records in order to hold the values of bound variables within a function body. In Alfalfa, since all functions are serial combinators, the node contains the full environment in which a function call is being executed.

Because a serial combinator may create calls to other serial combinators on remote processors, the system has to provide a mechanism for suspending evaluation of a function until the needed values return. This necesitates the creation of the activation records in a heap, and not in a stack, in order to preserve the "state" of the function call during suspension. The processor has to be free to use the stack for other purposes while a function call is suspending.

Due to the non-determinism in the order in which parallel function invocations complete, there must exist a mechanism to provide synchronization between functions that are returning values and the functions that called them. This is usually accomplished by updating various status bits when the values return. The suspending and resuming of a function invocation depend on the state of these bits. The overhead involved in resuming via the setting and testing the status bits is significant.

In many programs, however, there are sequences of function calls that contain no parallelism and require no mechanism for suspending. In these cases, heap allocation of activation records and the use of status bits for synchronization is unnecessary. The activation records for these function calls should be (and are in Alfalfa) allocated on the machine stack. The stack provides a sequential, deterministic, return mechanism at very low cost. A heap is still needed, however, to support lazy evaluation and higher order functions. The G-machine[11] is an example of a such a sequential graph reducer.

It is also true that in many cases, function is neither higher order nor is ever passed any evaluated arguments. If this is the case, no heap allocation is necessary and an invocation of the function can be allocated on the stack just as if it were part of a first order, applicative order, language.

Detecting which function invocations can be executed using sequential graph reduction may involve a sophisticated analysis such as an abstract or collecting interpretation[4,14,7]. Briefly, we say a function is sequential (and can be evaluated using a stack) if

- It makes no parallel function calls, i.e. it only calls one function at a time.
- It never "forks" a function call, i.e. it never proceeds without waiting for a value of a function call to return.
- It only calls functions that are themselves sequential.

Determining if a function, $f$, is sequential in a first order language involves solving a recursive set equation. In the higher order case, $f$ may make calls to 'unknown' functions and a collecting interpretation would be needed to determine if the 'unknown' function is is always sequential.

If a sequential function $f$ has arity $n$ then an application of $f$ to $n$ arguments can be executed using the sequential stack-based evaluation model if all arguments to the invocation have already been evaluated (assume that f may reference any of its arguments). In order to avoid constructing pieces of graph while executing $f$, $f$ cannot create a delayed expression nor a partially applied function. Even if $f$ does create one of these objects, it can still be executed on a stack but would have to allocate the closures in the heap.

The Alfalfa compiler currently solves the set equation for sequential function in the in the first order case. It gives up when encountering an unknown function call. For those functions $f_i$ that it has found to be sequential, it generates two definitions. One is for sequential stack based execution when the the compiler can determine that the actual parameters in a call to $f_i$ have been evaluated. A sequential graph reduction based definition is also generated for the case where a call to $f_i$ may involve unevaluated arguments.

### 3.1.4 Tail Recursion

Even if a function invocation is not able to be evaluated on the stack, the compiler can direct the creation of function invocations that are space efficient.

The naive implementation would allocate a new activation record in the heap for every function invocation. The heap space occupied by the activation record will eventually have to be reclaimed by the storage management system (Alfalfa uses a distributed reference counting system).

In fact, if the value that a function invocation $a$ returns is simply the result of a function invocation $b$, then the activation record for $b$ can be written over the space allocated for $a$. This is the very familiar optimization for tail recursion applied to the domain of graph reduction. This is possible in Alfalfa because all activation records are the same size. This tail recursion optimization saves heap space as well as the overhead of invoking the storage allocator for b.

# 4 Details of the Alfalfa system

## 4.1 Synchronization by serial combinators

Unlike Alfl functions, serial combinators contain constructs that specify the synchronization necessary for distributed execution. The algorithms used by the compiler to detect the inherent parallelism in an Alfl program and to generate the synchronization constructs necessary to exploit it is beyond the scope of this paper. A preliminary description can be found in [10] and a detailed explanation is forthcoming from the authors.

The basic synchronization constructs are *demand*, *wait*, and *spawn*. For ease of explanation, we will represent serial combinators using S-expression syntax much like that of Lisp.

Given a function $f$ represented in serial combinator intermediate form, the demand, wait, and spawn constructs can occur anywhere within the body of $f$ and are defined as follows.

(demand ($v_1 \ldots v_n$)
        body-exp)

indicates that at this point in the evaluation of $f$, the values of variables $v_1 \ldots v_n$ may be safely demanded (in parallel). Because serial combinators preserve laziness, we must be certain, using strictness analysis, that the values of $v_1 \ldots v_n$ will be needed at some point in the computation. It is important to note that $v_1 \ldots v_n$ do *not* have to return before body-exp is evaluated.

(wait ($v_1 \ldots v_n$)
body-exp)

indicates that the values of $v_1 \ldots v_n$ must return before the evaluation of body-exp can even begin. If any of $v_1 \ldots v_n$ are still being evaluated (perhaps by some remote processor) then evaluation of $f$ is suspended. Evaluation of $f$ is resumed when the needed values have returned. The assumption is made that $v_1 \ldots v_n$ had already been demanded or spawned (see below). Even though evaluation of a function is blocked, the processor is free to reduce any other segment of the graph whose value have been requested.

(spawn (($v_1$ $exp_1$) ... ($v_n$ $exp_n$)
body-exp)

indicates that for each $exp_i$, a new node $n_i$ is created in the graph (perhaps on a remote processor) that represents the expression $exp_i$. At the same time the value of $n_i$ is demanded. When $n_i$ becomes been evaluated, the value it returns is bound to the local variable $v_i$. Since the evaluation of $exp_i$ has already commenced, $v_i$ need not occur in a demand at any point in $f$. However, since evaluation of body-exp proceeds without blocking on the values of $v_1 \ldots v_n$, each $v_i$ must occur within a wait before its value is referenced within $f$.

(let (($v_1$ $exp_1$) ... ($v_n$ $exp_n$)
body-exp)

is similiar to spawn with the exception that the Alfalfa compiler has determined that each $exp_i$ is an expression that can be executed immediately on the stack and that every $exp_i$ should be evaluated on the local processor. The evaluation of body-exp continues as soon as the values have been computed. Naturally, no $v_i$ need appear in a demand or a wait.

In addition to the synchronization constructs describe above, serial combinators contain some information about where (i.e. on which processor) new nodes in the graph should be created. However, since the compiler is unassisted by the programmer, this information mainly consists of indicating whether the new node should be allocated locally or remotely. The Alfalfa run-time system is responsible for choosing a particular remote processor based upon system load information, etc.

Here is the divide and conquer factorial as it appears in Alfl and serial combinators.

```
{ pfac l h == l=h->l, { mid == (l+h)/2;
                        result pfac l mid + pfac (mid+1) h;
                        }
  result pfac 1 10;
  }
```

```
pfac l h == (demand (l h)
                (wait (l h)
                   (if (= l h) l
                      (let ((mid (/ (+ l h) 2)) (v1 (+ mid 1)) )
                         (spawn ((v2 (pfac l mid)) (v3 (pfac v1 h)))
                            (wait (v2 v3)
                               (+ v2 v3)))))))
```

## 4.2 The Alfalfa Run-time System

The Alfalfa run-time system is a completely distributed mechanism for distributing and performing the execution of the serial combinator code generated by the Alfalfa compiler. It is replicated indentically on each processor in the system and is solely responsible for reducing the portion of the graph residing in the local store. The major components of the run-time system, pictured in figure 1, are the graph reducer, message handler, task queue, diffusion scheduler, and storage manager.

## 4.3 The Graph Reducer

The synchronization constructs that serial combinators contain are simply calls to routines in the graph reducer component that perform the necessary transformations on the graph. Before discussing how these operations are performed, the data structures involved should be described.

### 4.3.1 Data Structures

A node in the graph is a contiguous block of bytes that contains the following fields:

- *State*: Either "unevaluated", "waiting" (which means that this node is waiting for the value of other nodes in the graph), or "evaluated".
- *Value*: If the node is evaluated, then its value. Otherwise the value field is a pointer to code that specifies the computation to be performed when the value of the node is requested.

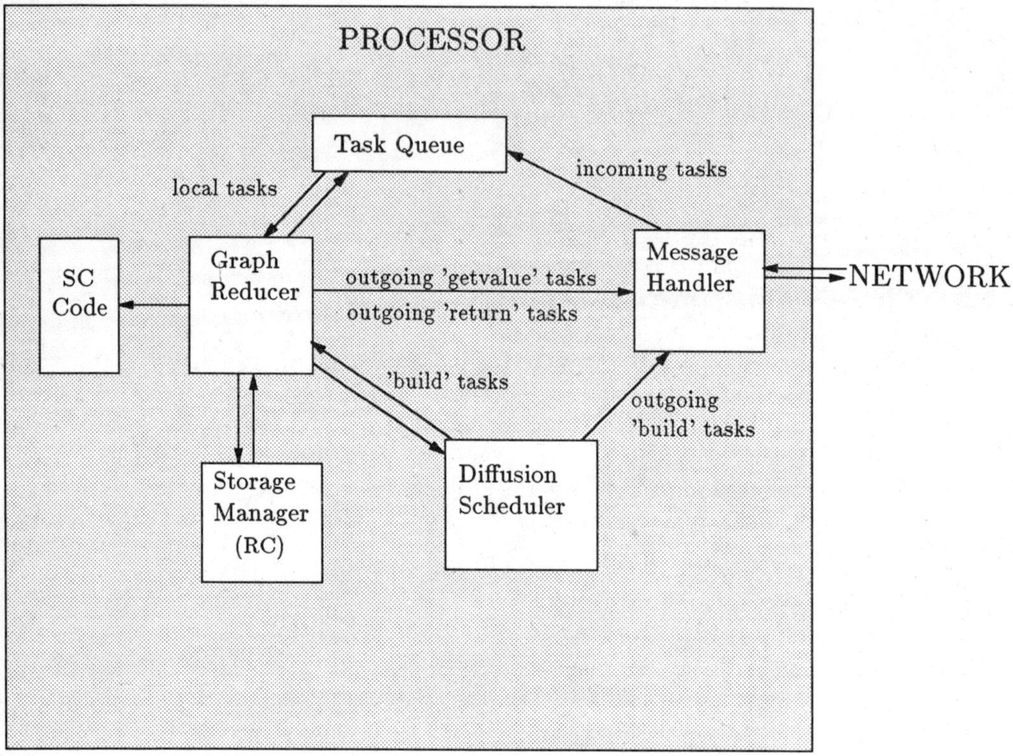


Figure 1: The Alfalfa run-time system

- *Args*: A vector containing the values of the arguments to the the function represented by this node. Each element contains either a base value, or a pointer to another node in the graph.
- *Requests*: A list of other nodes that have requested the value of this node.Because of sharing, there may be several nodes who have demanded the value of this one.
- *Evalfield*: A bitfield indicating the status of each element in the Args vector. If the ith bit of this bitfield is 1 then the ith argument has already been evaluated and contains a value. Otherwise the ith argument is a pointer to another node.
- *Waitmask*: A bitfield indicating which arguments must be evaluated before evaluation of this node proceeds. Evaluation can proceed when for every 1 in the waitmask, there is a corresponding 1 in the Evalfield.
- *RefCount*: The reference count of this node for storage reclamation purposes.

A pointer to a node in the graph has two fields:

- *Processor*: The address of the processor upon which the node resides.
- *Node*: The address of the node on its host processor.

A task is a structure that specifies a step for the run-time system to take in order to reduce the graph. Program execution proceeds by repeatedly removing tasks from the task queue for each node and performing the action specified by the task. There are three kinds of tasks:

- An *evaltask* contains pointers to a target node source node. It indicates that the value of the target node is being requested by the source node.
- A *returntask* contains a pointer to a target node and a value. It indicates that some other node has been evaluated and is returning its value to the target node. A returntask also contains and integer that indicates which element of the target node's args vector is to be updated.
- A *buildtask* contains a node description and a return pointer. It indicates that a new node in the graph should be built matching the node description and evaluated. The value of the new node is being requested by a node pointed to by the return pointer.

### 4.3.2 Execution

Execution begins by the creation of a collection of nodes representing the initial graph. An evaltask requesting the value of the root node of the graph is placed on the task queue of the processor upon which the root node resides. The run-time system on each processor then begins to remove tasks (if present) from the local task queue.

If an evaltask is encountered by the reducer and the target node, $n$, is unevaluated, then evaluation of $n$ proceeds by a jump to the code pointed to by $n$'s value field. This code, of course, is the code generated for a serial combinator by the Alfalfa compiler and contains calls to the reducer routines that support demand, wait, and spawn. When the serial combinator code is finished executing, a returntask is created to return the resulting value, $v$, to any requesting node. $n$'s state is modified to "evaluated" and its value field is overwritten with $v$.

If $n$'s state was already "evaluated" when the evaltask was encountered, then the reducer immediately creates a returntask with $n$'s value.

If a buildtask is encountered, then a subgraph matching the descriptor is created in the local graph space, and execution proceeds as though an evaltask was encountered for the root of the new subgraph.

We now describe what actions are taken when a demand, wait, or spawn is encountered in the code pointed to by node n. Figure 2 shows the state of a node before it is evaluated. The list of variables in a demand construct has been translated into a list of indices into $n$'s args vector. For each argument index, $i$, in the demand if the $i$th bit of $n$'s evalfield is 1 (i.e. the $i$th argument has already been evaluated) then no action is taken. Otherwise an evaltask is created in order to request the value of the node to which the $i$th argument points. As discussed above, execution of $n$'s code continues without blocking.

When a wait is encountered, $n$'s waitmask is modified such that if the $i$th argument occurs in the wait, then the $i$th bit of the waitmask is set to 1. $n$'s value field is modified so that it now points to a *continuation*, namely code that will be executed when the needed arguments return. If all the arguments occuring in the wait have already been evaluated, then execution proceeds via a jump to the continuation. Otherwise, control is relinquished to the run-time system to carry out the next task on the local task queue. Figure 3 shows the state of a node when all needed arguments have returned and it is about to resume.

When the run-time system encounters a returntask from any of the arguments that the $n$ is waiting for, it updates the appropriate element of the $n$'s argument vector and evalfield. If the evalfield now has a 1 in every bit position that the waitmask does, then the $n$ is ready to be awakened. Again, this is acomplished by simply jumping to the code pointed to by the $n$'s value field. Otherwise, no action is taken.

When a spawn is enountered, a buildtask is created for each expression in the spawn. For each element of the spawn, the run-time system invokes its dynamic scheduler (see

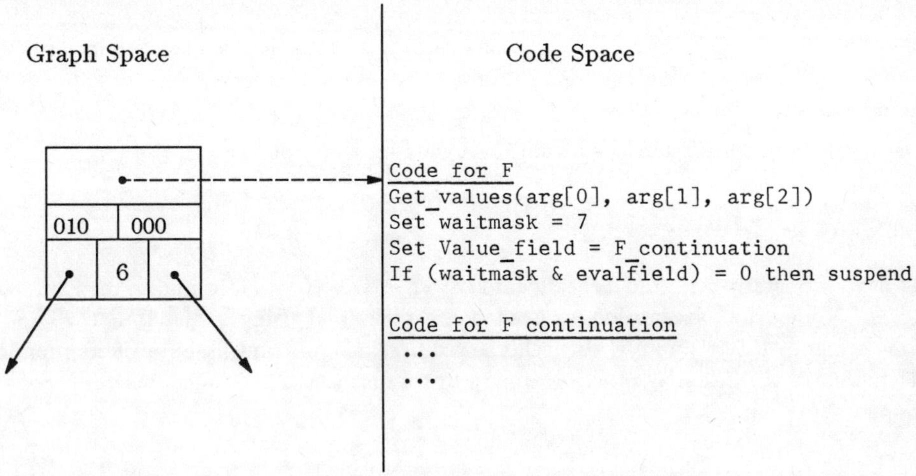


Figure 2: A node in its initial state

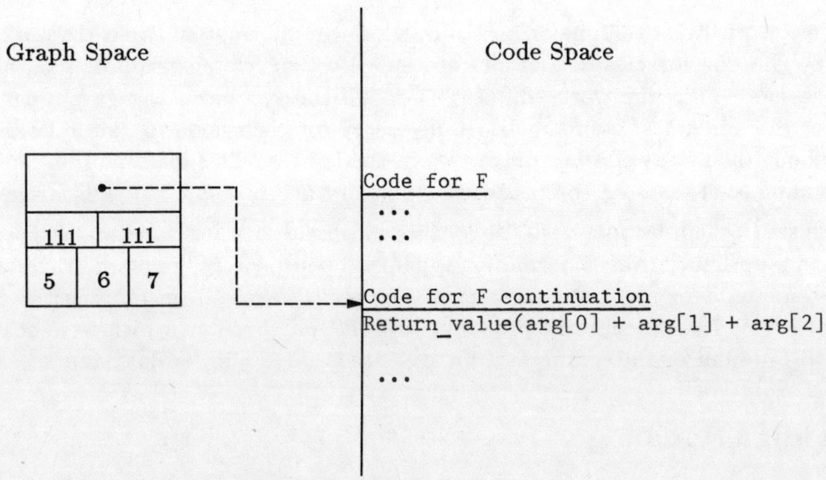


Figure 3: A suspended node about to resume

below) to select a processor upon which to create the new node. A description of the node, as specified by the compiler, is inserted into the task and the task is placed in the appropriate processors task queue. A slot in $n$'s arg vector is reserved for the value of the new node when it returns.

The result of the program is the value returned by the root node in the graph.

## 4.4 Storage Management

Nodes are a constant size and are allocated from a freelist that is maintained on each processor. Nodes that are no longer needed are reclaimed using a distributed reference counting scheme. Unfortunately, this scheme requires a significant amount of communication to maintain the integrity of the graph[8], but a discussion of it is beyond the scope of this paper.

## 4.5 Dynamic Scheduling

The function of Alfalfa's dynamic scheduler is choose a processor to allocate each new segment of the graph. A heuristic algorithm, called Diffusion Scheduling, is used which takes into account when selecting processors such information as system load characteristics and locality of references.

The diffusion scheduler running on each processor uses information about the work load of only its neighboring processors and allocates new portions of the graph only on neighboring processors. Thus the work "diffuses" through the system as the graph expands. This reduces the amount of communication necessary for each node to gain relevant information about the status of other processors in the system. This load information may include the number of tasks on the task queue, memory utilization, CPU utilization, etc.

The Diffusion Scheduler may also use locality information when selecting a processor for a new node, $n$. This locality information is gained by inspecting the pointers contained in $n$. For example, if many of the arguments to $n$ point to processor $P$, then the diffusion scheduler would likely choose a processor near $P$ or $P$ itself on which to allocate $n$. Hopefully, this minimizes the communication costs associated with evaluating $n$.

## 4.6 Message Handling

The message handler provides the interface to the network for each processor. It is responsible for parsing incoming messages and directing the information to the appropriate module in the run-time system. It is also responsible for packaging outgoing messages and directing them to the appropriate processor.

A messages can be one of several types:

- A *System Message* contains information about a processor in the system. Incoming system messages provide the local processor with load information about neighboring processors and the message handler passes the information to diffusion scheduler. The message handler sends out information about the local state as directed by the diffusion scheduler.
- A*Reducer Message* will be one of the tasks described above. Incoming tasks will be added to the local task queue. When a graph reduction operation is performed that requires communication, the message handler creates and sends the appropriate reducer message.
- A *Storage message* is a message that indicates an operation that changes the reference count of a node on the destination processor has occurred. Message passing is the only way to maintain consistent reference counts in a distributed memory system. Storage messages are generated when a reference to a node on a remote processor is either replicated or deleted.

# 5 Performance Measurements

We have used Alfalfa to test the effectiveness of some simple diffusion heuristics. In this section we show the results of a number of them applied to a single program which finds all solutions to the 8 queens problem (although in our case, it was 6 queens). In all the runs, the heuristics used only task queue length as a measure of a processor's load. The heuristics used in these experiments were of two basic types, communicating and non-communicating.

Since messages are extremely expensive on the iPSC we tested several algorithms that do not involve interprocessor communication. In these cases, the diffusion scheduler on each processor decides where to allocated a new node in the graph based only on some local information. The non-communicating heuristics we tested were:

- *Simple Round Robin*: If $l \geq T$ where $l$ is the length of the local task queue and $T$ is a constant, then the new node is allocated on a processor chosen using round robin. In figure 4 the execution time of Alfalfa is plotted using three different values for $T$. Notice that when $T$ is small, the time taken to execute on 2 processors is significantly greater than the execution time on one processor. This is due to the fact that the two processors are shipping nodes back and forth to each other and are paying a large communications penalty. When $T$ is large, performance on two processors is quite good. However a large $T$ restricts parallelism and the system is unable to take advantage of additional processors.
- *Dependent Round Robin*: If

$$l \geq 2^{(E-(C*d))}$$

FIG 4. Simple Round Robin Diffusion
Execution Time

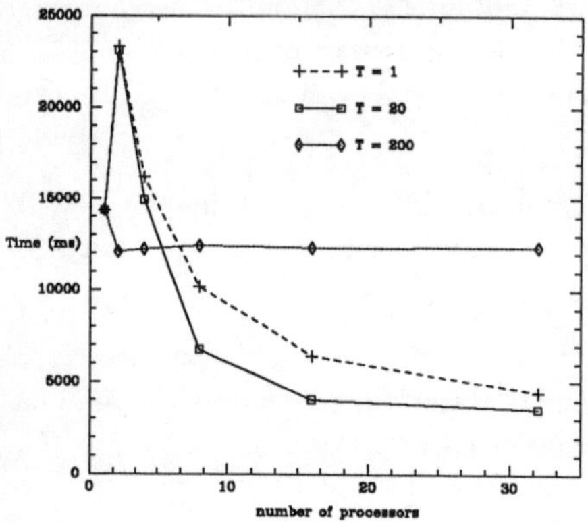


FIG 5. Dependent Round Robin Diffusion
Execution Time

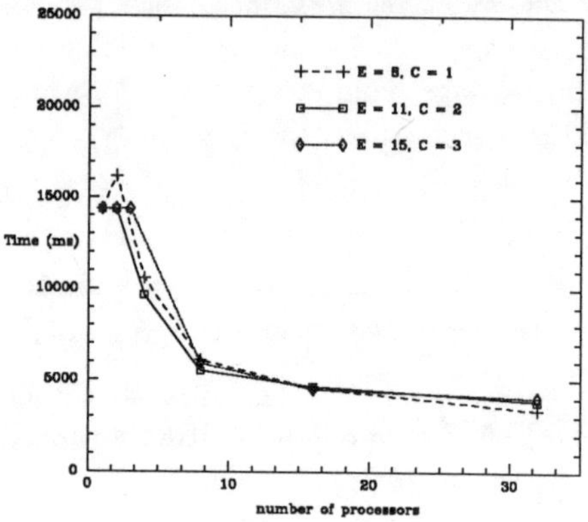


where $d$ is the dimension of the hypercube and $E$ and $C$ are constants, then the new node is allocated on a remote processor chosen using round robin. This heuristic is called dependent round robin because it is dependent on the number of processors in the system. The larger the value of E, the larger the number of nodes that will evaluated locally. If E is large, then the potential speedup is reduced, although in a system with a small number of processors, there is less likely to be wasted communication from sending nodes back and forth between busy processors. The larger the value of C, the faster the threshold decreases as processors are added. In figure 5 note the improvement in performance over the simple round robin heuristic.

We also tested a few simple heuristics that required inter-processor communication. A processor sends load information to neighboring processors to indicate how busy it is. When a new node is to be created, the local processor compares its own load with the reported load of its neighbors in order to choose an appropriate processor for the new node. Two issues that the communicating diffusion heuristics have to resolve is

- How often should load information be sent?
- Given the load information, what criteria are used to select a processor?

We measured the performance of two communicating heuristics.

- *Simple Communicating*: In this algorithm, load information is sent to neighboring processors whenever the length of the task queue, $l$, is different by at least factor of two from the previously reported value, *and* either the old value or new value is

FIG 6. Simple Communicating Diffusion Execution Time

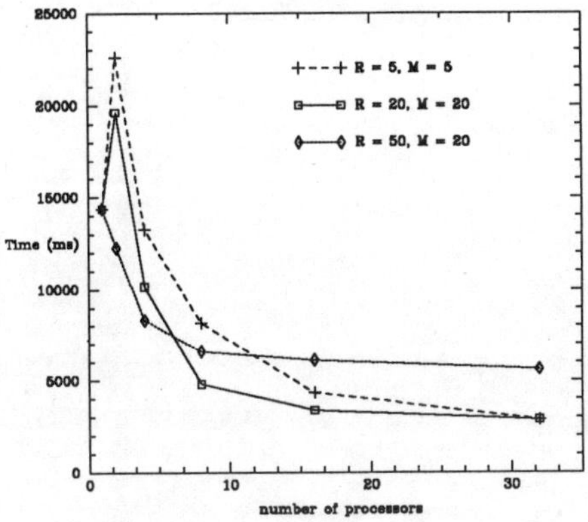


FIG 7. Dependent Communicating Diffusion Execution Time

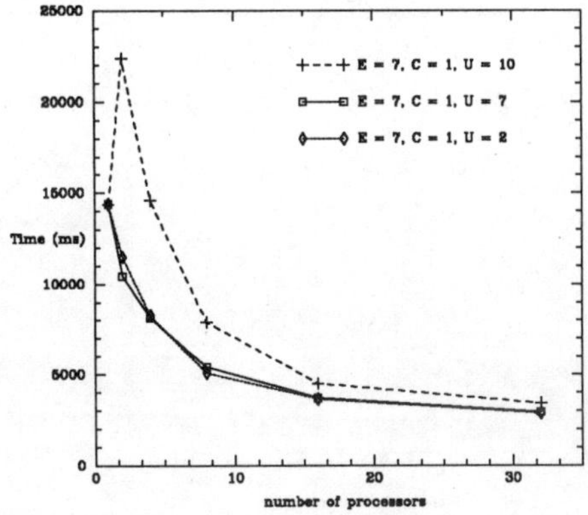


greater than some constant M. Since, a processor's task queue will often vary between a length of 0 and 1 or 1 and 2, M should be large enough to prevent messages being sent due to minor transitions.

When a new node is to be created, a remote processor, $p$, is chosen if

$$l_p \leq l - R$$

where $l_p$ is the lowest reported load of a remote processor, and $R$ is a constant. The execution times using this heuristic are plotted in figure 6.

- *Dependent Communicating*: This heuristic uses information about the number of processors in the system to decide when to send load information and when to create nodes remotely. It sends load information to neighboring processors when $l$ changes by factor of 2 and either the new value or the old value is greater than

$$2^{(U-(C*d))}$$

where U and C are constants and d is the dimension of the hypercube. In this case, the larger the value of U, the fewer load messages will be sent.

When a new node is to be created, a remote processor, $p$, is chosen if

$$l_p \leq 2^{(E-d)}$$

and $l_p$ is the lowest reported load of a remote processor and E is a constant. The execution times using this heuristic are plotted in figure 7.

FIG 8. Simple Round Robin Diffusion
Distribution of Tasks

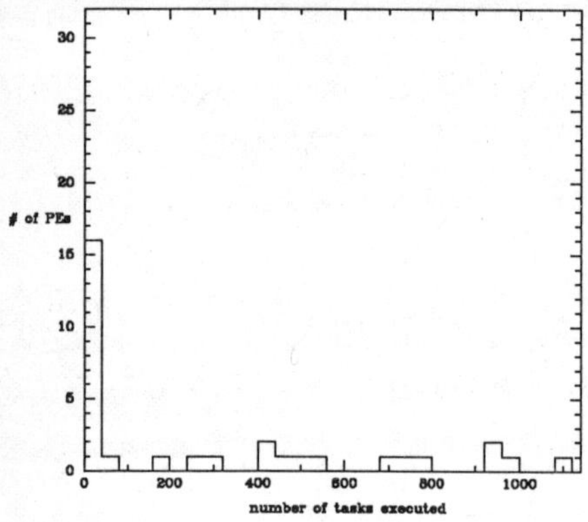


FIG 9. Dependent Round Robin Diffusion
Distribution of Tasks

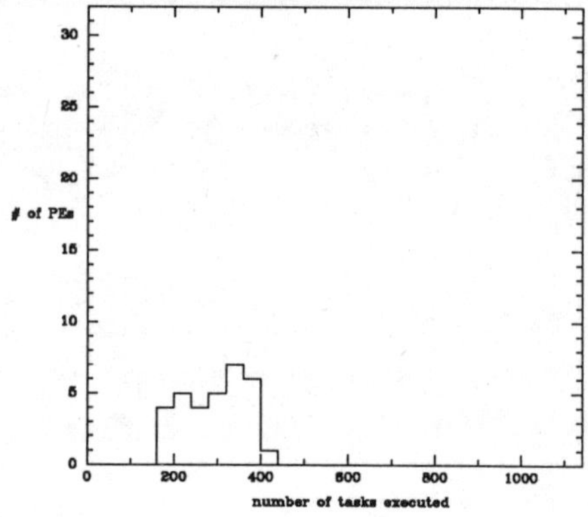


FIG 10. Simple Communicating Diffusion
Distribution of Tasks

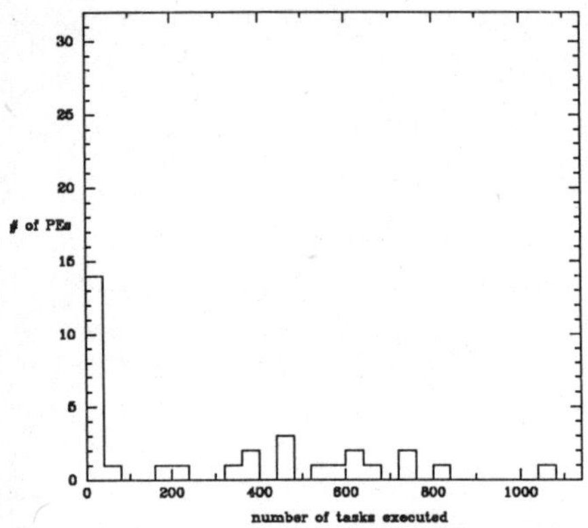


FIG 11. Dependent Communicating Diffusion
Distribution of Tasks

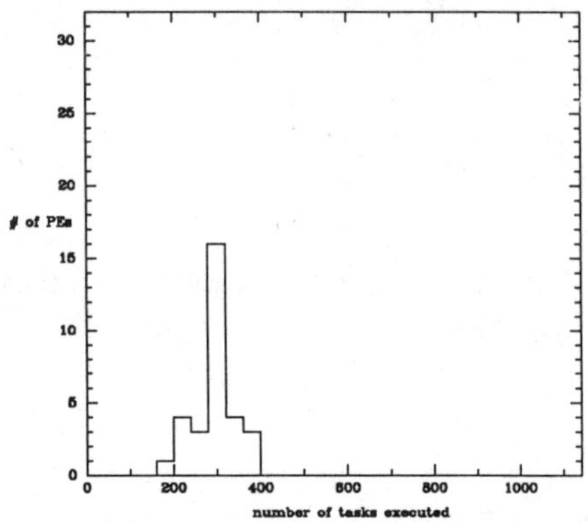

In addition to speedup, we are interested in determining how uniformly the processors in the system were utilized given a diffusion heuristic. In each case, we chose the parameters for the heuristic that gave the best performance and executed the program on a 32 processor system. In figures 8 through 11 we plot the distribution of processors according to the number of tasks that each processor executed. In is strikingly apparent that the dependent diffusion heuristics, both communicating and round robin, provided a much more even distribution of work than the simple diffusion heuristics. In figures 12 through 15, the distribution of the processors according to the number of reducer messages handled by each processor leads to precisely the same conclusion. Not suprisingly, we also found that the number of storage messages handled by each processor is directly proportional to the number of reducer messages it handles.

# References

[1] R.M. Burstall, D.B. MacQueen, and D.T. Sannella. Hope: an experimental applicative language. In *The 1980 LISP Conference*, pages 136–143, Stanford University, August 1980.

[2] T. Clarke, P. Gladstone, and Norman A. MacLean. Skim — the s, k, i reduction machine. In *The 1980 LISP Conference*, pages 128–135, Stanford University, August 1980.

[3] Intel Corporation. *iPSC User's Guide - Preliminary.* Technical Report 175455-001, Intel Corporation, July 1985.

[4] P. Cousot and R. Cousot. Abstract interpretation: a unified lattice model for static analysis of programs by construction or approximation of fixpoints. In *4th ACM Sym. on Prin. of Prog. Lang.*, pages 238–252, ACM, 1977.

[5] J. Darlington and M. Reeve. Alice: a multi-processor reduction machine for the parallel evaluation of applicative languages. In *Functional Programming Languages and Computer Architecture*, pages 65–76, ACM, October 1981.

[6] P. Hudak. *ALFL Reference Manual and Programmer's Guide.* Research Report YALEU/DCS/RR-322, Second Edition, Yale University, October 1984.

[7] P. Hudak. *Collecting interpretations of expressions.* Research Report 497, Yale University, Department of Computer Science, 1986. submitted to the Third Workshop on the Mathematical Foundations of Programming Language Semantics.

[8] P. Hudak. *Object and Task Reclamation in Distributed Applicative Processing Systems.* PhD thesis, University of Utah, July 1982.

FIG 12. Simple Round Robin Diffusion
Reducer Message Distribution: 32 PEs

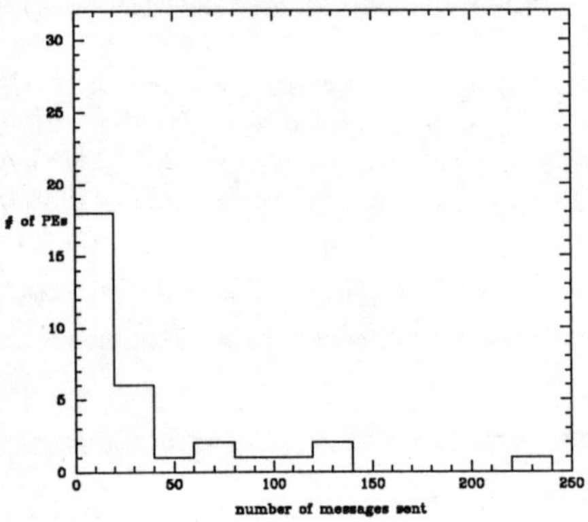


FIG 13. Dependent Round Robin Diffusion
Reducer Message Distribution: 32 PEs

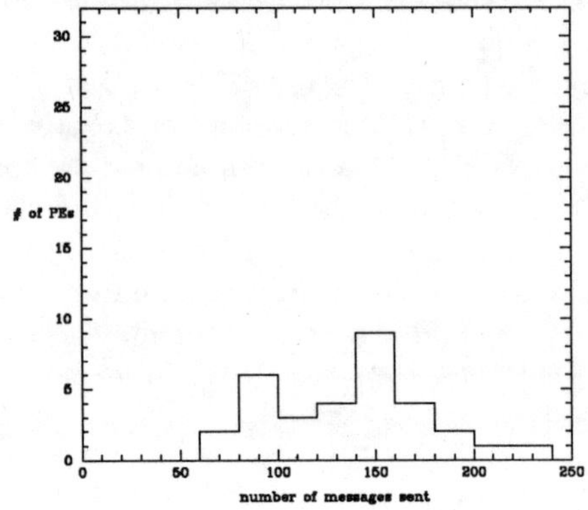


FIG 14. Simple Communicating Diffusion
Message Distribution: 32 PEs

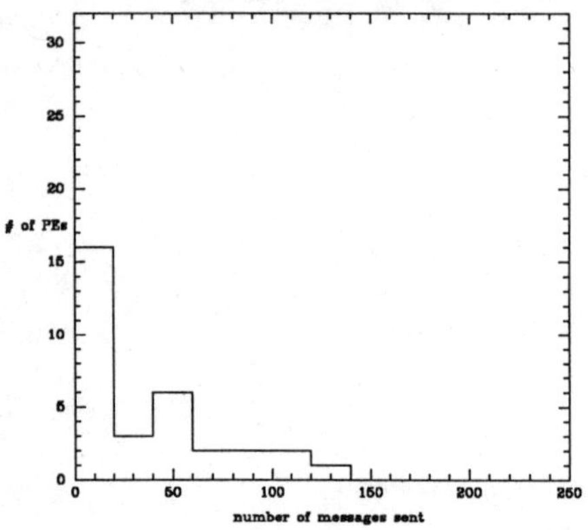


FIG 15. Dependent Communicating Diffusion
Message Distribution: 32 PEs

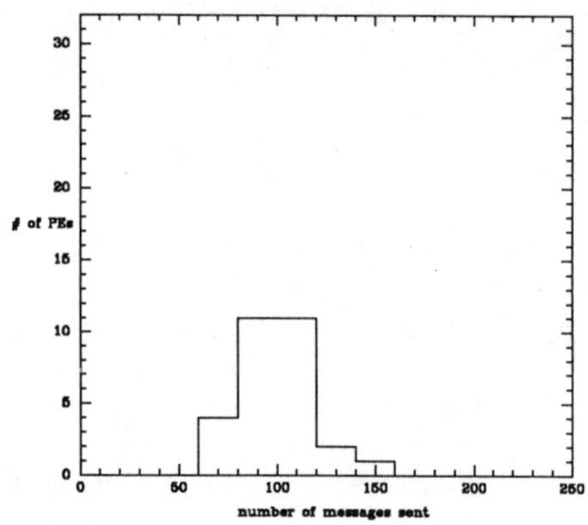

[9] P. Hudak and B. Goldberg. Distributed execution of functional programs using serial combinators. In *Proceedings of 1985 Int'l Conf. on Parallel Proc.*, pages 831–839, August 1985. Also appeared in IEEE Trans. on Computers, Vol C-34, No. 10, October 1985, pp. 881-891.

[10] P. Hudak and B. Goldberg. Serial combinators: "optimal" grains of parallelism. In *Functional Programming Languages and Computer Architecture*, pages 382–388, Springer-Verlag LNCS 201, September 1985.

[11] T. Johnsson. *The G-machine: an abstract machine for graph reduction.* Technical Report, PMG, Dept. of Computer Science, Chalmers Univ. of Tech., February 1985.

[12] R.B. Kieburtz. *The G-machine: a fast, graph-reduction evaluator.* Technical Report CS/E-85-002, Dept. of Computer Science, Oregon Graduate Center, January 1985.

[13] R. Milner. A proposal for standard ml. In *Proc. 1984 ACM Conf. on LISP and Functional Prog.*, pages 184–197, ACM, August 1984.

[14] A. Mycroft. *Abstract Interpretation and Optimizing Transformations for Applicative Programs.* PhD thesis, Univ. of Edinburgh, 1981.

[15] D.A. Turner. *SASL language manual.* Technical Report, University of St. Andrews, 1976.

# Parallel Graph Reduction on a Supercomputer: A Status Report

Randy Michelsen
Lauren Smith [1]
Elizabeth Williams
Bonnie Yantis

Computing and Communications Division[2]
Los Alamos National Laboratory
Los Alamos, NM 87545

## ABSTRACT

We describe an ongoing effort to develop a parallel graph reduction run-time system hosted on a multiprocessor supercomputer. This run-time system is presently augmented by the functional language compiler of the ALFALFA system. Admittedly, parallel graph reduction is hardly a novel idea. The interesting notion is the provision of a parallel execution environment sufficiently powerful to support development of large prototypical scientific and symbolic codes in a functional language. This will allow the investigation of the functional programming paradigm within these application domains in an empirical fashion. In this paper, we discuss the motivation for the effort, describe the basic elements of the implementation, and provide some preliminary insight distilled from our experience with an initial version of the run-time system.

## 1. INTRODUCTION

The continuing demand for increased computational power, especially in the realm of scientific computing, has spurred interest in parallel computation. Today, most application development for computationally intensive algorithms intended for high-performance computers is done in Fortran. Hence, there exists a considerable body of knowledge regarding its use and performance in such domains. We are interested in collecting similar data relating to use of the functional (or applicative) paradigm.

Toward this end, we are currently developing a parallel graph reduction run-time system for a supercomputer incorporating four high-speed processors. The code generation is performed by an existing functional language compiler resident on a traditional uniprocessor. Initially, the focus of the effort was the conversion, with minimal change, of a run-time system originally developed for a distributed memory multiprocessor to a shared memory multiprocessor. This strategy enabled us, in the near term, to rely upon an existing compiler and concentrate our energy instead on the development of the reduction run-time system. Subsequently, the emphasis will shift to performance tuning of the run-time system and the investigation of related compilation and scheduling issues.

Our motivation for undertaking this task was essentially twofold. A common complaint from potential users of functional languages outside of the language research community is the pragmatic impossibility of developing programs of a "realistic" size simply due to performance-related issues. The definition of realistic is of course open to interpretation, but it is worthy of note that typical production programs at Los Alamos National Laboratory (hereafter, the Laboratory) consist of 100,000

---

[1] currently at Department of Defense, 9800 Savage Road, Fort Meade, MD 20755, electronic mail address: lls@mimsy.umd.edu

[2] This work was performed under the auspices of the Department of Energy.

or more lines of Fortran. The performance deficiencies usually associated with functional languages may be attributed in part to the fact that most implementations currently available are intended for language research and lack the maturity (and implied fine tuning) of production compilers. Furthermore, the implementations generally exist on machines lacking the requisite processing and memory capacity for large-scale scientific computations. Our effort is particularly apropos given the latter circumstance.

We were also interested in exploring the performance ramifications of a large physical memory shared by cooperating reduction "engines." There have been graph reduction systems implemented on distributed memory machines or networks of machines, including several discussed in this volume. Yet, given the usual logical view of a graph space global to all reduction activity, a large shared memory architecture seems particularly well suited to parallel graph reduction.

As a result of these interests and practical concerns such as machine availability, we chose the CRAY X-MP computer as the current implementation target. This architecture combines a small number of high-performance processors with a large shared memory.

Section 2 contains background information and a brief discussion of the CRAY X-MP architecture. [2] Section 3 describes the design of the initial run-time system. Section 4 discusses early experience with the system and a recently initiated follow-up development effort. A summary is presented in Section 5.

## 2. BACKGROUND

### 2.1. The ALFALFA System

The work described here is closely related to the work on ALFALFA by Benjamin Goldberg and Paul Hudak of Yale University. We shall present a very brief description of the ALFALFA system, but the interested reader is referred to B. Goldberg's paper in this volume for a more complete treatment of the subject. ALFALFA consists of a compiler for the functional language ALFL [3] and an associated parallel graph reduction system resident on an Intel iPSC multiprocessor. A source program is parsed into an intermediate graph form called LIF (Lambda Intermediate Form). This graph is then annotated as a result of extensive type, sharing, and strictness analysis. The annotated graph is converted into a serial combinator [4] program graph. Finally, C procedures are generated from this graph for the combinators. This C program is then compiled and linked with the graph reduction system resident on the processing nodes of iPSC.

### 2.2. The CRAY X-MP

The CRAY X-MP is a multiprocessor derivative of the earlier CRAY-1 family of supercomputers. The basic architecture consists of an interleaved, multiport memory and input/output subsystems shared by several identical high-performance vector processors. Each processor has a major cycle time of 8.5 nanoseconds. Pipelining is utilized in the scalar and vector functional units, with parallel activity possible among the independent functional units. A parallel program has available to it shared registers - eight data registers, eight address registers, and 32 single-bit semaphore registers. A blocking test-and-set instruction can be performed on each semaphore register. Unlike some of its counterparts, the X-MP does not support virtual memory, but rather uses a contiguous allocation scheme for memory management. The X-MP is available in a variety of configurations, currently with up to four processors and sixteen million 64-bit words of main memory.

The primary use of modern supercomputers as "number crunchers" is reflected in the software environment of the Cray. The most mature and heavily used compiler available at the Laboratory is the vectorizing Fortran compiler developed by Cray Research Inc. Fortunately, a locally supported C compiler is also available. The C language was clearly more attractive in our development because of portability concerns and the availability of language features such as dynamically allocatable data structures.

The processors of a CRAY X-MP have traditionally been used to support multiple, independent job streams. However, a set of routines has been developed at the Laboratory to enable user program initiation and control of parallel tasks. These multitasking routines [1], written in Fortran and Cray assembly language, provide constructs for task creation, termination, and various styles of cooperation among asynchronous tasks. Among the latter are analogues of traditional synchronization mechanisms such as generalized counting semaphores, fetch/add instructions, locks, events, barriers, forks, and joins. Because of the real memory organization of the X-MP, the multitasking routines rely upon memory management utilities developed and in use at the Laboratory.

### 3. THE CRAY GRAPH REDUCTION RUN-TIME SYSTEM

Since one of our intentions was to provide access to Cray-class computational power for functional language programs in a timely fashion, the initial version of the Cray reduction system is a variant of the ALFALFA system developed for the Intel iPSC. Of course, given the disparate target architectures of the two efforts, there are considerable differences to accommodate. We are currently using the compiler from the ALFALFA system as the code generator for this reduction system. The C source code produced by the compiler undergoes source-to-source transformations to produce the appropriate Cray-compatible code. At this point, the source code is loaded onto the Cray. These combinator definitions are compiled and linked with the resident run-time system.

An alternative design for the run-time system was considered, but eventually rejected in favor of the design presented in this section. A primary goal in this initial implementation was to produce a run-time system highly compatible with the original ALFALFA compilation system, with some consideration given to performance factors. The candidate design was to associate with each node being reduced a separate reduction run-time system, with scheduling of these tasks performed by the host's operating system. This design was rejected because of the high overhead of task management relative to the actual work associated with the reduction activity.

In the selected design (refer to Figure 1), each physical processor executes a reduction task that is the run-time system. The term "task" is used in this context to designate an executing program image on the Cray host. The program graph and a queue of ready nodes, i.e., graph nodes that are candidates for immediate reduction, reside in shared memory. The common node queue is accessed by the reduction tasks whenever new work is desired. Access to this queue and the nodes of the program graph is controlled through the use of synchronization variables. Error handling and normal termination of the run-time system are accomplished via the notion of "synchronized meetings" provided by the multitasking routines. This construct allows a set of parallel tasks to synchronize on a single variable, denoted as the "meeting flag." After all the tasks meet at this synchronization point, a single task executes a sequential section of code. Upon the completion of this section of code, all tasks resume parallel execution. A variant of this design uses a fixed number of reduction tasks, which is strictly greater than the number of processors. When one reduction task must wait to gain access to a lock, another task may be able to reduce a node if one is on the ready queue; thus, all processors are executing if nodes are ready.

As indicated earlier, the run-time system is written primarily in C. Because of the need to interface the Fortran multitasking routines and the run-time system, a Fortran main program actually initiates the bootstrap routine for the run-time system. The C bootstrap routine initializes the ready node queue, the various synchronization variables, and constructs the initial program graph. When the initialization is complete, four parallel reduction tasks are created. One of the reduction tasks determines that it is the root (using a shared variable) and initiates the graph reduction. The tasks continue selecting nodes from the ready queue, mutating the program graph, and placing nodes on the ready queue until the graph is reduced to normal form or there is an abnormal termination.

I/O, because of its sequential nature, is accomplished through the use of a software "lock" on the I/O module. When one of the reduction tasks needs to perform I/O, it attempts to gain possession of the I/O lock. If the lock is unavailable (i.e., already possessed by another task), the requesting task waits until it is available. Memory allocation and deallocation are handled in a similar fashion.

Error handling is accomplished by a synchronized meeting. If a serious error is detected by one of the reduction tasks, the task sets the synchronization variable and initiates a meeting. The other

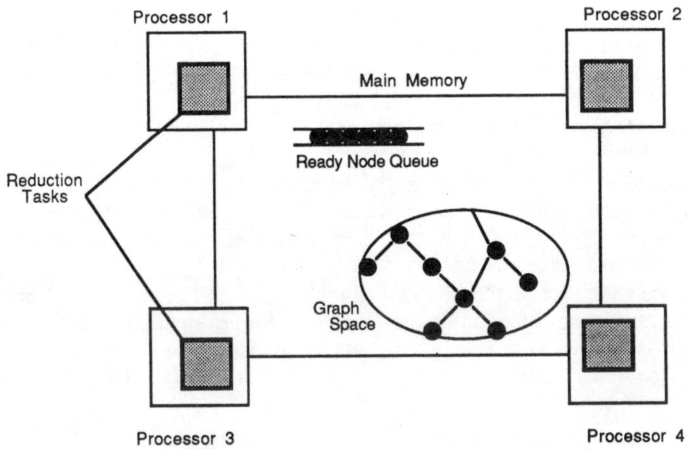


Figure 1. The Cray run-time system.

reducers will synchronize and the initiating task will execute the appropriate error handling. To ensure correct behavior in the advent of a meeting requested as a result of a fatal error, the tasks of the run-time system periodically inspect the meeting flags associated with this event. This inspection is critical in order to avoid potential deadlock as a result of tasks waiting for locks retained by abnormally terminated tasks. The occurrence of a fatal error will cause logging of state information, followed by the termination of the reduction tasks.

## 4. EARLY EXPERIENCE

At this point, the reduction system has been applied to simple ALFL programs. Several performance obstacles have been identified on the basis of our limited experience with the Cray environment, including the multitasking routines. The granularity of computation is far too small to be practical given the raw performance characteristics of the Cray processor. For example, the overhead associated with a procedure call is approximately 40 major cycles. This indicates a need to increase the granularity of the combinators generated by the compiler, as well as a need to significantly restructure the reduction system itself to reduce the cumulative performance impact of procedure invocation. Another obvious candidate for further scrutiny is synchronization. The tactic in this instance must be to reduce points of synchronization and to minimize the overhead associated with each synchronization.

These issues, among others, have been addressed in a subsequent development project targeted toward enhanced performance. In particular, the reduction system was redesigned to benefit more fully from the shared memory model, to more accurately reflect the performance characteristics of the Cray architecture, and to support measurement of performance and workload characteristics. Through careful design and utilization of a different mechanism, the minimum overhead incurred to control access to shared objects such as the ready node queue has been reduced by a factor of 7 to 15. Finally, an alternative scheme for internal data representation has been employed that is more amenable to possible future optimizations. This scheme supports memory allocation in a manner such that effective utilization of the Cray vector hardware is possible where appropriate.

## 5. SUMMARY

We briefly discussed the status of a parallel graph reduction system for the CRAY X-MP multiprocessor. A large portion of the initial effort was devoted to, out of sheer necessity, deciphering the interactions of several disparate software systems (e.g., memory management utilities). This effort will be fully appreciated by those acquainted with the rather meager software development support tools traditionally available on supercomputers.

The poor performance of the initial version of the run-time system prompted the development of a second redesigned system. This system is presently undergoing integration testing. This version has been tailored for increased performance on the host architecture. In particular, synchronization and other operational overheads have been significantly reduced.

In the future, we anticipate performing characterization studies of prototypical applications and additional performance tuning of the run-time system. The former should provide important data for the investigation of issues such as task granularity and scheduling. Finally, the entire question of vectorization has been ignored but must be addressed in any truly viable high-performance system.

## 6. REFERENCES


(1) Bobrowicz, F., "The Los Alamos Multitasking Control Library," Los Alamos National Laboratory unclassified release (in preparation).

(2) Cray Research, Inc., *CRAY X-MP Series Mainframe Reference Manual,* HR-0032 (1982), Cray Research, Inc.

(3) Hudak, P., "ALFL Reference Manual and Programmer's Guide," Research report YALEU/DCS/RR-322, 2nd ed., Computer Science Department, Yale University, October 1984.

(4) Hudak, P., and Goldberg, B., "Distributed execution of functional programs using serial combinators," *Proc. 1985 International Conference on Parallel Processing,* August 1985, pp. 831-839.

# Target code generation from G-machine code

Thomas Johnsson
Chalmers University of Technology
Göteborg, Sweden

**Abstract**

The G-machine is a stack machine for von-Neumann-like execution of lazy functional languages, using graph reduction. This paper describes how target code generation from G-machine code is performed in a compiler for Lazy ML. We briefly review the G-machine, and the compiler phases preceding the target code generation. Two different target code generation methods are described, as well as essential parts of the run-time system. The first target code generator is very simple, generating naïve code by means of 'macro expansion'. This code explicitly manipulates the stacks exactly as prescribed by the G-machine instructions. A simple peep-hole code improver can remove a fair amount of the most glaring inefficiencies of the naïve code. The second target code generator is less naïve, and takes a more systematic approach to avoid redundant operations in the resulting target code. This code generator can be neatly expressed as an attribute grammar over the G-machine code sequence.

# 1 Introduction

Many programming language researchers have long advocated the use of *functional languages*, in particular *lazy* ones, for their expressiveness and semantic simplicity — see e.g. [HM76], [Fri76], [Tur81]. At the same time, researchers in the field of computer architecture are working towards multiprocessor computers that take advantage of the parallelism inherent in such languages. But it may take many years before the envisioned novel architectures become available; and even when they do, the vast majority of computers in use will for a long time still be conventional ones. Therefore, if functional programming languages are to gain acceptance in a wider community it is necessary to develop efficient implementation techniques for conventional machines; such implementations have been notoriously lacking.

The *Lazy ML* (or *LML* for short) compiler project, pursued by Lennart Augustsson and myself, is an attempt to remedy the situation. Earlier papers have described some of the techniques used in this compiler: see [Aug84], [Joh84], [Aug85], and [Joh85]. Figure 1 shows the structure of the compiler.

In LML pattern matching is the primitive for scrutinizing values of concrete data types, both user defined and predefined ones (e.g. integers and lists). After parsing and type checking the compiler transforms pattern matching into *simple case expressions*; see [Aug85]. The program is then transformed into a set of global recursive functions, reminiscent of super-combinators [Hug82]. This transformation is called *lambda lifting* and is described in [Joh85]. Then each of the global functions are compiled into *G-machine* code. The *G-machine* is the abstract machine that embodies our paradigm for compiled graph reduction; it is described in [Joh84]. Finally, the G-machine code is translated into assembler code for the target machine (in two steps; see section 3.3). Currently the target machine is the VAX-11, but we also have preliminary target code generators for other machines; see section 3.3.

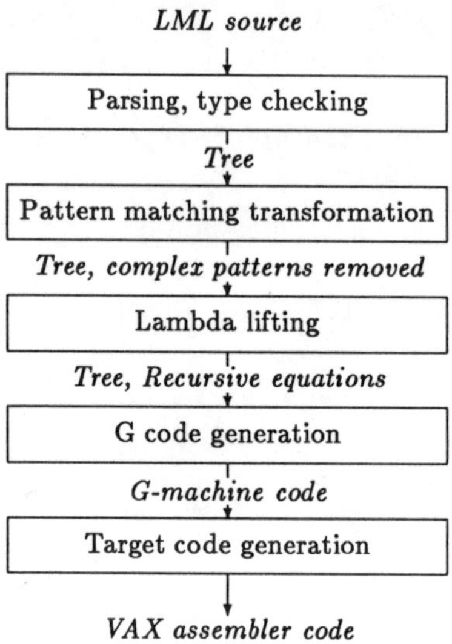


Figure 1: The *Lazy ML* compiler.

Figure 1 is in fact somewhat simplified: currently our LML compiler consists of 19 or so separate passes over the program. The compiler itself is written in LML (except for the parser which is written in Yacc and C), and consists of about 5600 lines of source code; it recompiles itself at a speed of 1 line/CPU second.

What makes our task different from that of generating target code for 'conventional' languages like FORTRAN and Pascal? The thing that determines the overall efficiency of the target code for any language is how well the most frequent operations in each language are implemented on the target machine.

Conventional languages mostly handle a "word" at a time, a word being an entity that can be conveniently handled by one machine instruction, and stored in a machine register or one memory location. Important things here are such things as register allocation for variables, and instruction selection for assignment statements and for code evaluating arithmetic and logical expressions—see for example [WJW*75].

In list oriented languages the main work is to handle lists and other data structures; arithmetic expressions are relatively rare. In a lazy functional language this is even more accentuated. Here we must also handle suspensions—data structures representing unevaluated expressions—build them, and later evaluate them. Therefore the prime consideration for us must be in achieving efficient heap space allocation and an efficient machinery for evaluating suspensions.

## 2 Review of the G-machine

The starting point of our work is the SKI-combinator reducer of Turner [Tur79], but instead of using the SKI combinators, we use the global functions of the form $f_i x_1 \cdots x_{m_i} = e_i$, which result from lambda lifting. These are used as rewrite rules for the graph.

*Naïve graph reduction* is performed by repeatedly rewriting left hand sides of definitions to right hand sides. Thus calling for the reduction of the graph for $f_i e'_1 \cdots e'_{m_i}$, this expression is rewritten into $e_i[x_1 = e'_1, \cdots, x_{m_i} = e'_{m_i}]$, i.e. the right hand side of the function definition with pointers to actual parameters substituted for formal ones; sharing of expression subgraphs may thus arise. Reduction will continue with this new expression graph. Figure 2 shows an example of this. With a definition $sqr\ x = mul\ x\ x$ (*mul* being the predefined primitive multiplication function), an expression graph for *sqr e* is rewritten into the graph for *mul e e*, sharing pointers to *e*. Next *mul* calls for reduction of its two arguments into numbers (which happens to be the same one because of the sharing). Finally, *mul* computes the product of the two numbers and remakes the top apply node into an integer node with the value of that product (see figure 2).

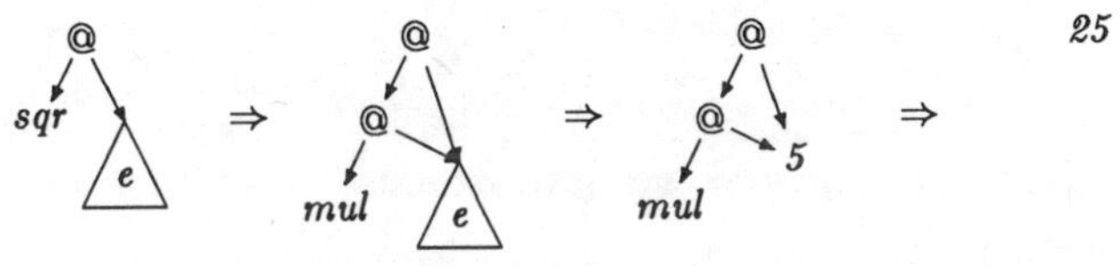


Figure 2: Naïve graph reduction of *sqr e*, with $sqr\ x = mul\ x\ x$.

An observation of paramount importance is that we always call for the reduction of an expression graph in order to get the value or *canonical form* of the expression. Thus in the example above reduction will not stop until the value of $e_i[x_1 = e'_1, \cdots, x_{m_i} = e'_{m_i}]$ has been reached. So instead of evaluating expressions by successive rewrites, more conventional expression evaluation techniques can be used. In the example of figure 2, this means that it is not necessary to construct the intermediate graph for *mul e e*. Instead, *sqr* should call for evaluation of its argument $x$, compute $x \times x$, and remake the apply node to an integer node with the value $x \times x$ directly; see figure 3. But although much of the graph manipulation can be bypassed in this manner, we still need to be able to construct expression graphs, for instance for lazy arguments to functions.

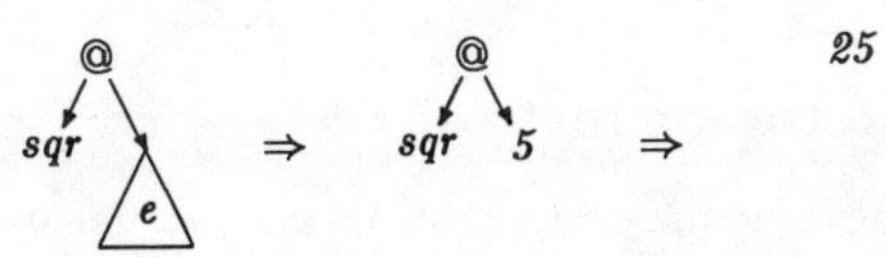


Figure 3: Non-naïve graph reduction of *sqr e*.

The *G-machine* is a stack machine, with instructions to construct and manipulate graphs. Reduction is started by executing an EVAL instruction, which calls for evaluation of the graph pointed to from the top of a stack of node pointers, the so called *S stack*. Each of the functions $f_i$ is compiled into G-machine code to compute the value of the right hand side, update the root node of the application graph, and return from the EVAL call. This code is executed when EVAL has been called to reduce the graph for an $f_i$ application. Pointers to the graphs of the arguments and the apply node to be updated are found on the S stack. For further details on the innards of the G-machine, see [Joh84].

In order for this paper to be self-contained, and also because the current approach in some respects differs from the one presented in [Joh84], appendix A lists a complete set of compilation rules for a small functional language. The G-machine instructions emitted by this compiler are the ones we will consider for target code generation in this paper. In appendix F we have given an example of compilation of the program *sieve of Erathostenes*, also showing the pattern matching transformation and the lambda lifting.

A state in the abstract G-machine can be described by a 7-tuple $\langle O, C, S, V, G, E, D \rangle$, where

$O$ is the output produced so far,

$C$ is the G-code sequence currently being executed,
$S$ is a stack of node names, i.e., pointers into the graph,
$V$ is a stack of basic values used for evaluation of arithmetic and logical expressions,
$G$ is a graph, i.e., a mapping from node names to nodes. In this paper we will consider the following node types:

| | |
|---|---|
| INT $i$ | integer nodes, |
| CONSTR $k\ n_1 \cdots n_m$ | constructor nodes, for constructor number $k$, having $m$ arguments, |
| AP $n_1\ n_2$ | application nodes, $n_1$ and $n_2$ are pointers to the function and argument graphs respectively, |
| FUN $f$ | a function node for the function named $f$, |
| HOLE | a node to be filled in later with a value, being used while constructing cyclic graphs for *letrec* expressions. |

$E$ is a global environment: a mapping from function names $f$ to pairs (*arity of f, code for f*),
$D$ is a dump used for recursive calls to EVAL; when calling EVAL the current $S$ and $C$ are dumped here.

In appendix B we give state transition rules for the G-machine instructions we consider in this paper.

# 3 The basic machinery of the G-machine implementation

In this section we describe the basic mechanisms of our implementation of the G-machine: the heap management scheme, the EVAL mechanism and a simple target code generator. In section 4 we describe an improved target code generator; the same run-time machinery and EVAL mechanisms are used in both cases.

## 3.1 Implementing the components of the abstract machine

In our implementation we implement the components of the abstract G-machine in a fairly simple and obvious way (see also figure 4):

$O$ the output, is simply printed on standard output.
$C$ the G-code, is the currently executing target code, i.e. the the code segment with the program counter, in the conventional manner.
$S$ the pointer stack, is mapped onto an auxiliary data area, and a machine register for a stack pointer, called *Sp*. The S stack contains the arguments and the local variables of the functions, and we do *not* attempt to allocate registers for these.
$V$ the basic value stack, is mapped onto the system stack together with the system stack pointer of the target machine, which we call *Vp*.
$G$ the graph, is mapped onto a heap area and a heap pointer register; see below on heap management scheme. Some nodes, which we would like to be 'non-volatile', e.g. function nodes and integer nodes, are put in a data area outside the heap, at compile time. This is to avoid allocating them on the heap every time they are needed.
$E$ is the compiled target code in the conventional manner; the arity information in the abstract machine is implemented with code to check the number of arguments present on the stack.
$D$ is mapped onto the system stack, with *Vp*, just as the V stack is. Dumping the current stack is implemented by pushing the current S stack pointer Sp. Dumping a code sequence correspond to saving the return address on the stack in a call to a subroutine.

The stacks in the abstract machine and in the implementation do not always correspond exactly to each other. For instance, some intermediate results of arithmetic computations which in the abstract machine are put on the V stack are held in machine registers in the implementation (see section 4). When a distinction is necessary, the V and S stacks in the implementation will be called the *concrete V stack* and the *concrete S stack* respectively; the counterparts in the abstract G-machine will be called the *abstract V stack* and the *abstract S stack*.

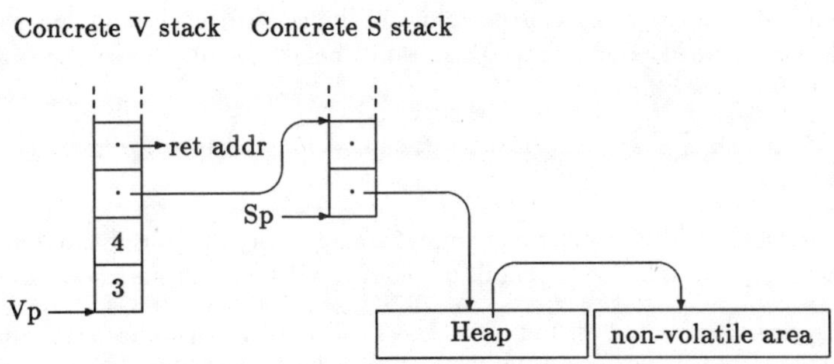


Figure 4: Stacks and the heap in the implementation.

## 3.2 Heap management scheme

It is very common when implementing list-oriented languages like Lisp and ML to use a linked list of free cells and a mark-scan garbage collector. We use a different scheme, which appears better suited for conventional computers, without hardware support for the heap; see figure 5. In this

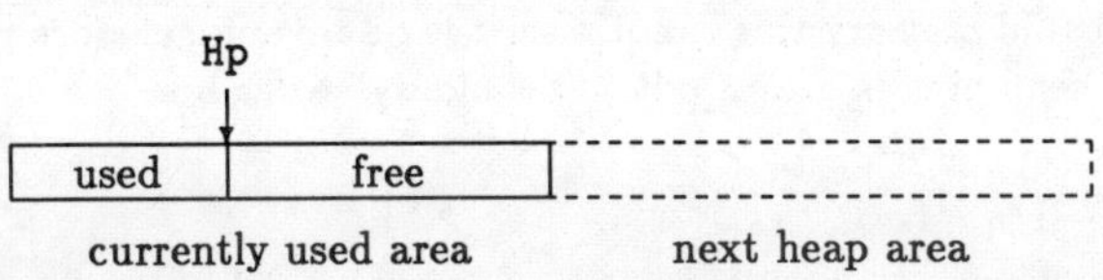


Figure 5: The heap.

scheme memory is allocated from left to right, by incrementing a heap pointer register. The heap space is subdivided into two heap areas, and one subarea at a time is used. When space runs out in the current subarea, the nodes in use are copied to the other heap area, leaving the garbage behind. The VAX and many other machines have an auto-incrementing indirect register addressing mode. Using this on the heap pointer register, an apply node or a cons node can be allocated and filled with its contents using only three machine instructions with three memory references—see figure 6.

This memory allocation scheme offers the following advantages.

- With our scheme short and efficient straight line code for graph construction is possible, as shown in figure 6 — a very important consideration, since graph reduction is very heap allocation intensive. Generally, faster code results if code for an operation is emitted in-line, rather than as a call to a subroutine, and not only because of the overhead of the call and return instructions. Jumps, calls and other instructions that break the current instruction

```
movl $AP,(Hp)+
movl function pointer,(Hp)+
movl argument pointer,(Hp)+
```

Figure 6: VAX code for constructing an apply node. Hp is an alias for the register used as heap pointer.

stream often are comparatively expensive on machines with an instruction prefetch queue: the queue has to be filled again and it takes a while before the first instruction can be decoded and executed after the jump.

- Nodes of different sizes need no special treatment — long vectors, for instance, can easily be accommodated.
- No temporary locations (registers, for example) are necessary for storing pointers to subgraphs while constructing a graph. They can all be referred via Hp at offsets known at compile-time.
- Before nodes are constructed on the heap, checks have to be made to ascertain that enough heap is available. This check can easily be done once for all node allocations in a long straight line sequence (see the target code in appendix F how this is done).
- A copying garbage collector compactifies the graph so that nodes close to each other in the graph will also be close to each other in the memory—an obvious advantage when using a machine with virtual memory. With the mark-scan scheme and list of free cells, nodes tend to become randomly distributed in the heap.
- In a copying garbage collector the copying time is proportional to the current size of the graph, making for a short garbage collection time for "small problems", even if a big heap is used.

A disadvantage of our scheme is that only half of the available heap memory can be used at a time. On machines with virtual memory this is not a serious disadvantage, since presumably the memory pages of the unused heap area migrates out to secondary storage.

## 3.3 M code

To abstract away from some of the details of target code generation, both in this exposition and in the actual compiler, instead of translating G-code to VAX machine code or whatever particular machine we consider, we translate the G-code into *M code*. The M code machine is intended to model most ordinary CPUs, in that it has a set of registers, an instruction set with addressing modes, etc. For historical reasons, the M code machine looks very much like the VAX — push and pop addressing modes, 3 operand instruction — so obtaining VAX assembler code from the M code is mostly only a matter of pretty printing the M code. But it has proved easy to obtain assembler code also for other machines; at the time of writing this we have code generators for the M68000, the NS32016 and the IBM360 architecture.

The M code machine is specialized for translating G-machine code. It has special addressing modes for pushing and popping the V and the S stacks, and moving a word to the heap using the Sp, Vp and Hp registers. Items are handled in units of *words*, an entity assumed big enough to be a pointer or an integer. Data memory, i.e. the stacks and the heap, is assumed to be word-addressed; data pointers always point to a word in memory. For instance, adding a number $n$ to a data pointer gives a new pointer pointing $n$ words ahead of the old one (independent of how many bytes there are in a word on the intended target machine). Appendix C summarizes the instructions and addressing modes of the M code.

## 3.4 Naïve target code generation

In this section we describe target code generation by means of 'macro expansion': each G-instruction is translated into a fixed sequence of machine instructions independent of the context. Just as in the abstract G-machine, in the code the operands are popped from the stacks and the result is pushed onto it. Inbetween the target code for each G-machine instruction, the concrete stacks corresponds exactly to the abstract stacks. This kind of code we call *naïve code*. There are several reasons for first showing naïve target code generation. Firstly, naïve target code generation is very simple to implement, and is thus of interest for that reason. Secondly, we do not want to obscure the exposition with details of more intricate code generation. Thirdly, when we describe an improved target generator in section 4, the naïve code provides a useful reference. The same runtime system and EVAL-unwind machinery will be used for both the naïve code and code less naïve.

We now give some examples of naïve target code generated from G-code. A complete list of rules for generating naïve code can be found in appendix D.

From the G-machine instruction SUB we generate naïve M code that pops the two operands from the V stack and pushes back the result onto the V stack, as follows:

```
sub popV,popV,pushV
```

In an M instruction (as well as in a VAX instruction) the operands are evaluated from left to right. The above M instruction corresponds to the VAX instruction

```
subl3 (sp)+,(sp)+,-(sp)
```

or

```
subl2 (sp)+,(sp).
```

The M code register Vp corresponds to the VAX stack pointer register sp.

From the MKAP instruction we generate the following naïve M code: (see also figures 4, 5 and 6)

```
move $AP,tohp
move popS,tohp
move popS,tohp
move $-3(Hp),pushS
```

First the tag word is moved to the heap, and then the function field and the argument fields are popped from the S stack into the heap; finally a pointer to the constructed node is pushed back into the S stack. The above M code corresponds to the following VAX code:

```
movl $AP,(Hp)+
movl (Sp)+,(Hp)+
movl (Sp)+,(Hp)+
moval -12(Hp),-(Sp)
```

where Hp and Sp are to be considered as aliases for the actual registers used.

## 3.5 Implementation of EVAL, unwind and RET

The EVAL instruction takes the graph pointed to from the top of the (abstract) S stack and reduces it to its value. If this node is already a value, eg. an INT node or a CONSTR node, then EVAL is a no-operation. If the node is an apply node, the G-machine enters an unwind loop to follow the function pointers until reaching the function node of the application, pushing the pointers to the apply nodes on the way.

In the target code and the runtime system for EVAL we want to do as little as possible if the node is a constructor. Our implementation method uses an object oriented view of node types and operations. For each node type we associate a table of pointers to code that implements the various operations we want to perform on them, like EVAL, unwind, and some others; see figure 7. Thus in the first position in the table for an apply node, for instance, we have a pointer to code to evaluate an apply node, in the second position code for unwind an apply node, etc. A node tag is simply a pointer to such a table.

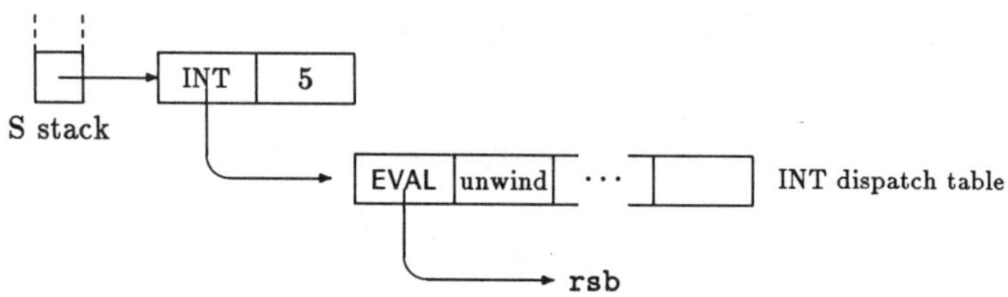


Figure 7: An INT node and its dispatch table. Since EVAL is a no-operation on an INT node, the eval entry in the table, the first one, points to an `rsb` (return from subroutine) instruction.

It may seem a waste to use an entire word for the tag when ordinarily a few bits or a byte would suffice, if the tag were instead a small number. The space waste may however turn to our advantage, since now all objects on the heap are words, and thus all accesses to the heap can be word-aligned. On the VAX, for instance, word accesses unaligned to word boundaries are possible, but slower than aligned ones. On many other machines unaligned accesses are not even possible. Another advantage of this scheme is that nodetypes can be added at will without changing the emitted code.

EVAL is thus implemented as a subroutine call via the node tag and the first entry in the table. Figure 8 shows naïve M and VAX code sequences for for the EVAL instruction according to this scheme. The instructions to pop into `r0` and the push into the S stack can most of the times be removed by a peep hole code improver (section 3.10) or can be avoided altogether by a less naïve code generator. If the node is a constructor, like an INT node, CONSTR etc, the eval subroutine call lands on a return instruction immediately. If on the other hand the node is an apply node, the eval entry points to code for starting the evaluation of an application. The unwind operation is implemented similarly.

Figure 9 shows the M code in the runtime system for EVAL and unwind for an apply node. The first thing to do when evaluating an apply node is to push the Sp register onto the V stack, which in the abstract machine corresponds to pushing the entire current S stack on the dump. For the unwind step the same method of jumping via the tag is used. This jump may either end up in `AP_unwind again`, or in the unwind entry for a function node.

We represent function nodes by two words: The FUN tag pointing to a dispatch table for FUN describing function nodes in general, and a *function description vector* which contains information about a particular function:

*M code for* EVAL:

```
move popS,r0
calltab eval,0(r0)
move r0,pushS
```

*VAX code for* EVAL:

```
movl (Sp)+,r0       Pop node pointer into r0
movl (r0),r1        Get tag, i.e. table pointer into r1
jsb *(r1)           Do a subroutine call to the eval code
movl r0,-(Sp)       Push back node pointer.
```

Figure 8: Target code for EVAL.

```
AP_eval:    move Sp,pushV
AP_unwind:  move r0,pushS
            move 1(r0),r0
            jmptab unwind,0(r0)
```

Figure 9: EVAL and unwind for an apply node.

- The arity of the function, i.e. the number of arguments necessary to perform reduction.
- A pointer to a 'non-volatile' function node for this function, emitted together with the code for the function. The code for PUSHFUN refers to this node; also, to save heap space, the garbage collector replaces all pointers to function nodes in the heap by pointers to these common function nodes.
- A pointer to code for the number-of-arguments check and stack rearrangement after unwind; ordinarily the arity information should suffice together with general code in the runtime system for the task. But to speed up this operation there is also straight line code for this operation for the most common small arities.
- A pointer to code for evaluating a vector application node; see section 3.7.
- A pointer to the target code for the function.

The naïve M code for FUNSTART is given in figure 10. Figure 11 shows the situation when unwind

```
FUNSTART f n  ⇒  Cf:  word FUN,Vf;
                 Vf:  word n, Cf, unwn, vunwn, Ff;
                 Ff:
```

Figure 10: The naïve M code for FUNSTART.

has reached a function node. (In this figure the function takes two arguments and returns a function which is applied to one argument; an example of this would be the application *K (add 2) 3 5* with

$K\ x\ y = x$ and $add\ x\ y = x+y$.) The code at `FUN_unwind` fetches the unwind code from the function description vector and jumps via it. This code in turn checks the number of arguments by comparing the saved Sp with the current one, and returns prematurely if there are to few arguments; otherwise rearranges the S stack, and ends by jumping to the target code for the function, whose pointer is also found in the function description vector.

The task of the function code is to compute the value of of the right hand side of the function definition given the actual arguments on the stack, update the root of the redex with the value, pop off the arguments, and exit from the function code. Since in general the value could be a function, whose argument may be found 'above' the just reduced redex, the code for RET must perform the unwind operation on the updated redex node. If the value node happened to be eg. an INT node after all, then the actual return from EVAL is done by the unwind code for INT; see figure 12. It is here assumed that the executed program is type correct, in the sense that an argument must only be applied to a thing which is a function.

## 3.6 Tail calls and direct calls: JFUN and CALLFUN

To return the value of a function application ($\mathcal{R}[\![f\ e_1 \cdots e_m]\!]r\ n$ in appendix A), an initial approach could be to use a "pure graph reduction" strategy. $\mathcal{R}[\![f\ e_1 \cdots e_m]\!]r\ n$ would then be code to build the graph of the application, update the apply node for the return value with the application graph, pop off the arguments, and execute a RET instruction which will start the unwinding again. Such a pure graph reduction strategy gives a loop-like behaviour for tail recursive calls, since unwind and the subsequent function code will reuse the old stack space.

Now assume that the called function $f$ has arity $m$, i.e., the same as the number of passed arguments in the tail call. Then after unwind and argument rearrangement we will have a stack with the $m$ arguments of $f$ on top of the stack, and a pointer to the apply node to update with the result of the application. However, this happens to be the same apply node that was updated with the graph of the application to begin with! So if the arity of the called function is actually known at compile-time to be the same as the number of applied arguments, returning the value of the application can be compiled into code that replaces the current content of the S stack with the new arguments of $f$ (this is what the compilation scheme $\mathcal{S}$ does, referred to in appendix A), followed by a jump to the code for $f$ (instruction JGLOBAL in the code example in appendix F).

To enable a nice behaviour at runtime for tail calls even when the arity of the called function match the number of arguments but the arity of the function is not known at compile time (for instance when the function is passed as an argument or the function is defined in a different separately compiled module), the stack is set up under the optimistic assumption that the arities will actually match. The tail call itself is then performed by a `jmptab jfun` instruction, which at run-time will compare the arity with the actual number of arguments and take appropriate action. Almost exactly the same is done for a direct call using the CALLFUN instruction (as a result of $\mathcal{E}[\![i\ e_1 \cdots e_m]\!]r\ n$, see appendix A), the only difference being that in the emitted target code before the `jmptab` instruction the return address and an appropriate S stack pointer is pushed onto the V stack. Figure 13 shows naïve target code for JFUN and CALLFUN. This call can then wind up in the `jfun` entry of a FUN node in the run-time system, where the number of passed arguments is compared with the actual arity of the function, as shown in figure 14. If they are equal, which is expected to happen most of the time, nothing extra has to be done and an immediate jump to the code for the function is performed. However, if the arities do not match the conservative action of building the application graph described above is taken. That action is also taken if the `jmptab jfun` winds up in the `jfun` entry for apply nodes.

Further refinements of the run-time system code for tail calls are possible. A trivial one would be to replace the arguments with pointers to the $m$ apply nodes at the same time as constructing them, thereby short-circuiting the $m$ unwind steps that would otherwise follow. A less trivial one is that

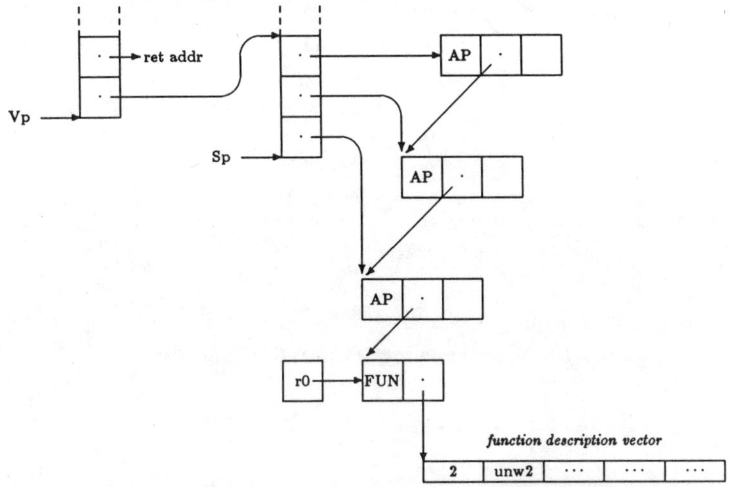


```
FUN_unwind:         move    1(r0),r3
                    jmptab  VFunwind,r3
                    :
unw2:               comp    $2(Sp),0(Vp)
                    jgt     premature_return
                    move    0(Sp),r0
                    move    2(r0),pushS
                    move    2(Sp),r0
                    move    2(r0),1(Sp)
                    jmptab  VFfuncode,r3
                    :
premature_return:   move    popV,Sp
                    move    -1(Sp),r0
                    return
```

Figure 11: The situation when unwind has reached a function node.

```
RET ⇒                move popS,r0
                     jmptab unwind,0(r0)

canonical_unwind:    move popV,Sp
                     return
```

Figure 12: Naïve target code for RET and the unwind code in the runtime system for a canonical node.

```
JFUN n ⇒        move    popS,r0
                move    $n,r2
                jmptab  jfun,0(r0)

CALLFUN n ⇒     move    popS,r0
                move    $n,r2
                move    $l,pushV
                move    $(n+1)(Sp),pushV
                jmptab  jfun,0(r0)
           l:
```

Figure 13: Naïve code for JFUN and CALLFUN.

| | | | |
|---|---|---|---|
| FUN_jfun: | move | $1(r0),r3 | Pointer to fun descr vector into r3 |
| | comp | r2,VFarity(r3) | no of args = arity? |
| | jneq | AP_jfun | if no, take conservative action |
| | jmptab | VFfuncode,r3 | if yes, go to function code. |
| do_mkap: | move | $AP,tohp | Begin MKAP: tag |
| | move | r0,tohp | function part from r0 |
| | move | popS,tohp | argument part from S stack |
| | move | $-3(Hp),r0 | new function pointer into r0. |
| AP_jfun: | sub | $1,r2,r2 | Decrement MKAP counter |
| | comp | r2,$0 | MKAP counter = 0 ? |
| | jneq | do_mkap | if not, go MKAP again. |
| | move | 1(Sp),r1 | pointer to node to update into r1 |
| | move | $AP,0(r1) | remake into AP node: tag |
| | move | r0,1(r1) | function part |
| | move | popS,2(r1) | argument part from stack. |
| | jmptab | unwind,0(r0) | go to the unwind state. |

Figure 14: Run-time system code for tail calls and direct calls.

we need not always be as drastic as to rebuild the apply graph completely and unwind, some minor adjustment of the stack and the graph might be enough. This is illustrated in figure 15. Here $f$ has been tail-called with two arguments (with one pending argument $a_3$), (a) shows the stack just after the execution of the JFUN instruction. If the arity of $f$ is only one the argument $a_2$ is transferred to the apply node and a new node is allocated for the return value, entering configuration (b). If on the other hand the arity of $f$ is three a pointer to the third argument is copied from the apply node for the pending argument into the stack, as indicated in (c); however, this is not enough! We must also actually build the graph $f\ a_1\ a_2$ and update, in order to preserve proper sharing. This is discussed in further detail in [Pey87, chapter 21].

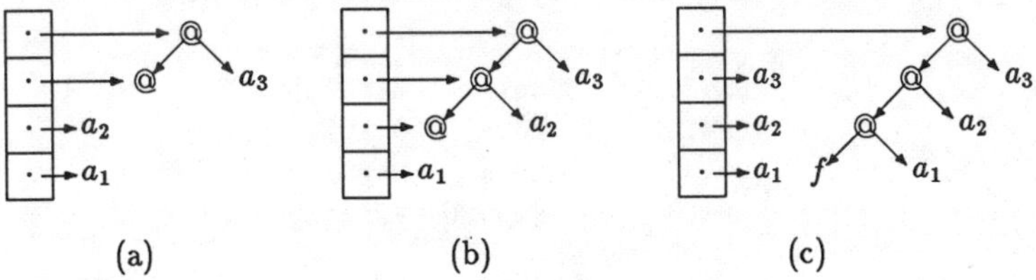


Figure 15: Stack and graph rearrangement at a tail call with one argument too many, (a)⇒(b), and one argument too few, (a)⇒(c).

## 3.7 Vector apply nodes

In this section we consider a more efficient representation of graphs of function applications.

Building the graph of the application, $f\ e_1\ e_2\ e_3$. say, the standard representation will be that of figure 16(a). If the arity of $f$ is three, the apply node containing $e_3$ will be updated with the value of the application when the code for $f$ is executed. Except for serving as place holders for the arguments $e_1$ and $e_2$, the other two apply nodes have been built in vain! Figure 16(b) shows an alternative representation of the same application, being a single node, called VAP (vector apply), containing all the applied arguments. It will be used *only* with number of arguments equal to the arity. This restriction simplifies the code involved immensely, while still covering a large number of common cases. If there are more arguments in an application then the extra ones are put in apply nodes as usual.

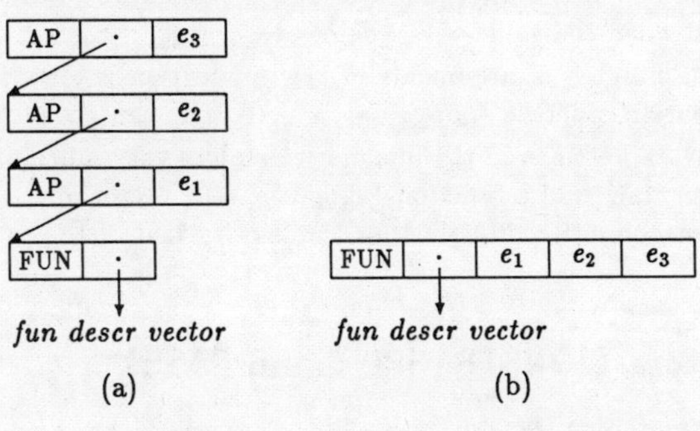


Figure 16: Standard (a) and alternative (b) representation of $f\ e_1\ e_2\ e_3$.

The VAP representation is cheaper to build, since fewer words have to be allocated on the heap. A VAP application node can also be reduced more efficiently. Figure 17 shows run-time system code for eval and unwind of a VAP node. A pointer to an appropriate vunw*n* entry is contained in the function description vector (see figure 10). The arity check (as in `FUN_unwind`) is not necessary here, since exactly the right number of arguments are known to be present. Instead of the ordinary unwind and rearrangement necessary with the standard representation, the stack can be set up immediately to the desired configuration.

```
VAP_eval:     move     Sp,pushV
VAP_unwind:   move     r0,pushS
              move     1(r0),r3
              jmptab   VFvapunwind,r3
              :
vunw3:        move     4(r0),pushS
vunw2:        move     3(r0),pushS
vunw1:        move     2(r0),pushS
vunw0:        jmptab   VFfuncode,r3
```

Figure 17: Run-time system code for EVAL and unwind of vector apply nodes.

Code to build a vector apply node can be emitted easily only if the function application is of the form $f\ e_1 \cdots e_m$, where $f$ is a global function with a known arity of $\leq m$. The lambda-lifting algorithm is designed to make sure that this is the case as often as possible; see [Joh85].

In [Joh84] it was stated that to construct the graph for an expression involving primitive operators $+$, $\times$ etc, we would build a graph with applications to the corresponding primitive functions; for example, to build the graph for $x + y \times x$ we would then build the application $add\ x\ (mul\ y\ x)$. This had the effect that evaluating this expression, by calling EVAL, caused evaluation to work in an 'interpretation mode', the eval machinery interpreting the graph to compute the value—hardly a very efficient evaluation method, compared to 'real' code for the evaluation of the same expression.

We can do much better by 'wrapping up' the expression in a function, i.e., transforming the expression $e$, whose graph is to be built, into the application $f\ v_1 \cdots v_m$, where $f\ v_1 \cdots v_m = e$ and $v_1 \cdots v_m$ are the free variables of $e$ (this can be done during lambda-lifting). For instance, the expression given above would be transformed into the application $f\ x\ y$, and the function definition $f\ x\ y = x + y \times x$. Thus, instead of building a potentially large graph of $e$ containing applications to the primitive functions *add*, *if* etc., (as described in [Joh84]) $e$ will now be represented by 'pure code' being executed when the value of $e$ is demanded.

Further, since the number of arguments in the application is always equal to the arity (being the number of free variables in the expression), a VAP node can always be built in this case.

Representing lazy expressions in this manner resembles very much an environment-based technique for the implementation of a functional language (the vector $v_1 \cdots v_m$ is a tiny environment!), e.g. the lazy SECD machine [Hen80,Lan64] or the lazy variant of the categorical abstract machine (CAM) [CCM85].

## 3.8 Constructors: CONSTR, SPLIT and CASE

Prior to the pattern matching transformation in the compiler (see [Aug85] for details) the concrete data type declarations are processed, and the constructors of each type are assigned ***constructor numbers*** from 0 and upwards. Expressions of the form $i\ e_1 \cdots e_n$ where $i$ is a constructor identifier, are transformed into $constr_k\ e_1 \cdots e_n$. For constructing an object of a concrete data type there is a G-machine instruction CONSTR $k$ $n$ which constructs an object with constructor number $k$, with $n$

arguments taken from the S stack. The inverse operation is performed by the instruction SPLIT $k$ $n$ which pops a pointer to a constructor node and pushes the $n$ constituent parts.

Since apply nodes can be updated with value nodes, i.e. INT nodes and nodes representing concrete data types, no node representing a value must be bigger than an apply node. Alas, were it not for this we could conveniently represent all constructor nodes with a vector of sufficient size. Instead we can introduce a root node for the constructor, and regard the CONSTR $k$ $n$ as a shorthand for the two-instruction sequence CVEC $n$; CUNIT $k$ which together build the nodes

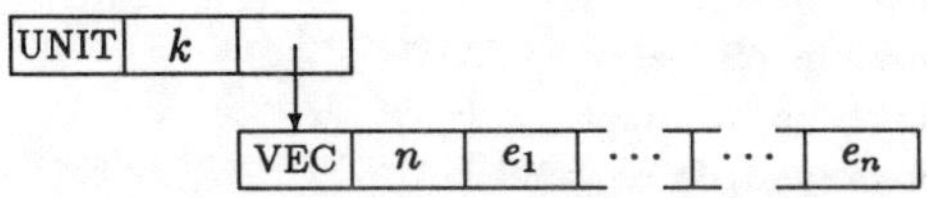


Correspondingly, SPLIT $k$ $n$ would then be a shorthand for SPLITUNIT; SPLITVEC $n$.

The above is a very spacious representation for contructors with one or two arguments, and we would like to use a more compact one whenever possible. Now consider a more standard representation of a list cons node:

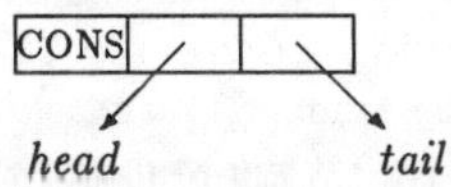


We would like this more compact representation also for binary constructors defined by the user.

However, the predefined CONS constructor has a specific constructor number (namely 1). Since in general a type can have many binary constructors we need CONS constructors with different constructor numbers associated to them. To implement this we have a set of pair tags PAIR$k$, for $0 \leq k \leq K$, with the corresponding constructor numbers associated to them. (Predefined CONS thus uses PAIR1.) Most concrete data types have a small number of constructors, so a fairly small $K$ should cover most cases; in our implementation we have chosen $K = 4$. If $k > K$ then the UNIT node is used as a root instead, as in the initial example. Various combinations of NIL, UNIT, PAIR and VEC nodes are used to optimize different cases, as shown in table 1. Figure 18 shows an example, together with the naïve target code for the resulting CONSTR and SPLIT variants.

| | | | |
|---|---|---|---|
| | *all k and* $n \leq 1$: | | |
| CONSTR $k$ 0 | $\Rightarrow$ CNIL $k$ | SPLIT $k$ 0 | $\Rightarrow$ *nothing* |
| CONSTR $k$ 1 | $\Rightarrow$ CUNIT $k$ | SPLIT $k$ 1 | $\Rightarrow$ SPLITUNIT |
| | $0 \leq k \leq K$ *and* $n \geq 2$: | | |
| CONSTR $k$ 2 | $\Rightarrow$ CPAIR $k$ | SPLIT $k$ 2 | $\Rightarrow$ SPLITPAIR |
| CONSTR $k$ 3 | $\Rightarrow$ CPAIR 0; CPAIR $k$ | SPLIT $k$ 3 | $\Rightarrow$ SPLITPAIR ; SPLITPAIR |
| CONSTR $k$ $n$ | $\Rightarrow$ CVEC $n-1$; CPAIR $k$ | SPLIT $k$ $n$ | $\Rightarrow$ SPLITPAIR ; SPLITVEC $n-1$ |
| | $k > K$ *and* $n \geq 2$: | | |
| CONSTR $k$ 2 | $\Rightarrow$ CPAIR 0; CUNIT $k$ | SPLIT $k$ 2 | $\Rightarrow$ SPLITUNIT ; SPLITPAIR |
| CONSTR $k$ $n$ | $\Rightarrow$ CVEC $n$; CUNIT $k$ | SPLIT $k$ $n$ | $\Rightarrow$ SPLITUNIT ; SPLITVEC $n$ |

Table 1: Coding of the CONSTR and SPLIT instructions.

The CASE $(k_1.l_1).\cdots.(k_m.l_m).l_d$ instruction pops a node pointer from the S stack and extracts the constructor number $k$ from this node. If $k = k_i$ then a jump to the label $l_i$ is performed, else if none of $k_i$ is equal to $k$ a jump to the default label $l_d$ is taken. A problem with the elaborate space-saving

| CONSTR 2 4 ⇒ | | SPLIT 2 4 ⇒ | |
|---|---|---|---|
| CVEC 3 | `move $VEC,tohp` | SPLITPAIR | `move popS,r0` |
| | `move $3,tohp` | | `move 2(r0),pushS` |
| | `move popS,tohp` | | `move 1(r0),pushS` |
| | `move popS,tohp` | SPLITVEC 3 | `move popS,r0` |
| | `move popS,tohp` | | `move 4(r0),pushS` |
| | `move $-5(Hp),pushS` | | `move 3(r0),pushS` |
| CPAIR 2 | `move $PAIR2,tohp` | | `move 2(r0),pushS` |
| | `move popS,tohp` | | |
| | `move popS,tohp` | | |
| | `move $-3(Hp),pushS` | | |

Figure 18: An example of CONSTR and SPLIT.

representation scheme presented above is that it might be hard to extract the constructor number from the representation. Here the tag dispatch mechanism described in section 3.5 comes to our rescue. The code for the CASE instruction executes a `calltab getconstrno,0(r0)` which for each node type has to bring the constructor number to a machine register, as shown in figure 19. Since the *case* expression can be used also for integer values, `getconstrno` code is also needed for INT nodes. See also appendix D for the naïve code of the CASE instruction and the variants of the CONSTR and SPLIT instructions.

```
CASE (0,L0),(1,L1),L2 ⇒  move     popS,r0
                         calltab  getconstrno,0(r0)
                         case     (0,L0),(1,L1),L2

                In runtime system:
PAIR0_getconstrno:  move    $0,r2
                    return
PAIR1_getconstrno:  move    $1,r2
                    return
                      ⋮
INT_getconstrno:
UNIT_getconstrno:   move    1(r0),r2
                    return
```

Figure 19: CASE code example, and the entries in the runtime system for extracting the constructor number.

## 3.9 Pairing of G-machine instructions

It is useful to let the target code generator consider some pairs or longer sequences of G-machine instructions, instead of generating code for each of them independently of each other.

Consider the sequence MKAP; UPDATE $n$. The MKAP instruction constructs an apply node, which is copied by the UPDATE instruction onto another node. Hence the memory allocated for the apply node is wasted, we might as well fill in the update nodes directly with the AP tag, and function and argument fields taken from the S stack. Thus:

$$\text{MKAP; UPDATE } n \Rightarrow \begin{array}{l}\texttt{move } (n-1)\texttt{(Sp),r0} \\ \texttt{move \$AP,0(r0)} \\ \texttt{move popS,1(r0)} \\ \texttt{move popS,2(r0)}\end{array}$$

This method is of course also applicable for a CONSTR or an MKINT followed by an UPDATE instruction. Even greater savings can be done if the node updated is the return value node, updated with a constructor e.g. an integer, and the update is followed by POP and RET. We then know beforehand how RET will behave: since the scrutinized node is a canonical one, the unwind operation will cause an immediate return. Therefore we can emit a `return` instruction instead of a `jmptab unwind`.

## 3.10 Peep-hole code improvement of the naïve code

*Peep hole code improvement* is a standard technique for locally improving the target code. It works by looking at the target code through a small *peep hole*, and whenever possible replacing the instructions in the peep hole with a shorter and faster equivalent. In general each replacement may give rise to possibilities for further improvements both before and after the current position of the peep hole. Therefore, in general a single pass over over the target code is not sufficient; to get the most out of it, either multiple scans must be done, or we could have a scheme where the peep hole moves both backwards and forwards in the target code.

Looking at the code emitted by our naïve target code generator, it is immediately apparent that it has a lot of rough edges. The most obvious imperfection is the large number of redundant pushes and pops; things are pushed onto the concrete V or S stack, only to be popped immediately in the next instruction, or very soon after. It is of course desirable to get rid of these redundancies, not only to reduce the number of target code instructions, but also to reduce the number of memory references which sometimes is a more limiting factor for speed than the number of executed instructions.

Despite its deficiencies surprisingly efficient code can be obtained by first generating naïve code and then improve it with a very simple peep hole code improver, even one that make only a single pass over the code. The worst rough edges can be taken away from the naïve code with a fairly small number of patterns, recognizing the most common cases. Figure 20 shows a representative example.

Carrying this approach to its limit, we soon find that patterns become increasingly specialized, and furthermore an increasing amount of computation has to go into recognizing pattern instances in the code. Another major drawback is that it is very hard to make efficient use of registers as temporary locations. Thus a different and more general approach to code generation is needed, and we present one in the next section.

# 4 A better code generator

In this section we present a target code generator that from the start avoids generating the redundant push and pop operations that the naïve target code generator produces and the peep hole code improver can remove to some extent.

The idea behind its operation is that instead of emitting code to push an operand on the concrete V or S stack, the operand is pushed on a simulated V or S stack internal to the code generator. Likewise, if an operation requires something to be popped from a stack and the corresponding simulated stack is nonempty then the operand is taken from the simulated stack; otherwise a popV or popS operand is used. A simulated stack will always correspond to an initial segment of the an abstract stack. It is convenient to have the simulated stack containing M code addressing modes, so that such an operand can be placed directly in an instruction in the emitted target code.

$\mathcal{B}[\![$x-1$]\!]$ [x=1] 1 =

| *G code* | *naïve M code* | *improved M code* |
|---|---|---|
| PUSH 0 | `move 0(Sp),pushS` | |
| EVAL | `move popS,r0` | `move 0(Sp),r0` |
| | `calltab eval,0(r0)` | `calltab eval,0(r0)` |
| | `move r0,pushS` | |
| GET | `move popS,r0` | |
| | `move 1(r0),pushV` | |
| PUSHBASIC 1 | `move $1,pushV` | |
| SUB | `sub popV,popV,pushV` | `sub $1,1(r0),pushV` |

*Patterns used:*

| | | |
|---|---|---|
| `move` *a*`,pushS; move popS,`*a* | ⇒ | ∅ |
| `move` *a*`,pushS; move popS,`*b* | ⇒ | `move` *a*`,`*b* |
| `move` *a*`,pushV; move` *b*`,pushV; sub popV,popV,`*c* | ⇒ | `sub` *b*`,`*a*`,`*c* |

Figure 20: An example of what a simple peep hole improver can do with a code sequence.

Figure 21 shows an introductory example where a simulated S stack is used for constructing graph construction code. Here the simulated stack is assumed to be empty initially. (The resulting simulated stack is shown on the same line as the G or M code instruction producing it.) In the emitted target code for MKAP both operands are taken from the simulated stack; the result is a reference to the heap which is pushed onto the simulated stack. Finally, assuming this pointer is needed on the concrete S stack, the simulated S stack is 'flushed', pushing the heap reference onto the concrete S stack.

| *G code* | *sim. S stack* | *target code* |
|---|---|---|
| | [] | |
| PUSHFUN g | [$g] | |
| PUSHFUN f | [$f,$g] | |
| MKAP | | `move $AP,tohp` |
| | | `move $f,tohp` |
| | [$-3(Hp)] | `move $g,tohp` |
| | [] | `move $-3(Hp),pushS` |

Figure 21: Target code generation using a simulated S stack.

The technique of simulating the stack operations on a compile-time stack is not new; it has been use in for instance [LS78].

In the G-code references to the elements in the S stack (eg. in the PUSH and UPDATE instructions) refers via the top of the *abstract* S stack; in the target code the corresponding *n*`(Sp)` operands refer via the top of the *concrete* S stack. Thus if the simulated S stack is non-empty the offset have to be adjusted accordingly. Figure 22 shows an example of this: here the PUSH 1 instruction refers to the second topmost element in the abstract S stack, but since the simulated S stack at that point is one element deep, it will refer to the topmost element of the concrete S stack.

If the offset of the PUSH instruction is such that it refers to an operand in the simulated S stack, then it will copy that operand onto the top of simulated S stack; figure 23 shows an example of

| *G code* | *sim. S stack* |
|---|---|
| | [] |
| PUSH 0 | [0(Sp)] |
| PUSH 1 | [0(Sp),0(Sp)] |

Figure 22: Adjusting the offsets of stack references.

this: here the PUSH 1 instruction refers to the bottommost element in the simulated S stack, and is therefore copied onto the top of the simulated stack.

| *G code* | *sim. S stack* |
|---|---|
| | [] |
| PUSHFUN f | [$f] |
| PUSHFUN g | [$g,$f] |
| PUSH 1 | [$f,$g,$f] |

Figure 23: Pushing by copying an element in the simulated stack.

A similar snag occurs when referring via Hp to newly constructed nodes on the heap. The offset of an operand `$n(Hp)` changes when new items are moved to the heap, so the offsets of such operands on the simulated S stack have to be adjusted accordingly. Figure 24 illustrates this. Here, after the `move $AP,tohp` of the second MKAP, the heap reference `$-3(Hp)` changes to `$-4(Hp)` on the simulated stack. This figure is also a further illustration of the handling of the *n*`(Sp)` references (in PUSH 2).

$\mathcal{C}$[[f x y]] [x=0,y=1] 1 =

| *G code* | *sim. S stack* | *target code* |
|---|---|---|
| | [] | |
| PUSH 0 | [0(Sp)] | |
| PUSH 2 | [1(Sp),0(Sp)] | |
| PUSHFUN f | [$f,1(Sp),0(Sp)] | |
| MKAP | | move $AP,tohp |
| | [1(Sp),0(Sp)] | move $f,tohp |
| | [0(Sp)] | move 1(Sp),tohp |
| | [$-3(Hp),0(Sp)] | |
| MKAP | [$-4(Hp),0(Sp)] | move $AP,tohp |
| | [0(Sp)] | move $-4(Hp),tohp |
| | [] | move 0(Sp),tohp |
| | [$-3(Hp)] | |

Figure 24: Handling of heap references on the simulated stack.

The operations on the simulated V stack can be dealt with similarly, as is illustrated in figure 25. For the result of the subtraction we allocate a register and put it onto the simulated V stack.

$\mathcal{B}[\![(4\text{-}3)\text{-}(2\text{-}1)]\!]\ r\ n =$

| G code | sim. V stack | target code |
|---|---|---|
| | [] | |
| PUSHBASIC 4 | [$4] | |
| PUSHBASIC 3 | [$3,$4] | |
| SUB | [r2] | sub $3,$4,r2 |
| PUSHBASIC 2 | [$2,r2] | |
| PUSHBASIC 1 | [$1,$2,r2] | |
| SUB | [r3,r2] | sub $1,$2,r3 |
| SUB | [r2] | sub r3,r2,r2 |

Figure 25: Simulation of the content of the V stack.

## 4.1 Breaks in the code

So far we have considered only straight line code sequences manipulating the stacks. We must now consider what to do with the simulated stacks when reaching a break in the code, a JMP or an EVAL for instance. Figures 26 illustrate some complications.

When reaching a join in the code sequence — a place where different execution paths meet — the different assumptions about the locations of the operands of the abstract stacks must be 'synchronised', i.e. code must be emitted to make the simulated stacks equal; see figure 26 for an example. Here, at the end of the two branches of the **if** expression the two simulated V stacks look different, and they must be synchronised in order to continue after LABEL 1.

$\mathcal{B}[\![$**if** $\cdots$ **then** 0 **else** x-1$]\!]\ r\ n =$

| G code | sim. V stack | target code |
|---|---|---|
| ... | | |
| PUSHBASIC 0 | [0] | |
| JMP L1 | ? | |
| LABEL L2 | | |
| ⋮ | ⋮ | |
| PUSHBASIC 1 | [$1,1(r0)] | |
| SUB | [r2] | sub $1,1(r0),r2 |
| | ? | ? |
| LABEL L1 | | |

Figure 26: Synchronisation problems with the V stack.

Sometimes calling conventions require that some abstract stack elements are put in particular locations . We will not attempt to put function arguments in registers, so before a CALLFUN the arguments and the pointer to the HOLE node for the return value must be put onto the concrete S stack where the called function expects them to be.

We must also consider what happens to the global state when we call a function, via EVAL or CALLFUN, and how the global state may have changed when the call returns. Firstly, values of registers for temporary values may be corrupted, invalidating register references on the simulated stacks. Secondly, very likely heap memory has been allocated by the called function, thus invalidating `$n(Hp)` references on the simulated S stack.

In standard compiler technology two different conventions are used when dealing with problems like the ones above. In the *callee saves* convention, the called function upon entry save the registers and other locations it need to use for its own purpose, and restores the old values of these locations upon exit from the function. In the *caller saves* convention, before calling a function, the caller saves the registers etc. that it wants to keep the values of.

Using the *callee saves* convention would be complicated by the fact that jumps to other functions may occur in the function code to implement tail recursion, and it may be hard to determine which registers need to be saved. Also, since the Hp register cannot be saved and restored, the problem with the $*n*(Hp) references would remain.

We will therefore use a the *caller saves* convention. Moreover, as a simple solution to all potential problems exhibited above, we will simply require that at the end of a basic block *all* the elements of the abstract stacks are to reside in their corresponding places in the concrete stacks, with the exception that EVAL, CALLFUN and JFUN require the top of the abstract S stack in r0. This strategy is of course much less than optimal in situations like the one in figure 26; here a move $0,r2 ending the *then*-branch would suffice to synchronize the V stacks. But to obtain this a global register allocation scheme would be required.

## 4.2 How to flush the simulated stacks

Having thus decided that we want the simulated stacks to be empty at the end of a basic block, the question then is how and when to 'flush' them. A simple strategy would be to simply flush the simulated stacks onto the real ones, by emitting a sequence of instructions which push the elements of the simulated stacks, as shown in figure 27(a). The two expressions 5-4 and (3-2)-1 are computed, and their values have been put in registers r1 and r2 respectively, and they thus need to be flushed out onto the real V stack. To avoid tying up precious registers (and thus risk running out of them),

| $\mathcal{B}[\![5\text{-}4]\!]\ r\ n;\ \mathcal{B}[\![(3\text{-}2)\text{-}1]\!]\ r\ n =$ | | | | |
|---|---|---|---|---|
| *G-code* | *sim. V* | *target code* | *sim. V* | *target code* |
| PUSHBASIC 5 | [$5] | | [$5] | |
| PUSHBASIC 4 | [$4,$5] | | [$4,$5] | |
| SUB | [r1] | sub $4,$5,r1 | [] | sub $4,$5,pushV |
| PUSHBASIC 3 | [$3,r1] | | [$3] | |
| PUSHBASIC 2 | [$2,$3,r1] | | [$2,$3] | |
| SUB | [r2,r1] | sub $2,$3,r2 | [r2] | sub $2,$3,r2 |
| PUSHBASIC 1 | [$1,r2,r1] | | [$1,r2] | |
| SUB | [r2,r1] | sub $1,r2,r2 | [] | sub $1,r2,pushV |
| | | move r1,pushV | | |
| | [] | move r2,pushV | | |
| | (*a*): *late* | | (*b*): *early* | |

Figure 27: Late and early flushing of the V stack.

and also to take advantage of the 3-operand instructions, it would be better to push items onto the stack as soon as possible if they would be flushed later on anyway. This is shown in figure 27(b); the resulting target code sequence is both shorter and require one register less than the corresponding code in figure 27(a).

For items in the simulated S stack the situation is not quite as complicated. There is no harm in delaying a flush of the simulated S stack since so far there has been no need for temporary pointer registers — all references in the simulated S stack are either a global function constant $\$f$, a reference to the heap via Hp, a reference to a variable in the S stack, e.g. $n$(Sp), or the special register r0. Nor can we shorten the instruction stream by earlier flushing of the S stack, as we could with the V stack; the reason it is possible for the V stack is the 3-operand instructions. Nevertheless, for uniformity if nothing else, we will flush also the simulated S stack as early as possible, in exactly the same manner as for the simulated V stack.

## 4.3 Looking ahead to see how operands are used

For each G-machine instruction translated to target code we want to know whether the result of the operation is going to be used again before the end of the basic block. If that is the case then we put the result in an as convenient place as possible for later access, e.g. in a register, and put the register reference or whatever on the simulated stack. Otherwise the operand is pushed immediately onto the concrete stack.

To obtain this information, we now develop a measure $U(c)$, an non-negative integer, being the number of operands that the G-code sequence $c$ will consume from the top of the abstract stack before the end of $c$. In the following discussion, we will only look at the V stack, with instructions PUSHBASIC and SUB operating on it.

Now consider generating target code for SUB.$c$, i.e., a SUB instruction followed by a G-code sequence $c$. If $U(c) > 0$ the result of the SUB instruction is going to be used later on, before the end of the basic block. In this case we should emit a target code instruction `sub` $a$,$b$,$rr$, and push $rr$ onto the simulated V stack. If on the other hand $U(c) = 0$ then the result of the subtraction will not be used before the end of the basic block. In this case we should emit `sub` $a$,$b$,`pushV`, and leave the simulated V stack unchanged (i.e. empty)—this implements the early flushing exemplified in figure 27(b). Similarly, for the PUSHBASIC $x$ instruction we either either push $\$x$ onto the simulated V stack if $U(c) > 0$, or emit an instruction `move $x,pushV` if $U(c) = 0$.

$U(c)$ can be computed as follows:

$$\begin{aligned} U(\emptyset) &= 0 \\ U(\mathsf{PUSHBASIC}\ x.c) &= dec(U(c)) \\ U(\mathsf{SUB}.c) &= 2 + dec(U(c)) \\ &\quad \text{where } dec(n) = \begin{cases} 0 & \text{if } n = 0 \\ n-1 & \text{if } n > 0 \end{cases} \end{aligned}$$

Guided by $U(c)$ in this manner, the intention is to have the simulated V stack empty at the end of the basic block, and we must convince ourselves that this will always be the case. Now consider what would happen if, before generating code for $c$, the depth of the simulated stack $d$ were greater than $U(c)$. Even if the target code for all of $c$ would push their results onto the concrete V stack, the simulated V stack could not be empty, instead it would have a non-zero depth of $d - U(c)$. Therefore, the code generation rules will obey the condition $d \leq U(c)$ at all times, provided that the simulated stack is empty at the beginning of the basic block. Further, our method requires that the simulated stack corresponds to an initial segment of the abstract stack. Therefore, the simulated stack must be empty when operands are pushed onto the concrete stack; the condition $d \leq U(c)$ ensures this.

We are now going to change the representation of the number $U(c)$ into a more convenient (and more illustrative) form; it also allow some generalisation, described later. This new form can also be extended to deal with register allocation of function arguments, in a manner described later. The new version of $U(c)$ has the value of a list of elements **use** and **push**. A number $n$ corresponds to $n$ **use** elements prepended in front of an infinite list of **push** elements. $U(c)$ can now be defined as follows:

$$
\begin{aligned}
U(\emptyset) &= \textit{allpush} \text{ where } \textit{allpush} = \textbf{push}.\textit{allpush} \\
U(\mathsf{PUSHBASIC}\ x.c) &= \textit{tl}(U(c)) \\
U(\mathsf{SUB}.c) &= \textbf{use}.\textbf{use}.\textit{tl}(U(c)) \\
&\quad \text{where } \textit{tl}(x.l) = l
\end{aligned}
$$

We will call such a list a ***use info stack***. Each element in a use info stack relates to a certain element in the simulated or concrete stack. A **use** entry indicates that the corresponding element of the abstract stack will be used later on, and should thus be in the simulated stack; a **push** entry indicates that the corresponding stack element will not be used before the end of the basic block and so should be pushed onto the concrete stack.

Thus, in general for each G-instruction the target code generator now takes two use info stacks and two simulated stacks, and produces two new simulated stacks and target code instructions. The new simulated stacks are used when generating target code for the next G-instruction — i.e. this information is propagated forwards in the G-code stream. The new use info stacks, however, are passed upwards and are used when generating target code for the preceding G-instruction — i.e. this information is propagated backwards in the code stream. In essence, our new target code generator is a two pass algorithm, processing a basic block at a time. In the first scan, backwards, the use info stacks are generated. In the second scan, forwards, the target code is generated based on the use info stacks and the simulated stacks.

Again, for our introductory example we will consider a sequence of PUSHBASIC and SUB instructions, and we will only look at the simulated V stack and its corresponding use info stack. For the SUB instruction the use info stack stack is used in the following manner when generating target code. First assume the topmost entry in the use info stack is **push**. This indicates that the result *must* be pushed onto the concrete V stack. The proper target code instruction to emit is then `sub` *a*,*b*,`pushV`, where *a* and *b* are the two source operands. The **push** entry is removed and replaced by two **use** entries, hinting to the translation of the preceding instruction that the SUB instruction pops two operands and wants them for as fast access as possible.

If on the other hand the topmost use info stack entry is **use** this indicates that the result is going to be used again before the end of the basic block, and it is apt to put the result in a machine register, and the instruction to emit is `sub` *a*,*b*,`r`*r* where *r* is a newly allocated register. Again, the entry on the use info stack is removed and replaces by two **use** entries.

The situation is much simpler for the PUSHBASIC instruction. Figure 28 summarizes the rules for SUB and PUSHBASIC. In the second rule for SUB, *r* is a suitably chosen scratch register. In each case the resulting use info stack is shown above the old one, whereas the resulting simulated stack is shown below the old one. Figure 29 shows an example of target code generation using use info stacks.

| *G-code* | *use V info stack* | *sim. V stack* | *target code* |
|---|---|---|---|
| | **use.use.**··· | *a.b.*··· | |
| SUB | **push.**··· | ··· | `sub` *a*,*b*,`pushV` |
| | **use.use.**··· | *a.b.*··· | |
| SUB | **use.**··· | `r`*r*.··· | `sub` *a*,*b*,`r`*r* |
| | ··· | ··· | |
| PUSHBASIC *x* | **push.**··· | ··· | `move $`*x*,`pushV` |
| | ··· | ··· | |
| PUSHBASIC *x* | **use.**··· | `$`*x*.··· | |

Figure 28: The target code generation cases for SUB and PUSHBASIC.

| *G-code* | *use info stack* | *sim. V stack* | *target code* |
|---|---|---|---|
| | ... | [] | |
| PUSHBASIC 5 | **use.**··· | [$5] | |
| PUSHBASIC 4 | **use.use.**··· | [$4,$5] | |
| SUB | **push.**··· | [] | `sub $4,$5,pushV` |
| PUSHBASIC 3 | **use.push.**··· | [ $3] | |
| PUSHBASIC 2 | **use.use.push.**··· | [$2,$3] | |
| SUB | **use.push.**··· | [r1] | `sub $4,$5,r1` |
| PUSHBASIC 1 | **use.use.push.**··· | [$1,r1] | |
| SUB | **push.push.**··· | [] | `sub $1,r1,pushV` |

Figure 29: An example of using a V use info stack.

The use info stacks can be extended to deal with register allocation of function arguments in the following manner. We introduce a new use info stack element: **reg** $r$, indicating that the corresponding result of an operation must be put in register $r$. For example, suppose that the register allocator has determined that the two arguments of $f$, say, are to be passed in registers `r3` and `r4`; then the use info stack to be passed upwards from e.g. a tail call to $f$ is **reg** 3.**reg** 4.**push.push.**···.

## 4.4 Programming the code generator

In the target code generator just outlined, information flows both backwards and forwards in the G-code stream. *Attribute grammars* [Knu68] provide a convenient formalism for expressing this kind of information flow. We now describe how the target code generator can be expressed as an attribute grammar. A complete attribute grammar for generating code for all the G-code instructions considered in this paper can be found in appendix E.

Attribute grammars deal with context free grammars and parse trees; they express how to decorate the interior nodes of the parse tree with *semantic values*. Our idea is to regard the a G-code stream as a parse tree for a grammar generating the sequences of G-machine instructions, as shown in figure 30. The G-code instructions correspond to the terminal symbols of the grammar.

We will write 'grammar rules' as for instance

$$C = \text{PUSH } n \text{ . } c$$

i.e. in a manner reminiscent of pattern matching in functional languages. The backwards (i.e. upwards) flow of information corresponds to *synthesized* attributes. In our case the backwards propagating information is the two use info stacks. Also the generated target code will be handled as a synthesized attribute. For each 'grammar rule' we must define the synthesized attribute of the left hand 'nonterminal' symbol $C$; they will be written as $C$!*code*, $C$!*Suse* and $C$!*Vuse*. Similarly, the two simulated stacks constitute the forwards flow of information, and they will appear as *inherited* attributes; they must be defined for the right hand 'nonterminal' $c$. They will be written as $c$!*Ssim* and $c$!*Vsim*.

We will now describe rules for some of the G-machine instructions in detail.

### 4.4.1 Attribute grammar rules for PUSHFUN

The 'grammar rule' for the PUSHFUN case is

$$C = \text{PUSHFUN } f \text{ . } c$$

*Grammar:*

$$
\begin{array}{lcl}
\text{gcode} & \rightarrow & \epsilon \\
 & | & \textsf{EVAL}\ \text{gcode} \\
 & | & \textsf{GET}\ \text{gcode} \\
 & | & \textsf{SUB}\ \text{gcode} \\
 & \vdots & \\
 & \textit{etc.} &
\end{array}
$$

*A G-code stream:*

PUSH 1; EVAL; GET

*The corresponding parse tree:*

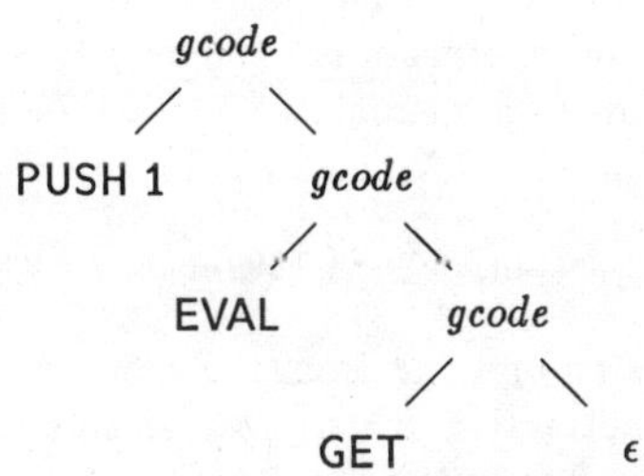


Figure 30: Regarding the G-code stream as a context free grammar.

with $f$ standing for a global function identifier i.e. the G-code. The S use info stack (i.e. list) is split into the top element and the rest:

$$u.C!Suse = c!Suse$$

i.e., $u$ becomes the head and $C!Suse$ becomes the tail for $c!Suse$.

A note about notation: in the attribute grammar rules code sequences and addressing mode operands will be written as strings, eg. "`move r0,pushS`". Subsitutions in such strings will be written by simply placing a variable in italic in such a string: eg. "`move` *a*`,pushS`" which has the value "`move r0,pushS`" if $a$="`r0`". Semicolon will be used for concatenation of such code sequences.

Next, we scrutinize $u$ to see if code is to emitted for the push, or if the operand is to be pushed onto the simulated stack instead.

$$
\begin{array}{lcl}
C!code & = & \textbf{case}\ u\ \textbf{in}\ \text{use: } c!code \parallel \text{push: ``}\texttt{move \$C}f\texttt{,pushS}\text{''};\ c!code\ \textbf{end} \\
c!Ssim & = & \textbf{case}\ u\ \textbf{in}\ \text{use: ``}\texttt{\$C}f\text{''}.C!Ssim \parallel \text{push: } C!Ssim\ \textbf{end}
\end{array}
$$

The above pattern is so common that we will define a function *newS* for it, with the following two cases:

$$
\begin{array}{l}
\text{newS use } a\ c\ S = (c,\ a.S) \\
\text{newS push } a\ c\ S = (\text{``}\texttt{move } a\texttt{,pushS}\text{''};c\ ,\ S)
\end{array}
$$

Using newS, $c!code$ and $c!Ssim$ can now be defined by

$$(C!code,\ c!Ssim) = \text{newS } u\ \text{``}\texttt{\$}f\text{''}\ c!code\ C!Ssim$$

The rest of the attributes are just trivially propagated up or down. The entire attribute grammar rule for PUSHFUN thus reads as follows.

C =PUSHFUN $f$ . c :

$$\begin{aligned} u.C!Suse &= c!Suse \\ C!Vuse &= c!Vuse \\ (C!code,c!Ssim) &= \text{newS } u \text{ "\$C}f\text{" } c!code\ C!Ssim \\ c!Vsim &= C!Vsim \end{aligned}$$

### 4.4.2 Attribute grammar rules for SUB

The V use info stack is split into the top element and the rest:

$$u.U = c!Vuse$$

Two source operands are to be chosen for the `sub` instruction. They are to be taken either from the simulated V stack if it is non-empty, of from the concrete V stack using a `popV` operand. This is conveniently handled by letting an empty simulated V stack be represented by an infinite list of `popV` operands, with the ordinary simulated V stack put on top of it. Thus

$$a.b.V = C!Vsim$$

will bind to $a$ and $b$ either `popV` or an operand from the ordinary simulated V stack. Further, we may want to put the result of the subtraction in a register, so an unused register has to be allocated. A register is unused if it is not referred to in the simulated stacks, so the register allocation function simply picks one that is not referred to there.

$$r_d = \text{getdatareg } V\ C!Ssim$$

The entire attribute grammar rule for SUB now reads as below.

$C$ = SUB . $c$ :

$$\begin{aligned} u.U &= c!Vuse \\ a.b.V &= C!Vsim \\ r_d &= \text{getdatareg } V\ C!Ssim \\ C!Vuse &= \text{use.use.}U \\ C!Suse &= c!Suse \\ (C!code,c!Vsim) &= \textbf{case } u \textbf{ in} \\ &\quad \text{use: ("sub } a,b,r_d\text{", } r_d.V) \\ &\quad \text{push: ("sub } a,b,\text{pushV", "popV".}V) \\ &\quad \textbf{end} \\ c!Ssim &= C!Ssim \end{aligned}$$

Here we cannot use the function newS, because it would have emitted a `move` intruction for the push onto the V stack; we want that to be done in the `sub` instruction.

### 4.4.3 Attribute grammar rules for EVAL

Before calling EVAL we require the simulated V and S stack stack to be empty, except that we want the top of the abstract S stack in `r0`. $C!Vuse$ is therefore set to the value of an infinite list of **push** objects. The same is done with $C!Suse$, except that we did not want the top of the abstract S stack to be in the concrete stack; therefore the top of $C!Suse$ is **use**. When we move this operand into `r0` using the function intoreg0 there is no risk that r0 is used for something else, since at this point the rest of the simulated stacks are empty. Also after the call the simulated stacks are empty — this we represent with infinite lists of popV and popS respectively — except for the top of the abstract S stack which is in r0.

$C$ =EVAL . $c$ :

$$
\begin{aligned}
u.U_s &= c!Suse \\
a.S &= C!Ssim \\
C!Vuse &= \text{allpush} \\
C!Suse &= \text{use.allpush} \\
C!code &= \text{intoreg0 } a \text{ ; ``}\texttt{calltab eval,0(r0)}\text{'' ; } c' \\
(c', c!Ssim) &= \text{newS } u \text{ ``r0'' } c!code \text{ allpopS} \\
c!Vsim &= \text{allpopV}
\end{aligned}
$$

### 4.4.4 Programming the attribute evaluator

Implementing the compiler in a lazy functional language, there is a simple way to perform the attribute evaluation. Attributes can be evaluated in a single pass over the data structure, in this case the G-machine code, with a function taking the inherited attributes as arguments and returning a tuple of the synthesized attributes. Using this method here gives the target code generator the following structure.

```
mcode (PUSHFUN f . c) CSsim CVsim =
    letrec (ccode,cSuse,cVuse)  =  mcode c cSsim cVsim
                       u.CSuse  =  cSuse
                         CVuse  =  cVuse
                 (Ccode,cSsim)  =  newS u (idlit f) ccode CSsim
                         cVsim  =  CVsim
    in (Ccode, CSuse, CVuse)
mcode (EVAL . c) CSsim CVsim =
  ⋮
```

Attribute evaluation using lazy functional languages is described in further detail in [Joh86].

# 5 Concluding remarks

In this paper we have described two different methods for generating target code from G-machine code. The first method described consists of generating naïve target code which explicitly manipulates the stacks. As is evident from the size of appendix D, such a code generator is almost trivial to implement, and is the method of choice for getting an implementation running quickly. A simple peep hole code improver can remove a fair amount of the most glaring inefficiencies in the form of redundant pushes and pops.

The second code generation method described is more systematic as to the avoidance redundant pushes and pops in the target code. The resulting algorithm can be expressed neatly as an attribute grammar over the G-machine code list. Further, attribute evaluation can be performed in a straightforward way in a lazy functional language by rewriting the attribute grammar into an evaluation function.

The attribute-based target code generator allocates machine registers only for a very special purpose: as storage locations for intermediate values of arithmetic expressions. We may speculate about the possible benefits of a more systematic use of machine registers, e.g. for function arguments and other variables (instead of having them in the concrete S stack). For conventional languages, using global register allocation, a reduction of execution time of 50% is not unusual (see e.g. [CH84]). On one hand, compared to functional languages prospects look better for us because functional languages have no side effects, allowing us to put variables into registers more often; in an imperative language a possible side effect on a variable e.g. via a pointer makes it necessary to have that variable in memory. On the other hand, when building a suspended function application, its argument must put in memory (i.e. the heap) anyway until the suspension is being evaluated. Further, the G-machine code for each function is usually well-sprinkled with EVAL instructions, so not knowing what function code each EVAL might cause to execute, worst-case assumptions must be made about how EVAL corrupts registers etc. Therefore, it seem likely that attempting such register allocation would be rather pointless without also accompanying it with data-flow analysis (which might determine the possible functions executed due to a certain EVAL), or strictness analysis (which could eliminate many EVALs altogether).

# 6 Acknowledgements

The LML compiler has been developed together with Lennart Augustsson; in particular, he implemented the code generators for M68000, the NS32016 and the IBM 360. Thanks are also due to the members of the Programming Methodology Group for help and encouragement of this work; in particular I would like to thank Kent Karlsson and Phil Wadler for careful proofreading and numerous comments on this paper. The LML compiler project has been financially supported in part by the Swedish Board for Technical Development (STU).

# References

[Aug84] L. Augustsson. A compiler for lazy ML. In *Proceedings of the 1984 ACM Symposium on Lisp and Functional Programming*, pages 218–227, Austin, 1984.

[Aug85] L. Augustsson. Compiling pattern matching. In *Proceedings 1985 Conference on Functional Programming Languages and Computer Architecture*, Nancy, France, 1985.

[CCM85] G. Cousineau, P.L. Curien, and M. Mauny. The categorical abstract machine. In *Proceedings 1985 Conference on Functional Programming Languages and Computer Architecture*, pages 50–64, Nancy, France, 1985.

[CH84] F. Chow and J. Hennessy. Register allocation by priority-based coloring. In *Proceedings of the SIGPLAN '84 Symposium on Compiler Construction*, pages 222–232, Montreal, 1984.

[Fri76] D. P. Friedman. *Cons should not evaluate its arguments*, pages 257–284. Edinburgh University Press, 1976. In the book *Automata, languages and Programming.*

[Hen80] P. Henderson. *Functional Programming.* Prentice-Hall, 1980.

[HM76] P. Henderson and J. H. Morris. A lazy evaluator. In *3'rd conference on The Principles of Programming Languages*, pages 95–103, Atlanta, Georgia, January 1976.

[Hug82] J. Hughes. Super combinators - a new implementation method for applicative languages. In *Proceedings of the 1982 ACM Symposium on Lisp and Functional Programming*, pages 1–10, Pittsburgh, 1982.

[Joh84] T. Johnsson. Efficient compilation of lazy evaluation. In *Proceedings of the SIGPLAN '84 Symposium on Compiler Construction*, pages 58–69, Montreal, 1984.

[Joh85] T. Johnsson. Lambda lifting: transforming programs to recursive equations. In *Proceedings 1985 Conference on Functional Programming Languages and Computer Architecture*, Nancy, France, 1985.

[Joh86] T. Johnsson. *Attribute grammars and functional programming.* Draft paper, Programming Methodology Group, Chalmers University of Technology, 1986.

[Knu68] Donald E. Knuth. Semantics of context-free languages. *Math. Systems Theory*, 2:127–145, 1968.

[Lan64] P. J. Landin. The mechanical evaluation of expressions. *Computer Journal*, (6):308–320, 1964.

[LS78] H. Lindström and J. Skansholm. *Code Generation in the MARY compiler for the IBM 360.* Technical Report 78.1, Department of Computer Science, Chalmers University of Technology, Göteborg, Sweden, 1978.

[Pey87] S. Peyton-Jones. *The Implementation of Functional Languages.* Prentice Hall, 1987.

[Tur79] D. A. Turner. A new implementation technique for applicative languages. *Software - Practice and Experience*, 9:31–49, 1979.

[Tur81] D. A. Turner. The semantic elegance of applicative languages. In *Proceedings 1981 Conference on Functional Languages and Computer Architecture*, Wentworth-by-the-Sea, Portsmouth, New Hampshire, 1981.

[WJW*75] W. Wulf, R.K. Johnsson, C.B. Weinstock, S.O. Hobbs, and C.M. Geschke. *The Design of an Optimizing Compiler.* Elsevier Computer Science Library, New York, 1975.

# Appendixes

## A G-machine compilation rules

In the rules given below, $r$ denotes an 'environment', a mapping telling where on the S stack the local variables are found. $n$ denotes the current stack depth. $f$ denotes a global function, $x$ and $v$ denotes local variables, and $i$ denotes any identifier.

---

$\mathcal{F}$ gives code for an entire function definition:

$\mathcal{F}[\![f\ x_1 \cdots x_n = e]\!]$ = FUNSTART $f\ n$; $\mathcal{R}[\![e]\!]\ [x_1{=}n, \cdots x_n{=}1]\ n$

---

$\mathcal{R}[\![e]\!]\ r\ n$ gives code to return the value of $e$, i.e. compute the value of $e$, update update the root apply node of the application, and return from EVAL. $\mathcal{R}$ has the particular property that it turns tail calls into jumps in the G-machine code.

$\mathcal{R}[\![$**case** $e$ **in** $\cdots \|c_{ki}\!: v_i \cdots v_{mi}\!: e_i \| \cdots \|v_d\!:\!e_d]\!]\ r\ n =$
    $\mathcal{E}[\![e]\!]\ r\ n$; PUSH 0; CASE $(k_1,l_1) \cdots l_d$;
        ⋮
    LABEL $l_i$; PUSH 0; SPLIT $k_i\ m_i$; $\mathcal{R}[\![e_i]\!]\ r[v_m{=}n{+}2, \cdots v_1{=}n{+}1{+}m_i]\ (n{+}1{+}m_i)$;

$\vdots$

LABEL $l_d$; $\mathcal{R}[\![e_d]\!]\ r[v_d{=}n{+}1]\ (n{+}1)$

Labels $l_1$, $l_2$ etc and $l_d$ are assumed unique

$\mathcal{R}[\![$**letrec** $v_1{=}e_1$ **and** $\cdots\ v_m{=}e_m$ **in** $e]\!]\ r\ n =$

$Cletrec[\![v_1{=}e_1, \cdots, v_m{=}e_m]\!]\ r\ n$; $\mathcal{R}[\![e]\!]\ r\ n$; $r[v_1{=}n{+}1, \cdots v_m{=}n{+}m]\ (n{+}m)$

$\mathcal{R}[\![i\ e_1 \cdots e_m]\!]\ r\ n{=}\ \mathcal{S}[\![i\ e_1 \cdots e_m]\!]\ r\ n$; JFUN $m$

*otherwise:*

$\mathcal{R}[\![e]\!]\ r\ n = \mathcal{E}[\![e]\!]\ r\ n$; UPDATE $(n{+}1)$; POP $n$; RET

---

$\mathcal{S}[\![i\ e_1 \cdots e_m]\!]\ r\ n$ rearranges the stack for a tail call. $\mathcal{S}$ is rather complex, so we refrain from giving it here.

---

$\mathcal{E}[\![e]\!]\ r\ n$ gives code that computes the value of $e$ and leaves a pointer to the result on top of the S stack.

$\mathcal{E}[\![$**case** $e$ **in** $\cdots \|c_{ki}\colon v_i \cdots v_{mi}\colon e_i \| \cdots \|v_d{:}e_d]\!]\ r\ n =$

$\mathcal{E}[\![e]\!]\ r\ n$; PUSH 0; CASE $(k_1,l_1) \cdots l_d$;

$\vdots$

LABEL $l_i$; PUSH 0; SPLIT $k_i\ m_i$; $\mathcal{E}[\![e_i]\!]\ r[v_m{=}n{+}2, \cdots v_1{=}n{+}1{+}m_i]\ (n{+}1{+}m_i)$;
MOVE $(m_i{+}2)$; POP $(m_i{+}1)$; JMP $l_{out}$;

$\vdots$

LABEL $l_d$; $\mathcal{E}[\![e_d]\!]\ r[v_d{=}n{+}1]\ (n{+}1)$; MOVE $(m_i{+}2)$; POP $(m_i{+}1)$;
LABEL $l_{out}$

Labels $l_1$, $l_2$ etc, $l_d$ and $l_{out}$ are assumed unique

$\mathcal{E}[\![$**failmatch**$]\!]\ r\ n =$ POP ($n$-*failstackdepth*); JMP *faillabel*

*faillabel* is the label of the closest surrounding default entry.
*failstackdepth* is the S stack depth at *faillabel*.
(see also remark below)

$\mathcal{E}[\![x]\!]\ r\ n =$ PUSH $n$-$r(x)$; EVAL

$\mathcal{E}[\![f]\!]\ r\ n =$ PUSHFUN $f$

$\mathcal{E}[\![$**letrec** $v_1{=}e_1$ **and** $\cdots\ v_m{=}e_m$ **in** $e]\!]\ r\ n =$

$Cletrec[\![v_1{=}e_1, \cdots, v_m{=}e_m]\!]\ r\ n$; $\mathcal{E}[\![e]\!]\ r\ n$; $r[v_1{=}n{+}1, \cdots v_m{=}n{+}m]\ (n{+}m)$; MOVE $(m{+}1)$; POP $m$

$\mathcal{E}[\![constr_k\ e_1 \cdots e_m]\!]\ r\ n = \mathcal{C}[\![constr_k\ e_1 \cdots e_m]\!]\ r\ n$

$\mathcal{E}[\![i\ e_1 \cdots e_m]\!]\ r\ n =$ ALLOC; $\mathcal{C}[\![e_m]\!]\ r\ (n{+}1)$; $\cdots\ \mathcal{C}[\![e_1]\!]\ r\ (n{+}m)$; $\mathcal{C}[\![i]\!]\ r\ (n{+}1{+}m)$; CALLFUN $m$

$\mathcal{E}[\![e_1\text{-}e_2]\!]\ r\ n = \mathcal{B}[\![e_1\text{-}e_2]\!]\ r\ n$; MKINT

---

$\mathcal{B}[\![e]\!]\ r\ n$ gives code that computes the ***basic value*** of $e$ and leaves the result on top of the basic value stack V (see below).

$\mathcal{B}[\![$**case** $e$ **in** $\cdots \|c_{ki}\colon v_i \cdots v_{mi}\colon e_i \| \cdots \|v_d{:}e_d]\!]\ r\ n =$

$\mathcal{E}[\![e]\!]\ r\ n$; PUSH 0; CASE $(k_1,l_1) \cdots l_d$;

$\vdots$

LABEL $l_i$; PUSH 0; SPLIT $k_i\ m_i$; $\mathcal{B}[\![e_i]\!]\ r[v_m{=}n{+}2, \cdots v_1{=}n{+}1{+}m_i]\ (n{+}1{+}m_i)$;
POP $(m_i{+}1)$; JMP $l_{out}$;

$\vdots$

LABEL $l_d$; $\mathcal{B}[\![e_d]\!]\ r[v_d{=}n{+}1]\ (n{+}1)$; POP $(m_i{+}1)$;
LABEL $l_{out}$

Labels $l_1$, $l_2$ etc, $l_d$ and $l_{out}$ are assumed unique

$\mathcal{B}[\![$**failmatch**$]\!]\ r\ n =$ POP ($n$-*failstackdepth*); JMP *faillabel*

*faillabel* is the label of the closest surrounding default entry.
*failstackdepth* is the S stack depth at *faillabel*.
(see also remark below)

$\mathcal{B}[\![\textbf{letrec } v_1{=}e_1 \textbf{ and } \cdots v_m{=}e_m \textbf{ in } e]\!]\ r\ n =$

$\quad Cletrec[\![v_1{=}e_1, \cdots, v_m{=}e_m]\!]\ r\ n;\ \mathcal{B}[\![e]\!]\ r\ n;\ r[v_1{=}n{+}1, \cdots v_m{=}n{+}m]\ (n{+}m);$ POP $m$

$\mathcal{B}[\![constr_k]\!]\ r\ n =$ PUSHBASIC $k$

$\mathcal{B}[\![e_1\text{-}e_2]\!]\ r\ n = \mathcal{B}[\![e_1]\!]\ r\ n;\ \mathcal{B}[\![e_2]\!]\ r\ n;$ SUB

*Otherwise*

$\mathcal{B}[\![e]\!]\ r\ n = \mathcal{E}[\![e]\!]\ r\ n;$ GET

---

$\mathcal{C}[\![e]\!]\ r\ n$ gives code that constructs the graph for $e$ and leaves a pointer to it on top of the S stack.

$\mathcal{C}[\![x]\!]\ r\ n =$ PUSH $(n\text{-}r(x))$

$\mathcal{C}[\![f]\!]\ r\ n =$ PUSHFUN $f$

$\mathcal{C}[\![constr_k\ e_1 \cdots e_m]\!]\ r\ n = \mathcal{C}[\![e_m]\!]\ r\ n; \cdots \mathcal{C}[\![e_1]\!]\ r\ (n{+}m\text{-}1);$ CONSTR $k\ m$

$\mathcal{C}[\![\textbf{letrec } v_1{=}e_1 \textbf{ and } \cdots v_m{=}e_m \textbf{ in } e]\!]\ r\ n =$

$\quad Cletrec[\![v_1{=}e_1, \cdots, v_m{=}e_m]\!]\ r\ n;\ \mathcal{C}[\![e]\!]\ r\ n;\ r[v_1{=}n{+}1, \cdots v_m{=}n{+}m]\ (n{+}m);$ MOVE $(m{+}1)$; POP $m$

$\mathcal{C}[\![i\ e_1 \cdots e_m]\!]\ r\ n = \mathcal{C}[\![e_m]\!]\ r\ n; \cdots \mathcal{C}[\![e_1]\!]\ r\ (n{+}m\text{-}1);\ \mathcal{C}[\![i]\!]\ r\ (n{+}m);\ \underbrace{\text{MKAP}; \cdots \text{MKAP}}_{m \text{ times}}$

$Cletrec[\![v_1{=}e_1, \cdots, v_m{=}e_m]\!]\ r\ n =$

$\quad \underbrace{\text{ALLOC}; \cdots \text{ALLOC}}_{m \text{ times}};\ \mathcal{C}[\![e_1]\!]\ r'\ n';$ UPDATE $m$; $\cdots\ \mathcal{C}[\![e_m]\!]\ r'\ n';$ UPDATE 1

where $r' = r[v_1{=}n{+}1, \cdots v_m{=}n{+}m]$ and $n' = n{+}m$.

---

Remark: In the code for **failmatch** the two entities *faillabel* and *failstackdepth* are to be considered as extra parameters to the G-code generation schemes, but they have been left out in order not to clutter up the above table. These parameters are trivially propagated down everywhere, except in the handling of the case expression. Below we show this for $\mathcal{B}[\![\textbf{case}\cdots]\!]$; $\mathcal{E}[\![\textbf{case}\cdots]\!]$ is similar.

$\mathcal{B}[\![\textbf{case } e \textbf{ in } \cdots \|c_{ki}{:}\ v_i \cdots v_{mi}{:}\ e_i \| \cdots \|v_d{:}e_d]\!]\ r\ n\ faillabel\ failstackdepth =$

$\quad \mathcal{E}[\![e]\!]\ r\ n;$ PUSH 0; CASE $(k_1,l_1) \cdots l_d;$

$\quad \vdots$

$\quad$ LABEL $l_i$; PUSH 0; SPLIT $k_i\ m_i;\ \mathcal{B}[\![e_i]\!]\ r[v_m{=}n{+}2, \cdots v_1{=}n{+}1{+}m_i]\ (n{+}1{+}m_i)\ l_d\ (n{+}1)$

$\quad$ POP $(m_i{+}1)$; JMP $l_{out}$;

$\quad \vdots$

$\quad$ LABEL $l_d$; $\mathcal{B}[\![e_d]\!]\ r[v_d{=}n{+}1]\ (n{+}1)\ faillabel\ failstackdepth$; POP $(m_i{+}1)$;

$\quad$ LABEL $l_{out}$

# B State transition rules for G-machine instructions

1. $\langle o,$ PRINT$.c,\ n.s,\ v,\ G[n = \text{INT } i],\ E,\ D\rangle \Rightarrow \langle o;i,\ c,\ s,\ v,\ G[n = \text{INT } i],\ E,\ D\rangle$
2. $\langle o,$ PRINT$.c,\ n.s,\ v,\ G[n = \text{CONSTR } k\ n_1 \cdots n_m],\ E,\ D\rangle \Rightarrow$
   $\langle o,\ \underbrace{\text{EVAL.PRINT.}\cdots\text{.EVAL.PRINT}}_{m \text{ times}}.c,\ n_1.\cdots.n_m.s,\ v,\ G[n = \text{CONSTR } k\ n_1 \cdots n_m],\ E,\ D\rangle$
3. $\langle o,$ EVAL$.c,\ n.s,\ v,\ G[n = \text{AP } n_1\ n_2],\ E,\ D\rangle \Rightarrow$
   $\langle o,$ UNWIND$.(),\ n.(),\ v,\ G[n = \text{AP } n_1\ n_2],\ E,\ (c,s).D\rangle$
4. $\langle o,$ EVAL$.c,\ n.s,\ v,\ G[n = \text{INT } i],\ E,\ D\rangle \Rightarrow \langle o,\ c,\ n.s,\ v,\ G[n = \text{INT } i],\ E,\ D\rangle,$
   *similarly for nodes* CONSTR $k\ n_1 \cdots n_m$ *and* FUN $f$.
5. $\langle o,$ UNWIND$.(),\ n.s,\ v,\ G[n = \text{AP } n_1\ n_2],\ E,\ D\rangle \Rightarrow \langle o,$ UNWIND$.(),\ n_1.n.s,\ v,\ G[n = \text{AP } n_1\ n_2],\ E,\ D\rangle$
6. $\langle o,$ UNWIND$.(),\ n_0.n_1 \cdots n_k.s,\ v,\ G[n_0 = \text{FUN } f,$
   $n_1 = \text{AP } n_1'\ n_1'', \cdots n_k = \text{AP } n_k'\ n_k''],\ E[f = (k,c)],\ D\rangle \Rightarrow$
   $\langle o,\ c,\ n_1'' \cdots n_k''.n_k.s,\ v,\ G[n_0 = \text{FUN } f,\ n_1 = \text{AP } n_1'\ n_1'', \cdots n_k = \text{AP } n_k'\ n_k''],\ E[f = (k,c')],\ D\rangle$

7. $\langle o, \text{UNWIND}.(), n_0.n_1 \cdots n_k.(), v, G[n_0 = \text{FUN } f], E[f = (a,c')], (c',s').D\rangle$ *and* $k < a \Rightarrow$
$\langle o, c', n_k.s', v, G[n_0 = \text{FUN } f], E[f = (k,c')], D\rangle$

8. $\langle o, \text{JFUN } m.c, n.s, v, G[n = \text{FUN } f], E[f = (m,c')], D\rangle \Rightarrow$
$\langle o, c', s, v, G[n = \text{FUN } f], E[f = (m,c')], D\rangle$

9. $\langle o, \text{JFUN } m.c, n.s, v, G, E, D\rangle \Rightarrow$
$\langle o, \underbrace{\text{MKAP}.\cdots.\text{MKAP}}_{m \text{ times}}.\text{UPDATE 1.RET}.(), s, v, G, E, D\rangle$ *otherwise*

10. $\langle o, \text{CALLFUN } m.c, n_0.n_1.\cdots.n_m.s, v, G, E, D\rangle \Rightarrow \langle o, \text{JFUN } m.c, n_0.n_1.\cdots.n_m.(), v, G, E, (c,s).D\rangle$

11. $\langle o, \text{RET}.c, n.(), v, G[n = \text{INT } i], E, (c',s').D\rangle \Rightarrow \langle o, c', n.s', v, G[n = \text{INT } i], E, D\rangle$,
*similarly for* $n = \text{CONSTR } k\ n_1 \cdots n_m$.

12. $\langle o, \text{RET}.c, n.s, v, G[n = \text{AP } n_1\ n_2], E, D\rangle \Rightarrow \langle o, \text{UNWIND}.(), n.s, v, G[n = \text{AP } n_1\ n_2], E, D\rangle$
*similarly for* $n = \text{FUN } f$.

13. $\langle o, \text{PUSH } m.c, n_0.\cdots.n_m.s, v, G, E, D\rangle \Rightarrow \langle o, c, n_m.n_0.\cdots n_m.s, v, G, E, D\rangle$

14. $\langle o, \text{PUSHFUN } f.c, s, v, G, E, D\rangle \Rightarrow \langle o, c, n'.s, v, G[n' = \text{FUN } f], E, D\rangle$

15. $\langle o, \text{PUSHBASIC } i.c, s, v, G, E, D\rangle \Rightarrow \langle o, c, s, i.v, G, E, D\rangle$

16. $\langle o, \text{SUB}.c, s, i_2.i_1.v, G, E, D\rangle \Rightarrow \langle o, c, s, (i_1 - i_2).v, G, E, D\rangle$

17. $\langle o, \text{MOVE } m.c, n_0.\cdots.n_{m-1}.n_m.s, v, G, E, D\rangle \Rightarrow \langle o, c, n_1.\cdots.n_{m-1}.n_0.s, v, G, E, D\rangle$

18. $\langle o, \text{POP } m.c, n_1.\cdots.n_m.s, v, G, E, D\rangle \Rightarrow \langle o, c, s, v, G, E, D\rangle$

19. $\langle o, \text{GET}.c, n.s, v, G[n = \text{INT } i], E, D\rangle \Rightarrow \langle o, c, s, i.v, G[n = \text{INT } i], E, D\rangle$

20. $\langle o, \text{UPDATE } m.c, n_0.\cdots n_m.s, v, G[n_0 = N_0\ , n_m = N_m], E, D\rangle \Rightarrow$
$\langle o, c, n_1.\cdots n_m.s, v, G[n_0 = N_0\ , n_m = N_0], E, D\rangle$

21. $\langle o, \text{LABEL } l.c, s, v, G, E, D\rangle \Rightarrow \langle o, c, s, v, G, E, D\rangle$

22. $\langle o, \text{JMP } l.\cdots \text{LABEL } l.c, s, v, G, E, D\rangle \Rightarrow \langle o, c, s, v, G, E, D\rangle$

23. $\langle o, \text{ALLOC}.c, s, v, G, E, D\rangle \Rightarrow \langle o, c, n'.s, v, G[n' = \text{HOLE}], E, D\rangle$

24. $\langle o, \text{MKAP}.c, n_1.n_2.s, v, G, E, D\rangle \Rightarrow \langle o, c, n'.s, v, G[n' = \text{AP } n_1\ n_2], E, D\rangle$

25. $\langle o, \text{MKINT}.c, s, i.v, G, E, D\rangle \Rightarrow \langle o, c, n'.s, v, G[n' = \text{INT } i], E, D\rangle$

26. $\langle o, \text{CASE } \cdots(k_i.l_i)\cdots.l_d.c, n.s, v, G[n = \text{CONSTR } k\ n_1 \cdots n_m], E, D\rangle \Rightarrow$
$\langle o, \text{JMP } l_i.c, s, v, G[n = \text{CONSTR } k\ n_1 \cdots n_m], E, D\rangle$ *if* $k = k_i$

27. $\langle o, \text{CASE } \cdots(k_i.l_i)\cdots.l_d.c, n.s, v, G[n = \text{CONSTR } k\ n_1 \cdots n_m], E, D\rangle \Rightarrow$
$\langle o, \text{JMP } l_d.c, s, v, G[n = \text{CONSTR } k\ n_1 \cdots n_m], E, D\rangle$ *otherwise*

28. $\langle o, \text{CONSTR } k\ m.c, n_1.\cdots.n_m.s, v, G, E, D\rangle \Rightarrow$
$\langle o, c, n'.s, v, G[n' = \text{CONSTR } k\ n_1 \cdots n_m], E, D\rangle$

29. $\langle o, \text{SPLIT } k\ m.c, n.s, v, G[n = \text{CONSTR } k\ n_1 \cdots n_m]\ E, D\rangle \Rightarrow$
$\langle o, c, n_1.\cdots.n_m.s, v, G[n = \text{CONSTR } k\ n_1 \cdots n_m]\ E, D\rangle$

# C The M code

The M code machine is an abstraction of conventional CPUs with registers, addressing modes, condition codes etc; it is similar to the VAX-11 architecture. M code instructions handle data in units of words, an entity assumed big enough to hold an integer or a pointer. The memory, i.e. the heap, the stacks and the special data segment, is assumed to be word-addressed. Addressing modes can have side-effects, so the evaluation order of the instruction operands are specified to be from left to right. In the lists below, $a$, $b$, $s$ and $d$ stand for addressing modes, $n$ and $m$ stand for numbers, and $r$ stands for a register number. See also section 3.3.

## Instructions:

| | |
|---|---|
| `move` *s*,*d* | Moves source operand $s$ to destination location $d$. |
| `sub` *a*,*b*,*d* | Computes $b - a$ and stores the result in $d$. Similarly for `add`, `mul` etc. |
| `comp` *a*,*b* | Compares operands, and sets condition code. It is always tested immediately in the following jump-conditional instruction. |
| `jlt` *l*, `jne` *l* *etc* | If condition code set by the preceding `comp` *a*,*b* instruction indicates that $a < b$ etc, then jump to global label $l$. When comparing adresses, it is assumed that the stacks grow towards low addresses and the heap towards high addresses. |
| `jmptab` *offset*,*n* | and |

| | |
|---|---|
| `calltab` *offset*, $a$ | A table-indexed jump or call, to implement a case mechanism on node tags (see section 3.5); *offset* is a symbolic name denoting an offset in a dispatch table, and $a$ contains the address of the table. In a call, the return address is pushed onto the V stack. *offset* is one of the names listed below: |

| | |
|---|---|
| `eval` | To implement the EVAL instruction. |
| `unwind` | To implement the unwind operation. |
| `jfun` | To perform a tail call or a 'direct' call. |
| `getconstrno` | To get the constructor number of a constructor. |
| `gc` | To copy a node during garbage collection. |
| `VFunwind` | The unwind-rearrange code for a FUN node. |
| `VFfuncode` | The function code associated to a FUN node. |
| `VFvapunwind` | The unwind-pushargs code for a VAP node (see section 3.7). |

| | |
|---|---|
| `case` $a$, $(k_1, l_1) \cdots (k_m, l_m), l_d$ | If the value of $a$ is $k_1$ the jump to label $l_1$, etc. If none of the constants $k_i$ is equal to $a$ then a jump is made to the default location $l_d$. $a$ is only computed once. |
| `jmp` $a$ | An ordinary jump to global label $a$. |
| `call` $a$ | An ordinary call to a global label $a$. The return address is pushed onto the V stack. |
| `return` | Returns by popping the return address from the V stack. |
| $l$: | A label in the code or data segment. |
| `.data` | Ordinarily M code instructions are emitted in a code segment. The `.data` directive specifies that the instructions that follows are to be put in a data segment instead. In this segment we only put `words` and labels. |
| `word` $a_1, a_2 \cdots$ | A secuence of word constants, $a_i$ is either a global label or a number. |
| `.code` | Switch back to the code segment. |

## Addressing modes:

| | |
|---|---|
| `Sp` | The S stack pointer register. |
| `pushS` | Push a word onto the S stack; can only be a destination operand. |
| `popS` | Pop a word from the S stack; can only be a source operand. |
| $n$`(Sp)` | The $n$'th S stack location relative to the top of the S stack. |
| `$`$n$`(Sp)` | A pointer to an S stack location. |
| `Vp` | The value pointer register. |
| `pushV` | Push a word onto the V stack; can only be a destination operand. |
| `popV` | Pop a word from the V stack; can only be a source operand. |
| `Hp` | The heap pointer register. |
| `tohp` | Move a word to the heap, and increment `Hp`. |
| `$`$n$`(Hp)` | The pointer value `Hp`+$n$. |
| *rr* | A register location; here we leave it unspecified the actual number of registers. For simplicy's sake, we assume that there are always 'enough' registers. In practice, we want to parameterize this number w.r.t. the actual target machine. |
| $n$`(rr)` | Indexed register, i.e. the location given by the address $n$+*content-of(*`rr`*)*. |
| `$`$i$ | A global memory location. |
| `$`$x$ | $x$ is either a number or, an identifier i.e. a pointer to a fixed memory location given by the symbolic label $x$ in which case $x$ must appear as a label somewhere. |

# D A naïve target code generator

| | |
|---|---|
| FUNSTART $f$ $n$ | C$f$: word FUN,V$f$;<br>V$f$: word $n$, C$f$, unw$n$, vunw$n$, F$f$;<br>F$f$: |
| EVAL | move popS,r0; calltab eval,0(r0); move r0,pushS |
| JFUN $n$ | move popS,r0; move \$$n$,r2; jmptab jfun,0(r0) |
| CALLFUN $n$ | move popS,r0; move \$1,pushV; move \$$(n+1)$(Sp); move \$$n$,r2; jmptab jfun,0(r0); $l$: |
| RET | move popS,r0; jmptab unwind,0(r0) |
| PUSH $n$ | move $n$(Sp),pushS |
| PUSHFUN $f$ | move \$$f$,pushS |
| PUSHBASIC $x$ | move \$$x$,pushV |
| SUB | sub popV,popV,pushV |
| MOVE $n$ | move popS,$(n-1)$(Sp) |
| POP $n$ | move \$$n$(Sp),Sp |
| GET | move popS,r0 |
| UPDATE $n$ | move popS,r0; move $(n-1)$(Sp),r1; move 0(r0),0(r1);<br>move 1(r0),1(r1); move 2(r0),2(r1) |
| LABEL $l$ | L$l$: |
| JMP $l$ | jmp L$l$ |
| ALLOC | move Hp,pushS; move \$HOLE,tohp; move \$2(Hp),Hp |
| MKAP | move \$AP,tohp; move popS,tohp; move popS,tohp; move \$-3(Hp),pushS |
| MKINT | move \$INT,tohp; move popV,tohp ; move \$-2(Hp),pushS |
| CASE $(k_1,l_1)\cdots(k_m,l_m)$ $l_d$ | move 0(Sp),r0; calltab getconstrno,0(r0); case r2,$(k_1,l_1)$, $\cdots$ $(k_m,l_m)$,$l_d$ |
| CNIL $m$ | .data; L$l_u$:; word NIL; word $m$; .code; move \$L$l_u$,pushS |
| CUNIT $m$ | move \$UNIT,tohp; move \$$m$,tohp; move popS,tohp; move -3(Hp),pushS; |
| CPAIR $m$ | move \$PAIR$m$,tohp; move popS,tohp; move popS,tohp; move -3(Hp),pushS |
| CVEC $n$ | move \$VEC,tohp; move \$$n$,tohp; $\underbrace{\texttt{move popS,tohp;}}_{n \text{ times}}$ move $-(n+2)$(Hp),pushS |
| SPLITUNIT | move popS,r0; move 2(r0),pushS |
| SPLITPAIR | move popS,r0; move 2(r0),pushS; move 1(r0),pushS |
| SPLITVEC $n$ | move popS,r0; move $(n+1)$(r0),pushS;$\cdots$move 2(r0),pushS |

# E An attribute-based target code generator

This appendix gives attribute grammar rules generating target code for all G-machine instructions considered in this paper. Utility functions and common definitions used here are defined at the end of this appendix.

$C$ = FUNSTART $f$ $n$ . $c$ :

| | | |
|---|---|---|
| $C$!*Vuse* | = | allpush |
| $C$!*Suse* | = | allpush |
| $C$!*code* | = | "C$f$:"; "word FUN,V$f$";<br>"V$f$:"; "word $n$, C$f$, unw$n$, vunw$n$, F$f$";<br>"F$f$:" |
| $c$!*Ssim* | = | allpopS |
| $c$!*Vsim* | = | allpopV |

$C$ = EVAL . $c$ :

| | | |
|---|---|---|
| $u.U$ | = | $c$!*Suse* |
| $a.S$ | = | $C$!*Ssim* |
| $C$!*Vuse* | = | allpush |
| $C$!*Suse* | = | use.allpush |
| $C$!*code* | = | intoreg0 $a$; "calltab eval,0(r0)"; $c'$ |
| ($c'$,$c$!*Ssim*) | = | newS $u$ "r0" $c$!*code* allpopS |
| $c$!*Vsim* | = | allpopV |

$C = \mathsf{JFUN}\ n\ .\ c:$

$$\begin{aligned} a.S &= C!Ssim \\ C!Vuse &= \text{allpush} \\ C!Suse &= \text{use.allpush} \\ C!code &= \text{intoreg0 } a;\ \text{“}\texttt{move \$}n\texttt{,r2}\text{”};\ \text{“}\texttt{jmptab jfun,0(r0)}\text{”};\ c!code \\ c!Ssim &= \text{allpopS} \\ c!Vsim &= \text{allpopV} \end{aligned}$$

$C = \mathsf{CALLFUN}\ n\ .\ c:$

$$\begin{aligned} u.U &= c!Suse \\ a.S &= C!Ssim \\ l &= \textit{a unique label} \\ C!Vuse &= \text{allpush} \\ C!Suse &= \text{use.allpush} \\ m_1 &= m+1 \\ C!code &= \text{intoreg0 } a;\ \text{“}\texttt{move \$L}l\texttt{,pushV}\text{”};\ \text{“}\texttt{move \$}m_1\texttt{(Sp),pushV}\text{”};\ \text{“}\texttt{move \$}n\texttt{,r2}\text{”}; \\ &\quad \text{“}\texttt{jmptab jfun,0(r0)}\text{”};\ c' \\ (c', c!Ssim) &= \text{newS } u\ \text{“}\texttt{r0}\text{”}\ c!code\ \text{allpopS} \\ c!Vsim &= \text{allpopV} \end{aligned}$$

$C = \mathsf{RET}\ .\ c:$

$$\begin{aligned} a.S &= C!Ssim \\ C!Vuse &= \text{allpush} \\ C!Suse &= \text{use.allpush} \\ C!code &= \text{intoreg0 } a;\ \text{“}\texttt{jmptab unwind,0(r0)}\text{”};\ c!code \\ c!Ssim &= \text{allpopS} \\ c!Vsim &= \text{allpopV} \end{aligned}$$

$C = \mathsf{PUSH}\ n\ .\ c:$

$$\begin{aligned} u.C!Suse &= c!Suse \\ C!Vuse &= c!Vuse \\ (C!code, c!Ssim) &= \text{newS } u\ (\text{indS } n\ C!Ssim)\ c!code\ C!Ssim \\ c!Vsim &= C!Vsim \end{aligned}$$

$C = \mathsf{PUSHFUN}\ f\ .\ c:$

$$\begin{aligned} u.C!Suse &= c!Suse \\ C!Vuse &= c!Vuse \\ (C!code, c!Ssim) &= \text{newS } u\ \text{“}\texttt{\$C}f\text{”}\ c!code\ C!Ssim \\ c!Vsim &= C!Vsim \end{aligned}$$

$C = \mathsf{PUSHBASIC}\ x\ .\ c:$

$$\begin{aligned} u.C!Vuse &= c!Vuse \\ C!Suse &= c!Suse \\ (C!code, c!Vsim) &= \text{newV } u\ \text{“}\texttt{\$}x\text{”}\ c!code\ C!Vsim \\ c!Ssim &= C!Ssim \end{aligned}$$

$C = \mathsf{SUB}\ .\ c:$

$$\begin{aligned} u.U &= c!Vuse \\ a.b.V &= C!Vsim \\ r_d &= \text{getdatareg } V\ C!Ssim \\ C!Vuse &= \text{use.use.}U_v \\ C!Suse &= c!Suse \\ (C!code, c!Vsim) &= \textbf{case } u \textbf{ in} \\ &\quad \text{use: (“}\texttt{sub } a\texttt{,}b\texttt{,}r_d\text{”};\ c!code,\ r_d.V)\ \| \\ &\quad \text{push: (“}\texttt{sub } a\texttt{,}b\texttt{,pushV}\text{”};\ c!code,\ \text{“}\texttt{popV}\text{”}.V) \\ &\quad \textbf{end} \\ c!Ssim &= C!Ssim \end{aligned}$$

$C = \mathsf{MOVE}\ n\ .\ c:$

$$\begin{array}{rcl}
a.S & = & C!Ssim \\
n' & = & n - 1 - \text{depthSsim}\ S \\
(C!code, c!Ssim) & = & \begin{cases} n' < 0: & (c!code,\ \text{replaceSsim}\ n\ a\ S) \\ n' \geq 0 \wedge \text{isinSsim}\ \text{“}n'\texttt{(Sp)}\text{”}\ S: & (\text{“}\texttt{move}\ n'\texttt{(sp)},r\text{”};\text{“}\texttt{move}\ a\texttt{,}n'\texttt{(Sp)}\text{”};c!code, \\ & \quad \text{substSsim}\ \text{“}n'\texttt{(Sp)}\text{”}\ r\ S) \\ n' \geq 0 \wedge \neg\text{isinSsim}\ \text{“}n'\texttt{(Sp)}\text{”}\ S: & (\text{“}\texttt{move}\ a\texttt{,}n'\texttt{(Sp)}\text{”};c!code,\ S) \end{cases} \\
c!Vsim & = & C!Vsim \\
C!Suse & = & \text{use}.c!Suse \\
C!Vuse & = & c!Vuse
\end{array}$$

$C = \mathsf{POP}\ n\ .\ c:$

$$\begin{array}{rcl}
n' & = & n - \text{simdepth}\ S \\
(C!code, c!Ssim) & = & \begin{cases} n' \leq 0: & (c!code,\ \text{tail}\ n\ S) \\ n' > 0: & (\text{“}\texttt{move}\ n'\texttt{(Sp),Sp}\text{”};\ c!code,\ \text{allpopS}) \end{cases} \\
C!Vsim & = & c!Vsim \\
C!Suse & = & \underbrace{\text{use}.\cdots.\text{use}}_{n\ \text{times}}.c!Suse \\
C!Vuse & = & c!Vuse
\end{array}$$

$C = \mathsf{GET}\ .\ c:$

$$\begin{array}{rcl}
u.C!Vuse & = & c!Vuse \\
a.S & = & C!Ssim \\
C!Suse & = & \text{use}.c!Suse \\
(c_a, r_a) & = & \text{intoaddrreg}\ a\ S\ C!Vsim \\
C!code & = & c_a;\ c' \\
(c', c!Vsim) & = & \text{newV}\ u\ \text{“}\texttt{1(}r_a\texttt{)}\text{”}\ c!code\ C!Vsim \\
c!Ssim & = & C!Ssim
\end{array}$$

$C = \mathsf{UPDATE}\ n\ .\ c:$

$$\begin{array}{rcl}
a.c!Ssim & = & C!Ssim \\
(c_a, r_a) & = & \text{intoaddrreg}\ a\ C!Vsim\ S \\
(c_b, r_b) & = & \text{intoaddrreg}\ (\text{indS}\ (n-1)\ S)\ C!Vsim\ (r_a.S) \\
C!Suse & = & \text{use}.c!Suse \\
C!Vuse & = & c!Vuse \\
C!code & = & c_a;\ c_b;\ \text{“}\texttt{move 0(}r_a\texttt{),0(}r_b\texttt{)}\text{”};\ \text{“}\texttt{move 1(}r_a\texttt{),1(}r_b\texttt{)}\text{”};\ \text{“}\texttt{move 2(}r_a\texttt{),2(}r_b\texttt{)}\text{”};\ c!code \\
c!Vsim & = & C!Vsim
\end{array}$$

$C = \mathsf{LABEL}\ l\ .\ c:$

$$\begin{array}{rcl}
C!Vuse & = & \text{allpush} \\
C!Suse & = & \text{allpush} \\
C!code & = & \text{“}\texttt{L}l\texttt{:}\text{”} \\
c!Ssim & = & \text{allpopS} \\
c!Vsim & = & \text{allpopV}
\end{array}$$

$C = \mathsf{JMP}\ l\ .\ c:$

$$\begin{array}{rcl}
C!Vuse & = & \text{allpush} \\
C!Suse & = & \text{allpush} \\
C!code & = & \text{“}\texttt{jmp L}l\text{”} \\
c!Ssim & = & \text{allpopS} \\
c!Vsim & = & \text{allpopV}
\end{array}$$

$C = \mathsf{ALLOC}\ .\ c:$

$$\begin{array}{rcl}
u.C!Suse & = & c!Suse \\
C!Vuse & = & c!Vuse \\
C!code & = & \text{“}\texttt{move \$HOLE}m\texttt{,tohp}\text{”};\ \text{“}\texttt{move \$2(Hp),Hp}\text{”};\ c' \\
(c', c!Ssim) & = & \text{newS}\ u\ \text{“}\texttt{\$-3(Hp)}\text{”}\ c!code\ (\text{map}\ (\text{H}\ 3)\ S) \\
c!Vsim & = & C!Vsim
\end{array}$$

$C = \mathsf{MKAP}\ .\ c$ :

$$
\begin{aligned}
u.U_s &= c!Suse \\
a.b.S &= C!Ssim \\
C!Vuse &= c!Vuse \\
C!Suse &= \text{use.use.}U_s \\
C!code &= \texttt{"move \$AP,tohp"; "move }a_1\texttt{,tohp"; "move }b_2\texttt{,tohp"; }c' \\
&\qquad \textbf{where } a_1 = \text{H 1 } a \textbf{ and } b_2 = \text{H 2 } b \\
(c',c!Ssim) &= \text{newS } u\ \texttt{"\$-3(Hp)"}\ c!code\ (\text{map (H 3) } S) \\
c!Vsim &= C!Vsim
\end{aligned}
$$

$C = \mathsf{MKINT}\ .\ c$ :

$$
\begin{aligned}
u.U &= c!Suse \\
a.V &= C!Vsim \\
C!Vuse &= c!Vuse \\
C!Suse &= \text{use.}U \\
C!code &= \texttt{"move \$INT,tohp"; "move }a\texttt{,tohp"; }c' \\
(c',c!Ssim) &= \text{newS } u\ \texttt{"\$-2(Hp)"}\ c!code\ (\text{map (H 2) } C!Ssim) \\
c!Vsim &= C!Vsim
\end{aligned}
$$

$C = \mathsf{CASE}\ (k_1,l_1) \cdots (k_m,l_m)\ l_d\ .\ c$ :

$$
\begin{aligned}
a.S &= C!Ssim \\
C!Vuse &= \text{allpush} \\
C!Suse &= \text{use.allpush} \quad \text{CASE \textit{pops its arg}} \\
C!code &= \text{intoreg0 } a\texttt{; "calltab getconstrno,0(r0)";} \\
&\qquad \texttt{"case r2,(}k_1\texttt{,}l_1\texttt{),}\cdots\texttt{,(}k_m\texttt{,}l_m\texttt{),}l_d\texttt{"; }c!code \\
c!Ssim &= \text{allpopS} \\
c!Vsim &= \text{allpopV}
\end{aligned}
$$

$C = \mathsf{CNIL}\ k\ .\ c$ :

$$
\begin{aligned}
u.U &= c!Suse \\
C!Vuse &= c!Vuse \\
C!Suse &= U_s \\
l &= \textit{a unique label} \\
C!code &= \texttt{".data"; "L}l\texttt{:"; "word NIL"; "word }k\texttt{"; ".code"; }c' \\
(c',cCcstk) &= \text{newS } u\ \texttt{"L}l\texttt{"}\ c!code\ C!Ssim \\
c!Vsim &= C!Vsim
\end{aligned}
$$

$C = \mathsf{CUNIT}\ k\ .\ c$ :

$$
\begin{aligned}
u.U &= c!Suse \\
a.S &= C!Ssim \\
C!Vuse &= c!Vuse \\
C!Suse &= \text{use.}U \\
C!code &= \texttt{"move \$UNIT,tohp"; "move \$}m\texttt{,tohp"; "move }a_2\texttt{,tohp"; }c' \\
&\qquad \textbf{where } a_2 = \text{H 2 a} \\
(c',c!Ssim) &= \text{newS } u\ \texttt{"\$-3(Hp)"}\ c!code\ (\text{map (H 3) } S) \\
c!Vsim &= C!Vsim
\end{aligned}
$$

$C = \mathsf{CPAIR}\ k\ .\ c$ :

$$
\begin{aligned}
u.U_s &= c!Suse \\
a.b.S &= C!Ssim \\
C!Vuse &= c!Vuse \\
C!Suse &= \text{use.use.}U_s \\
C!code &= \texttt{"move \$PAIR}k\texttt{,tohp"; "move }a_1\texttt{,tohp"; "move }b_2\texttt{,tohp"; }c' \\
&\qquad \textbf{where } a_1 = \text{H 1 } a \textbf{ and } b_2 = \text{H 2 } b \\
(c',c!Ssim) &= \text{newS } u\ \texttt{"\$-3(Hp)"}\ c!code\ (\text{map (H 3) } S) \\
c!Vsim &= C!Vsim
\end{aligned}
$$

$C = \text{CVEC}\ n\ .\ c$ :

$$
\begin{aligned}
u.U_s &= c!Suse \\
a_1.\cdots.a_n.S &= C!Ssim \\
C!Vuse &= c!Vuse \\
C!Suse &= \underbrace{\text{use}.\cdots.\text{use}}_{n \text{ times}}.U \\
C!code &= \texttt{"move \$VEC,tohp"; "move \$}n\texttt{,tohp";} \\
&\quad \texttt{"move (}H\ 2\ a_1\texttt{),tohp";}\cdots\texttt{"move (}H\ (n+1)\ a_n\texttt{),tohp"};\ c' \\
(c', c!Ssim) &= \text{newS}\ u\ \texttt{"\$-(}n+2\texttt{)(Hp)"}\ c!code\ (\text{map H}\ (n+2)\ S) \\
c!Vsim &= C!Vsim
\end{aligned}
$$

$C = \text{SPLITUNIT}\ .\ c$ :

$$
\begin{aligned}
u.U_s &= c!Suse \\
a.S &= C!Ssim \\
(c_a, r_a) &= \text{intoaddrreg}\ C!Vsim\ S \\
C!Suse &= \text{use}.U_s \\
C!Vuse &= c!Vuse \\
C!code &= c_a;\ c' \\
(c', c!Ssim) &= \text{newS}\ u\ \texttt{"2(}r_a\texttt{)"}\ c!code\ S \\
c!Vsim &= C!Vsim
\end{aligned}
$$

$C = \text{SPLITPAIR}\ .\ c$ :

$$
\begin{aligned}
u_1.u_2.U_s &= c!Suse \\
a.S &= C!Ssim \\
(c_a, r_a) &= \text{intoaddrreg}\ C!Vsim\ S \\
C!Suse &= \text{use}.U_s \\
C!Vuse &= c!Vuse \\
C!code &= c_a;\ c_1 \\
(c_2, S_2) &= \text{newS}\ u_2\ \texttt{"2(}r_a\texttt{)"}\ c!code\ S \\
(c_1, c!Ssim) &= \text{newS}\ u_1\ \texttt{"1(}r_a\texttt{)"}\ c_2\ S_2 \\
c!Vsim &= C!Vsim
\end{aligned}
$$

$C = \text{SPLITVEC}\ n\ .\ c$ :

$$
\begin{aligned}
u_1.\cdots.u_n.U_s &= c!Suse \\
a.S &= C!Ssim \\
(c_a, r_a) &= \text{intoaddrreg}\ C!VsimS \\
C!Suse &= \text{use}.U_s \\
C!Vuse &= c!Vuse \\
C!code &= c_a;\ c_1 \\
(c_n, S_n) &= \text{newS}\ u_n\ \texttt{"(}n+1\texttt{)(}r_a\texttt{)"}\ c!code\ S \\
(c_{n-1}, S_{n-1}) &= \text{newS}\ u_{n-1}\ \texttt{"}n\texttt{(}r_a\texttt{)"}\ c_n\ S_n \\
&\ \vdots \\
(c_2, S_2) &= \text{newS}\ u_2\ \texttt{"3(}r_a\texttt{)"}\ c_3\ S_3 \\
(c_1, c!Ssim) &= \text{newS}\ u_1\ \texttt{"2(}r_a\texttt{)"}\ c_2\ S_2 \\
c!Vsim &= C!Vsim
\end{aligned}
$$

## Utility functions:

allpush = push.allpush
allpopS = popS.allpopS
allpopV = popV.allpopV
getdatareg $V\ S$ = getreg *possible_data_regs* $V\ S$
getaddrreg $V\ S$ = getreg *possible_addr_regs* $V\ S$
getreg $(r.regs)\ V\ S$ = **if** $\neg$(used $r\ S\ \vee$ used $r\ V$) **then** $r$ **else** getreg *regs* $V\ S$

used $r$ ("rp".$s$) = $r = p\ \vee$ rused $r\ s$
used $r$ ("$n(p)$".$s$) = $r = p\ \vee$ rused $r\ s$
used $r$ ("\$$n(p)$".$s$) = $r = p\ \vee$ rused $r\ s$
used $r$ (popV.$s$) = false
used $r$ (popS.$s$) = false

```
        used r (a.s) = used r (a.s)
intoreg0 "r0" = ""
intoreg0 a = "move  a,r0"
intoaddrreg "rr" V S = ("", "rr")
intoaddrreg a V S = ("move  a,r", r)
                              where r = getaddrreg V S
newS use a c S = (c, a.S)
newS push a c S = ("move  a,pushS";c, S)
indS n (popS.S) = "n(Sp)"
indS 0 (x.S) = x
indS n (x.S) = indS (n - 1) S
H n "$m(Hp)" = "$(m - n)(Hp)"
H n a = a
depthSsim (popS.S) = 0
depthSsim (a.S) = 1 + depthSsim S
isinSsim x popS.S = false
isinSsim x a.S = x = a∨ isinSsim x S
replaceSsim 1 a (x.S) = a.S
replaceSsim n a (x.S) = x.replaceSsim (n - 1) a S
substSsim a b S = map (λx.if x = a then b else x) S)
tail 0 l = l
tail n (x.l) = tail (n - 1) l
map f [] = []
map f (x.l) = f l.map f l
```

# F A compilation example

**Input program:**

```
let rec from n = n.from (n+1)
  and sieve (p.l) = let rec filterp (x.L) = if x % p = 0 then filterp L else x.filterp L
                    in sieve (filterp l)
```

**After pattern matching transformation:**

```
Pmain input =
let rec from n = n.from (n+1)
  and sieve A = case A in
                cons p l: let rec filterp B = case B in
                                              cons x L: if x % p = 0 then filterp L else x.filterp L
                                              $: Pfail "No match in filterp"
                                              end
                          in p. sieve(filterp l)
                $: Pfail "No match in sieve"
                end
in sieve (from 2)
```

**After lambda-lifting:**

```
Pmain input = sieve (from 2)
from n = n.from (f n)
f n = n+1
sieve A = case A in
          cons p l: p.sieve (filterp p l)
          $: Pfail "No match in sieve"
          end
filterp p B = case B in
              cons x L: if x % p = 0 then filterp p L else x.filterp p L
              $: Pfail "No match in filterp"
              end
```

## G-machine code:

| *G-machine code:* | *completes the expression:* |
|---|---|
| FUNSTART Pmain 1; PUSHINT 2; PUSHFUN _from; MKAP; | $\mathcal{C}$⟦ from 2 ⟧ [input=1] 1 |
| MOVE 1; JGLOBAL _sieve; | $\mathcal{R}$⟦ sieve(from 2) ⟧ [n=1] 1 |
| FUNSTART _from 1; PUSH 0; PUSHFUN f; MKAP; | $\mathcal{C}$⟦ f n ⟧ [n=1] 1 |
| PUSHFUN _from; MKAP; | $\mathcal{C}$⟦ from(f n) ⟧ [n=1] 1 |
| PUSH 1; CONSTR 1 2; | $\mathcal{C}$⟦ n.from(f n) ⟧ [n=1] 1 |
| UPDATE 2; POP 1; RET; | $\mathcal{R}$⟦ n.from(f n) ⟧ [n=1] 1 |
| FUNSTART f 1; PUSH 0; EVAL; GET; | $\mathcal{B}$⟦ n ⟧ [n=1] 1 |
| PUSHBASIC 1; ADD; | $\mathcal{B}$⟦ n+1 ⟧ [n=1] 1 |
| MKINT; UPDATE 2; POP 1; RET; | $\mathcal{R}$⟦ n+1 ⟧ [n=1] 1 |
| FUNSTART _sieve 1; PUSH 0; EVAL; | $\mathcal{E}$⟦ A ⟧ [A=1] 1 |
| PUSH 0; CASE (1,L3) L1; | *case on that value* |
| L1: PUSHSTRING "No match in sieve"; | $\mathcal{C}$⟦ "No match in sieve" ⟧ [A=1] 2 |
| MOVE 2; POP 1; JGLOBAL Pfail; | $\mathcal{R}$⟦ Pfail "No match in sieve" ⟧ [A=1] 2 |
| L3: PUSH 0; SPLIT 1 2; | *push* l *and* p *on the S stack* |
| PUSH 1; PUSH 1; PUSHFUN _filterp; MKAP; MKAP; | $\mathcal{C}$⟦ filterp p l ⟧ [A=1,l=3,p=4] 4 |
| PUSHFUN _sieve; MKAP; | $\mathcal{C}$⟦ sieve(filterp p l) ⟧ [A=1,l=3,p=4] 4 |
| PUSH 1; CONSTR 1 2; | $\mathcal{E}$⟦p.sieve(filterp p l) ⟧ [A=1,l=3,p=4] 4 |
| UPDATE 5; POP 4; RET; | $\mathcal{R}$⟦p.sieve(filterp p l) ⟧ [A=1,l=3,p=4] 4 |
| FUNSTART _filterp 2; PUSH 1; EVAL; | $\mathcal{E}$⟦ B ⟧ [B=1,p=2] 2 |
| PUSH 0; CASE (1,L6) L4; | *case on it* |
| L4: PUSHSTRING "No match in filterp"; | $\mathcal{C}$⟦ " ... "⟧ [B=1,p=2] 3 |
| MOVE 3; POP 2; JGLOBAL Pfail; | $\mathcal{R}$⟦ Pfail "No match in filterp" ⟧ [B=1,p=2] 3 |
| L6: PUSH 0; SPLIT 1 2; | *push* L *and* p *onto the S stack* |
| PUSH 0; EVAL; GET; | $\mathcal{B}$⟦ x ⟧ [B=1,p=2,L=4,x=5] 5 |
| PUSH 3; EVAL; GET; | $\mathcal{B}$⟦ p ⟧ [B=1,p=2,L=4,x=5] 5 |
| MOD; PUSHBASIC 0; EQ; | $\mathcal{B}$⟦ x % p = 0 ⟧ [B=1,p=2,L=4,x=5] 5 |
| JFALSE L7; | *jump if false* |
| PUSH 1; MOVE 5; POP 3; JGLOBAL _filterp; | $\mathcal{R}$⟦ filterp p L ⟧ [B=1,p=2,L=4,x=5] 2 |
| L7: PUSH 1; PUSH 4; PUSHFUN _filterp; MKAP; MKAP; | $\mathcal{C}$⟦ filterp p l ⟧ [B=1,p=2,L=4,x=5] 2 |
| PUSH 1; CONSTR 1 2; UPDATE 6; POP 5; RET. | $\mathcal{R}$⟦ p.filterp p l ⟧ [B=1,p=2,L=4,x=5] 2 |

## M code for function filterp:

Below we show target code generation for the function *filterp*, showing the S and V use info stacks and simulated stacks. To save space, u and p have been used instead of **use** and **push**. p... denotes an infinite list of **push**es, so p... = p.p... etc; sometimes the latter is used instead of the former to show how use info stack elements are consumed. Note that the use info stacks resulting from one G-machine instruction are shown in the line for the preceding G-machine instruction, as in figure 29.

| *G-code* | *S use* | *S sim* | *V use* | *V sim* | *M code* |
|---|---|---|---|---|---|
| FUNSTART _filterp 2 | p... | [] | p... | [] | `C_filterp:`<br>`word FUN,V_filterp`<br>`V_filterp:`<br>`word 2,C_filterp,unw2,vunw2,F_filterp`<br>`F_filterp:` |
| PUSH 1 | u.p... | [1(Sp)] | u.p... | [] | |
| EVAL | p... | [r0] | p... | [] | `move 1(Sp),r0`<br>`calltab eval,0(r0)`<br>`move r0,pushS`[1] |
| PUSH 0 | u.p... | [1(Sp)] | p... | [] | |
| CASE (1,L6),L4 | p... | [] | p... | [] | `move 0(Sp),r0`<br>`calltab getconstrno,0(r0)`<br>`case r2,(1,L6),L4` |
| L4: | p... | [] | p... | [] | `L4:` |
| PUSHSTRING "No ..."[2] | u.p... | [$Lt1] | p... | [] | `.data`<br>`Lt1:`<br>`word STRING,Lt2,0`<br>`Lt2:`<br>`string "No match in filterp"`<br>`.code` |

| | | | | | |
|---|---|---|---|---|---|
| MOVE 3 | u.u.p... | [] | p... | [] | move $Lt1,2(Sp) |
| POP 2 | p... | [] | p... | [] | move $2(sp),Sp |
| JGLOBAL Pfail[3] | p... | [] | p... | [] | jmp FPfail |
| L6: | p... | [] | p... | [] | L6: |
| PUSH 0 | u.p... | [0(Sp)] | p... | [] | |
| SPLIT 1 2[4] | p.p.p... | [] | p... | [] | move 0(Sp),r3<br>move 2(r3),pushS<br>move 1(r3),pushS |
| PUSH 0 | u.p.p.p... | [0(Sp)] | p... | [] | |
| EVAL | u.p... | [r0] | p... | [] | move 0(Sp),r0<br>calltab eval,0(r0) |
| GET | p... | [] | p.p... | [] | move 1(r0),pushV |
| PUSH 3 | p... | [3(Sp)] | p.p... | [] | |
| EVAL | u.p... | [r0] | u.p... | [] | move 3(Sp),r0<br>calltab eval,0(r0) |
| GET | p... | [] | u.u.p... | [1(r0)] | |
| MOD | p... | [] | u.p... | [r3] | mod 1(r0),popV,r3 |
| PUSHBASIC 0 | p... | [] | u.u.p... | [$0,r3] | |
| EQ; JFALSE L7 | p... | [] | p... | [] | comp $0,r3<br>jne L7 |
| PUSH 1 | u.p... | [1(Sp)] | p... | [] | |
| MOVE 5 | u.u.p... | [] | p... | [] | move 1(Sp),4(Sp) |
| POP 3 | u.u.u.p... | [] | p... | [] | move $3(sp),Sp |
| JGLOBAL _filterp | p... | [] | p... | [] | jmp F_filterp |
| L7: | u.u.u.u.u.u.p... | [] | p... | [] | L7: |
| PUSH 1 | u.u.u.u.u.u.u.p... | [1(Sp)] | p... | [] | |
| PUSH 4 | u.u.u.u.u.u.u.u.p... | [3(Sp),1(Sp)] | p... | [] | |
| MKVAP _filterp 2[5] | u.u.u.u.u.u.p... | [$-4(Hp)] | p... | [] | comp Hp,_hplim<br>jlt Lt3<br>call GC<br>Lt3:<br>move $VAP,tohp<br>move V_filterp<br>move 3(Sp),tohp<br>move 1(Sp),tohp |
| PUSH 1 | u.u.u.u.u.u.u.p... | [0(Sp),$-4(Hp)] | p... | [] | |
| RETCONSTR 1 2 5[6] | | [] | p... | [] | move popV,Sp<br>move -1(Sp),r0<br>move $PAIR1,0(r0)<br>move -6(Sp),1(r0)<br>move $-4(Hp),2(r0)<br>return |

## Notes:

1. This instruction is redundant.
2. The PUSHSTRING instruction pushes a pointer to a string, i.e. a list of characters. This is implemented by building a 'non-volative' STRING node, which has pointer to the bytes of the string. The STRING node must be evaluated to become canonical.
3. The JGLOBAL instruction is a simplification of the JFUN instruction, used when arities are known at compile-time to match.
4. The SPLIT 1 2 instruction becomes a SPLITPAIR.
5. MKVAP _filterp 2 is a shorthand for PUSHFUN _filterp; MKAP; MKAP.
6. RETCONSTR 1 2 5 is a shorthand for CONSTR 1 2; UPDATE 6; POP 5; RET.

# Toward The Design of a Parallel Graph Reduction Machine

## The MaRS Project*

Michel Castan**
Guy Durrieu
Bernard Lecussan
Michel Lemaître

Alessandro Contessa
Eric Cousin
Paulino Ng

ONERA-CERT - Department of Computer Science
2, avenue Edouard Belin, B.P. 4025
31055, Toulouse CEDEX (FRANCE)

*This research is supported by the DGA/STPA under contract STPA n° 8595004.07
**UUCP: ..!seismo!mcvax!inria!tls-cs!castan

---

### Introduction

Research in the aeronautical field requires more and more non vectorial computation for aircraft design. This non-vectorial computation is either numeric or symbolic. For many years, the computer architecture group at ONERA-CERT has been involved in the design of parallel machines for such purposes. For non-vectorial numeric computation, a project of a Data-Flow machine was started in 1973 and ended in 1980 : the LAU project [Berger80]. Between 1982 and 1985, a revised version of LAU machine was designed : LAURA [Durrieu86].

In 1984, we started a new project of a parallel machine for symbolic computation : MaRS project. Preliminary studies set functional languages as the best suited programming languages for such a machine, and parallel graph reduction as the best execution model. The MaRS machine presents three main characteristics :

- The machine language is semantically close to functional languages in order to facilitate compilation and simplify the execution model. Combinator code as proposed by Turner

[Turner79] has been chosen as a basis for the machine language.

- The execution model takes into account the parallelism intrinsically found in functional languages. However, annotations in the source code will be authorized in order to increase the performance of this machine.
- The machine is to be implemented in VLSI. Four kinds of devices are used in this system, respectively for storage, reduction, communication and input-output.

Some specific aspects of MaRS appear to exist in other projects. SKIM2 [Stoye84], DAPS [Hudak84], COBWEB [Hankin85], ALICE [Darlington81], REDIFLOW [Keller84] are all, in some sense, close to MaRS. However, MaRS differs in the way these mechanisms are implemented. This paper first presents our approach in the characterization of the general mechanisms involved in MaRS architecture (chapter 1). Then we outline some typical aspects of MaRS, such as the set of combinators, or the different types of processors (chapter 2). Finally, we conclude this presentation by the overall planning of the MaRS project, actually under development at CERT-DERI (Toulouse-France).

## 1. Characterization of the basic mechanisms

Functional languages are based on lambda-calculus, originally introduced by A. Church [Church41]. Lisp, the first one, was introduced by Mac Carthy in 1960 [McCarthy60]. One very important feature of functional languages is the "referential transparency", which broadly states that the value of a given expression depends only on its textual context and not on the past history of the computation. This is very important in terms of parallel execution. A consequence of this property is that the application of a function to an argument, and the result of this application, can be considered as different representations of the same object.

Since 1960, many functional languages have been designed. Each one has its own syntax and semantics, but every purely functional language can be translated into pure Lisp in a straightforward way. Thus, we have chosen pure Lisp as the primary programming language of MaRS.

Among the mechanisms that characterize a parallel machine, we can define two classes : mechanisms induced by the language and mechanisms induced by architectural considerations.

### 1.1 Mechanisms induced by the language

#### 1.1.1 Choice of the execution model

There are three main models for the execution of a functional program : Control Flow, Data Flow and Reduction [Treleaven82]. Control Flow and Data Flow models make a distinction between program and data. Control can be associated either to the program (Control Flow) or to the data (Data Flow). Roughly, Control Flow model considers instructions as an active agents that transforms the passive data, whereas Data Flow model considers data as active agents moving through the passive code graph [Vegdhal84]. These two models have in common a static representation of programs during execution.

Reduction model does not make any distinction between program and data. Instructions and operands are both expressions. The application of an operator to operands and the result of this application are two differents forms of the same object. Based upon a rewriting mechanism, the Reduction model appears to be very close to the theoric model, the lambda calculus. Finally, the Reduction model can handle with ease and in an homogeneous manner such concepts as infinite data structures, recursivity and higher order functions [Henderson80].

Reduction model considers programs as expressions that are rewritten during execution. Considering the nature of the expression, one can distinguish two kinds of reductions: string reduction where expressions are formed of litterals and values, and graph reduction, where expressions are constituted of litterals and references. In our opinion, graph reduction is more interesting for the following reasons : because of the notion of reference, graph reduction allows the share of computation, but it also provides an efficient way for manipulating large data structures. Graphs are uniform representations of suspensions (closures) as well as of shared data structures. For these reasons we have chosen Graph Reduction as the execution model of MaRS.

### 1.1.2 Program and data representation

The first contact between programmers and the programming language is through the syntax of the language and the objects it handles. In the case of Lisp, there is a great homogeneity between program representation and data representation. On the one hand, we have atomic objects like numbers or atoms, and on the other we have non-atomic objects like lists.

In Lisp, every expression is evaluable. For the evaluator, it is necessary to distinguish the type of a given object. This distinction can be done by software, for example with a partition of the address space [Greussay77], or by hardware, for example with the use of extra bits. The idea of typing all information by hardware appeared first in 1973 in an article of E.A. Feustel [Feustel73] and has been used in most Lisp-machines [Bawden77, Castan80, Chu81]. This leads naturally to the so called self-defined object, which is an object containing all the information needed to identify it. In MaRS, an object is made out of a number of parts: one to define the type of the object and the rest containing one or more values. This organisation remains transparent to the user.

In a reduction context, the basic composed object is either the application of a single function to an argument, or the cons-cell. This implies the binary cell as the elementary composed object. The binary cell is designed by a self-defined object whose mark bits point out its type (application or cons-cell) and whose value is the binary cell address. Because the same object can refer to an application of a function to an argument, as well as to the result of this application, it is necessary to add extra bits in the memory cell, in order to encode the status of the cell : at least an unreduced state and a reduced state.

An important and useful characteristic of functional languages, is the possibility to create new objects dynamically by allocation of memory cells. This implies a dynamic management of the memory, in particular concerning the collection of unused cells. Many methods can be used for such a purpose. All of them require some extra bits intended for use by the garbage collector, the GC field in the memory cell.

1.1.3 Execution function

We set up two states associated with the memorization of an object : the initial state (or un-reduced state) in which the object contains the application of a function to an argument, and the final state (or reduced state) in which the object contains the result of this application. Changing the state of a memory cell is done by a new object, namely the process. The existence of a process associated with a memory cell is indicated by a third state of the memory cell : the "under reduction" state.

One question is : do we need one process for all objects, or one process for each object, or an intermediate solution like one process for a group of objects? As Church-Rosser's first theorem points out, no functional expression can be completely reduced into two different results. In other words, two finite reductions of the same initial expression always give the same result. Thus, parallelism during reduction is safely allowed. It will be achieved by a multiprocess execution model.

Figure 1 : Memory-Process relation

Until this point, we have introduced two kinds of objects: binary cell and process. The relations between these two objects can be summarized in figure 1. Considering the time a process remains associated with a cell, many strategies can be used. Roughly, we consider two main approaches: the data-driven approach, where a process is associated to a cell as soon as this cell is created, and the demand-driven approach where the association occurs as soon as a read access to an un-reduced cell occurs. Demand-driven strategy is necessary to take advantage of the so-called lazy evaluation. However, this implies some hardware mechanisms for operators like suspended-CONS [Friedman76] or for delayed-evaluation [Henderson80]. The main advantage of the data driven approach is a higher parallelism. On the other hand, some care must be taken in controlling the parallelism, like garbage collection for the process [Grit81].

The advantage of a graph reduction versus string reduction is the possibility of sharing computation between processes. However, we must avoid a situation where several processes reduce the same cell at the same time. A mechanism for synchronization of accesses to a cell must be provided : each read access to a cell in the state "under reduction" will be delayed until the cell reaches the reduced state. A waiting queue must be associated to each cell whose state is under reduction.

### 1.1.4 The beta-reduction

Beta-reduction is fundamental in lambda-calculus, because it formalizes the parameter-passing operation during a function call. During the execution of the beta-reduction, each occurrence of a formal parameter of a function is eliminated by substitution of the corresponding argument. In other words, the beta-reduction consists in replacing the body of the function by a new copy in which all formal parameters have been replaced by the effective parameters.

Many approaches are possible to realize the substitution of the formals parameters. The substitution can be done as soon as possible, that is to say when the function is called. The substitution can also be done as late as possible, when the value of a parameter is needed (for example, by using an association list). An intermediate approach consists in making a substitution in several steps, using combinators. This method has several advantages: there is no notion of environment, the transformation of an expression in combinatory code remains very simple, and the combinatory code retains all functional properties of the source language.

## 1.2 Mechanisms induced by the architecture

### 1.2.1 General model

As a result of the mechanisms induced by the language, we can elaborate a first description of a functional parallel machine. Roughly, it is a multiprocess machine, based on graph reduction. Processes apply reduction rules to the nodes of the graph. Beta-reduction is done incrementally by using combinatory code. The strategy for process creation is mainly demand-driven with some features of the data-driven approach in order to increase the parallelism of the execution. A hardware mechanism is used for synchronisation of memory accesses, in order to prevent more than one process reducing the same node at the same time.

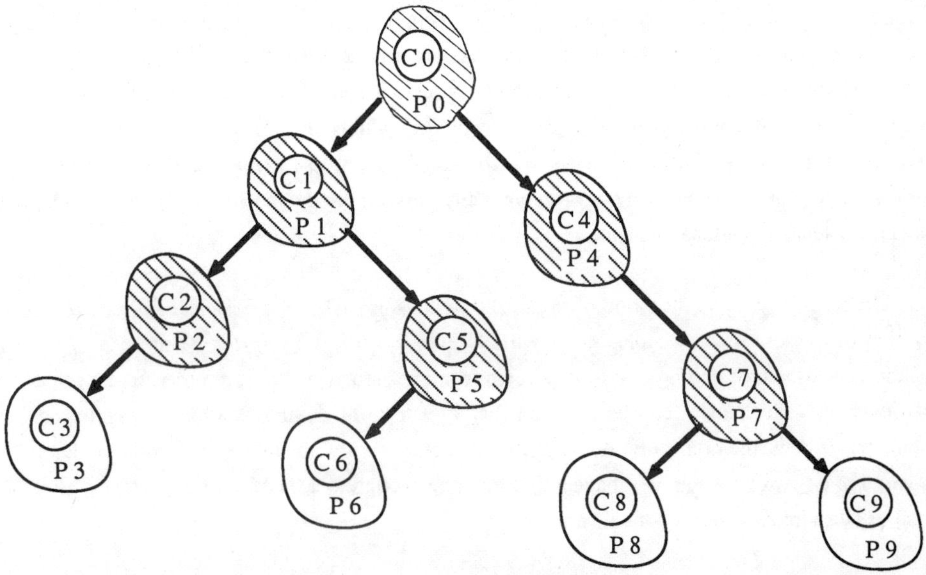


Figure 2 : multiprocess organization

1.2.2 Memory-process separation

A process requires some resources such as storage for arguments, evaluation stack, etc. Whatever execution model is adopted, these resources will be much bigger than the memory space required for one node of the graph. Therefore, it is not at all realistic to associate a process with only one memory cell during the whole cell lifetime (there is a big disparity between the cell lifetime and the time the cell state is under reduction). Process and node must be two independent entities. Then it becomes natural to have two kinds of processors, one for processes (reduction processor) and one for nodes (memory processor). At this point of the discussion, what we call processor is a unit able to manage a set of objects sequentially (one object at a time).

After we have decided on the logical separation between processes and nodes, we must define the relationship between processes and nodes. One question can be, for example : does a process have to remain associated with a cell until the reduction of that cell is completed, or can a process be dissociated from a cell in order to reduce some other cells which have higher priority ? In the first case, let's examine both the relation between processes and the association process-cell at a given time during a parallel execution. These relations can be sketched as in the figure 2. In that figure, a link between two processes denotes a data dependency : for example, process P2 is waiting for completion of process P3. With the hypothesis of a machine having limited process resources, it is obvious that such organisation can't be envisaged because it is impossible to predict which resources will be needed for a given program. A better organization consists in associating one process to many cells at the same time, as suggested in figure 3. In this way, it is theoretically possible for the machine to support an application having a potentially high parallelism.

Unfortunately, this latter kind of organization does not avoid the existence, at a certain time, of waiting processes. For example, as shown in figure 3, P0 is waiting for P1 and P2. An obvious solution consists in detaching a waiting process from the associated cells, in order to reduce new cells. However, this mechanism can be very expensive in term of communication between process and memory. In fact, the configuration as shown in figure 3 only appears when a process leads to the creation of a new one which will have a lifetime long enough to cause the waiting state. This problem shall be taken into account at compile time.

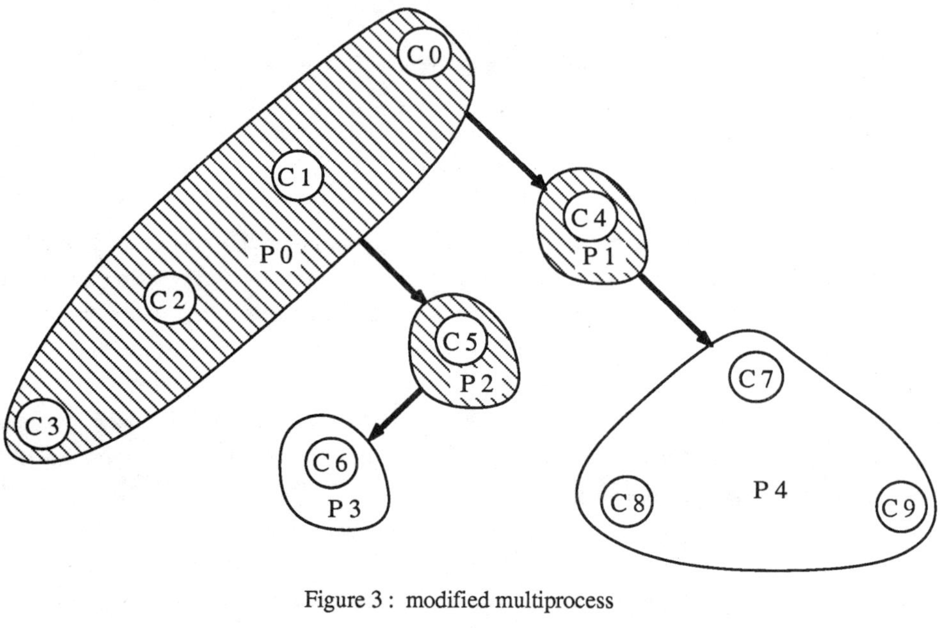


Figure 3 : modified multiprocess

The various processors must exchange information. Communication by message passing seems to be the best solution, as this model conceptually allows a complete separation between sender and receiver. In this way, it is possible to recover the communication time as well as the memory access time during the execution, provided the instantaneous degree of parallelism is sufficient. Furthermore, we avoid some kind of busy waiting due to the access conflicts on the data path.

### 1.2.3 Implications on the organization

We have now defined our machine in terms of mechanisms required by functional programs. It is a multiprocessor machine with storage processors and reduction processors communicating by message passing through a communication network.

With several processors, the communication latency time, that is to say the time between the emission of a request and the reception of the corresponding response, can be rather long. For a full use of the reduction processor, it is necessary to recover the communication time by allowing a number of requests before the reception of a response [Arvind83]. This characteristic can be obtained through multiprocess management in each reduction processor, a process swap occuring each time a response is being awaited. However, since this operation is crucial with regard to time, a specialized hardware mechanism is necessary.

The dynamic aspect of the process creation and the memory cell creation implies a mechanism for the distribution of these objects among the different processors. In order to avoid bottlenecks in the machine, this mechanism must be totally distributed. The communication network is a good candidate to support the load balancing of the machine.

Because of the constraints of a VLSI implementation, the number of processes managed by each reduction processor is low, say 8 processes. It becomes necessary to have a mechanism to avoid deadlock when the instant parallelism reaches the machine saturation value. The first approach consists in slowing down the expansion of parallelism, for example by changing the execution model from a "breadth first" to a "depth first" model [Keller84]. However this method does not avoid saturation in all cases. A better mechanism must totally stop the expansion of parallelism before overloading occurs. The simplest one consists in changing the execution model from a parallel execution to a sequential execution for each processor. In order to implement this mechanism, we must provide a signal to indicate to each reduction processor the execution model to be used. Once again, to avoid bottlenecks, this signal must be generated by the communication network.

A last consequence of this implementation concerns the possibility of the non-deterministic parallel operators, such as the parallel-AND and the parallel-OR. In order to take full advantage of the power of such a machine, we should have all of the reduction processors busy all the time. Then, when the machine reaches saturation, the parallel execution model is switched to a sequential one. This implies that AND and OR operators cannot have a true parallel semantics (in fact, true parallel semantics theoretically implies a number of available processes equal to the number of the memory cells). In any case, allowing the use of truly parallel operators would imply a definite jump out of the purely functional model : a given expression could have different results depending on the machine state or configuration. This has to be avoided if we want to keep the good properties of functional languages, specially as regards debugging.

Related to this problem is the status of the MERGE operator needed for input-output management (see for example [Henderson82] or [Stoye86]). This non-deterministic MERGE can be proved to be theoretically equivalent to a non-deterministic parallel OR or AND (that is to say, if we have one, we can build the other). A solution is, like that proposed in [Stoye86], to transfer this operator out of the proper reduction machinery. In MaRS, it is envisaged to entrust input-output processors with the execution of MERGEs.

## 2. The MaRS project

### 2.1 The set of combinators

The machine language of MaRS is based on a set of indexed combinators. One advantage of this solution concerns the expansion rate (ratio of the length of the produced object code to the length of the source code). It has been proved to be smaller than two, for a set of combinators close to the one presented below [Castan85], assuming integers are stored in constant space.

#### 2.1.1 Basic combinators

In this section, we describe the functionality of the MaRS combinators, giving their reduction rules. Notice that compilation steps are merely reduction steps in reverse order. We use five basic combinators : I, $M_n$, $W_n$, $K_n$, and $L_n$ :

I is the identity, as usual :

$$I\ e \rightarrow e$$

$M_n$ is the "move" combinator :

$$M_n\ e_1\ e_2\ \dots\ e_n \rightarrow e_1\ e_n\ e_2\ \dots\ e_{n-1} \qquad \text{, for } n \geq 3$$

$W_n$ is the "copy and move" combinator :

$$W_n\ e_1\ e_2\ \dots\ e_n \rightarrow e_1\ e_n\ e_2\ \dots\ e_{n-1}\ e_n \qquad \text{, for } n \geq 2$$

$K_n$ is the "cancell" combinator :

$$K_n\ e_1\ e_2\ \dots\ e_n \rightarrow e_1\ e_2\ \dots\ e_{n-1} \qquad \text{, for } n \geq 2$$

$L_n$ creates n-2 application nodes at once :

$$L_n\ e_1\ e_2\ \ldots e_n \rightarrow\ e_1\ (e_2\ \ldots\ e_n) \qquad , \text{for } n \geq 3$$

Indexes show the range of the combinators, i.e. the number of arguments involved in the associated rewriting rule. These arguments are needed to trigger the reduction of the combinator. One has to think of these rewriting rules as moves in an execution stack, the combinator being on the top of it, and a continuation being just under the top. Reducing a combinator just moves the arguments of the continuation to right places. The new application nodes created by the $L_n$ combinator are created out of the stack, in the shared memory. There is one important exception to this rule, when the constructed expresion ( $e_2\ \ldots\ e_n$ ) can be reduced immediatly. In this case, there is no need to build the expression in shared memory. This is called sequentialization and is discussed below in more details.

These combinators are generalizations of some well-known combinators. For example, $L_n$ is the generalization of the "compositor" combinator :

$$B\ x\ y\ z \rightarrow x\ (y\ z)$$

For the sake of optimization, some additional combinators are used :

$$N_n\ e_1\ e_2\ \ldots e_n \rightarrow\ e_1\ (e_2\ e_n)\ e_3\ \ldots e_{n\ 1} \qquad , \text{for } n \geq 3$$

$$Q_n\ e_1\ e_2\ \ldots e_n \rightarrow\ e_1\ (e_2\ e_n)\ e_3\ \ldots e_{n-1}\ e_n \qquad , \text{for } n \geq 3$$

The effect of $N_n$ (respectively $Q_n$) can be generally obtained by a combination of $L_3$ and $M_{n-1}$ (respectively $L_3$ and $W_{n-1}$).

Another combinator for optimization purpose, the "swap" combinator, is defined by :

$$S_n\ e_1\ e_2\ \ldots e_n \rightarrow\ e_1\ e_n\ e_3\ \ldots e_{n-1}\ e_2 \qquad , \text{for } n \geq 3$$

### 2.1.2 Annotating the combinators

As stated in the first part of this paper, each node of the graph being reduced can be in three different states: unreduced, under reduction, or reduced. These states are internally coded in nodes, and are used for inter-process synchronization : a reduced node can be read by any process, without any restriction; an unreduced node can be read by any process, but the first access will change its state to "under reduction", thus preventing other accesses to the node until it is reduced.

One can often detect, at compile time, the fact that a node created by a $L_n$ combinator is irreductible. Here is a simple example : suppose we have in an execution stack the following components :

$$L_4\ *\ +\ x\ y\ z$$

The reduction of $L_4$ will result in :

$$*\ (+\ x\ y)\ z$$

The newly created expression, (+ x y ) , consists of two new application nodes, namely + applied to x, and this applied to y. The first one, + applied to x, is obviously irreducible, so that here is no need to devote a process to its reduction. This fact can be detected at compile time, and associated to the combinator creating the node as an annotation.

The other newly created node of the previous example is the root node of the subgraph (+ x y ). It will certainly need to be reduced, because it corresponds to an argument of a strict operator, namely *. Here we have two options at execution time :

1) create and trigger a new process to reduce the sub-expression (+ x y ) at once
2) let the current process do the job, starting a sequential sub-process, to be achieved before the * can proceed. In this case, no new node is created outside the stack.

A static analysis, conducted at compile time, involving stricticity, time cost and user annotations in the source code, fixes a choice among the previous options. This choice is materialized by another annotation of the $L_4$ combinator. In our example, the combinator will be annoted $LP_4$, if we want a new process, or $LS_4$ if we want to sequentialize the computation. In the first case, $LP_4$, the first created node (+ x ) will be marked as being in reduced state.

In the general case, the reduction of a $L_n$ obeys the following rules:

$$LX_n\ e_1\ e_2\ \dots\ e_n \rightarrow e_1\ (e_2\ \dots\ e_n)$$

if X=S, then $e_2\ \dots\ e_n$ remains in the stack and ( $e_2\ \dots\ e_n$ ) is sequentially reduced by a sub-process of the current process.

if X=P, then ( $e_2\ \dots\ e_n$ ) is created in shared memory, and a new process is activated to reduce this sub-graph unless the machine is saturated. In that case, $LP_n$ is treated as $LS_n$.

if X=R, then ( $e_2\ \dots\ e_n$ ) is created in shared memory, with all nodes marked "reduced".

if X=N, then ( $e_2\ \dots\ e_n$ ) is created in shared memory, with all nodes marked "reduced" except the root node marked "unreduced". A process will be created only when an access request to this root node occurs.

Such annotations are also available for the combinators $N_n$ and $Q_n$.

Another example serves to clarify the discussion. Let F be the function defined by :

F = ( lambda ( x y z ) ( * ( + z x ) ( - z y ) ) )

We show below the different transformations performed by the compiler in order to produce the combinatory code corresponding to the definition of F, first without annotations :

$$
\begin{aligned}
& * \; (+ z x) \; (- z y) \\
= \quad & L_4 * + z x \; (- z y) \\
= \quad & M_4 \; (L_4 * +) \; x \; (- z y) \; z \\
= \quad & M_3 \; (M_4 \; (L_4 * +)) \; (- z y) \; x z \\
= \quad & L_4 \; (M_3 \; (M_4 \; (L_4 * +))) \; - z y \; x z \\
= \quad & W_4 \; (L_4 \; (M_3 \; (M_4 \; (L_4 * +))) \; -) \; y \; x z \\
= \quad & M_3 \; (W_4 \; (L_4 \; (M_3 \; (M_4 \; (L_4 * +))) \; -)) \; x y z
\end{aligned}
$$

Hence, $F = M_3 \; ( W_4 \; ( L_4 \; ( M_3 \; ( M_4 \; ( L_4 * + ))) - ))$. Now, there are four possibilities when applying F to its three arguments :

1) the whole computation is achieved sequentially, using only one process.

2) two processes are used, one for + and *, and the second one for -, in parallel with the first.

3) two processes, one for - and *, the second for +.

4) three processes, one for each operator.

All these cases are distinguished by different annotations of $L_n$ combinator. The first case (sequential) is obtained when all $L_n$ are $LS_n$. Here is an example of a complete reduction corresponding to this case :

$$
\begin{aligned}
& F \; 3 \; 4 \; 5 \\
= \quad & M_3 \; (W_4 \; (LS_4 \; (M_3 \; (M_4 \; ( LS_4 * +))) \; -)) 3 \; 4 \; 5 \\
\rightarrow \quad & W_4 \; (LS_4 \; (M_3 \; (M_4 \; (LS_4 * +))) \; -) \; 4 \; 3 \; 5 \\
\rightarrow \quad & LS_4 \; (M_3 \; (M_4 \; (LS_4 * +))) \; - \; 5 \; 4 \; 3 \; 5
\end{aligned}
$$

The reduction of $LS_4$ causes $(- \; 5 \; 4)$ to be reduced directly on the stack, giving :

$$
\begin{aligned}
\rightarrow \quad & M_3 \; (M_4 \; (LS_4 * +)) \; 1 \; 3 \; 5 \\
\rightarrow \quad & M_4 \; (LS_4 * +) \; 3 \; 1 \; 5 \\
\rightarrow \quad & LS_4 * + \; 5 \; 3 \; 1
\end{aligned}
$$

Here again $(+ \; 5 \; 3)$ is reduced on the stack :

$$
\begin{aligned}
\rightarrow \quad & * \; 8 \; 1 \\
\rightarrow \quad & 8
\end{aligned}
$$

In this reduction example, it should be understood that non-atomic expressions like

$$( W_4 \; ( LS_4 \; ( M_3 \; ( M_4 \; ( LS_4 * + ))) \; - )))$$

do not exist in the execution stack, but are represented in the stack by a pointer to the corresponding sub-graph in the shared memory.

Other control strategies are supported by the following annotations. Case 2) and 3) are :

$$F = M_3 ( W_4 ( LP_4 ( M_3 ( M_4 ( LS_4 * + ) ) ) - ) )$$

and $$F = M_3 ( W_4 ( LS_4 ( M_3 ( M_4 ( LP_4 * + ) ) ) - ) ).$$

The last case 4) is of course :

$$F = M_3 ( W_4 ( LP_4 ( M_3 ( M_4 ( LP_4 * + ) ) ) - ) ).$$

Let us follow the reduction of the second case :

$$F \; 3 \; 4 \; 5$$

$$= \quad M_3 (W_4 (LP_4 (M_3 (M_4 (LS_4 * +))) -)) \; 3 \; 4 \; 5$$

$$\rightarrow \quad W_4 (LP_4 (M_3 (M_4 (LS_4 * +))) -) \; 4 \; 3 \; 5$$

$$\rightarrow \quad LP_4 (M_3 (M_4 (LS_4 * +))) - \; 5 \; 4 \; 3 \; 5$$

The reduction of $LP_4$ causes the sub-graph ( - 5 4 ) to be created in the shared memory, with a new parallel process to reduce it.

$$\rightarrow \quad M_3 (M_4 (LS_4 * +)) \; p \; 3 \; 5$$

where p is a pointer to the future result being computed. The current process goes on :

$$\rightarrow \quad M_4 (LS_4 * +) \; 3 \; p \; 5$$

$$\rightarrow \quad LS_4 \; * \; + \; 5 \; 3 \; p$$

$$\rightarrow \quad * \; 8 \; p$$

p is now fetched and the associated result value is waited for :

$$\rightarrow \quad * \; 8 \; 1$$

$$\rightarrow \quad 8$$

31 … 0

| integer | S | value | |
|---|---|---|---|
| ptr. | processor n° | node address | |
| comb. | | combinator n° | index |

Figure 4 : The elementary 32-bit word

## 2.2 The processors

The MaRS machine is composed of four different processors. All processors are able to manipulate the same indivisible information : a 32-bit wide word. Figure 4 gives the possible organization of this word : it has a tag that indicates the semantics of the rest of the word (integer, pointer, combinator, etc...); The rest of the word can have different organizations : sign and value, processor address and node or process address, combinator number and index value, etc.

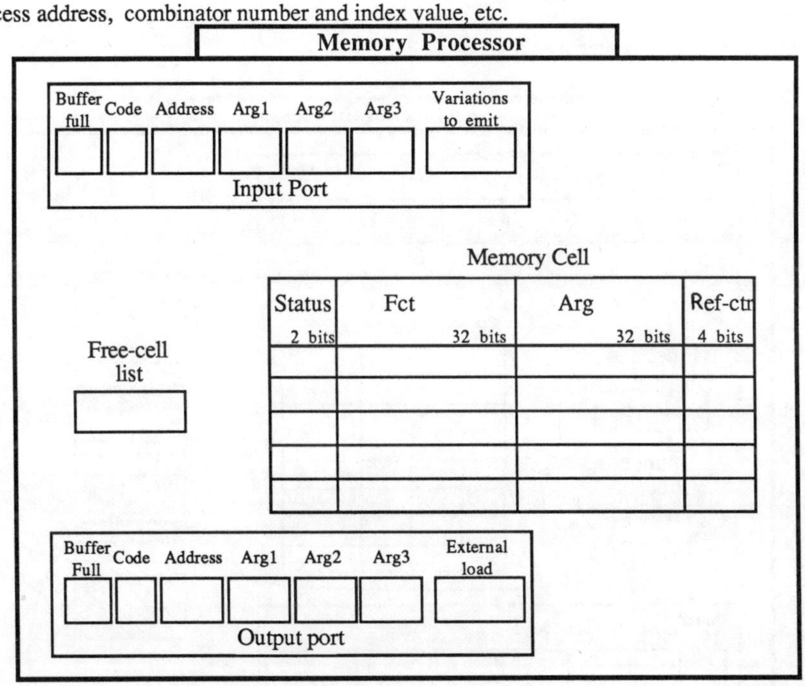


Figure 5 : The Memory processor

2.2.1 The memory processor

The memory processor manages a set of memory cells. Figure 5 gives an overview of the principal elements of a memory processor. A memory cell consists of four parts. Status word encodes the state of the cell. Fct and Arg words encode respectively the function part and the argument part of an application. These two words are 32 bits long as shown in figure 4. Ref-ctr word is used to memorize the number of references to the corresponding cell for garbage collecting purpose.

In addition to the set of cells managed by a processor, there are two communication ports, one for receiving messages (the input port) and one for sending messages (the output port). Each port comprises two main parts : one is related to the message buffer (buffer full, code, address, arg1, arg2 and arg3) ; the second is used for the load management. The "variation to emit" register contains the variation of the memory load of the processor. The memory load of a processor dynamically gives the number of allocated cell in that processor. This information is used for load balancing purpose. The "external load"

register contains the memory load of the machine, viewed through the output port of the processor. This information is used by the memory processor
to slow down creation of new processes (for example, to treat a creation of unreduced node with process message as a creation of unreduced node message).

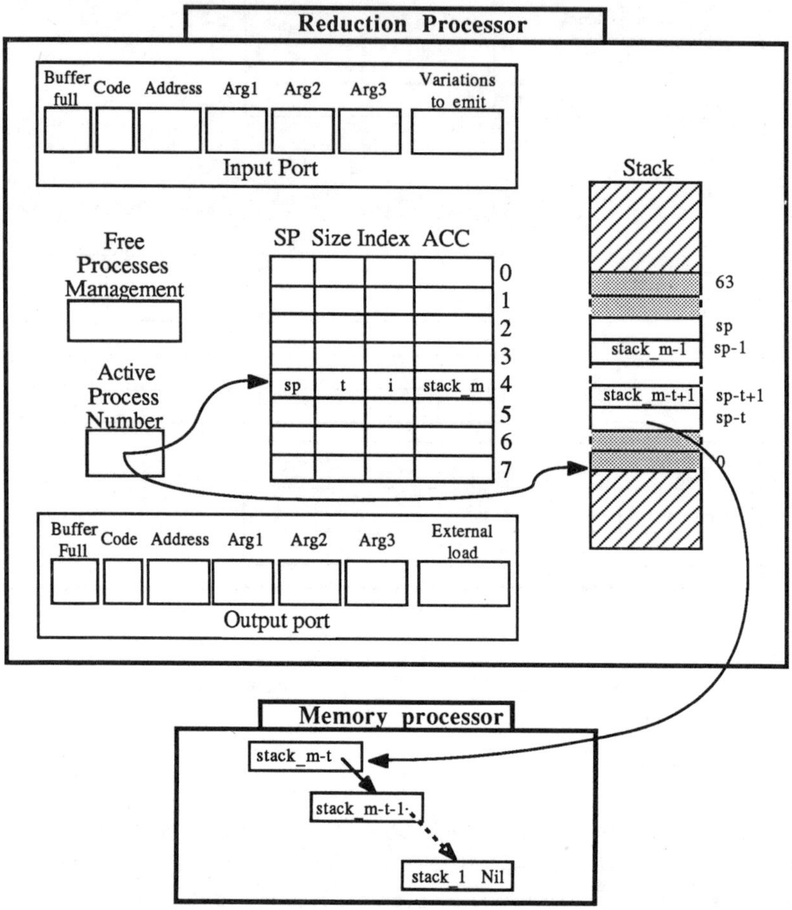


Figure 6: The reduction processor

2.2.2 The reduction processor

The principal elements of a reduction processor are presented in figure 6. As with the memory processor, we have two communication ports, an input port and an output port. The purpose of the reduction load management on each port is just the same as for the memory processor. The reduction load sent through the input port is the number of allocated process in the processor, whereas the load information viewed through the ouput port is used by the processor to decide which execution model is needed for the evaluation of a $LP_n$ combinator : sequential or parallel.

Due to the VLSI implementation, the reduction processor will manage up to 8 processes. Only one process per processor is active at a given time. Its number is stored in the active process register. The free processes
are indicated by the register free processes management. The reduction processor has a memory used to implement the stack associated to each process. In fact, only 64 words are allocated to each process. These words only represent the top of the execution stack of each process. The bottom of the stack resides in the shared memory as a graph. In addition to these 64 words, each process has a context process comprising the stack pointer (SP), the number of words of the execution stack stored in the local memory, an index register used to address the local memory, and an accumulator register that contains the last word written into the stack. Figure 6 gives the organization of the execution stack of process n° 4 : this execution stack contains m words; only t words are in the local memory of the reduction processor, while the rest resides in shared memory as a list of m-t cells.

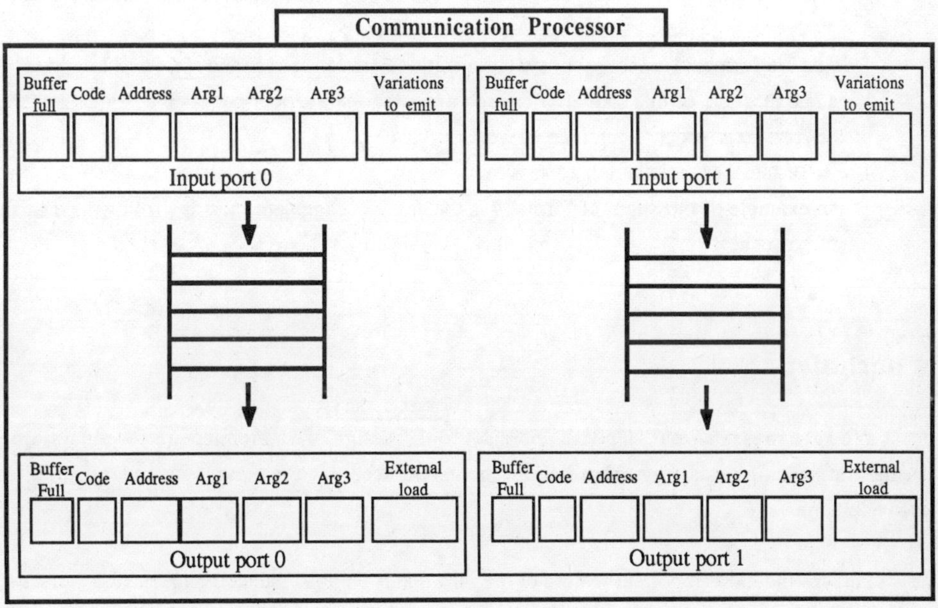


Figure 7 : The communication processor

### 2.2.3 The communication processor

Figure 7 gives the composition of a communication processor. As can be seen, there are two input ports, two output ports and two queues to store messages. The purpose of a communication processor is double. First, it is used to direct messages from an input port to an output port. Second, it is used to send backward a load value through the input ports, this load being computed from the load information received at its two output ports.

The communication processor handles messages between reduction processors and memory processors. A message may have one of the four following formats :

1- < code-message, destination address, function, argument >

This is a message to be sent to a destination address. There is no sender address because no response is required by such a sender. This format is used for the message sent by a process to write its result on the corresponding node.

2- < code-message, emitter address, function, argument >

This is a message sent without a destination address. It can be a message for either a process allocation or a node allocation. This message will be sent to the less charged output port, considering the load corresponding to the code message (execution load or memory load).

3- < code-message, destination address, emitter address >

This is a format of message for synchronization between sender and receiver. Increment ref-ctr is one message of this format. For such a message, the sender needs an acknoledgement from the receiver in order to continue its work.

4- < code-message, destination address >

An example of messages of format 4 is the acknowledgement message following a increment ref-ctr request.

## Conclusion

We have presented our stepwise approach to the design of an architecture for a multiprocessor reduction machine, starting from requirements induced by the language and by architectural considerations.

Perhaps the most important feature of the machine is the load balancing mechanism, that includes the decentralized computation and distribution of an instantaneous load information, allowing tasks to be equally distributed among processors. Furthermore, this load information allows each reduction processor to dynamically change its execution model, thus contributing to the regulation of the instantaneous parallelism in the machine. We also saw how information concerning management of potential parallelism can be conveyed by combinators themselves.

More details of this architecture, such as the network organization, garbage collector, etc., and details concerning compilation, can be found in [Castan85, Castan86a, Castan86b, Lemaître86].

## Planning of the MaRS project

The global aim of the project is to build a prototype of a multiprocessor reduction machine by 1989. The MaRS project started in 1984. So far, work has been done on the following topics : evaluation of available parallelism in functional programs; translation of functional languages into combinatory code; definition of an abstract multi-processor machine with shared memory and message communication; and simulation of the abstract machine.

Present and future work includes : precise definition of a source language; optimized compilation taking into account potential parallelism in programs; "fine grain" execution model and its simulation; study of a minimum programming environment for the machine; collection of parallel symbolic functional programs for tests and validation; VLSI design using CAD systems.

It is planned to have the communication processor available as a VLSI chip in 1987, and the other processors by 1988. Integration, tests and final evaluation performance should be conducted by 1989.

## References

[Arvind83] Arvind, R.A. Iannucci
Two fundamental issues in multiprocessing : The Data Flow solution
MIT/LCS/TM-241, September 1983.

[Bawden77] A. Bawden, R. Greenblatt, J. Holloway
Lisp machine progress report, MIT memo ≠444, 1977.

[Berger80] Ph. Berger, D. Comte, N. Hifdi, B. Pelois, J.C. Syre
Mise en oeuvre d'un prototype de multiprocesseur en assignation unique
Rapport final, TEAU 18, ONERA-CERT, 1980.

[Castan80] M. Castan
Conception et réalisation d'une machine spécialisée dans le traitement des formes arborescentes.
Thèse de troisième cycle, Octobre 1980, UPS Toulouse, France.

[Castan85] M. Castan, M.H. Durand, M. Lemaître
A set of combinators for abstraction in linear space
September 1985, to be published in Information Processing Letters

[Castan86a] M. Castan, G. Durrieu, B. Lecussan, M. Lemaître
Etude sur le parallélisme non vectoriel dans les applications aéronautiques : mécanismes de base de la machine MaRS, machine multiprocesseurs à réduction symbolique.
Rapport final n° 1/3244/DERI, Volume 2, Janvier 86.

[Castan86b] M. Castan, M. H. Durand, G. Durrieu, B. Lecussan, M. Lemaître
MaRS: a multiprocessor machine for parallel graph reduction
Hawaii International Conference on System Sciences, Honolulu, January 1986.

[Chu81] Y. Chu, M. Abrams
Programming languages and direct execution computer architecture, Computer, July 1981.

[Church41] A. Church
The calculi of lambda-conversion
Ann. of Math. studies, Princeton, N.J. 1941, 2nd ed. 1951.

[Darlington81] J. Darlington, M. Reeve
ALICE, a multiprocessor reduction machine for the parallel evaluation of applicative languages
Proc. of the Conference on functional programming languages and computer architecture.
October 1981, Portsmouth, New Hampshire.

[Durrieu86] G. Durrieu
LAURA : a parallel non vectorial data driven processor for aeronautic workstations
Hawaii International Conference on System Sciences, Honolulu, January 1986.

[Feustel73] E.A. Feustel
On the advantages of the tagged architecture
IEEE Transactions on Computer, Vol 22, N° 4, 1973

[Friedman78] D.P. Friedman, D.S. Wise
CONS should not evaluate its arguments
Michaelson and Milnes Eds., Automata, Languages and Programming, Edindurgh University Press, 1976.

[Greussay77] P. Greussay
Contribution à la définition interprétative et à l'implémentation des lambda languages
Thèse d'état, Université de Paris VI, Novembre 1977.

[Grit81] D.H. Grit, R.L. Page
Deleting irrelevant tasks in an expression oriented multiprocess or system
ACM Transaction on programming languages and systems, 3(1), pp 49-59, January 1981.

[Hankin85] C.L. Hankin, P.E. Osmon, M.J. Shute
COBWEB: A combinator reduction architecture
Functional Languages and Computer Architecture, Nancy France, September 1985.

[Henderson80] P. Henderson
Functional Programming, application and implementation
Prentice Hall International, 1980

[Henderson82] P. Henderson
Purely Functional Operating Systems, in Functional programming and its applications, Darlington, Henderson, Turner Eds.
Cambridge University Press, 1982.

[Hudak84] P. Hudak, B. Goldberg
Experiments in diffused combinator reduction
ACM Symposium on Lisp and Functional Programming, Austin Texas, August 1984.

[Keller84] R.M. Keller, F.C.H. Lin, J. Tanaka
Rediflow Multiprocessing
Proc. IEEE Compcon, February 1984

[Lemaître86] M. Lemaître, M. Castan, M.H. Durand, G. Durrieu, B. Lecussan
Mechanism for efficient multiprocessor combinator reduction
ACM Symposium on Lisp and Functional Programming, Cambridge Massachusetts, August 1986

[McCarthy60] J. Mc Carthy

Recursive functions of symbolic expressions and their computation by machines
CACM Vol 3, n°4, April 1960

[Stoye84] W.R. Stoye, T.J. Clarke, A.C. Norman

Some practical methods for rapid combinator reduction
ACM Symposium on Lisp and Functional Programming, Austin Texas, August 1984

[Stoye86] W.R. Stoye

Message Based Functional Operating Systems
Science of Computer Programming, 6 (1986).

[Treleaven82] P.C. Treleaven

Computer architecture for functional programming, in Functional Programming and its Application, J. Darlington, P. Henderson, D.A. Turner (eds), Cambridge University Press, 1982

[Turner79] D.A. Turner

A new implementation technique for applicative languages
Software practice and experience, vol. 9, 1979.

[Vegdhal84] S.R. Vegdhal

A survey of proposed architectures for the execution of functional languages
IEEE Transaction on computers, C-33(12), December 1984.

---

# THE PARALLEL GRAPH REDUCTION MACHINE, ALICE

P. G. Harrison and M. J. Reeve
Department of Computing
Imperial College
London

## Abstract

Traditionally the design of programming languages has been compromised in order to exploit the most efficient machine architecture available, which has usually been of the von Neumann type. However, we believe that *functional languages* provide the most effective means for producing software and that the right approach is to develop a *customised architecture* for the implementation of the most suitable computational model for these languages.

In this paper we adopt *graph reduction* as the most natural computational model for functional languages and show how to represent it in a packet-based abstract computer architecture, ALICE. This architecture is highly parallel in nature, with a collection of processing agents performing redex reductions concurrently and asynchronously on the one hand, and the packets that represent the nodes in a function expression graph being distributed over several memory segments on the other. We suggest how a concrete parallel machine can be built to realise this abstract architecture, and describe the design of an experimental prototype machine. This prototype is a hardware emulator of the ideal concrete architecture which is to be built in customised VLSI, and is constructed using the INMOS Transputer as the basic building block. The design utilised a performance model to assist in choosing the optimal organisation of machine components, such as the ratio of processor to memory boards, and the model's predictions for the performance of various configurations of the machine are presented, along with some preliminary measurements made on the prototype. Finally, we make some suggestions for future design enhancements based on our experiences to date.

## 1. Introduction

We advocate the design of a machine tailored to executing programs written in *functional languages*. This contrasts with the historical approach where programming language design has been constrained by the von Neumann computer architecture. A natural computational model for the evaluation of functional expressions is *graph reduction*, in which the graph representing an expression is repeatedly reduced, by transforming sub-graphs which represent *redexes* (reducible expressions). At each stage there may be one or more redexes which can be reduced, the choice of the next redex or redexes being determined by the evaluation order (computation rule). Thus we may adopt for our customised architecture a parallel evaluator in which the tasks to be computed simultaneously are identified with the redex rewrites which occur in the graph reduction computational model. In this way the architecture can exploit the parallelism inherent in functional languages, and formal transformation offers the prospect of providing techniques to find equivalent functions for which parallelism can be exploited to a greater extent.

In the next section we explain the concept of graph reduction in the context of the functional language HOPE [BM80], and show how a packet-based machine architecture can be used to realise this computational model. This leads to an outline, at the abstract level, of the ALICE architecture, [DR81]. In section 3 we describe the design of the transputer-based prototype machine, before discussing its parameterisation, which employed an analytical model, in section 4. Performance issues and current status are addressed in section 5 before the paper concludes with an outline for future research and a summary in section 6.

## 2. A packet-based abstract architecture for graph reduction

### 2.1 Evaluation of first-order expressions

Graph reduction, corresponding to the evaluation of first-order functional expressions, proceeds by transforming the graph representing an expression with respect to some *redex* node which represents a function application (β-reduction in the case of λ-expressions). The node is then *overwritten* with the sub-graph which represents the result of the application. There are three cases to consider:

(i) If the associated function is primitive, and all of its required argument sub-graphs have been evaluated, i.e. cannot be further reduced, then the application is evaluated and the redex node is replaced by the root-node of its result. In the case of a primitive arithmetic function, for example, all the arguments must be evaluated, and the result-sub-graph consists of the single root-node which contains an item of atomic data. But in the case of the conditional function, only the predicate must be evaluated, and the result is the appropriate argument-sub-graph.

(ii) If the associated function is primitive, but not all of the arguments it requires have been evaluated, evaluation of the next required argument (specified in the delta rules of the primitive function) is initiated.

(iii) If the associated function is programmer-defined, the redex node is replaced by the sub-graph formed from that representing the function's body (i.e. its defining expression), but with each arc that is connected to a node representing a formal parameter redirected to the corresponding sub-graph representing the actual argument. In other words, the body is essentially copied with its formal parameters instantiated by argument expressions in the usual way.

In fact (iii) is a little more complex if the reduction machine performs *pattern matching* , as is the case with ALICE. Pattern matching requires that certain arguments have been evaluated at least far enough to permit comparison with the patterns. When they have not, analogues of the delta rules of primitive functions initiate such evaluation prior to reduction.

The selection of the next redex to reduce corresponds to the evaluation order (or computation rule), for example Normal Order Reduction (which is semantically safe) or Parallel Reduction. We will return to this point later in this section, but it should be noted that Parallel Reduction gives Normal Order Semantics (apart from the problems of halting unwanted non-terminating computations, if such are initiated).

We now give an example of a functional program, written in HOPE, which will be used to illustrate the operation of the ALICE machine and then show how an expression-graph can be represented as a collection of packets in the memory of an abstract architecture. Consider the data type "tree", and suppose we wish to write a function, "size", which takes a tree as argument and returns the number of data items held in the tree as its result. In HOPE this might be written as:

**data** tree($\alpha$) == tip($\alpha$) ++ node(tree($\alpha$) x $\alpha$ x tree($\alpha$))

**dec** size : tree($\alpha$) -> NUM

--- size(tip(i)) <= 1

--- size(node($t_1$, i, $t_2$)) <= plus (1, plus(size($t_1$), size ($t_2$)))

In other words, the size of a tree which is a single tip is one, and the size of a tree with a data item at its root is one more than the sum of the sizes of its sub-trees. For example,

size (node (tip(i2), i1, tip(i3))) = 3

A picture of the argument is the tree:

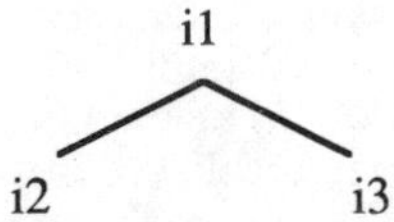


Underlining the redex(es) chosen for re-writing at each stage, reduction of the graph representing this expression proceeds as follows:

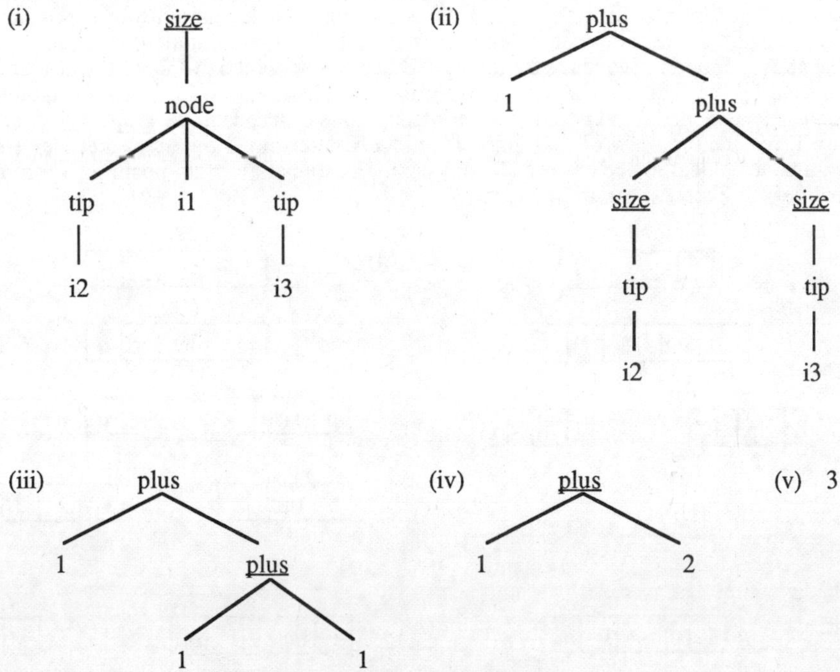


Figure 1 Graph reduction of the expression size(node(tip(i2), i1, tip(i3)))

## 2.2 Representation by packets

A node in an expression graph may be represented by a *packet* which is a record divided into fields of the form

| Identifier | Function | Argument List | Secondary Fields |
|---|---|---|---|

The Identifier field simply labels the packet uniquely so that it can be referenced by other packets (see below). It is typically implemented as the *address* of the location in computer storage of the packet, thereby not appearing explicitly in the packet itself. The Function field identifies the function associated with the node represented by the packet, and in practice would be the address of the code of a user defined function or the identifier of a primitive function. The Argument List represents the outgoing arcs from the node, and so consists of a list of identifiers of other packets (representing the top nodes of the argument sub-graphs). The length of the list is variable, so that conceptually packets too have variable length. (In practice packets would normally be of fixed length, providing for a given maximum number of arguments, m say. Function applications with more than m arguments would be implemented by linking together a sufficient number of packets to hold the extra argument-packet addresses). The secondary fields will be discussed later as they become necessary; they contain control information used by the evaluation mechanism.

We are now in a position to show how a sequence of graph reductions may be implemented by the creation and updating of packets. Since we have a one-to-one representation of nodes and arcs by a collection of packets, any graph transformation (corresponding to the rewriting of a sub-expression) can be expressed equally well and uniquely in terms of packets. Thus the above reduction sequence for the HOPE expression which evaluates the size of a tree is implemented by the sequence of packet collections shown in figure 2. The packets within any collection are positioned to reflect the corresponding node positions in the graph reduction shown in figure 1, but the topology of that graph is uniquely determined by the packet identifiers and their argument fields; the packets may be physically located at arbitrary positions in storage, i.e. the identifiers (addresses) are arbitrary.

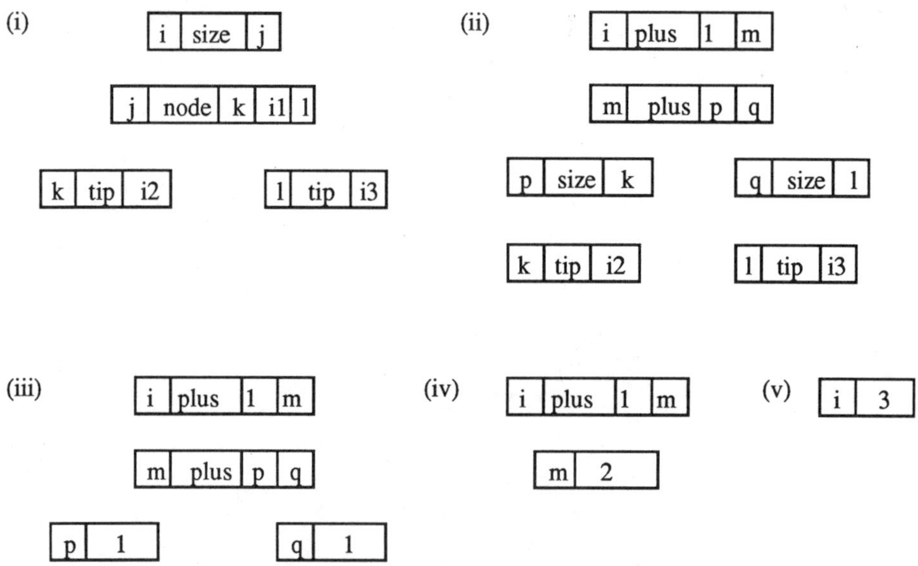


Figure 2 Packet representation of graph reduction

In the main part of this paper we do not consider higher-order functions, but they are supported quite simply in our packet-based model. Higher-order functions are handled at two levels. First, all higher-order functions are translated by the compiler into *partial applications* of a function to a smaller number of arguments than it has formal parameters. This is done by replacing any references to parameters of "outer" functions in a function's defining expression by *new* parameters of this function. (This is equivalent to *lambda-lifting*, [JO85]). Secondly, ALICE supports partial applications by creating *suspensions*, which are packets that indicate the number of missing arguments together with the arguments it has already collected by application of previous suspensions. The application of a suspension to an argument either just appends the argument to its argument list (for later evaluation) or else causes a normal packet rewrite if the argument is its last one.

## 2.3 Packet communication

Clearly the reduction sequence shown in figure 2 will evaluate the size of the tree correctly, but we have not yet specified how this particular sequence can be generated in the machine executing it. To do this we need a mechanism for indicating the packet (or packets) which corresponds to the next redex (or set of redexes) selected by the chosen evaluation order. A selected redex-packet is easily indicated by a status bit (and/or attaching the identifier of the packet to a "job list" of rewritable packets maintained somewhere in the machine's control system), and then only rewritable packets will be selected for reduction. It therefore remains to determine and mark these packets at each stage in the reduction sequence.

Now, an application of a user-defined function is selected as a redex according to the specified evaluation order, and to show the potential for parallelism, we use an "all redexes in parallel" rule. This will reduce a redex to its Normal form, but may also initiate some non terminating computation, which we will not worry about for the present. For an application of a primitive function, the corresponding delta rule will determine which arguments are needed to permit rewriting. The *number* of arguments needed is stored in the Pending Signals secondary field of the packet representing the primitive function's application, call this packet P, and only the packets representing the required argument-expressions are marked as rewritable in their status fields. After evaluation of an expression, represented by packet Q say, if this is an awaited argument of the primitive function application represented by P, the packet P must be "signalled" that one of its awaited arguments has been evaluated. This is done by having included the identifier of packet P in the Signal Set secondary field of packet Q at the time Q was generated (to cause the communication), and by reducing the Pending Signals field in P when the signal from Q is received. If this becomes zero, the packet P is marked rewritable. Note that the number of packet identifiers in the signal-set is variable; another reason why packets are conceptually of variable length.

The second configuration in the reduction sequence of Figure 2 is elaborated to show the signalling information in Figure 3.

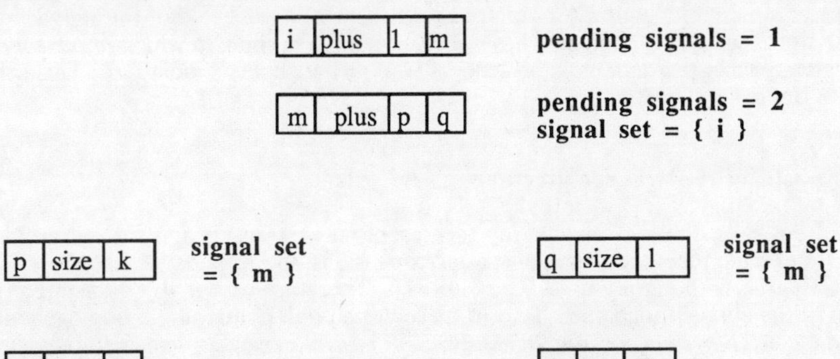


Figure 3 Detecting Re-writable Packets

The signalling mechanism as so far described implies a *control flow* scheme, whereby a signal merely indicates the availability of the result of an argument evaluation. The result itself remains in the signaling packet as per figure 4(a). (Of course if the result is a large structure, access to it would be indirect through the packet which would hold only a pointer.)

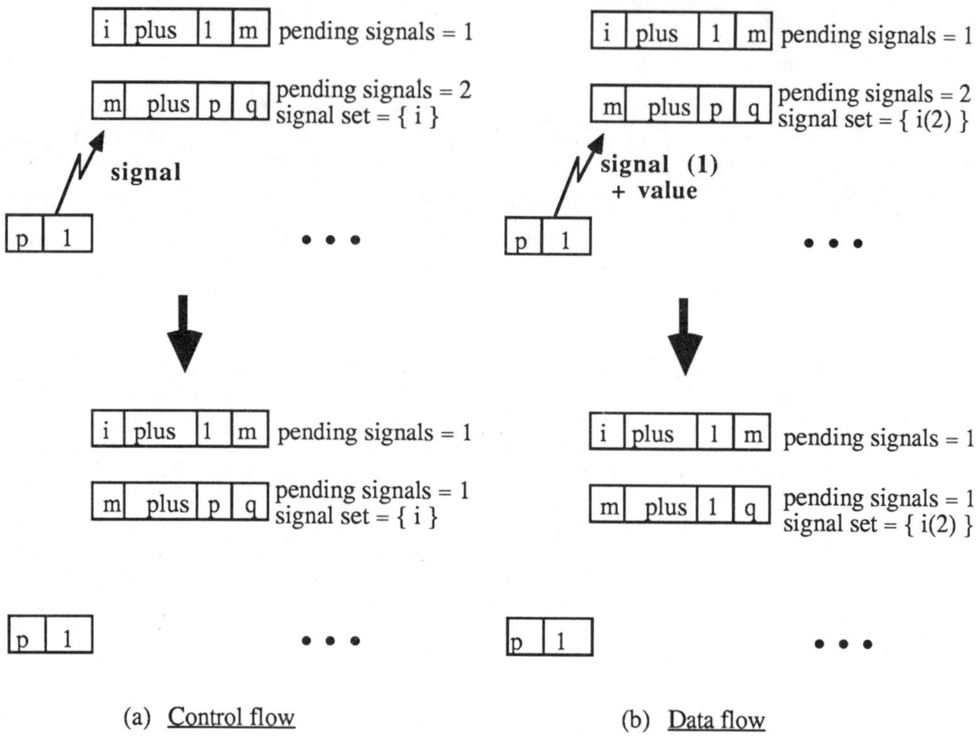


Figure 4 Signalling Mechanism

A more efficient scheme uses a *data flow* mechanism, whereby the data which is the result of the argument evaluation (or pointer to a structure) is sent to the waiting packet as the signal. The signal set information now has to be augmented to show to which *argument position* in the signalled packet the data must be sent, as well as the packet's identifier. This scheme is illustrated in figure 4(b).

## 2.4 An abstract ALICE architecture

We now have enough information to present an abstract design for our architecture, customised to execute functional, or applicative, expressions: the Applicative Language Idealised Computing Engine with acronym ALICE, [DR81]. There are essentially only three types of component in the abstract machine. Two of these are a *pool* containing a homogenous set of packets and a *collection of processing agents*, each of which repeatedly selects (or is assigned by some server) any rewritable packet from the pool, rewrites it and returns the updated redex-packet to the pool along with any newly generated packets. The whole process proceeds asynchronously until eventually the packet which originally represented the redex of the top-level expression has been fully evaluated and contains the result. This architecture is illustrated in figure 5.

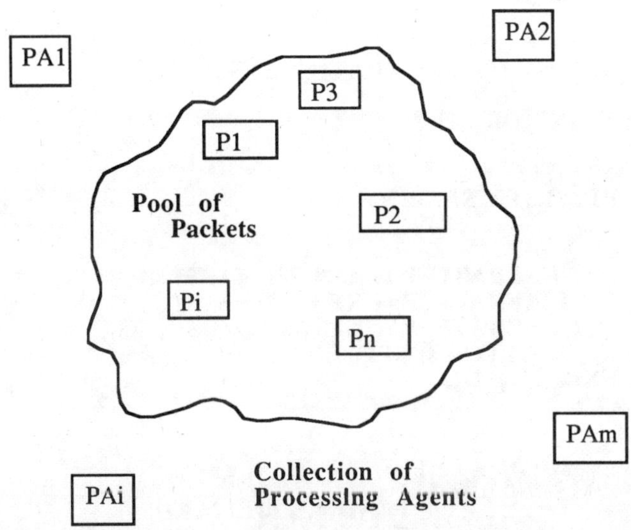


Figure 5 Abstract ALICE Architecture

The third component is implicit in the figure and provides the required *communication*; between packets, as described in the previous section (as well as for load sharing and other housekeeping) and between the packet pool and processing agents as just discussed above. The way that this abstract architecture has been realised in a prototype computer is the subject of the next two sections. Clearly many implementation problems have not been considered in this abstract model, for example the representations of function bodies and of structured data such as lists, garbage collection etc. Before moving on to consider the practical problems, we close this section by briefly describing the language which is used to control the operation of ALICE.

## 2.5 THE ALICE Compiler Target Language (ALICE CTL)

The ALICE Compiler Target Language (ALICE CTL) is the lowest level of user programming access to the machine and the target code for all compilers. Its most important feature is that it is independent of the underlying implementation. This has proved to be invaluable because it has allowed us to provide a *stable base* for software development that has decoupled the system software and compiler writers from the many changes that have been made to the physical architecture of the machine during the course of the project.

An ALICE CTL program is built from *constructor operator declarations* and *rewriteable operator definitions*. The former are analogous to the data statements found in a Hope program and the latter to the collections of equations that define a function. A rewriteable operator definition specifies the actions to be performed by a processing agent in order to process a packet whose Function field contains that operator.

A sample ALICE CTL program which rewrites applications of the size function considered in our previous examples follows, the various primitive instructions being explained below.

```
CONSTRUCTOR   tip($1), node ($1, $2, $3)

REWRITEABLE   size($1)

SEQ_DO
   SNAPSHOT_PACKET($1: OPERATOR, OPERAND())
   SEQ_ALT
      IS_CONSTRUCTOR($1)
         ALT
            $1 IS tip($T1)
               REWRITE_$PBP(!1)
            $1 IS node($T1, _, $T2)
               SEQ_DO
                  GENERATE_PACKET(&1: size($T1))
                  GENERATE_PACKET(&2: size($T2))
                  GENERATE_PACKET(&3: plus(&1, &2))
                  REWRITE_$PBP(plus(!1, &3))
               END_SEQ_DO
         END_ALT
      TRUE
         DO
            UPDATE_PACKET($1:
                                 PENDING_DEMANDS = --1,
                                 SIGNAL_SET = ++ (CONTROL_FLOW($PBP,
                                                  WHEN_CONSTRUCTOR)
                                                  )
                                 )
            RESTORE_$PBP(PENDING_SIGNALS = 1)
         END_DO
   END_SEQ_ALT
END_SEQ_DO
```

The 'brackets' ALT, END_ALT and SEQ_ALT, END_SEQ_ALT enclose alternative actions, only one of which is performed. Each alternative is "guarded" by a boolean expression. In an ALT group, all the guards are evaluated (in any order or indeed concurrently) and if more than one evaluates to true a programmer-defined arbitration method chooses just one alternative. In a SEQ_ALT the guards are evaluated in their textual order and the first alternative whose guard evaluates to true is chosen.

The brackets DO, END_DO and SEQ_DO, END_SEQ_DO enclose actions to be performed. Actions enclosed by DO and END_DO may be performed in any order or indeed concurrently. Those enclosed by SEQ_DO and END_SEQ_DO must be performed in the order specified.

Names beginning with the '$' symbol refer to arguments of the packet being processed, including the components of arguments which are composite data. The special name '$PBP' refers to the packet being processed itself. Names beginning with the '&' symbol refer to new packets generated during the course of the rewrite.

In section 3.4 below we describe how a 'size' packet is processed. The reader will be able to relate the steps taken to the above ALICE CTL very easily.

## 3. Design of the prototype machine

In this section we describe a physical realisation of the abstract architecture given in the previous section; a *concrete architecture*. The ideal machine would ultimately be constructed with customised VLSI chips, but a prototype, built from off-the-shelf programmable components, to emulate this architecture was favoured partly in order to shorten the delivery time, partly for reasons of cost and funding, but more importantly to achieve a high degree of flexibility in the design and control of the machine. In this way, much post-construction experimentation becomes possible, and indeed advisable, from which the design of a future VLSI machine would be greatly assisted. The prototype comprises three distinct types of boards of electronic components, corresponding to the logical components of the abstract machine, viz. *agent boards* (to perform asynchronous packet rewriting), *memory boards* (to provide a distributed packet pool) and an *interconnection network* (to provide fast, reliable communication between the agents and memories. The concrete architecture, showing these three basic constituents, is illustrated in figure 6.

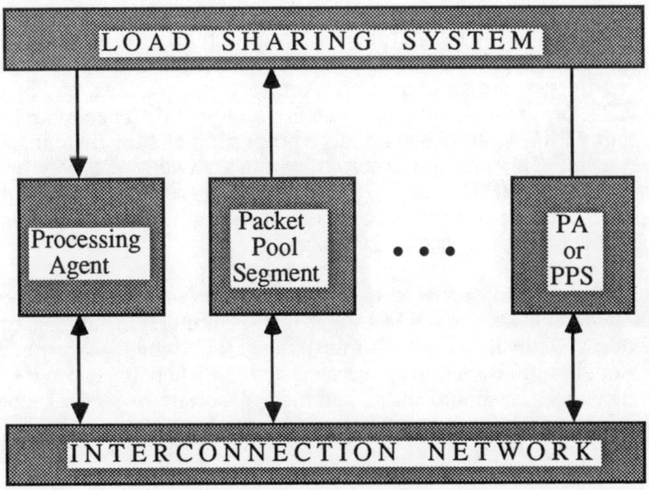


Figure 6 Concrete ALICE Architecture

A further type of communication is in general necessary, in order to implement a *load sharing system*, and at present this is also provided for by the interconnection network. This system is used to distribute work amongst the processing agents and the packet pool segments according to the current load on each such element. The additional communication is necessary in order to maintain and make available the appropriate information, although if load sharing is done on a purely *random* basis, no such communication is necessary; this is precisely what has been implemented in the initial version of the prototype. For expressions with high levels of potential concurrency, for which most processors can be kept busy for most of the time, one would expect the random algorithm to be efficient. This, we shall see in section 6, was found to be the case. However, the converse was also found to hold, expressions evaluating with a large degree of sequential behaviour giving poor performance without improved load sharing.

The designs of each of the three types of physical component are discussed in the following sub-sections, and section 3 closes with a description of the sequence of tasks performed during the evaluation of a functional expression, to illustrate the machine's principles of operation.

### 3.1 Agent boards

The processing agent component of the abstract architecture is realised by five INMOS transputers on a single board in the prototype machine. Two of these perform the packet rewriting (packet rewriting units, abbreviated to PRUs); one manages the local cache storage which holds the function definitions required by each PRU (function definition management unit, abbreviated to FDU); one serves as an input buffering device, handling all incoming transactions from the network and routing them to the appropriate PRU; and one serves as an output buffering device similarly. The logical organisation of these devices is shown in figure 7, the supporting hardware being omitted.

Each PRU has 64k of RAM and is devoted to the task of rewriting packets, according to their ALICE CTL code which is guaranteed to be in the local memory of the PRU by the FDU (see below). The ALICE CTL code exists in compiled form, in a lower level "Implementation Specific Language " or ALICE ISL. This language represents an idealised instruction set for a PRU, and is what is actually used in the prototype; the transputers, running OCCAM programs, *emulate* a PRU and *interpret* ISL code. Of course this is not the most efficient implementation, but it is the most suitable for the purposes of prototyping. Now, the cycle involved in the rewriting of a packet involves a number of *network transactions* ; to fetch a rewritable packet and argument packets, to generate an updated packet, to signal packets, etc, as may be seen in the sample ALICE CTL program of section 2.5 and in section 3.4. The amount of local processing is fairly small, and a PRU would spend a large proportion of time awaiting responses from the packet pool segments. This problem is avoided in the conventional way by *multiprocessing* a number of *virtual PRUs* (VPRUs or "virtual agents") in the PRU, so that when one VPRU is awaiting a response, another may be processing. If the number of VPRUs multiprocessed is v, the buffering transputers each multiplex 2v processes.

The processors which provide the network interfaces are the External Request Unit transputer (ERU) which has 8k of RAM and buffers incoming transactions, decoupling the PRU pair, and the External Transmission Unit transputer (ETU) which performs the complementary task of handling outgoing transactions generated by the PRU pair. It is the ETU which initiates and manages the network communication, and the appropriate *process* executing in it is marked explicitly as NCU (Network Connection Unit) in Figure 7.

Finally, the FDU is responsible for managing the function definition caches of the two PRUs. When a packet arrives in the ERU destined for a PRU, call it A, it is first passed to the FDU which inspects the function field and checks if there is a copy in A's 64k of RAM. If there is, then the packet is passed on to A. If there is a copy in the other PRU's cache, a copy of this is written into A's cache. Otherwise further network transactions are scheduled through the ETU to fetch the function definition from the central function store, residing in the packet pool memory. Now, the ALICE CTL comprising a function-definition is generated by a compiler, written in HOPE for example, as an ordinary data structure in terms of constructor functions. It is therefore compiled into packets in the usual way, and so a function-definition also may be distributed in any way over the packet pool segments. Thus to obtain a function's definition, the FDU schedules a sequence of packet-fetches. When the packets containing the code defining the rewrite rules have all arrived (through the ERU), and been written into A's cache, the original redex-packet is sent to A (by the FDU) which can then perform the rewriting. The states of the caches of the two PRUs are held in tables in the FDU's 8k of RAM.

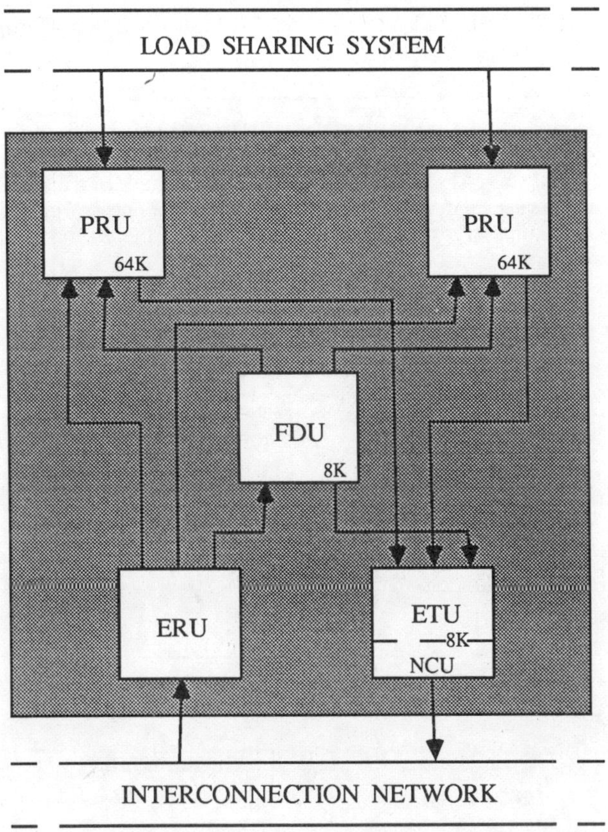


Figure 7 Processing Agent Board

## 3.2 Packet Pool Segment Boards

The abstract packet pool is distributed over a number of *packet pool segments* (PPSs) in the prototype machine, each segment being built on one board. The segments form one address space. The most significant bits of a packet's identifier are used to specify the segment in which it is stored and the remaining bits provide the offset within that segment. The processing performed by a packet pool segment is rather simpler than that of an agent, comprising in the main the transmission of network transactions in response to incoming messages from PRUs. The major part of the board is allocated to memory chips, providing two megabytes for each segment, controlled by a transputer called the packet management unit (PMU). This transputer also handles the buffering of inputs from the network via a high priority process which has access to enough input buffer space to accommodate one message from *every* virtual agent. Thus incoming network transactions can never be blocked and deadlock cannot occur. As with the processing agent boards, there is a separate ETU transputer to buffer the outgoing transactions and decouple the PMU. The incorporation of input buffering into the PMU avoids the need for a separate ERU, so saving one transputer per PPS board. This is feasible because the computing tasks required of a PMU are sufficiently small, whereas the same is not true of the PRUs, which would become overloaded if they also had to support high-priority buffering processes. The layout of a PPS board is shown in Figure 8.

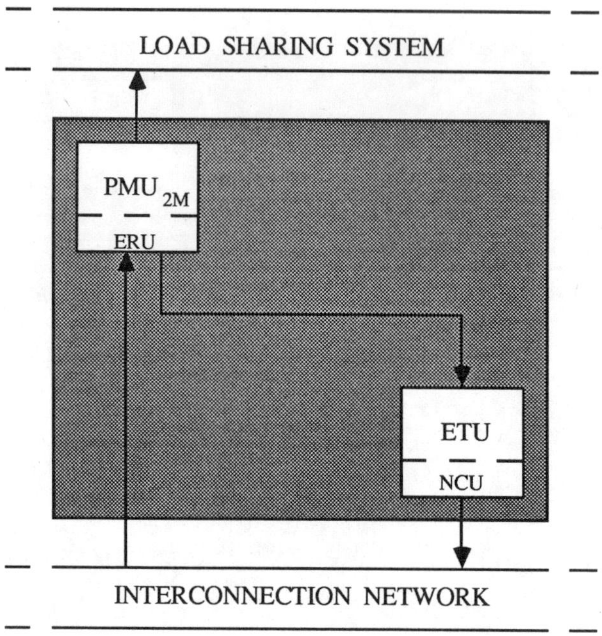


Figure 8 Packet Pool Segment Board

### 3.3 Communication and the interconnection network

The interconnection network provides bi-directional communication between any processing agent and any packet pool segment. Its ideal implementation would be a *crossbar switch* but such are impractical for all but small numbers of inputs and outputs, their complexity being proportional to the square of the size. A *multi-stage switching network* with a *generalised-cube* topology still allows any input to be connected to any output but is far more cost effective; when there are N input and output ports the complexity is proportional to NlogN. Such networks are the subject of [SI85]. Their disadvantage is that the routes from two different inputs to two different outputs may share some intermediate link, thus forcing one communication to have to wait for the other; this is known as *blocking*. There are many types of shuffle exchange networks, the one used in the prototype machine is a *delta network* [Pa79]. The basic building block is a b x b crossbar switch, in our case an in-house designed 4 x 4 switch implemented in Emitter Coupled Logic (ECL) and known as the XS1 [FC85]. A network with $b^n$ inputs and outputs is built from n *stages* of $b^{n-1}$ switches. The links between the switches in successive stages are arranged so that each of the $b^n$ inputs to the network "sees" a complete 1-to-$b^n$ decode tree encompassing all the $b^n$ outputs.

In order to keep the engineering complexity of the network within managable bounds we use *bit-serial* communication. The XS1 chip can switch any input to any output in 85nS and will support data rates of up to 150Mbits/sec, but in the prototype machine, communication is via Transputer serial links which operate at 10Mbits/sec. The XS1 is fully *asynchronous*, allowing paths to be set-up at any time and held for any period of time. The network in the prototype machine uses a *circuit switching* communication protocol; i.e. when a source processor wishes to transmit a message, it first establishes a path through the network that links its ETU (output buffer) to the destination processor's ERU (input buffer) and then initiates a direct end-to-end transfer of the data. During the path-building phase, in the event of a clash at an intermediate switch, the partial path established is *held* rather than released, although the latter alternative is provided for in the hardware. A blocked message essentially 'waits' at an input pin of the blocking switch which is polled on every cycle of a 4-phase, non-overlapping clock within the switch to prevent more than one message being allocated the same output simultaneously. When the output required by a blocked message becomes free, the message can be connected to the appropriate input pin of the next stage. Thus all contention is internal in switches, there is no queueing, and deadlock cannot occur.

Communication over the network is monitored by fault detection mechanisms that check for misrouted or corrupted messages. After six months of continuous use we have yet to see any such errors during normal operation. The network in the prototype machine has 64 inputs and outputs. Thus there are 3 stages of 16 XS1s. The 48 chips and their supporting electronic components such as buffers and drivers are distributed over 8 boards.

### 3.4 Principles of operation

We have seen how functional expressions are evaluated in the abstract architecture, and how the primitive logical elements of this architecture can be realised in a prototype. It is now an easy task to describe the operational principles of ALICE in terms of the *network transactions* generated in the rewriting of a reducible packet residing in one of the PPSs.

First, a packet is *fetched* by any of the idle virtual agents (VPRUs) which issues a request for a reducible packet to one of the PPSs, chosen by the VPRU *at random* (recall we have only random load sharing at present). A reducible packet, if one is present in the PPS, is sent back to the VPRU as a response - otherwise the response indicates that there are no reducible packets, and the transactions are wasted if there are reducible packets elsewhere. Such wasted transactions may or may not significantly affect performance (as we discuss in section 5), but intuitively would be less frequent and have less effect in evaluations having more parallelism, where more processing elements are likely to be busy.

Once a rewritable packet arrives at the ERU of its destination agent board it is passed on to the FDU which ensures that the code defining the function is present in the requesting PRU's cache, by fetching the code via further network transactions if necessary, as discussed in section 3.1. The FDU then sends the redex-packet to the appropriate PRU, addressed to the requesting VPRU. The remainder of this VPRU's operations and network transactions are fully defined by the ALICE CTL associated with the function appearing in the packet. This may involve fetching argument-packets from given addresses (i.e. the PPS and location within its memory), updating and re-storing the redex packet at its original address and generating new packets. A new packet is stored in the RAM of a PPS which is selected at random by the VPRU, at an address which is sent back to the VPRU as an acknowledgement by the PPS. (If the RAM happened to be full, an error code would be returned and the VPRU would try another PPS.) The only other main operation that the VPRU performs is to rewrite a packet with a *primitive function* in its function field.

An illustration of these operations is given by the ALICE CTL code for the rewrite rules of the size function in section 2.5. Having received a size packet to rewrite, the VPRU first fetches the argument-packet, $1. If this is a constructor, the argument has been evaluated (in a lazy implementation) and the packet can be re-written. If the constructor is "tip", the result is the constant 1 and if it is "node" two new packets are generated to evaluate the sizes of the sub-trees. Their addresses, &1 and &2, are bound when the acknowledgements are received from the chosen PPSs. A third packet is then generated which adds the results of packets &1 and &2, and has its address bound to &3. The redex-packet ($PBP) is then updated so as to evaluate the sum of 1 and the result of packet &3. If the argument-packet is not a constructor, then its evaluation must be initiated. This is accomplished by first decrementing its Pending Demands field (this will have been 1 in the lazy implementation), then inserting the identifier of the redex-packet into its Signal Set together with an indication that the signal is to be transmitted when its function field becomes a constructor (i.e. when the packet has been evaluated), and then re-storing it in the packet pool. The control flow signalling scheme is used in this example. The redex-packet is then re-stored with its Pending Signals field set to 1; to be awoken on receipt of the signal from its newly updated argument-redex-packet.

### 3.5 Other Features

There are a number of supporting features which should be mentioned. These include the diagnostic and monitoring system, Input-Output facilities and garbage collection. The diagnostic and monitoring system provides sophisticated fault detection and location tools and enables detailed data on the behaviour of the system to be collected for analysis; this data will be used in the design of any successor machine. The Input-Output system provides an Ethernet link to the machine, which is used as a high-speed application load and monitor data dump route.

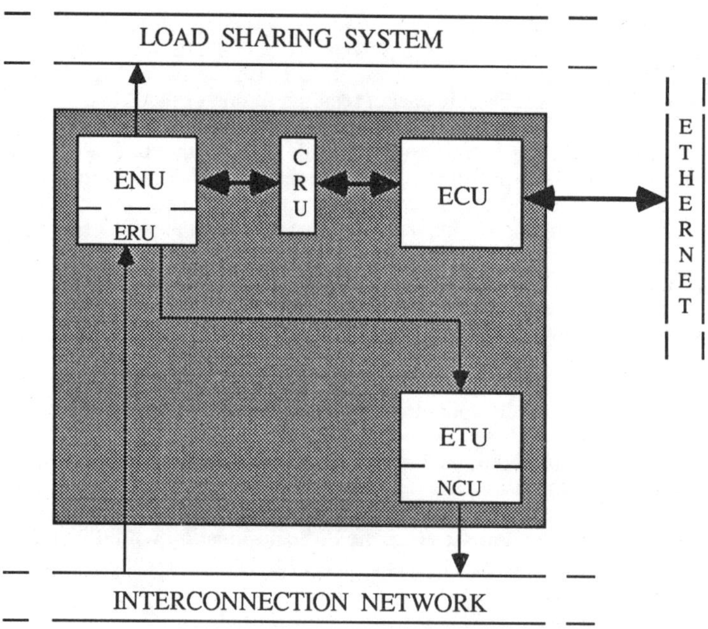


Figure 9 Input-Output Board

As a computation proceeds, many intermediate results are generated. When these are finished with, i.e. are no longer arguments of any function application, the packets they occupy can be reused. This reclaiming process is known as *garbage-collection*. A count, known as the *reference count*, is maintained in all packets, and records how many times the value held in a packet is used by other parts of the computation, i.e. how many times a packet's identifier appears in the Argument List field of other packets. Each time a new reference to a packet is generated the reference count of the packet is incremented by one. Similarly, each time a redex evaluation is completed, the references to its argument-packets disappear and their reference counts are decremented by one. If the reference count of a packet reaches zero, no part of the computation requires its value and so the packet can be reclaimed. This is known as *reference count garbage collection*, and on ALICE this is performed concurrently with expression evaluation. Reference count updating is performed by PRUs as part of the packet rewriting process, but is not shown in the ALICE CTL code given in section 2.5 for the function size.

## 4. Parameterisation by an analytical performance model

From the discussion of the previous section, it should be clear that a correctly operating prototype ALICE machine could be constructed using any numbers of agent boards, say p, and PPS boards, say m, together with a large enough delta network, i.e. one that provides at least (p + m)-way communication, and supporting hardware. A major design problem, therefore, is to determine the optimum ratio, p : m, given a constraint on the total size of the machine. In practice the constraint was essentially the size of the rack, which could in fact accommodate 40 processor and memory boards, i.e. p + m = 40. The total number of transputers, the most expensive of the electronic components, was also limited.

A second design issue concerned the degree of multiprocessing of VPRUs in each PRU, the parameter v in section 3.1. Assuming for the moment that memory size is unlimited, the mean packet rewriting rate will always increase with v, provided that there are sufficiently many reducible packets, i.e. at least 2pv, the total number of VPRUs. However, since at most 2p PRUs can be processing at any time, the gains in performance will diminish as v continues to be increased beyond a certain level, and an asymptote will be approached. If now memory is also taken into account, the more VPRUs multiprogrammed, the less local RAM will be available for the function definition cache, and so the greater will be the "miss rate" encountered by the FDU. Thus the number of remote function store accesses will increase, *degrading* performance. There is also the 'microscopic' storage problem of *addressing* of the VPRUs - whenever v increases through a power of 2, an extra addressing bit is required. So what value should we choose for the number of VPRUs in each PRU transputer? Other questions which arise are 'What will be the throughput of a given configuration (in terms of p, m and v as well as other parameters like device processing rates)?'. 'Where will the bottlenecks occur?'. 'How many PRUs should be supported on a single agent board?'.

Rough intuitive guesses could be made for all of these questions, based on the experience of the engineers and using timings of the OCCAM code in which the controlling firmware was written. These guesses suggested that p should be a bit less than m, v should be about 4, throughput might be around 10-20k rewrites of the Fibonnacci function per second (the "NFIB" rate discussed in section 5) and that either the PRUs or PMUs would form the bottleneck. It was possible to be a little more precise regarding the further question of the number of PRUs, n say, to put on an agent board. At low utilisations, one is enough. In a high-loading situation, recall that all of the incoming work for each PRU first passes through the ERU (and FDU in the case of a new redex-packet). Therefore if the mean time delay for an incoming message in the ERU is $T_e$ and in a PRU is $T_p$, then we should aim to have $T_p = nT_e$ so that the throughputs of the ERU and the combined set of PRUs match (and any bottleneck is then shared between the ERU and PRUs, giving best performance). If $n > T_p/T_e$, performance will not degrade, but the cost of the machine will increase. If $n < T_p/T_e$ performance will suffer.

In order to make precise, reliable predictions for the other quantities of interest, a *performance model* was constructed. In view of the number of components in the concrete architecture, and other disadvantages, a simulation model was rejected, and an analytical model was adopted, [HF86]. An analytical model which represented every transputer individually would also be expensive to run, and a method of aggregating like components was developed using decomposition methods [CO77]. This resulted in a queueing network model with five servers having queue length dependent service rates, which was solved using the standard Convolution Algorithm, see for example [SC81]. The five different servers represented the corresponding five collections of all (a) PMUs; (b) PPS board ETUs; (c) combined ERUs and FDUs on agent boards; (d) PRUs; (e) agent board ETUs. The service rate functions of the aggregate servers were calibrated using the mean service times estimated by OCCAM timings for the associated transputers, and mean DMA transmission times. In addition, each service rate depends on the number of devices aggregated, and number of processes multiprogrammed in them in cases (c), (d) and (e). Thus, formulae for the service rate functions of the five aggregate-servers were given in terms of the mean service times referred to above and respectively m, m, (p and 2v), (2p and v) and (p and 2v) for a prototype with p agent boards and m PPS boards. In each case, the service rate for any queue length is proportional to the rate of an individual server in the aggregate, given by the reciprocal of the estimated mean service time. When the queue length is $n$, the service rate function, $R(n)$, assuming that individual transputer service rates are unity, is as follows (see [HF86]):

Cases (a) and (b): $R(n) = mn/(m+n-1)$

Cases (c) and (e): $R(1) = 1, \; R(n) = \{(2(p-1)v-n+1)/(2pv-n+1)\}R(n-1) + 2pv/(2pv-n+1) \quad (n\geq 1)$

Case (d): $R(1) = 1, \; R(n) = \{((2p-1)v-n+1)/(2pv-n+1)\}R(n-1) + 2pv/(2pv-n+1) \quad (n\geq 1)$

The model gave predictions for (amongst other quantities) throughput, for various parameterisations, assuming highly parallel execution in which all virtual agents are always busy. A graph of throughput versus number of agent boards, p, where p + m = 40, shown in Figure 10, indicated that little more performance improvement can be obtained for v > 8, optimum performance then being obtained for p = 16 and m = 24, giving throughput of approximately 120k network transactions per second. The service rate of a single PRU transputer was estimated to be 4, so that this performance prediction is quite close to the theoretical maximum of p.2.4 = 128 transactions per second, obtained when no PRU is idle at any time.

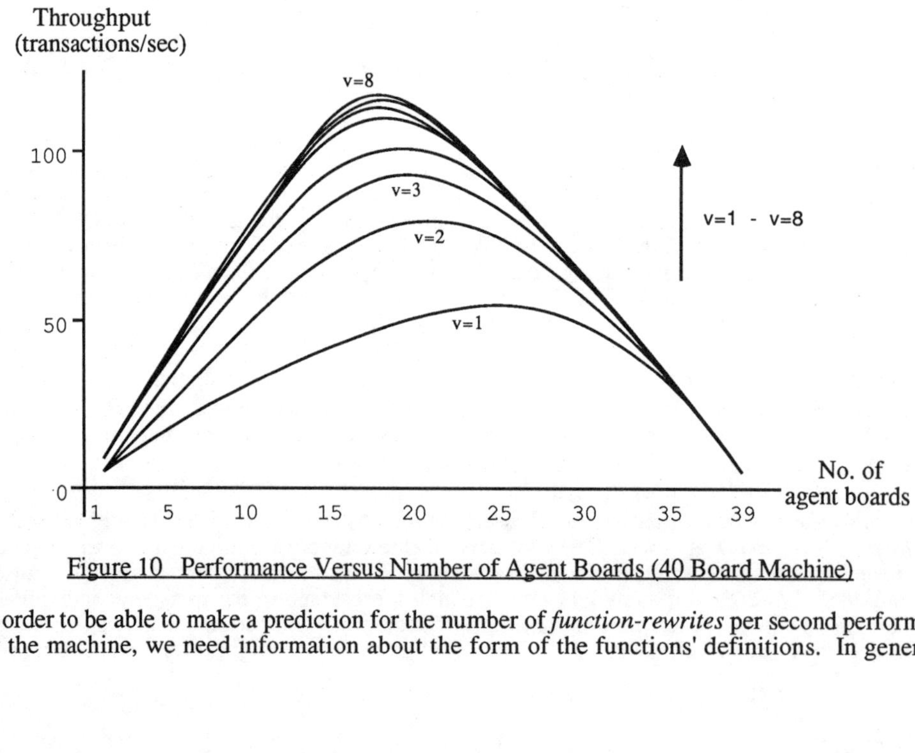


Figure 10 Performance Versus Number of Agent Boards (40 Board Machine)

In order to be able to make a prediction for the number of *function-rewrites* per second performed by the machine, we need information about the form of the functions' definitions. In general, this requires a *workload model also*, but for the purposes of benchmarking we can just inspect the ALICE CTL of the functions concerned to find the number of transactions involved in a rewrite. Our main benchmark is based on the fibonnacci function and has HOPE defining equations

--- NFIB (1) <= 1;
--- NFIB (2) <= 1;
--- NFIB (succ(succ(n))) <= 1 + NFIB (succ(n)) + NFIB(n);

This function has the property that the total number of function calls made to NFIB in the evaluation of an application of NFIB is equal to the result (the addition of 1 includes the top-level call). This property follows immediately by induction on the argument, using the recursive equations. Thus, mean NFIB-function-rewrite rate can be estimated by applying NFIB to any argument and dividing the result by the measured execution time - we call this the 'nfib-rate'. However, this will only give the same answer for every argument if the ratios of the numbers of calls to "plus", NFIB-base-case (NFIB(1) or NFIB(2)) and NFIB-non-base-case are always the same. Moreover, in computing the total number of network transactions involved, we need to know separately the numbers of these different types of call. These results can be derived quite simply as follows. First define the class of functions $f_{ij}$ $(i,j\geq 0)$ by

$$f_{ij}(1) = f_{ij}(2) = i, \text{ and } f_{ij}(n+2) = j + f_{ij}(n+1) + f_{ij}(n) \quad (n\geq 1)$$

Thus $f_{10}$ is the normal Fibonnacci function, *fib* say, and it follows at once that for all $n\geq 1$, $i,j\geq 0$, $f_{i0}(n) = i.f_{10}(n)$ and $f_{ij}(n) = f_{i+j,0}(n) - j$. Now fixing the argument $n$ = N, and writing F to denote *fib*(N), in the evaluation of the expression NFIB(N) = $f_{11}$(N) = 2F - 1, we have that

(i) The number of base-case calls to NFIB is F

(ii) The number of non-base-case calls to NFIB is $f_{01}$(N) = F - 1

(iii) The number of calls to "plus" is $f_{02}$(N) = 2F - 2

These results again follow by simple induction, and we note as a check that the total number of function calls must be $f_{13}$(N) = 4F - 3, by similar reasoning, remembering to count two pluses and the top-level call in the recursive defining equation. This total is the same as the sum of (i), (ii) and (iii) as required. We have therefore established that the numbers of calls to "plus" , base-case-NFIB and non-base-case-NFIB are effectively equal to 2F, F and F respectively for even moderately large N, and that our nfib-rate will be the same for all N.

Since applications of NFIB have well-balanced expression graphs, NFIB is 'highly parallel' in that for a sufficiently large argument, there will be enough redexes to ensure that none of the virtual agents need be idle throughout most of the computation. Thus the assumptions of the model are acceptable for this benchmark. We assume use of the *eager* evaluation order for NFIB(N) in which every sub-expression that *can* be rewritten is marked reducible, and the data-flow mode of signalling between packets. In this way, all rewritable packets have their arguments in *immediate* form. Now, for the rewrite of a non-base-case-NFIB packet, there are 5 network transactions: fetch a reducible packet; generate two packets to evaluate NFIB applied to the smaller arguments; generate a packet to add the results of these two packets; and rewrite the redex-packet to add the immediate data value 1 to the result of this packet. For the rewrites of a "plus" or a base-case NFIB packet, there are 3 transactions: fetch a reducible packet; send the result as data to the packet with identifier in its signal set; and rewrite the redex-packet (although an optimisation would omit this transaction which is not necessary in the data-flow mode). Thus the total number of network transactions in the evaluation of NFIB(N) is 5(F-1) + 3(2F-2) + 3F

≅ 14F, and so the predicted maximum nfib-rate is approximately 2F/(14F/120,000 ) ≅ 17k rewrites per second.

The model predicts further that near the optimal parameterisation of the system, a 10% increase (or decrease) in the service rate of the PRUs results in a decrease (or increase) of 1 in the optimum value for the number of agent boards, p (when the total number of boards, p + m = 40). These predictions are displayed in the graph of Figure 11, which shows the performance curves near the optimum number of agent boards for individual PRU transputer service rates between 3.6 and 4.6. In fact at the optimum point, we really had p ≅ 17, but it is expected that the PRU code can be optimised significantly, increasing the service rate of the PRU server. Hence the conclusion that the ideal numbers for p and m are 16 and 24 respectively.

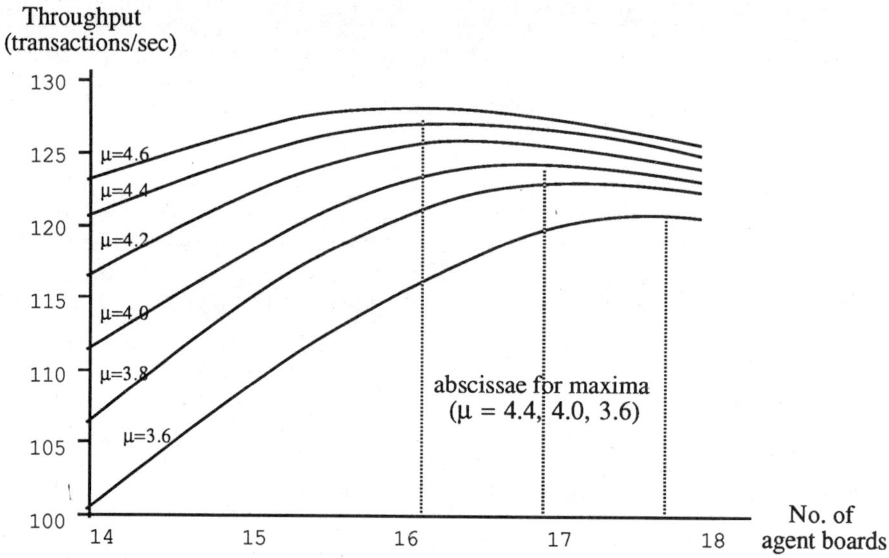


Figure 11 Sensitivity Curves Near the Optimal Parameterisation for PRU service rates between 3.6 and 4.6

The model also predicted the mean queue lengths and utilisations of the transputers, and the result of the bottleneck analysis was that with the optimal parameterisation, the PRU and PMU transputers are *necessarily* joint bottlenecks.

## 5. ALICE performance and current status

The ALICE prototype is now operational. At present, RAM images of a computation's initial packets (the top-level expression, its data and the ALICE ISL of the associated functions) are compiled off-line and loaded by the Ethernet link into storage addresses generated by the compiler. However, it is intended to mount a small operating system so that the machine can be made stand-alone. Currently application-level input-output is limited to a keyboard and screen, but it is planned that this will be extended to include a high-bandwidth link via the Ethernet. The source language used is HOPE, which is compiled into ALICE CTL, and a second compilation translates ALICE CTL code into ALICE ISL. It is expected that the parallel logic programming language, PARLOG, will also be compiled into ALICE CTL soon, providing an alternative application language, [DR83].

A number of functional programs have been monitored during their execution on ALICE. The main benchmark is the NFIB function, defined by the HOPE equations given in Section 4. As already noted there, applications of this function have well-balanced graph representations and so great potential for exploiting parallelism. Thus, when applied to a large enough argument, all VPRUs can be kept busy for the large majority of the execution time, and so the prototype machine can be expected to come close to giving its maximum performance. Measurements indicate this to be about 16000 NFIB-rewrites per second, or 32000 function rewrites per second if the contributions from the additions are also included. It has also been observed that, as the machine is scaled down, using fewer than 40 boards in total but always in the optimum ratio, performance decreases linearly with size - i.e. the NFIB rate is about 1k rewrites per second per agent board, e.g. 2k when there are two agents and three PPSs. In fact we also observe 1k when there is one agent and two or more PPSs. This follows since any PPSs in excess of 1.5 are wasted in that the PRUs are then the bottlenecks and become saturated. The linearity of performance with machine size observed for evaluation of NFIB applications is simply a consequence of the high degree of parallelism. In the next example, as the machine size increases, not all VPRUs (and PRUs) can be kept busy and performance does not increase significantly.

The second test function solves an optimal route planning problem called "Mike's tour" after Mike Reeve's travelling exploits. This evaluates in a very sequential fashion, and originally exhibited the anomalous property that as machine size increased, execution time *also increased*! The cause was soon discovered to be a limitation of the very naive implementation of "random load sharing". On becoming idle, a VPRU immediately issues a request to a (randomly selected) PPS for a packet to rewrite. If the PPS has none, it indicates thus in its response. The VPRU then chooses a PPS randomly again and the cycle will repeat until a reducible packet is located. Unfortunately a function application may have inherently sequential execution characteristics, i.e. may have a biased rather than balanced reduction graph; consider the factorial function, for example. In such a case, there will be few reducible packets at any time, and large numbers of network transactions will be generated as most VPRUs repeatedly request packets and receive the null response. "Mike's tour" is such a function, and the problem is compounded by the use of lazy evaluation which limits the number of rewritable packets to a minimum. Thus the DMA channels associated with network transactions saturate, and the PRUs and PMUs doing useful work become underutilized. In queueing model terms, the bottleneck has moved due to the sudden decrease in the mean service times of the PRU and PMU servers; although this results in *increased* transaction rate, most transactions do not represent useful work.

The problem of the saturation of the DMA channels with useless messages is easily solved. When a PPS receives a request for a packet to rewrite, but has none, it should *not* send a response until it does have a reducible packet, which it *then* sends. The useless message traffic is thereby removed, and the only problem now occurs when a VPRU is blocked waiting for its destination PPS to acquire a reducible packet when there *are* such packets in other PPSs. However, in a low-load situation, other VPRUs will rewrite these packets, and in a high-load situation a reducible packet will soon arrive in any given PPS. The solution therefore appears effective.

The final function application we consider illustrates the non-deterministic properties of parallel implementations. The example is the function that merges two lists, and can be written non deterministically in a HOPE style using equations with overlapping left hand sides thus:

```
--- merge (x :: y, z) <= x :: merge(y, z);
--- merge (y, x :: z) <= x :: merge(y, z);
```

The symbol :: denotes CONS, and the type declarations and base case equations have been omitted. Two lists will merge non deterministically in ALICE if the above equations generate new packets in a *parallel* ALT block in the ALICE CTL defining the rewrite rules for merge. Alternatively, by using SEQ_ALT, biased or alternating merges can be achieved.

## 6. Future plans and conclusions

A great deal of experimentation and enhancement is planned for the ALICE prototype in the near future, with the particular intention of aiding the design of a customised VLSI-version. This work will be considered in the categories of modelling, experimentation and machine enhancements.

There is an important role for modelling in the design of complex architectures, as we saw in section 4 for example. The development of reliable analytical models requires systematic validation, and the availability of a flexible experimental machine provides a rare opportunity for this. Conversely, the availability of a reliable model greatly enhances the prospects for the successful design and implementation of significant enhancements, as well as for the design of future machines. Indeed we expect to employ such models in the design of the customised successor to ALICE, so that ALICE herself will not only aid this design through direct experimentation, but also indirectly by assisting in the development of modelling techniques.

Regarding machine enhancements we have already noted in previous sections that a more sophisticated load sharing system than the random one should be developed, with the aim of allocating newly generated packets to the less utilised PPSs and ensuring that requests from VPRUs for reducible packets are addressed to PPSs which have them. A further improvement would then ensure that the reducible subset of packets is also evenly distributed amongst the PPSs in order to reduce the queueing of packet requests at overloaded PPSs. Any improved load sharing will require information to be available about the distribution of work in the system, which itself will require inter-PPS communication, which can be provided by the delta network. A more radical suggestion is to have the packet rewriting cycles initiated by the PMUs, which would choose VPRUs to which to send redex packets. This, some claim, facilitates more balanced load sharing.

An anticipated problem which is inherent in the pure abstract architecture is that of a *grain size* which is too small. In other words, the amount of computation devoted to expression evaluation performed by a VPRU between network transactions is small compared with the latency of the transactions, i.e. the round-trip response time associated with requests sent to the PPSs. In order to recover this lost performance, a very high degree of parallelism may be required. However, excessive parallelism will overload the set of VPRUs which are limited in number to 2pv, or 256 in ALICE. Thus we should exploit *locality* where possible, for example by increasing the sizes of function bodies. Other ways of reducing the number of network transactions would be to allocate packets for rewriting to PRUs which already have the required function definitions in their caches, and for a VPRU to retain at least one packet that it generates in *reducible* form for rewriting later. The second technique should also increase the effect of the first; consider a recursive function definition.

A further machine enhancement we mention concerns the delta network. Although this does *detect* faults it is not fault-tolerant. However, extra stages could be added to provide alternative routes in the normal way, see for example [SI85].

There are also a number of possible language-related enhancements. The prototype machine has been organised to provide for the parallel implementation of a number of languages. However, it is planned to optimise the transputer OCCAM control programs to handle purely functional expression evaluations only, with maximum efficiency. This might also lead to *compiling* ALICE CTL into OCCAM which would execute directly in the PRUs, rather than into ALICE ISL which is currently interpreted.

Finally, we briefly address the conflict between laziness and strictness. A strict (call-by-value) implementation can generate much more parallelism than the lazy scheme, but may also generate non-terminating computations. A solution under investigation is to provide a *strictness analyser* in the compiler which will identify strict arguments in function applications, i.e. those which will *definitely* be needed and so can be pre-evaluated. This would be the perfect solution if *all* such arguments could be detected. However, this cannot be done in general, but current research is very encouraging and it may be possible to identify an extensive class of strict arguments without too much compile-time effort.

To conclude, the experience gained by all concerned in the development of ALICE, from an ideal for the advancement of functional programming to an operational prototype machine offering great flexibility for experimentation, has been invaluable. The ultimate long-term goals are still some way off, and apart from building the successor machine in VLSI, we are also investing a substantial effort in the design of an enhanced functional language, $HOPE^{+}$, and in the development of a transformation-based programming support environment. But in the arrival of ALICE, we have reached a significant milestone on the way to the attainment of a functional-language-based system for solving the complex computer applications of today and tomorrow.

## Acknowledgements

There have been several others involved in the work described in this paper, and we wish to acknowledge the efforts of the members of the Functional Programming Research Section led by Professor John Darlington in the Department of Computing, Imperial College, and of the ALICE Team at ICL's Mainframe Systems Division, West Gorton.

## References

[BM80] R. M. Burstall, D. B. MacQueen, D. T. Sannella, "HOPE: An Experimental Applicative Language", 1st International LISP Conference, 1980.

[CF86] M. D. Cripps, A. J. Field, M. J. Reeve, "The Design and Implementation of ALICE: A Parallel Graph Reduction Machine", In 'Functional Programming: Languages, Tools and Architectures', ed. S. Eisenbach. Ellis Horwood Limited, 1986.

[CO77] P. J. Courtois, "Decomposability", Academic Press, 1977.

[DR81] J. Darlington, M. J. Reeve, "ALICE : A Multiprocessor Reduction Machine for the Parallel Evaluation of Applicative Languages", ACM/MIT Conference on Functional Programming Languages and Computer Architecture, 1981.

[DR83] J. Darlington, M. J. Reeve, "ALICE and the Parallel Evaluation of Logic Programs", 10th Annual Symposium on Computer Architecture, 1983.

[FC85] A. J. Field, M. D. Cripps, "Self-Clocking Networks", International Conference on Parallel Processing, 1985.

[SI85] H. J. Siegel, "Interconnection Networks for Large-Scale Parallel Processing: Theory and Case Studies", Lexington Books, 1985.

[GH84] H. W. Glaser, C. L. Hankin, D. Till, "Principles of Functional Programming", Prentice-Hall, 1984.

[HF86] P. G. Harrison, A. J. Field, "Performance Modelling of Parallel Computer Systems", PERFORMANCE '86/ACM SIGMETRICS, 1986.

[JO85] T. Johnsson, "Lambda Lifting: Transforming Programs to Recursive Equations", Conference on Functional Programming and Computer Architecture, 1985.

[KO79] R. A. Kowalski, "Logic for Problem Solving", North Holland, 1979.

[MO82] I. W. Moor, "An Applicative Compiler for a Parallel Machine", SIGNPLAN Symposium on Compiler Construction, 1982.

[PA79] J. H. Patel, "Processor-Memory Interconnections for Multi-Processors", 6th Annual Symposium on Computer Architecture, 1979.

[RE81] M. J. Reeve, "An Introduction to the ALICE Compiler Target Language", Research Report, Dept. of Computing, Imperial College, 1981.

[SC81] C. H. Sauer, K. M. Chandy, "Computer Systems Performance Modelling", Prentice-Hall, 1981.

# Overview of Rediflow II Development†

Robert M. Keller
Quintus Computer Systems, Inc.
Mountain View, CA 94041

Jon W. Slater
ESL, Inc.
Sunnyvale, CA 94088

Kevin T. Likes
University of Utah
Salt Lake City, UT 84112

**Abstract** A status report is provided on the design of *Rediflow II*, a proposed multiprocessor architecture based on graph reduction. We emphasize recent work on hardware design to support graph reduction, communication, and load balancing, as well as indicating an evaluation model oriented toward larger grain computation than in our previously-reported work.

## Introduction

Rediflow is a proposed multiprocessing system based on the concept that parallelism should be achieved in as transparent a fashion as possible. This is accomplished by using functional languages as a base for programming, normal order evaluation, and a dynamic load balancing technique [Keller 84a-b, 85a-b, Lin 85]. It is the most recent in a line of development based on graph reduction which began with work reported in [Keller 79]. The reader is invited to review the papers mentioned in this paragraph to motivate the work, as we do not do so here.

Parallel normal order evaluation is implemented by suspending and resuming functions based on the availability of arguments to the functions. The advantages to normal order evaluation it facilitates asynchronous parallelism, only the code necessary for evaluation is executed, and a natural and efficient implementation of streams and other infinite data structures is supported [Keller 85a].

Normal order evaluation may be implemented using a "graph reduction" approach, in the sense that nodes representing unevaluated expressions are replaced by their value. However, the approach being used currently does not take this idea to its ultimate limit, as was proposed in [Keller 79]. Rather, graph reduction is used only selectively, with sequential code being used where graph reduction is not really needed. This dichotomy is achieved by compile-time preprocessing. The sequential code model also appears to provide a more natural approach toward the inclusion of sequential processes communicating via streams ("Kahn processes"[Kahn 74]) than did an earlier proposal for Rediflow [Keller 84a].

The "II" in the title indicates that, in the past year, three performance improvements have been made to the Rediflow design: the Redilisp language, an approach to graph reduction called PSCED, and the Redilink architecture [Slater 86]. This paper gives an overview of these developments. Like its predecessor, Rediflow II employs a *dynamic load-balancing* mechanism which serves the purpose of evenly distributing the available work among the processors, using the *gradient model* for task distribution [Lin 85]. It also uses saturation control in an attempt to retain locality of migrational tasks [Keller 84b].

---

† This work was performed in part while the authors were resident at the University of Utah, where it was supported in part by a grant from the IBM corporation. We also wish to thank Sperry, Inc. for providing a prototyping facility and parts for the prototype of the Redilink intercommunication switch module.

## Redilisp

Redilisp is a Lisp dialect used to program Rediflow. It descended from our earlier FEL [Keller 82], but has Lisp, rather than "English" [Landin 64], syntax. It is a lexically-scoped language with higher-order functions and procedures, with emphasis on a declarative programming style. Parallel evaluation occurs by default, rather than by required introduction of special constructs, but the latter may also be introduced by the programmer for greater control. An objective of Redilisp is to allow the programmer to integrate functional, imperative, and object-oriented styles of programming into one language. As such, Redilisp provides destructive operators, however the user is encouraged to program in a fully functional subset.

## PSCED

In order to improve performance over pure graph-reduction, a PSCED approach has been adopted in the new simulator and architectural specification. This also allows an efficient method for the integration of sequential code. This idea is similar to the *Multilisp* implementation and the *SECD-m* effort [Halstead 86, Abramsky 85]. In this scheme, we can achieve normal-order evaluation using what seems to be a much more efficient approach than our former, more-interpretive, graph reduction model. With this decreased complexity, it becomes possible to construct support hardware for handling run time function demands and suspend processes.

Landin's original "SECD" abstract machine derived from four registers in an abstract virtual machine, called "stack", "environment", "control", and "dump" [Landin 64]. We began using the modified name "SCED" as a more pronounceable acronym ("sked"), which evolved to "PSCED" (still pronounceable as "sked") to emphasize the added feature of parallel evaluation, and also to suggest a *process* identifier component. Meanwhile, the stack component has become a *register file*, for greater efficiency in accessing local values, but the component is still identified as "S". The PSCED approach of multiplexing among various states is consistent with the multiple register windows being proposed for certain RISC architectures [Katevenis85]. Likewise, we have made an effort to keep the complexity of the architecture down to a small number of machine instructions, and the heavy use of vectors has been helpful in this regard.

As mentioned above, the components of a "state" of the machine are

**P** The index of the process (for descriptive purposes).

**S** A register file for holding temporary values. Every S is fixed in its number of registers at process-creation time. We currently assume no upper bound on the size of S among processes, although this might have to be modified for certain processor architectures.

**C** A Control string, which specifies a sequence of machine instructions to be executed. In the current implementation, this is a linked list of instructions. New control strings are installed as the result of certain instructions, such as those for application and conditional branching.

**E** An environment, which is an implementation of a mapping from variable identifiers to values. E is implemented so that adding new environment layers, as necessary when creating new functions, is easy to effect. In a parallel model such as this one, it is also desirable to achieve a high degree of sharing and locality of access. Such considerations led us to choose a linked-list of vectors as the environment implementation. Each vector is a "frame" which comes into existence as the result of an application or related construct. For example, the parameters of a procedure are constructed within such a frame, which becomes the top layer of the environment when the procedure code is installed.

Although the method described imparts a sequential component to the access of a frame, such components are optimized by storing (pointers to) selected frames within local registers. This capability was part of the motivation behind use of a register file, rather than a stack.

**D** The Dump is used as a means for saving state for later return. Effectively the dump embodies a continuation, and can be used to implement continuation objects in the language.

Although we alluded to a PSCED value as a "state", it is really a collection of such values which forms the entire state of the parallel machine, each individual PSCED being the state of a *process*, of which there may be an arbitrary number. The evaluation of suspensions (code for expressions to be evaluated on demand) gives rise to the *spawning* of new processes (represented by their states). In this way, asynchronous parallelism is achieved.

We pause here to explain the suspension concept to the reader unfamiliar with it. If evaluation were sequential, we would have the notion described in [Friedman 76]: each value of interest is represented by a memory cell which can hold a *tag* in addition to the value itself. If the tag indicates "ready", the value is as present in the cell. If the tag indicates "undemanded", the value is really that of a suspension, *i.e.* a pointer to code and environment, the execution of which will provide the value. Since the tag is set to *ready* once the value is computed, the code will only be executed once.

For parallel evaluation [Keller 79], we need a third tag value in addition to *ready* and *undemanded*, which says that the value has been demanded, but the evaluation is not yet complete. In this case, the value is that of a pointer to list of recipients of the value when the value is computed. In Rediflow, a distributed system, these recipients are represented by words in memories which are possibly remote to the cell itself.

## Compilation

Compilation consists of several phases. Included among them is a fairly traditional code-generation phase similar to the one described by in [Henderson80], with differences noted below. This phase is continuation-based, for inclusion of run-time continuation-capture primitives. Other phases are necessary to process the input for parallel execution, and are done at the source level. Here we are supplanting much of the work done at run-time in graph-reduction implementations. The phases are:

> *clustering*: here "important" sub-expressions, ones of sufficiently-large granularity to be worthy of parallel processing, are identified. Likewise, common sub-expressions of functional forms are identified. The sub-expressions are identified by incorporating equations for them within a "let" construct [Landin 64], using what can be regarded as synthesized identifiers which are ultimately compiled away.
>
> *demand-propagation*: based upon tabulated properties of primitive operators, certain sub-expressions are included within "demand*" special forms:

The value of the Redilisp expression

(demand* <varlist> <expr>)

is that of <expr>, but the values of variables in <varlist> are demanded in parallel prior to the start of the evaluation of <expr>. During the evaluation of <expr>, whenever the value of a variable in <varlist> is needed, a wait automatically occurs if the value is not yet ready.

The user is free to use demand*, but the purpose of the demand propagation phase is to infer where it may be inserted (keeping in mind that the previous clustering phase has named important sub-expressions as variables).

As an example, given the following definition for the "parallel factorial" function:

```
(def pfac
 (lambda (m n)
    (if (< m n)
       (let ((half (/ (+ m n) 2)))
          (let ((s1 (pfac m half))
                (s2 (pfac (+ half 1) n)))
             (* s1 s2)))
       m))))
```

the output of the propagation phase is

```
(def pfac
  (lambda (m n)
    (demand* (m n)
      (if (< m n)
        (let ((half (/ (+ m n) 2)))
          (demand* ()
            (let ((s1 (pfac m half))
                  (s2 (pfac (+ half 1) n)))
              (demand* (s1 s2)
                (* s1 s2)))))
        m))))
```

## PSCED Instruction Set

In order to present examples of PSCED execution, we present a part of the virtual-machine instruction set. As well as generalizing the notion of "suspending cons", as presented in [Friedman 76], to adapt to parallelism, it also extends it to *vectors* of suspensions. Thus, a cons cell is a vector of length 2. Implicit in what follows, each component of such a vector is *tagged* to indicate whether the contents represent a datum or a suspension which will compute the datum. This is discussed further below.

| | |
|---|---|
| [ldc c1 r2] | Loads constant c1 into register r2 |
| [mv r1 r2] | Copy the contents of register r1 into register r2. |
| [Op r1 r2 r3] | where Op = +, *, minus, /, <, >, eq?, cons, *etc.*, performs X on contents of registers r1, r2, leaving result in r3 |
| [if r1 c2] | If contents of register r1 is not nil, continues with code strip c2. Otherwise continues as usual. |
| [close c1 r2] | Makes a *closure* from code strip c1 and the current environment, and stores it into r2. |
| [find-frame c1 r2] | Loads address of frame at lexical level c1 (counting innermost as 0) into register r2. |
| [vec-dem-im r1 c2 r3] | Register r1 is assumed to contain a (pointer to a) vector (e.g. a frame). The constant c2, an integer 0, 1, 2, ... is used to index the vector, and the corresponding value is demanded if tagged undemanded. Demanding takes place through the construction of a *fetch* packet, as described later. The demanded component contains a suspension, the computation of which will execute in parallel with the present instruction stream as a new process. The tag is set to WAIT meanwhile, and an address associated with the demander is left in a waiting list in place of the value itself. The new tag value insures that there is no second initiation of a process to compute the same value. When the computation is complete, the value is put in the vector component indexed, after the waiting list is removed. The tag READY is then put into the component to signify completion. As later described, *forward* packets are sent to addresses on the waiting list. |
| [reg-wait r1] | If register r1 + 1 indicates that the value in register r1 is not ready, blocks the current process until the data value for the register is received in the form of a *forward* packet (see previous instruction, and later discussion). A pointer to the state-vector (PSCED) of the blocked process is saved in register r1, so that the process can be unblocked. |
| [ap r1 r2 r3] | Register r1 must contain a closure. It is applied to the argument frame in r2, and the result is stored in r3. |

[tap r1 r2 r3] Like [ap r1 r2 r3] except the application is tail-recursive, meaning that no new dump is installed. If possible, the current register file is reused.

[rtn r1] Used within a procedure, the current procedure is exited. The dump is used to specify the return state to be loaded. The value currently in r1 is stored into the register specified in the return state's register 0.

[rtn-susp r1] The current process, presumed to be executing a suspension, is exited. The dump is assumed to contain a pointer to a cell in which there is a *waiting list* of process states, which is extracted, and the value in r1 is stored into the cell. A *forward* packet is sent to the processes on the waiting list at their respective processors. In particular, the packet is targeted at the vector component which contained the suspension originally giving rise to the current process. The storing of the value unblocks the original process by placing it on the task-queue. (The processes would normally then re-try the instruction which caused them to suspend.)

[add-frame r1] Register r1 is assumed to contain (a pointer to) a vector. This is pushed onto the current environment. (For *let*, this is done after the frame is loaded with values or suspensions. For *letrec*, it is done before. In the latter case, it is important that the frame remain in a register so that its contents can be established.)

[pop-frame] The top frame of the current environment is discarded. (Used to leave let or letrec.)

[mkvector-im c1 r2] Constant c1 specifies a vector size, which is created and stored into r2. (Used to make environment frames.)

[vec-store-susp r1 c2 c3] Register r1 is assumed to contain a vector. A suspension is created from the code strip c3, and is stored in the position of the vector indexed by c2.

## Parallel evaluation

Parallel evaluation is achieved by a sequence of 1 or more *vec-dem-im* instructions, which demand the evaluation of code contained within suspension objects which are stored in the demanded components of the vector argument. The result of such a suspension evaluation will ultimately appear in both the vector component and the register nominated in the instruction. It will appear in the vector component by virtue of the process doing the evaluation.

The compiler allocates the register following the recipient specified in *vec-dem-im* as a companion register which holds a flag indicating whether the evaluation of the suspension is complete, and hence whether the value in the register is valid data. When the process needs the data (normally some time after it has demanded it), it does the *reg-wait* instruction, causing itself to block if the datum is not ready. This puts (the address of) the descriptor of the blocked process in the next adjacent register, by convention. Meanwhile, in the vector component demanded, the address of the context block for the process and the register index has been saved. When the evaluation of the corresponding suspension completes, the value is stored into the vector and the demanding process resumed, fetching the value into the register.

As an example, consider the function *pfac* discussed earlier. The inner *let* expression defines the values of variables s1 and s2. At run-time, there is a frame vector allocated for this *let*, with suspension for the values of s1 and s2. The code which issues the demands and waits for parallel evaluations to be complete appears as follows, assuming that r0 contains the address of the frame vector:

```
[vec-dem-im r0 0 r1] % Demand the value of s1, starting a process
[vec-dem-im r0 1 r3] % Demand the value of s2, starting a process
                     % The sub-evaluations proceed in parallel
[reg-wait r1]        % Wait for s1 to be ready
[reg-wait r3]        % Wait for s2
[+ r1 r3 r5]         % Perform the addition
```

## Redilink

Redilink [Slater 86] is a switch design for Rediflow Xputers (the term we use to denote the replicated node containing processor, memory, and switch) intended to implement three major functions of the Rediflow system. These are interprocessor communication, load balancing, and the PSCED graph reduction technique. They are assisted by the following specialized modules: *Interconnection Switch Module*, *Apply Task Queue Module*, and *Graph Reduction Memory Support Module*.

Considerations of scalability require that the Xputers have a bound on the number of physical links between nodes. It is thus necessary to traverse some number of switches when one processor communicates with another. Each such traversal introduces, in an unloaded network, latency equal to the number of traversed links times a constant, the latter being the amount of time it takes for a message to enter and exit a switch. It is desired to minimize this constant for all messages. This is the primary goal of the **Interconnection Switch Module**.

A common approach to switch development involves designing for the highest possible bandwidth to accommodate heavily loaded network conditions. However, high bandwidth alone does not yield the best possible performance. The key issue is that of *latency* through the switch, and the Interconnection Switch Module is designed to keep this to a minimum. We use a simple time-division multiplexing design which lends itself to a full hardware (non-microcoded) implementation.

To obtain a high bandwidth, yet allow a reasonable cost and degree of physical interconnection, our prototype of Redilink uses two 8 bit links (one in each direction) between neighboring nodes, and allows a total of four neighbors. The internal processor-switch connection has a still-higher bandwidth to minimize congestion from several simultaneous incoming messages. This is accomplished by two 32 bit links for the processor-switch connection. The peak capacity of the switch is 800 megabits per second, 160 on each of five through channels, the maximum number which can be active.

The Interconnection Switch Module is a *store-and-forward* routing scheme utilizing *virtual cut-through* [Kermani 79] and *direct-mapping buffers* [Gerla 81]. Because of variable length packets, the switching component can be said to be a *datagram* style [Fujimoto 83]. We attempt to reduce contention by making the switch fast enough that long messages will not occupy links for an extended amount of time, and by employing FIFO buffers on both input and output ports. With buffers on both ports we achieve a 20% gain in bandwidth in a loaded network through elimination of handshaking.

The Interconnection Switch Module must support a number of features specifically required for the Rediflow multiprocessor: Migrational apply tasks must be routed dynamically based on the load distribution algorithm of the system. Routing is to be accomplished by a writable routing table implemented with a RAM loaded at initialization time according to the topology for the given configuration. This allows a wide variety of possible topologies and can also be used for routing around a defective Xputer, an important consideration in a fault tolerant system. The switch block diagram appears in Figure 1. It consists of five input and five output ports, and logic for arbitration, control, and routing memory.

In order to achieve a single logical global memory space, it is necessary to have header information associated with communication packets. The message header format is shown in Figure 2. The header contains the Xputer node address, word count (length of the message), and the type of packet. The Rediflow system has three of types packets for interprocessor communication: the apply packet, the forward packet, and the fetch packet [Keller 79]. The apply packet represents a migrational Rediflow task. The Xputer address of the apply packet is normally unspecified, due to the fact that load balancing is accomplished dynamically by the switch. Alternatively, the programmer can specifiy the destination of a task, which would result in the specification of an Xputer address. The data part of the apply packet consists of a closure and an argument pointer.

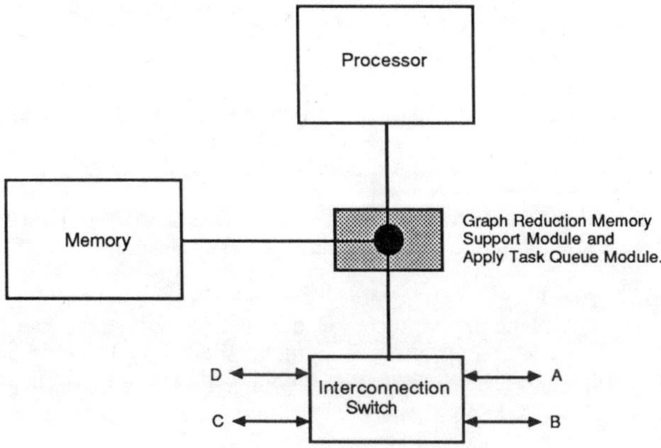


Figure 1

**Apply Packet**

| Node Address 8 bits | Word Count 8 bits | Packet Type 4 bits | Unused 12 bits |
|---|---|---|---|

| Data |
|---|

**Forward Packet**

| Node Address 8 bits | Word Count 8 bits | Packet Type 4 bits | Unused 12 bits |
|---|---|---|---|

| Destination Address 20 bits | Unused 12 bits |
|---|---|

| Data |
|---|

**Fetch Packet**

| Node Address 8 bits | Word Count 8 bits | Packet Type 4 bits | Unused 12 bits |
|---|---|---|---|

| Destination Address 20 bits | Unused 12 bits |
|---|---|

| Node Address 8 bits | Word Count 8 bits | Packet Type 4 bits | Unused 12 bits |
|---|---|---|---|

| Source Address 20 bits | Unused 12 bits |
|---|---|

Figure 2

In Figure 3, we see that the switch controller has 5 external inputs that specify "pressure" (the measure of processor load) direction. The dynamic load balancer specifies where to send the migrational apply tasks. Upon seeing an apply task (specified in the type field), the switch uses the pressure direction to route the apply packet. Pressure direction comes from the co-processor which is in contact with the nearest neighbor Xputers via another serial communication channel. This co-processor obtains the pressure of its neighbors and determines a gradient in which to route the message. The switch then forwards the apply task in the direction specified by the co-processor. All other packets simply use the routing memory to determine which output port to take. Variable length messages are supported by latching the length field of the header of incoming packets into a counter which decrements on each transfer of a word. When the count is equal to zero, the input port is considered free and will use the next incoming word as a header.

The **Apply Task Queue Module** consists of a doubly ended queue that holds migrational tasks. Upon receiving a signal from the load balancing co-processor, this module will start to send tasks out through the switch into the network. These tasks normally will make their way to some other Xputer which is underloaded. The number of tasks in the apply queue is maintained internally and is periodically sampled by the co-processor in order to compute the local Xputer's pressure.

The purpose of the **Graph Reduction Memory Support Module** is twofold:

> It handles messages at the rate delivered by the interconnection switch.
>
> It reduces the overhead associated with context switching due to suspending function applications.

Figure 4 is a block diagram which specifies the partitioning of the Graph Reduction Memory Support Module. In addition to apply tasks, there are two other types of tasks which allow communication of globally accessible objects within the Rediflow system:

> A **Fetch task** results from demand of the component of a vector, as a consequence of a *vec-dem-im* instruction.
>
> A **Forward task** is the ultimate response to a fetch. It consists of the value of a vector, the address of a target process descriptor, and a register address indicating the register to which the value is to be sent.

Routing of these tasks is implemented by means of seven "handlers" implemented within the Rediflow switch. The processor communicates with the handlers as devices on its memory bus.

> The **In Handler** manages all incoming messages from other Xputers. It routes messages to one of the Apply, Fetch, or Forward handlers.
>
> The **Out Handler** manages all outgoing messages to other Xputers.
>
> The **Demand Handler** manages all demands invoked by its own Xputer.
>
> The **Apply Handler** manages all Apply Tasks on its own Xputer. It is a doubly ended queue which partially implements load balancing.
>
> The **Fetch Handler** manages all fetch requests targeted for its own Xputer.
>
> The **Forward Handler** manages all forward requests targeted for its own Xputer.
>
> The **Local Task Handler** manages all tasks which must be executed on its own Xputer.

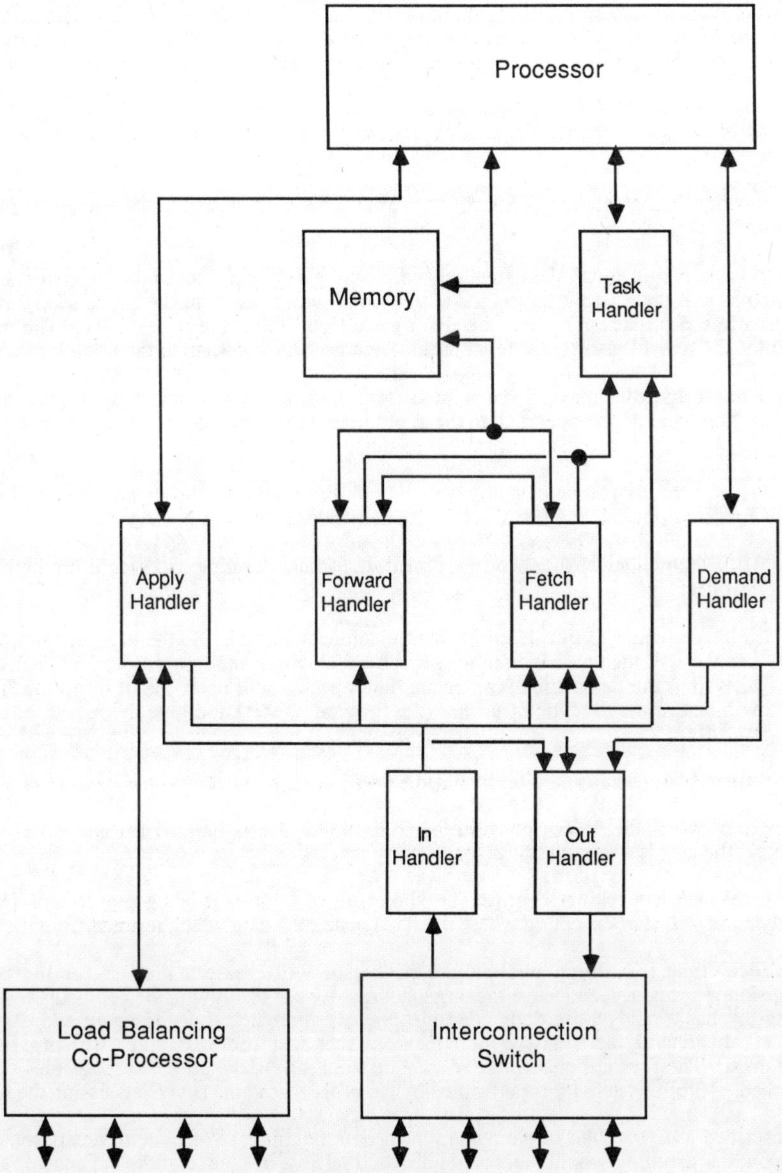
Processor
Memory
Task Handler
Apply Handler
Forward Handler
Fetch Handler
Demand Handler
In Handler
Out Handler
Load Balancing Co-Processor
Interconnection Switch
To Neighboring Switches

Figure 4

The following is a cursory description of the data flow among handlers. As a result of a *vec-dem-im* instruction, the processor performs a demand request, for a vector component value. The demand request sends two items to the **demand handler**: A pointer to the cell being demanded and the register to receive the value.

The **demand handler**, upon seeing a demand request, will build a **fetch packet**, and send it to one of two places based on locality.

> If the location being fetched is **local**, the demand handler will give the packet to the local **fetch handler**.
>
> If the location is **non-local**, the demand handler will give the packet to the local **out handler**. The out handler will then pass it on to the **switch** which will route it to the destination Xputer. Having arrived at the destination Xputer, the fetch packet will enter the **in handler**. The in handler will determine the type of packet (i.e. fetch packet) and give the message to the **fetch handler**.

Note that the **demand handler** makes local vs. non-local demands transparent to the system software. That is, both local and non-local demands end up in the appropriate **fetch handler**. Once in the **fetch handler**, the fetch packet will test the function call object for readiness:

> If the **datum** is **ready** (as indicated by the ready bit being set), the **fetch handler** will read the data, make a **forward packet**, and test whether the demander resides in this Xputer:
>
>> If the demander of the forward is **local**, the **fetch handler** will give it to the local **forward handler**.
>>
>> If the demander is **non-local**, the fetch handler will give it to the local **out handler**. The out handler will then pass it on to the **switch** which will route it to the destination Xputer. Having arrived at the destination Xputer, the fetch packet will enter the **in handler**. The in handler will determine the type of packet (i.e. forward packet) and give the message to the **forward handler**.
>
> If the **datum** is **not ready**, the **fetch handler** will pass the packet to the **local task handler**.

When it runs out of work, the processor will read from the local task handler for new tasks to perform. It must test whether the task is a restart request, or a fetch packet:

> If the **local task** is a **restart request** (i.e. the result of a forward being sent to a blocked process ), the processor will start executing where it left off, using the data which resulted from the demand.
>
> If the **local task** is a **fetch packet**, the processor will create a process for the corresponding suspension if necessary, and leave the target address for the result with the cell being requested. If the suspension has already been demanded, no process is created, but the result address still must be recorded. In general, there may be more than one such result address, and a **waiting list** of these is maintained. The forward chain is used as a distribution list when the suspension's value has been computed. All of this activity is performed by the *processor* in the act of processing the fetch packet.

Above we described a *distributed* implementation of data fetching. Were we to desire instead a shared-memory version, the handling would be considerably simplified, of course at the expense of sacrificing the scalability of the architecture. One item we glossed over here is the corresponding distributed handling of the *find-frame* instruction. Since this entails chasing a chain of pointers from one frame to the next, and thus possibly from one Xputer to the next, it is necessary to provide a mechanism to do the chasing and get the final frame address back to the originating Xputer. We leave the details to the reader's ingenuity.

## Current Status of Rediflow II

An initial version of the Redilisp compiler is complete, as well as is a simulator which executes the generated code. A TTL prototype of the Interconnection Switch Module has been designed and tested. A reduced VLSI prototype has also been designed and simulated. Work is currently under way to place a PSCED interpreter on an Inmos Transputer network. This seems to be the most economical way to demonstrate the Rediflow concept in a distributed environment, lacking funds to complete the hardware implementation work.

## References


[Abramsky 85] S. Abramsky and R. Sykes.
Secd-m: a virtual machine for applicative programming.
in J.-P. Jouannaud (ed.), Functional programming languages and computer architecture, Lecture Notes in Computer Science, 201, 81-98 (Sept. 1985).

[Friedman 76] D. Friedman and D. Wise.
CONS should not evaluate its arguments.
in S. Michaelson and R. Milner (eds.), Automata, Languages, and Programming, 257-284, Edinburgh University Press (1976).

[Fujimoto 83] R. Fujimoto.
VLSI communication components for multicomputer networks.
PhD Thesis, UC Berkeley, California (August, 1983).

[Gerla 81] M. Gerla.
Routing and flow control.
in F.F Kuo (ed)., Protocols and techniques for data communication networks 122-174, Prentice-Hall, Inc., Englewood Cliffs, New Jersey (1981).

[Halstead 86] R.H. Halstead.
Implementation of Multilisp: Lisp on a multiprocessor.
Symposium on Lisp and Functional Programming, 9-17, ACM (August 1984).

[Henderson 80] P. Henderson.
Functional programming: Application and implementation.
Prentice-Hall International, Inc., London, Great Britain (1980).

[Kahn 74] G. Kahn.
A simple language for parallel processing.
Proc. IFIP '74.

[Katevenis 85] M.G.H. Katevenis.
Reduced instruction set computer architectures for VLSI.
MIT Press (1983).

[Keller 79] R.M. Keller, G. Lindstrom, and S. Patil.
A loosely-coupled applicative multiprocessing system.
Proc. 1979 NCC, 613-622 (June 1979).

[Keller 82] R.M. Keller.
FEL (Function Equation Language) Programmer's Guide.
AMPS Technical Memorandum No. 7, University of Utah, Department of Computer Science, (1982).

[Keller 84a] R.M. Keller, F.C.H. Lin, and J. Tanaka.
Rediflow multiprocessing.
IEEE Compcon 1984, 410-417 (February, 1984).

[Keller 84b] R.M. Keller and F.C.H. Lin.
Simulated performance of a reduction based multiprocessor.
Computer 17(7):70-82, (July, 1984).

[Keller 85a] R.M. Keller.
Distributed computation by graph reduction.
Systems Research 2(4):285-295, (June, 1985).

[Keller 85b] R.M. Keller.
Rediflow architecture prospectus.
University of Utah, Department of Computer Science, Technical Report No. UUCS-85-105 (August, 1985).

[Kermani 79] P. Kermani and L. Kleinrock.
Virtual cut-through: A new computer communication switching technique.
Computer Networks 3(4):267-286 (Sept., 1979).

[Landin 64] P.J. Landin.
The mechanical evaluation of expressions.
Computer J. (1964).

[Lin 85] F.C.H. Lin.
Load balancing and fault tolerance in applicative systems.
PhD Thesis, University of Utah (August, 1985).

[Slater 86] J.W. Slater.
Redilink: An architecture for the Rediflow multiprocessor
M.S. Thesis Proposal, University of Utah (June, 1986).

# Specification of Reduction Strategies in Term Rewriting Systems.

**M.C.J.D. van Eekelen,**
**M.J. Plasmeijer.**

**Nijmegen University,**
**The Netherlands.**

## Abstract.

There is a growing interest in Term Rewriting Systems (TRS's), which are used as a conceptual basis for new programming languages such as functional languages and algebraic specification languages. TRS's serve as a computational model for (parallel) implementations of these languages. They also form the foundation for a calculus for Graph Rewriting Systems (GRS's). In Rewriting Systems reduction strategies play an important role because they control the actual rewriting process. Strategies determine the order of the rewriting and the rules to apply. Hence they have a great influence on the efficiency and the amount of parallelism in the computation. In ambiguous or non-deterministic TRS's, they even influence the outcome of the computation. Some of the reduction strategies used in TRS's are extremely complex algorithms. Unfortunately, there is no common formal specification method for reduction strategies yet.

In this paper three formal methods for specifying reduction strategies in TRS's are presented. In the first method the reduction strategy is encoded in the TRS itself. The original TRS is transformed to a so called annotation TRS in which the strategy is encoded using functions. This annotation TRS itself may use any normalizing reduction strategy. Unfortunately, compared with the number of rules of the original TRS, the annotation TRS may contain an exponential number of additional rules. In the second method this drawback is prevented, simply by using a priority TRS as annotation TRS. The desire to specify a strategy uniformly for all TRS's leads to the third method. A new TRS system is introduced that uses two basic primitives for matching and rewriting and that is build out of three separate TRS's. The use of this abstract-interpretation TRS is shown to be the most promising method.

## 0. Introduction.

A Term Rewriting System (TRS) consists of a *term* $\tau$ to rewrite and a set $\rho$ of *rewrite rules*. A *reduction strategy* $\sigma$ is an algorithm which determines in which order the rewritable subterms of the term (the so called redexes i.e. reducable expressions) have to be rewritten and which rules have to be applied. A TRS with a strategy will be called a term *reducer* or a term *rewriter*. A *redex* is a subterm which matches the left-hand-side (LHS) of a rewrite rule. TRS's consist of *variables* and *symbols*. Symbols start with an upper case character. All symbols that occur as left-most symbol of a rewrite rule are *functions*, other symbols are *constructors*.

---

This research was partly sponsored by the Dutch Parallel Reduction Machine project.

For instance when we have the following well-known set of rewrite rules:

| | | |
|---|---|---|
| Ap (Ap (Ap S x) y) z | → Ap (Ap x z) (Ap y z) | (S) |
| Ap (Ap K x) y | → x | (K) |

the term Ap (Ap (Ap K S) S) (Ap (Ap K K) K) can be rewritten to Ap S (Ap (Ap K K) K) and finally to Ap S K by applying two times the rule (K). In this example Ap is a function, S and K are constructors and x, y and z are variables.

A thorough introduction to TRS's is given in [KLO85]. TRS based languages like DACTL [GLA85] and Lean [BAR86a] are used as a computational model for implementations of new languages such as Miranda [TUR85] and OBJ [FUT84]. Currently various attempts ([BAR86b] based on [RAO84] and [BRO86] based on [EHR78]) are made to extend TRS's to a calculus for Graph Rewriting Systems (GRS's), in order to make them serve as a computational model of (parallel) graph reducing implementations.

A strategy in a TRS can be compared with the control flow in an ordinary imperative programming language. In the example above the strategy recursively takes the left-most redex in the term until the term after rewriting contains no redexes anymore and therefore it is in *normal form*. The outcome of the rewriting process may depend on the strategy followed. In particular this is the case if the TRS is ambiguous.

We distinguish the following forms of *ambiguity*:

- *Non-deterministic* type of rules, i.e. rules of which the LHS's match the same instance, such as

  | | |
  |---|---|
  | Choose x y | → x |
  | Choose x y | → y |

- *Strongly ambiguous* rules of which (an instance of) a subterm of a LHS is a redex, e.g.

  | | |
  |---|---|
  | G (K n 1) | → 1 |
  | K x y | → x |

  or, more subtle

  | | |
  |---|---|
  | F (F x) | → ... |

  which is ambiguous with itself.

For TRS's like these the outcome is depending on the strategy that is followed. Non-deterministic rules such as Choose might be useful but, of course, the strategy must support it by making a non-deterministic choice between the rules. Another example of a useful ambiguous definition is the following TRS which still has the *Church-Rosser* property (i.e. the system has an unique normal form):

```
Or True x        → True
Or x True        → True
Or False False   → False
```

Although in the case that the TRS is Church-Rosser, the outcome of the computation will not depend on the order in which redexes are chosen, it can happen that depending on the strategy followed the rewrite process may or may not *terminate*. Take for instance Berry's example which has a unique normal form:

```
F x 0 1    → x
F 0 x 0    → x
F 1 1 x    → x
```

Let's assume that of the actual arguments of F in the term there are two arguments which reduce to the indicated normal form while the other argument has no normal form at all. The most efficient strategy would not touch the latter. Unfortunately, on forehand we do not know which argument that is. Forcing argument evaluation if a pattern is specified on the corresponding position in a rule will start a non terminating computation. The only safe way to reach the normal form is to reduce each argument of F once to see if the term has become a redex (the parallel outermost strategy). This strategy is not always the most efficient one.

Another example which shows the importance of strategies is the following program with a pretty familiar appearance:

```
Fac 0    → 1                     (1)
Fac n    → * n (Fac (- n 1))     (2)
```

It only has the obvious semantics if the right strategy is chosen. In general the rules are ambiguous because Fac 0 matches both rules. Fac (- 1 1) even matches the wrong rule. However a valid definition of Fac can be obtained if the argument of Fac is evaluated on forehand and the first rule has priority over the second rule. Now Fac (- 1 1) will be reduced to Fac 0 and the priority of the rules guarantees that Fac 0 now matches only the first rule. The reduction strategies in most functional languages are of this type.

Hence we must conclude that, unfortunately, there is no best strategy for all TRS's. Safe strategies are not always in all cases efficient. Efficient strategies are not always in all cases safe. Some algorithms can be expressed more conveniently in one strategy than another. Also one could prefer some strategies for specific reasons, e.g. for a parallel architecture one would like to reduce as much redexes as possible in parallel. One could also imagine, as is proposed in Lean, that several strategies are mixed in one and the same TRS in order to optimize performance and descriptive power. All this gives rise to complex strategies which cannot be explained anymore in terms like "take the left-most redex". Hence there is a need to specify strategies formally.

Before we will specify reduction strategies we must first ask ourselves the question what the properties must be of a good strategy specification. Of course it must specify which redexes are to be rewritten in which order and which of the matching rules have to be applied. If we see the specification as an algorithm, the execution of that algorithm must reduce exactly the same redexes in exactly the same order as the original strategy does. There is however a problem in the case of *parallel* reduction strategies which is caused by the concept time. The question arises if two redexes, which according to the original strategy have to be reduced in parallel, also have to be executed at the same time when we execute the corresponding specification.

This consideration leads to two views on parallel reduction: *synchronous* and *asynchronous*. According to the synchronous view the strategy recursively determines one or more redexes which have to be reduced in parallel after which these redexes are all rewritten in one step. This view is often used in a theoretical framework but in practice there are only very few machines for which this view is applicable (e.g. [MAG80]). In most distributed environments the strategy algorithm is also implemented distributed and the asynchronous view is therefore more appropriate. In the asynchronous view redexes which are reduced in parallel are actually reduced in any (*possibly* parallel) order while the strategy may already have determined new redexes although not all previously assigned redexes were rewritten. In the asynchronous view we cannot state that the execution of the specification must rewrite *all* redexes in the same order as the original strategy would do. A specification can only model the dynamic behaviour. Though in this paper the specifications are used mainly in a sequential context, they can be easily extended to a parallel environment with the synchronous view as well as using the asynchronous view.

There are many ways in which strategies can be specified. In this paper three methods are introduced and a comparison between them is made. All methods create TRS's which can be reduced with any normalizing strategy.

1. Specification of the strategy by transforming the TRS $\tau$ to another one annotation-$\tau$; reducing the latter induces the wanted strategy on $\tau$.
2. Same as 1., but now in annotation-$\tau$ there is a priority in the rewrite rules (specificity).
3. The strategy is specified in a separate TRS on a different level.

For each method we will work out as a simple example the TRS consisting of the Curry combinators K and S (notated with explicit application functions). The strategy to be specified will be head reduction, i.e. left-most reduction to head normal form. A term is in head normal form if no reduction can possibly lead to a rewrite of the head of the term. We will refer to the example as Curry's example. For every method Curry's example will be proven correct. These proofs will be using O'Donnell's definition of a TRS simulating another TRS [O'DO85]. The methods are compared in terms of ease of correctness proving, ease of developing, ease of expressing efficient strategies etcetera.

## 1. Transforming a TRS to an Annotation TRS.

### 1.1 Description of the Method.

The method is inspired by the use of strictness annotations in functional programming languages (Miranda e.g.) which serve as compiler directives. The general idea is that we use such annotations, for instance a shriek "!", for labeling those redex(es) in the term that should be rewritten. Instead of directives we use functions. These functions have ordinary semantics which enables us to use the full TRS semantics to reason about their meaning. Therefore the original TRS is transformed to an annotation TRS. All rules in this annotation TRS start with these functions and therefore only those redexes of the original TRS which are "annotated", are also redexes in the annotation TRS. The annotation TRS itself can be reduced with any normalizing reduction strategy.

A small example: if the rules of the original TRS are:

| | | |
|---|---|---|
| Ap (Ap (Ap S x) y) z | → Ap (Ap x z) (Ap y z) | (S) |
| Ap (Ap K x) y | → x | (K) |

Then these rules are transformed to:

| | | |
|---|---|---|
| ! (Ap (Ap (Ap S x) y) z) | → Ap (Ap x z) (Ap y z) | (1) |
| ! (Ap (Ap K x) y) | → x | (2) |

This TRS has only one function; the function !. The transformation has as a consequence that functions in the original TRS (Ap in the example) now become constructors in the annotation TRS. Of course, also the term to reduce must be marked. For instance if Ap ((Ap (Ap K S) S) (Ap (Ap K K) K) is transformed to Ap (! (Ap (Ap K S) S)) (Ap (Ap K K) K) it has only one redex in the annotation TRS. Due to the applicative order reduction this redex will be reduced giving Ap S (Ap (Ap K K) K) as result. Hence only one of the redexes that is present in the original TRS, is reduced. This is the one which was marked.

In order to continue this process one also has to insert new markings for those redexes which have to be reduced next. When we specify a strategy in this way quite a lot of additional rules have to be added to deal with markings in the right way. Also the outcome of the computation must be the same as in the original TRS, i.e. all marking must disappear at the end. Let's try to make clear what kind of additional rules are needed by looking at our example.

### 1.2 Expressing Head Reduction for Curry's Example.

We start off with the rules of the ordinary TRS:

| | | |
|---|---|---|
| Ap (Ap (Ap S x) y) z | → Ap (Ap x z) (Ap y z) | (S) |
| Ap (Ap K x) y | → x | (K) |

We add exclamation marks to the left-hand-side of the rules of the original TRS in order to promote the original redexes to redexes in the new system.

| | | |
|---|---|---|
| ! (Ap (Ap (Ap S x) y) z) | → Ap (Ap x z) (Ap y z) | (1) |
| ! (Ap (Ap K x) y) | → x | (2) |

Note that the function ! has only one argument, hence the parentheses around the argument will be sometimes redundant. When there can be no confusion, we will leave them out.

We need a general propagation rule to achieve a left-most search for a redex.

| | | |
|---|---|---|
| ! (Ap (Ap (Ap (Ap v w) x) y) z) | → Ap (! (Ap (Ap (Ap v w) x) y)) z | (3) |

The propagation rule takes care of the propagation of the ! for the case that we have a pattern with four Ap's on the left. We also have to specify what happens for all other cases, i.e. when we have less than four Ap's. They can be divided in two classes:

1) The ! may have an argument which contains no redex at all. In this case the !-function must reduce to its argument in order to get the same normal form as in the original TRS. For each non-redex we have to add a rule. This gives us five additional rules.

| | | |
|---|---|---|
| ! S | → S | (4) |
| ! K | → K | (5) |
| ! (Ap K x) | → Ap K x | (6) |

! (Ap S x) → Ap S x (7)

! (Ap (Ap S x) y) → Ap (Ap S x) y (8)

Note that we could not write a more general rule like

! (Ap x y) → Ap x y

because this rule is ambiguous with the K rule. This would introduce the danger that ! (Ap (Ap K K) K) is reduced to Ap (Ap K K) K instead of being reduced to K.

2) The ! may have an argument of which a subterm is a redex. We call rules that are introduced to handle these cases *envelope-rules*. In this example only one envelope-rule is needed:

! (Ap (Ap (Ap K x) y) z) → Ap x z (9)

The nine rules we have given so far, model exactly **one** step of the head reduction strategy. In order to model the complete reduction to head normal form, the exclamation mark has to be created over and over in a driver rule like

* term → * (! term)

In order to let this reduction stop we must encode in the term whether or not a rewrite (according to the strategy on the original TRS) was done. We use a success constructor '+' to denote that a rewrite was done and a failure constructor '-' to denote that it wasn't. We add the following 'driver' rules

* (- term) → term (10)

* (+ term) → * (! term) (11)

The strategy stops if on the term no rewrite was done (10). Of course we must also change all the other rules to make them produce a result with the proper success or failure constructor (+/-). Furthermore we must also propagate these constructors back to the beginning of the term so that we can make the decision whether or not to stop the strategy. This is done by adding the following rules:

~ (Ap (- x) y) → - (Ap x y) (12)

~ (Ap (+ x) y) → + (Ap x y) (13)

and by changing the propagation rule to:

! (Ap (Ap (Ap (Ap v w) x) y) z) → ~ (Ap (! (Ap (Ap (Ap v w) x) y)) z) (3)

Finally the complete set of rules is:

| | | |
|---|---|---|
| ! (Ap (Ap (Ap S x) y) z) | → + (Ap (Ap x z) (Ap y z)) | (1) |
| ! (Ap (Ap K x) y) | → + x | (2) |
| ! (Ap (Ap (Ap (Ap v w) x) y) z) | → ~ (Ap (! (Ap (Ap (Ap v w) x) y)) z) | (3) |
| ! S | → - S | (4) |
| ! K | → - K | (5) |
| ! (Ap K x) | → - (Ap K x) | (6) |
| ! (Ap S x) | → - (Ap S x) | (7) |
| ! (Ap (Ap S x) y) | → - (Ap (Ap S x) y) | (8) |
| ! (Ap (Ap (Ap K x) y) z) | → + (Ap x z) | (9) |
| * (- term) | → term | (10) |
| * (+ term) | → * (! term) | (11) |
| ~ (Ap (- x) y) | → - (Ap x y) | (12) |
| ~ (Ap (+ x) y) | → + (Ap x y) | (13) |

The initial term is

* (! term).

Note that it was also possible to achieve the same result using only one annotation function in stead of three different functions. This was not done for reasons of clarity.

## 1.3 Correctness Proof.

In this section we first formally define what it means for a strategy to specify another strategy. This definition will also be used in the proofs of the other methods. Then we proof that Curry's example, as it was specified with this method, satisfies the corresponding conditions and we finish this section with some general remarks on proving specifications that are constructed with this method. We also introduce some new terminology for a special kind of TRS that we encounter.

### 1.3.1 Definition of Specification.

In his excellent book [O'DO85] Michael O'Donnell defines what it means for a TRS to simulate another TRS. When one considers a strategy of a TRS to be giving a restriction on the reduction relation, then restricting the reduction relation to what is specified by the strategy, his definition can be used directly as the definition for one term rewriter specifying another. This results in the following definition:

Let $<\tau_o, \rightarrow_\sigma>$ and $<\tau_s, \rightarrow_\pi>$ be reducers where $\rightarrow_\sigma$ and $\rightarrow_\pi$ are the reduction relations of $\tau_o$ (the original TRS) and $\tau_s$ (the specification TRS) restricted to the strategies $\sigma$ and $\pi$, then $\tau_{s\pi}$ *specifies* $\tau_{o\sigma}$ if there exist

an *encoding set* $E \subseteq \tau_s$,
a *decoding function* $d: \tau_s \rightarrow \tau_o \cup \{nil\}$ with $nil \notin \tau_o$,
a *computation relation* $\rightarrow_c \subseteq \rightarrow_s$,
such that

1. $d[E] = \tau_o \;\&\; d^{-1}[N\tau_{o\sigma}] \cap E \subseteq N\tau_{s\pi}$
   where $N\tau_{x\psi}$ is the set of normal forms of a TRS $\tau_x$ with respect to the strategy $\psi$,
2. $\forall \alpha, \beta \in \tau_s \;\; \alpha \rightarrow_c \beta \;\Rightarrow\; d(\alpha) \rightarrow_o d(\beta)$,
3. $\forall \alpha, \beta \in \tau_s \;\; \alpha\, (\rightarrow_{s\pi} - \rightarrow_c)\, \beta \;\Rightarrow\; d(\alpha) = d(\beta)$
   where $\rightarrow_{s\pi} - \rightarrow_c$ stands for any $s\pi$ reduction that is not a c reduction,
4. $\forall \alpha \in E, \beta \in \tau_o \;\; d(\alpha) \rightarrow_{o\sigma} \beta \;\Rightarrow\; \exists \delta \in E \;\; \alpha\, (\rightarrow_{s\pi} - \rightarrow_c)^* \rightarrow_c (\rightarrow_{s\pi} - \rightarrow_c)^*\, \delta \;\&\; d(\delta) = \beta$,
5. $\forall \alpha \in N\tau_{s\pi} \;\; d(\alpha) \in N\tau_{o\sigma} \cup \{nil\}$,
6. There is no infinite $(\rightarrow_{s\pi} - \rightarrow_c)$ reduction path and moreover there exists a bounded function $b : \tau_o \rightarrow \mathbb{N}$ such that $\forall \alpha, \beta \in E \;\; \alpha\, (\rightarrow_{s\pi} - \rightarrow_c)^m \rightarrow_c (\rightarrow_{s\pi} - \rightarrow_c)^n\, \beta \;\Rightarrow\; m < b(d(\alpha)) \;\&\; n < b(d(\alpha))$.

The specification is called *effective* if b, $\rightarrow_c$, d and E are all total computable.

Intuitively, this means that we encode terms into another TRS allowing multiple encodings. The first condition requires that decoding respect normal forms. The specifying TRS has computational reductions which mirror a rewrite in the original TRS and non-computational reductions which are internal book-keeping steps. Conditions 2,3 and 4 require that every original reduction is simulated by any number of book-keeping steps which do not change the encoded expression, followed by exactly one computational reduction to effect the change in the encoded expression. This reduction may again be followed by book-keeping steps. Condition 5 prevents dead ends in the specification. Condition 6 states that the number of book-keeping steps is not allowed to grow without bound, otherwise e.g. all possible original reduction sequences could be simulated before officially choosing one of them, which clearly is not what we want. All conditions together mean that we really simulate the steps of the original TRS and not only return the appropriate normal forms as a result.

For the function d we basically need to describe which term is encoded for all reducts of the elements of the encoding set, but we also have to define which terms are not reducts of elements of E and are therefore decoded to nil. Maybe it is possible to give a precise definition of a TRS simulating another one using a decoding function which is only defined on reducts of elements of E. In the future we would like to do research along this line to find out whether for our purposes it is possible to simplify this definition.

### 1.3.2 Proof of Curry's Example.

We define the encoding set E as $N\tau_{o\sigma} \cup \{ *(!\,t) \mid t \in \tau_o\text{-}N\tau_{o\sigma}\}$, where $\tau_o\text{-}N\tau_{o\sigma}$ stands for the set of all terms of $\tau_o$ that are not terms of $N\tau_{o\sigma}$. The set of constructors *,+,-,!,~ is called A. The decoding function d is defined as follows:

$x \in N\tau_{o\sigma}$ $\Rightarrow$ $d(x) = x$
x matches one of the rules of $\tau_s$ and the subterms matching variables do not contain elements of A $\Rightarrow$ $d(x) = x$ with all applications of elements of A skipped
in all other cases $d(x) = nil$.

The computation relation c is given by rules (1), (2) and (9). This leaves us to proof that all conditions are satisfied. When we are not interested in the strategy $\pi$ of $\tau_{s\pi}$ (i.e. the statement holds for any normalizing strategy of $\tau_s$), we will leave out the strategy suffix $\pi$.

1a. $d[E] = \tau_o$

Clear from the definitions.

1b. $d^{-1}[N\tau_{o\sigma}] \cap E \subseteq N\tau_s$

Suppose we have $\alpha \in E$ and $\beta \in N\tau_{o\sigma}$ with $d(\alpha) = \beta$, then $\alpha = \beta$ and because $\alpha$ does not contain any elements of A, $\alpha \in N\tau_s$.

2. $\forall \alpha, \beta \in \tau_s \;\; \alpha \rightarrow_c \beta \;\Rightarrow\; d(\alpha) \rightarrow_o d(\beta)$

Reductions according to the rules (1), (2) and (9) decode into S and K reductions.

3. $\forall \alpha, \beta \in \tau_s \;\; \alpha\, (\rightarrow_s\text{-}\rightarrow_c)\, \beta \;\Rightarrow\; d(\alpha) = d(\beta)$

Follows directly from the definition of d.

4. $\forall \alpha \in E, \beta \in \tau_o \;\; d(\alpha) \rightarrow_{o\sigma} \beta \;\Rightarrow\; \exists \delta \in E \;\; \alpha\, (\rightarrow_s\text{-}\rightarrow_c)^* \rightarrow_c (\rightarrow_s\text{-}\rightarrow_c)^* \,\delta \;\&\; d(\delta) = \beta$

Suppose we have $\alpha \in E$, $d(\alpha) \rightarrow_{o\sigma} \beta$; then $d(\alpha) \notin N\tau_{o\sigma}$, hence $\alpha$ is of the form $*(!\,t)$ with $t \notin N\tau_{o\sigma}$. So $d(\alpha) = t$ and t reduces in $\tau_{o\sigma}$ to $\beta$. We have to find a $\delta \in \tau_s$ such that $d(\delta) = \beta$ **and** $\alpha$ $(\rightarrow_s\text{-}\rightarrow_c)^* \rightarrow_c (\rightarrow_s\text{-}\rightarrow_c)^* \,\delta$; $t \notin N\tau_{o\sigma}$, so t has a head-redex r in $\tau_o$, hence t is of the form $Ap^1$ ( ...

$Ap^n$ (Ap ( Ap (Ap S x) y) z) $arg_n$) ... $arg_1$ or $Ap^1$ ( ...$Ap^n$ (Ap (Ap K x) y) $arg_n$) ... $arg_1$. Consequently, the form of $\beta$ is also known (apply S or K rule). Take $\delta = *$ (! $\beta$) then trivially $d(\delta) = \beta$.
Let us first assume that r is a S redex. If $n > 0$ then the only rule that is applicable on $\alpha$ is rule (3) which is not a computational rule. After one step this reduces to ~ ( $Ap^1$ ( ! ($Ap^2$ (... $Ap^n$ ( Ap ( Ap (Ap S x) y) z) $arg_n$) ... $arg_1$) and again if $n-1 > 0$ only rule (3) is applicable. So after n steps the result is ~ ($Ap^1$ ( ~ ($Ap^2$ (... ~ ($Ap^n$ (! ( Ap ( Ap (Ap S x) y) z)) $arg_n$) ... $arg_1$). These n steps are the internal steps of $\tau_S$ preceding the computational step which is applying rule (1), resulting into ~ ($Ap^1$ ( ~ ($Ap^2$ (... ~ ($Ap^n$ (+ ( Ap ( Ap x y) (Ap x z)) $arg_n$) ... $arg_1$). Now the only possible reductions are n applications of rule (13), which gives us * (+ $\beta$), which can only reduce to * (! $\beta$) = $\delta$.
The other case we look at, is r being a K redex. We now have to be careful because there can be two computational rules corresponding to a K reduction. If $n \geq 1$, then we can follow the same reasoning as in the S case. We have n-1 applications of rule (3) on application of rule (9) and n-1 applications of rule (13), followed by one application of rule (11); if n=0, then we can only apply directly the computational rule (2) and we have only one internal step from * (+ $\beta$) to * (! $\beta$).

5. $\forall \alpha \in N\tau_S \; d(\alpha) \in N\tau_{o\sigma} \cup \{nil\}$

Take $\alpha \in N\tau_S$ then either $\alpha$ is also a term of $N\tau_{o\sigma}$ and then $d(\alpha) = \alpha$ and $d(\alpha) \in N\tau_{o\sigma}$ or $\alpha$ is not a term of $N\tau_{o\sigma}$ and then $d(\alpha)$ is nil because $\alpha$ is not a redex in $\tau_S$.

6. There is no infinite $(\rightarrow_S\text{-}\rightarrow_c)$ reduction path and moreover there exists a bounded function $b : \tau_o \rightarrow \mathbb{N}$ such that $\forall \alpha, \beta \in E \;\; \alpha\, (\rightarrow_S\text{-}\rightarrow_c)^m \rightarrow_c (\rightarrow_S\text{-}\rightarrow_c)^n \beta \Rightarrow m < b(d(\alpha))\; \& \; n < b(d(\alpha))$.

In the proof of 4 it is shown implicitly that there is no infinite internal reduction path. We define b as $b(t) = 3 *$ (the number of Ap's on the spine) + 1. Take $\alpha, \beta \in E$ then $\beta$ can be an element of $N_{o\sigma}$ or not: $\beta \notin N\tau_{o\sigma} \Rightarrow \beta = *$ (! $d(\beta)$) and in the proof of 4 we saw that both m and n are less than the number of Ap's on the spine in respectively $\alpha$ and $\beta$. So certainly $m < b(d(\alpha))$ and $n < b(d(\beta))$; analogously to the reasoning in 4 one can easily show that $\beta \in N\tau_{o\sigma} \Rightarrow \alpha\, (\rightarrow_S\text{-}\rightarrow_c)^{m1} \rightarrow_c (\rightarrow_S\text{-}\rightarrow_c)^{n1}$ (* (+ $\beta$)) $(\rightarrow_S\text{-}\rightarrow_c)^{m2}\, \gamma\, (\rightarrow_S\text{-}\rightarrow_c)^{n2}$ (* (- $\beta$)) where $\gamma$ is a term not containing ! and n1, n2, m1 and m2 are all less or equal to the number of Ap's on the spine of $\alpha$, $\beta$ and $\gamma$ respectively. Proving this we use the fact that rules (1), (2), (3), (4), (5), (6) and (9) cover all possible terms ! t with t $\in \tau_{o\sigma}$.

The specification is *effective* because b, $\rightarrow_c$, d and E are all total computable.

### 1.3.3 Remarks

Several parts of the proof heavily relied on the specific rules, the specific structure of the term and of course on the specific strategy. We see no way to easily extend these proofs to other reducers.

Note that starting with terms * (! t) every possible reduct had at most one redex. We call a term with such a property *linear*, because the reduction path is linear. A TRS where all terms are linear is called a *linear* TRS. For those who think this is confusing considering the concept of left-linearity, i.e. having repeated variables on the left-hand-side, we suggest *left-comparing* as a possibly better name in stead of left-linear. It does not have the relation to polynomial terms (linear, quadratic etc.) and moreover it names the essential aspect of left-linearity which is that arguments have to be identical in order to be able to apply the rule. Linear TRS's have good prospects for efficient sequential execution, and left-comparing TRS's can cause trouble during execution when used with infinitely growing arguments.

### 1.4 Evaluation of the Method.

Although it is possible to convert the original TRS to an annotation TRS in which the strategy is explicitly encoded, the method has severe drawbacks.

First of all our annotation TRS has many more rules than the original TRS. Unfortunately the number of extra 'non-redex' rules per function can become exponential: the number of constants, that can occur, to the power of the width of the pattern. This results in too many rules. Some of these rules are very awkward such as the envelope rule. Furthermore the rules in the annotation TRS not only depend on the strategy but also on the original TRS. Consequently the correctness of the strategy can not be proven for all TRS's at once, but it must be proven for every TRS separately. Furthermore it is rather tedious work to give such a proof.

Concluding we can state that though it is very well possible to express a strategy in an annotation TRS, the number of rules and their complexity makes this method not very suitable for practical use.

## 2. Transforming the TRS to an Annotation TRS with Priority Rules.

### 2.1 Description of the Method.

The method we will describe in this section is very closely related to the way strategies are expressed in Dactl [GLA85]. It differs from the previous method in that we will use a different kind of annotation TRS namely an annotation TRS with priority rules [BAE86], a so called priority rewriting system (PRS). The difference with an ordinary TRS is that whenever a rule matches a given redex, it may only be chosen if none of the rules with higher priority can ever be applicable on the (internally rewritten) term.

## 2.2 Expressing Head Reduction for Curry's Example.

We start again with the rules of the ordinary TRS

| | | |
|---|---|---|
| Ap (Ap (Ap S x) y) z | → Ap (Ap x z) (Ap y z) | (A) |
| Ap (Ap K x) y | → x | (B) |

and we add exclamation marks to the rules

| | | |
|---|---|---|
| ! (Ap (Ap (Ap S x) y) z) | → Ap (Ap x z) (Ap y z) | (1) |
| ! (Ap (Ap K x) y) | → x | (2) |

Our propagation rule is much simpler now:

| | | |
|---|---|---|
| ! (Ap x y) | → Ap (! x) y | (3) |

The reason for this is that if a redex matches the third rule **and** one of the other two rules, the topmost rule (**not** rule 3) is taken because It has higher priority. We indicate this by putting an arrow in front of rules with decreasing priority. Rules with the same priority are indicated by adding a bar in front of them. Rules without any indication are not affected by the priority mechanism.

The non-redex rules are extremely simple now:

| | | |
|---|---|---|
| ! x | → x | (4) |

Anything that does not match one of the other rules, is not a redex. That's it! The somewhat strange envelope-rule of the previous section is also not necessary any more.

So we have elegantly modelled one step of this strategy. In order to model the strategy completely we still must encode whether or not a subterm was successfully rewritten. This is modelled by exactly the same changes and extra rules as in the previous section. Note that in general we must be careful with adding rules because our PRS could cause that they will never be applied. In this case we have no problems with that issue.

The complete set is now:

| | | |
|---|---|---|
| ↓ \| ! (Ap (Ap (Ap S x) y) z) | → + (Ap (Ap x z) (Ap y z)) | (1) |
| ↓ \| ! (Ap (Ap K x) y) | → + x | (2) |
| ↓ ! (Ap x y) | → ~ (Ap (! x) y) | (3) |
| ↓ ! x | → - x | (4) |
| * (- term) | → term | (5) |
| * (+ term) | → * (! term) | (6) |
| ~ (Ap (- x) y) | → - (Ap x y) | (7) |
| ~ (Ap (+ x) y) | → + (Ap x y) | (8) |

Again, the initial term must be of the form

* (! term)

Note that only the third and the fourth rule really rely on the order in which rules are matched.

## 2.3 Correctness Proof.

In order to prove this specification we first define what the semantics of a PRS is. Then we will show that in this case the semantics of the PRS is equivalent to the semantics of a TRS without priority. This TRS will turn out to be equivalent to the one we constructed using the previous method.

### 2.3.1 Semantics of a PRS.

A PRS has a unique semantics (is *well-defined*) if it has a unique sound and complete rewrite set. Unfortunately our underlying TRS is not strongly normalizing nor bounded. So we have to use the stabilization lemma formulated in [BAE86]. We have to prove that starting with $R^{\underline{0}} = R_{max}$ ($R_{max}$ being the set of all possible rewrites of $\tau_s$ i.e. all possible instances of all rules not considering the priority), there exists a $\underline{n}$ for which $R^{\underline{n}} = R^{\underline{n+1}}$, where $R^{\underline{n+1}}$ is defined as $(R^{\underline{n}})^c$ with $(R)^c$ being the set of all rewrites that is correct with respect to R. A rewrite $t \rightarrow_r s$ is *correct* if there is no internal R-reduction $t \rightarrow^* t'$ to an r'-rewrite $t' \rightarrow_{r'} s' \in R$ with $r' > r$ (r' has higher priority than r). R is *sound* if all rewrites which are an element of R, are correct w.r.t. R. R is *complete* if it contains all possible rewrites of $R_{max}$ which are correct w.r.t. R.

### 2.3.2 Proof.

Because this method resembles the previous one, we can take all definitions the same except for the computation relation c which will be given by rules 1 and 2 only. But before we can even start to say something about this solution, we must prove that this PRS has a unique semantics.

If we restrict ourselves to those terms t for which $d(t) \neq nil$, then one can easily see that the term has at most one redex for which several rules with different priorities can be applied. Of course, the intended semantics is that the rule with the highest priority is applied. Since there is always at most one redex, there are no internal reductions of more than zero steps. Yet there are trivial zero-step reductions using other instantiations for the variables. So we start with the set $R_{max}$ and compute $(R_{max})^c$. We prove that it is sound and complete, hence $((R_{max})^c)^c = (R_{max})^c$ and $(R_{max})^c$ is the semantics of our PRS.

We want to determine the set of rewrites that are correct w.r.t. $R_{max}$. As was already stated, the only internal redexes to be considered are other instantiations of variables. We know the structure of the terms, so all instantiations of the left-hand-side of rule (4) are ruled out by rule (3) except for the instantiations S and K. So the PRS is equivalent to another one where rule (4) is substituted by the rules:

| | | |
|---|---|---|
| ! S | →- S | (4') |
| ! K | →- K | (4") |

In the same way we can determine the patterns for which rule (3) is really applicable i.e. not ruled out by rule (1) or rule (2), giving the following rules replacing rule (3):

| | | |
|---|---|---|
| ! (Ap S y) | →~ (Ap (! S) y) | (3') |
| ! (Ap K y) | →~ (Ap (! K) y) | (3") |
| ! (Ap (Ap S a) y) | →~ (Ap (! (Ap S a)) y) | (3"') |
| ! (Ap (Ap (Ap K a) b) y) | →~ (Ap (! (Ap (Ap K a) b)) y) | (3"") |
| ! (Ap (Ap (Ap (Ap a b) c) d) y) | →~ (Ap (! (Ap (Ap (Ap a b) c) d)) y) | (3""') |

So the set $(R_{max})^c$ is the set determined by this new TRS without priority. This set is sound because evidently all rewrites in it are correct and it is complete because it also contains all rewrites which are correct with respect to it. One easily checks that the resulting rules that determine the semantics of the PRS, are equivalent to the rules of the previous example. So we have proven that our PRS model specifies Curry's example correctly.

### 2.3.3 Remarks.

Note that this proof more or less included the proof for the previous method and that we needed special analysis to determine the semantics of this PRS. In general there is not even always a unique semantics and it is not known whether a semantics always exists [BAE86].

### 2.4 Evaluation of the Method.

Clearly this method is a lot better than the previous one. There are more rules in the annotation TRS than in the original TRS, but the growth is no longer exponential because the non redex case can now be expressed with one rule (4). Also the funny envelope rules have disappeared.

Still this method has a severe disadvantage. The rules generated still heavily depend on the original TRS so if we want the same strategy for other TRS's we still have to change the rules of each one of them. The proof must be given for every TRS separately as was the case with the previous method. The complexity of the proof is even worse than the complexity of the previous proof because we also have to prove the well-definedness of the PRS. It would be better if we could define a strategy in such a way that the same description is valid for all TRS's.

## 3. Defining the Strategy Separately in an Abstract-interpretation TRS.

### 3.1 Description of the Method.

In order to be able to specify strategies uniformly for all TRS's we will consider three different conceptual levels of TRS's:

a) The first TRS is the *user-defined TRS*. This is the original TRS which now remains unchanged. An example of such a TRS is the S-K TRS.

b) The reduction strategy for the user-defined TRS is specified in a separate TRS called the *interpretation TRS*. Functions in the interpretation TRS may have a user-defined TRS or any subterm of such a TRS as argument. For instance one may write:

| | | |
|---|---|---|
| HeadReduce *S* | →*S* | (I-a) |
| HeadReduce *K* | →*K* | (I-b) |

The arguments of HeadReduce are written in *italic* style to indicate that they are part of a TRS on another level. Instead of the rules above one can write a more general rule which is valid for *all* combinators:

| | | |
|---|---|---|
| HeadReduce *c : Combinator* | →*c* | (I-c) |

In this rule *c* is a variable which will be bound to a (sub)term of a TRS. The suffix *Combinator* restricts the number of matching expressions. Such a suffix is called an *abstraction* .

c) These abstractions which are used as patterns in the interpretation TRS's, are defined by a third TRS: the *abstraction TRS*. This abstraction TRS is used to make an abstract syntax of the user-defined TRS available to the interpretation TRS. An example of such an abstraction TRS is:

| | | |
|---|---|---|
| *Combinator* | →*S* | (A-a) |
| *Combinator* | →*K* | (A-b) |

Now the extra restriction on *c* imposed by the suffix *Combinator* implies that the actual value of *c* must be a normal form of the term *Combinator* in the abstraction TRS. Hence only HeadReduce *S* and HeadReduce *K* will match rule (I-c). The abstraction level can be seen as a preprocessing level where things are done like syntactical categorizing, type checking, strictness analysis etcetera.

To make the specification of strategies really easy we introduce two primitive functions which can be used in the interpretation TRS:

Match *term rules*

which returns True if the *term* is a redex according to the *rules* and False otherwise.

Rewrite *term rules*

which returns the *term* after one rewrite according to the *rules* if the *term* matches the *rules* and the *term* itself otherwise.

The match function only checks whether or not the given term as a whole matches one of the given rules. The rewrite function will make a non-deterministic choice out of the matching rules. Although it is possible to define these primitive functions precisely in the interpretation TRS, the formal definition is rather tedious.

### 3.2 Expressing Head Reduction for Curry's Example.

First we give the rules of the user-defined TRS.:

| | | |
|---|---|---|
| Ap (Ap (Ap S x) y) z | → Ap (Ap x z) (Ap y z) | (U-S) |
| Ap (Ap K x) y | → x | (U-K) |

Furthermore we have to define those abstractions of the TRS that we need at the interpretation level. In this example these abstractions are extremely trivial. The abstraction TRS is:

| | | |
|---|---|---|
| *Combinator* | $\rightarrow S$ | (A-1) |
| *Combinator* | $\rightarrow K$ | (A-2) |

Finally we have to define the interpretation TRS. We will start with describing one step of the strategy: we will call that OneStepHead.

When the term is a redex we rewrite it and otherwise we search in the function part of apply for a redex (the propagation rule):

```
OneStepHead ( Ap f a ) rules        → Cond  ( Match ( Ap f a ) rules )
                                            ( Rewrite ( Ap f a ) rules )
                                            ( Ap ( OneStepHead f rules ) a )      (I-1)
```

In this rule Cond has the ordinary meaning. The result of OneStepHead applied to a function Ap with parameters is, if it matches as a whole, the rewrite of it, otherwise the result is the function Ap with as new parameters the result of OneStepHead applied to the function part. This surely terminates because of the next rule (I-2). Note that functions now also must have *rules* as parameter in order to be able to use the matching and rewriting primitives.

Again S and K cannot be rewritten, which we can now express in one rule:

```
OneStepHead c:Combinator rules      → c                                           (I-2)
```

The result of OneStepHead applied to a single combinator is that combinator itself.

We will now extend the one-step strategy to a complete strategy by adding another function:

```
Head term rules                     → Cond  ( ContainsAHeadRedex term rules )
                                            ( Head ( OneStepHead term rules ) rules )
                                            term                                  (I-3)
```

When our term contains a head-redex we do one left-most rewrite and continue, in the other case we stop and the result is the term itself.

The definition of ContainsAHeadRedex is very similar to the definition of OneStepHead. When there is a Rewrite in the definition of OneStepHead, we return True and when in OneStepHead we return the parameter as a result, we return False in ContainsAHeadRedex. Of course, recursive calls in OneStepHead are simply converted to recursive calls in ContainsAHeadRedex.

The definition of ContainsAHeadRedex is:

| | | |
|---|---|---|
| ContainsAHeadRedex *c:Combinator rules* | → False | (I-4) |
| ContainsAHeadRedex ( *Ap f a* ) *rules* | → Cond ( Match ( *Ap f a* ) *rules* ) | |
| | True | |
| | ( ContainsAHeadRedex *f rules* ) | (I-5) |

The complete set is now:

| | | |
|---|---|---|
| OneStepHead ( *Ap f a* ) *rules* | → Cond ( Match ( *Ap f a* ) *rules* ) | |
| | ( Rewrite ( *Ap f a* ) *rules* ) | |
| | ( *Ap* ( OneStepHead *f rules* ) *a* ) | (I-1) |
| OneStepHead *c:Combinator rules* | → *c* | (I-2) |
| Head *term rules* | → Cond ( ContainsAHeadRedex *term rules* ) | |
| | ( Head ( OneStepHead *term rules* ) *rules* ) | |
| | *term* | (I-3) |
| ContainsAHeadRedex *c:Combinator rules* | → False | (I-4) |
| ContainsAHeadRedex ( *Ap f a* ) *rules* | → Cond ( Match ( *Ap f a* ) *rules* ) | |
| | True | |
| | ( ContainsAHeadRedex *f rules* ) | (I-5) |

The initial term must be Head *Term SKRules*.

### 3.3 Correctness Proof.

In order to prove this specification we first define what the semantics of the abstract-interpretation TRS is. We will proof that the conditions are satisfied after proving some simple lemmas. These lemmas state general facts on the functions that are defined. These facts are easy to find because they cover the intention with which we constructed the functions. It will be rather simple to prove those lemmas.

#### 3.3.1 Semantics of the Abstract-interpretation TRS.

We start with settling the semantics of *c : Combinator* in the rules for OneStepHead and ContainsAHeadRedex. Because of the structure of the terms OneStepHead is equivalent to:

| | | |
|---|---|---|
| OneStepHead ( *Ap f a* ) *rules* | → Cond ( Match ( *Ap f a* ) *rules* ) | |
| | ( Rewrite ( *Ap f a* ) *rules* ) | |
| | ( *Ap* ( OneStepHead *f rules* ) *a* ) | (I-1) |
| OneStepHead *S rules* | → *S* | (I-$2_a$) |
| OneStepHead *K rules* | → *K* | (I-$2_b$) |

For our semantics this eliminates the abstraction TRS. Furthermore the user-defined and the interpretation TRS can be seen as one and the same TRS with restrictions on the construction of terms, such as: the first argument of OneStepHead is an element of $\tau_O$, the second argument are the rules, etcetera. During reduction reducts of terms that meet these restrictions also meet these restrictions. So we now have only one TRS with the ordinary meaning. However we do not deal with the full reduction relation but only with those reductions that are allowed by normalizing strategies. Strategies for this TRS are normalizing if they reduce the first argument of Cond and the Cond itself without reducing the other arguments and if they reduce the argument of Head to a term which is an element of $\tau_O$, before applying the Head rule again (this restriction is a consequence of uniting the two levels). So we simply remove those reductions out of the reduction relation. The reducer we get this way is called $\tau_S$ again and we will prove the specification following again O'Donnell's definitions.

### 3.3.2 Proof.

Before we proof that the specification satisfies O'Donnell's definitions we want to prove some lemmas about the functions.we have defined. The formulation of these lemmas is simple because of the way the functions were constructed.

**Lemma** The normal form of ContainsAHeadRedex x SKRules with $x \in \tau_O$ is True if x has a head-redex in $\tau_O$, False otherwise.

Proof: Recall the definitions:

Def: x contains a head-redex $\Leftrightarrow$ x matches a S or K rule or else if x = Ap a b, a contains a head-redex.

| | | |
|---|---|---|
| ContainsAHeadRedex *S rules* | → False | ($I\text{-}4_a$) |
| ContainsAHeadRedex *K rules* | → False | ($I\text{-}4_b$) |
| ContainsAHeadRedex ( *Ap f a* ) *rules* | → Cond ( Match ( *Ap f a* ) *rules* )<br>True<br>( ContainsAHeadRedex *f rules* ) | (I-5) |

Suppose $x \in \tau_O$ contains a head-redex then $x \neq S$ and $x \neq K$ (S and K themselves do not match the S or K rule) so x = Ap a b. So ContainsAHeadRedex x SKRules reduces in $\tau_S$ to Cond (Match (Ap a b) SKRules) True (ContainsAHeadRedex f SKRules). If x matches a S or K rule then obviously the normal form is True because we restricted the reduction relation to safe strategies. If x does not match a S or K rule then the left part of the application contains a head-redex which by induction and the absence of terms with an infinite spine results in the correct normal form. Proceeding this way it is very simple to prove the other parts of the lemma.

The following lemma is just as simple to proof :

**Lemma** OneStepHead x SKRules with $x \in \tau_o$ has as its normal form: if x has a head-redex in $\tau_o$, the reduct of x after reducing this head-redex and if x does not have a head-redex in $\tau_o$, its normal form is x.

We can use these lemmas in our proof of the specification. It will make everything very easy. We define E as: $x \in N\tau_o \Rightarrow E(x) = x$ ; $x \notin N\tau_o \Rightarrow E(x) =$ Head x SKRules. The computational relation is given by reductions of the primitive Rewrite. The definition of d is:

if x is an element of $N\tau_o$ then d(x) = x;

if x can be reduced via non computational reductions to a x' which is an element of $N\tau_o$ then also d(x) = x;

if x is not an element of $N\tau_o$ then we define d a bit special: if x can be reduced to x' via non computational reductions until the only possible reduction is a computational one and x' is of the form : $Ap^1$ ( ..... $Ap^n$ ( (Rewrite (Ap f a)) .....), perhaps surrounded by Head and SKRules then the decoding of x is defined as x' with applications of Head, SKRules and Rewrite skipped.

In other cases the decoding of x is nil.

Because we are only interested in reducts of elements of E, we restrict $\tau_s$ to those reducts.

1a. $d[E] = \tau_o$

Trivially true.

1b. $d^{-1}[N\tau_{o\sigma}] \cap E \subseteq N\tau_s$

Suppose $\alpha \in E$ and $\beta \in N\tau_o$ with $d(\alpha) = \beta$, then clearly $\alpha = \beta$, and since $N\tau_o \subseteq N\tau_s$, $\alpha$ is an element of $N\tau_s$.

2. $\forall \alpha, \beta \in \tau_s \ \alpha \rightarrow_c \beta \Rightarrow d(\alpha) \rightarrow_o d(\beta)$

Suppose $\alpha$ reduces to $\beta$ via a reduction of Rewrite then if $\alpha$ and $\beta$ are reducts of elements of E then clearly the decodings also reduce to each other.

3. $\forall \alpha, \beta \in \tau_s \ \alpha (\rightarrow_s - \rightarrow_c) \beta \Rightarrow d(\alpha) = d(\beta)$

Analogously to 2.

4. $\forall \alpha \in E, \beta \in \tau_o\ d(\alpha) \rightarrow_{o\sigma} \beta \Rightarrow \exists \delta \in E\ \alpha (\rightarrow_s\text{-}\rightarrow_c)^* \rightarrow_c (\rightarrow_s\text{-}\rightarrow_c)^* \delta\ \&\ d(\delta) = \beta$

Suppose $\alpha \in E$, $d(\alpha) \rightarrow_{o\sigma} \beta \Rightarrow \alpha \notin N\tau_o$, so $\alpha$ = Head $\alpha'$ SKRules with $\alpha' \in \tau_o$. Furthermore $d(\alpha) \rightarrow_{o\sigma} \beta$, so $\alpha'$ has a head-redex. Then OneStepHead $\alpha'$ SKRules reduces to $\delta$ with $d(\delta) = \beta$ and Head $\alpha'$ SKRules reduces to Head $\delta$ SKRules.

5. $\forall \alpha \in N\tau_s\ d(\alpha) \in N\tau_{o\sigma} \cup \{nil\}$

If $\alpha$ is a normal form of an element of E then $d(\alpha) \in N\tau_{o\sigma}$, $d(\alpha)$ is nil otherwise.

6. There is no infinite $(\rightarrow_s\text{-}\rightarrow_c)$ reduction path and moreover there exists a bounded function $b : \tau_o \rightarrow \mathbb{N}$ such that $\forall \alpha, \beta \in E\ \alpha (\rightarrow_s\text{-}\rightarrow_c)^m \rightarrow_c (\rightarrow_s\text{-}\rightarrow_c)^n \beta \Rightarrow m < b(d(\alpha))\ \&\ n < b(d(\alpha))$.

There is no infinite non computational path because we only use safe strategies. We define b to be: b(x) = 3*( the number of Ap's on the spine of x) + 3. We notate sp(x) for the number of Ap's on the spine of x. If $x \in \tau_o$ then ContainsAHeadRedex x SKRules has a number of non computational reductions which is bound by 3*sp(x). OneStepHead x SKRules is bound by 3*sp(x) + 1 because it has an extra Rewrite. So if $\beta \notin N\tau_s$ then m= 3*sp($\alpha$) + 1 and n = 1; if $\beta \in N\tau_s$ then we may need 3*sp($\beta$) + 2 non computational reductions to discover that we have a normal form.

The specification is *effective* because b, $\rightarrow_c$, d and E are all total computable.

### 3.3.3 Remarks.

The proof was easy because the most essential aspects were already covered by the lemmas which were themselves simple and easy to proof. The use of the primitives Match and Rewrite made it possible to reason easily about the TRS system.

## 3.4 Transformations of the Specification.

The description we have developed is a very nice one but operationally it is not the same as in the other two sections, because a call of ContainsAHeadRedex will induce an extra sweep through the term. If we want our description to mirror exactly the operations that were performed by the other methods, then we must change the rules and make them deliver a composite result consisting of the term and a boolean indicating whether the term was rewritten. The resulting rules are:

| | | |
|---|---|---|
| OneStepHead ( *Ap f a* ) *rules* | → Cond ( Match ( *Ap f a* ) *rules* ) | |
| | ( Pair True ( Rewrite ( *Ap f a* ) *rules* )) | |
| | ( Pair | |
| | (First ( OneStepHead *f rules* )) | |
| | ( *Ap* (Second ( OneStepHead *f rules* )) *a* )) | (I-1') |
| OneStepHead *c:Combinator rules* | → Pair False *c* | (I-2') |
| Head (Pair False *term*) *rules* | → *term* | (I-3') |
| Head (Pair True *term*) *rules* | → Head ( OneStepHead *term rules* ) *rules* | (I-4') |

with as initial term: Head (OneStepHead *Term SKRules*) *SKRules*.

Probably it will not be very difficult to prove this specification to be equivalent to the one without the booleans. Though this specification can now be considered to be just as efficient as the specifications in the other sections, it does not exactly mirror the same actions. In the other sections annotation functions were used in stead of booleans. Of course it is also possible to give a similar description with the abstract-interpretation method. We will rename Head to * and OneStepHead to !. We will also introduce success and failure constructors +/- and a propagate function called ~. The set of corresponding rules is:

| | | |
|---|---|---|
| ! ( *Ap f a* ) *rules* | → Cond ( Match ( *Ap f a* ) *rules* ) | |
| | ( + ( Rewrite ( *Ap f a* ) *rules* )) | |
| | ( ~ ( *Ap* ( ! *f rules* ) *a* )) | (I-1") |
| ! *c:Combinator rules* | → - *c* | (I-2") |
| * ( - *term* ) *rules* | → *term* | (I-3") |
| * ( + *term* ) *rules* | → * ( ! *term rules* ) *rules* | (I-4") |
| ~ ( *Ap* ( + *x* ) *y* ) *rules* | → + ( *Ap x y* ) | (I-5") |
| ~ ( *Ap* ( - *x* ) *y* ) *rules* | → - ( *Ap x y* ) | (I-6") |

If this specification is compared with that of section 2.2 we see that the redex rules and the propagation rule are now all covered by rule (I-1") thanks to the power of the match and rewrite primitives. The non-redex rules are covered by rule (I-2"). The last four rules are the same.

We have shown that it is relatively easy to express a reduction strategy using the abstract-interpretation TRS with several algorithms and that it is also simple to transform one specification to another.

### 3.5 Evaluation of the Method.

This method enables us to write elegant strategy specifications. The non-redex cases are easily handled using the Match and Rewrite primitives. The different actions for syntactically different terms are conveniently dealt with using the abstraction mechanism. The fact that we write our strategy description on another level helps us in specifying strategies more generally.

The abstract-interpretation TRS system makes it possible to specify strategies formally in such a way that the specification holds for a large class of TRS's. For instance the description given at the interpretation level in this section is a valid head reduction specification for all TRS's using explicit application functions. Only a very trivial adaptation to the abstraction TRS is necessary in order to summarize all combinators. The proof then also holds for this large class of TRS's. Besides that, the proof was less tedious because we were led immediately to some general lemmas which themselves were easy to proof using the properties of the primitives Match and Rewrite. This specification is very short and readable and it appeals to our intuition. It enables us to easily give transformations of one specification to other specifications which have specific properties.

Concluding we can state that we have a promising facility with a great expressive power for describing strategies independently of the TRS.

## 4. Conclusions and Further Research.

All of the three methods introduced in this paper are suitable for the specification of reduction strategies. Using an ordinary TRS gives rise to an exponential number of rewrite rules. This drawback disappears when a PRS is used. However the proof of the specification using this PRS was even more difficult than the proof with an ordinary TRS.

The most readable and general specification can be obtained by using an abstract-interpretation TRS extended with special primitives for matching and rewriting. The structure of the specification made it possible to give a simple proof and to construct transformations of the specifications in order to get alternative specifications with special desirable properties. Although this new TRS system is specially designed for the specification of reduction strategies, we think that its descriptive power is suitable for the specification of abstract interpretations [BUR85] in general.

In the near future we will search for simplifications of the definition of specification and we will investigate the use of the abstract-interpretation TRS in its full strength for the specification of strategies in graph rewriting systems, for giving correctness proofs of more complex strategies and for the investigation of the descriptive power for other domains.

## 5. Acknowledgements.

We would like to thank Ronan Sleep, John Glauert and Richard Kennaway of the University of East-Anglia for the many fruitful discussions about reduction strategies and we also thank Jan-Willem Klop of the Centre for Mathematics and Computer Science in Amsterdam for his explanations. Most of all we are grateful to Henk Barendregt of the University of Nijmegen for several important observations and valuable improvements.

## References.

[BAE86]
Baeten, J.C.M., Bergstra J.A., Klop J.W., "Term Rewriting Systems with Priorities", University of Amsterdam, Report FVI 86-03, 1986.

[BAR86a]
Barendregt, H.P., Eekelen, M.C.J.D. van, Glauert, J.R.W., Kennaway, J.R., Plasmeijer, M.J., Sleep, M.R., "Towards an Intermediate Language based on Graph Rewriting", University of East-Anglia and University of Nijmegen, Nijmegen internal report 88.

[BAR86b]
Barendregt, H.P., Eekelen, M.C.J.D. van, Glauert, J.R.W., Kennaway, J.R., Plasmeijer, M.J., Sleep, M.R., "Term Graph Rewriting", University of East-Anglia and University of Nijmegen, Nijmegen internal report 87.

[BRO86]
Broek, P.M. van den, Hoeven, G.F. van der, "Combinatorgraph Reduction and the Church-Rosser Property", internal report INF 86-15, Technical University of Twente, june 1986.

[BUR85]
Burn, G.L., Hankin, C.L., Abramsky, S., "The Theory and Practice of Strictness Analysis for Higher Order Functions", Research Report DoC 85/6, Imperial College, London 1985.

[EHR78]
Ehrig, H., "Introduction to the Algebraic Theory of Graph Grammars", Lecture Notes in Computer Science 73, Springer-Verlag Berlin, Aug. 1978.

[FUT84]
Futatsugi, K., Goguen, J.A., Jouannaud, J.P., Messaguer, J., "Principles of OBJ2", Twelfth Annual ACM Symposium on Principles of Programming Languages, Jan 1984, pp. 52-66.

[GLA85]
Glauert, J.R.W., Holt, N.P., Kennaway, J.R., Reeve, M.J., Sleep, M.R., Watson, I., "DACTLO: A Computational Model and an associated Compiler Target Language", University of East-Anglia 1985, internal report.

[KLO85]
Klop, J.W., "Term rewriting systems", Notes for the Seminar on Reduction Machines, Ustica 1985, to appear.

[MAG80]
Magó, G.A., "A Cellular Computer for Functional Programming", digest of Papers, IEEE Comp. Soc. COMPCON, Spring 1980, pp. 179-187.

[O'DO85]
O'Donnell, M.J., "Equational Logic as a Programming Language", Foundations of Computing Series, MIT Press, 1985.

[RAO84]
Raoult, J.C., "On Graph Rewritings", Theor. Comp. Sc. 32, North-Holland, 1984, pp. 1-24.

[TUR85]
Turner, D.A., "Miranda: A non-strict functional language with polymorphic types", Proceedings, Conference On Functional Programming Languages and Computer Architecture, Nancy 1985, Jouannaud, J.P., Ed., Lecture Notes in Computer Science 201, Springer-Verlag, Berlin Heidelberg New York Tokyo, pp. 1-16.

## CONTROLLING REDUCTION PARTIAL ORDER IN FUNCTIONAL PARALLEL PROGRAMS*

by

F. Warren Burton
Department of Computer Science
3160 MEB
University of Utah
Salt Lake City, Utah 84112
U. S. A.

## Abstract

In a functional parallel programming language a programmer may wish to control the reduction partial order to increase parallelism, reduce storage requirements, or reduce the total amount of work performed.

We show by a series of example that the reduction partial order can be easily controlled by using a combination of call-by-value, call-by-name, and call-by-speculation (an eager form of call-by-need).

## 1. Introduction

Functional programs are usually easier than procedural programs to read, write, verify, transform, and modify. With software costs increasing dominating hardware cost, it is often reasonable to use a less efficient language if programmer productivity can be increased as a result. This is becoming increasingly true with parallel systems, which are cheaper per unit of computation but harder to program. However, it is almost never reasonable to use a language for serious programming if the overheads are more than constant factor.

In a functional language, the reduction partial order may be controlled in order to:

- Increase parallelism,
- Reduce storage requirements, or
- Reduce the total amount of work performed.

The reduction partial order of a program can be controlled by using a combination of three parameter passing mechanisms. These mechanisms are:

- Call-by-value,
- Call-by-name, and
- Call-by-speculation.

*This material is based upon work supported by the National Science Foundation under Grant No. DMC-8514946.

Call-by-speculation is an eager form of call by need. If a processor would otherwise be idle, it may evaluate a call-by-speculation parameter before it is know to be needed. The system speculates that the value of the parameter will be needed later. If right, some time is saved. If wrong, some work is wasted. A more detailed explanation of these parameter passing mechanisms and their interactions may be found in [3, 7].

By using an appropriate combination of the above parameter passing mechanism, together with program transformation, a programmer can write programs which are both simple and efficient. Common behavior patterns often can be encapsulated in standard functions. These functions my be kept in a library.

Since changing the parameter passing mechanism can not change the result produced by a functional program, the indication of a particular parameter passing mechanism can be regarded as an annotation that does not effect the semantics of a program. (Poor annotations can result in nontermination, but we like to regard this as an extreme form of inefficiency.)

In this paper we will be primarily concerned with controlling the reduction partial order in the context of two computational models. These are the single processor (sequential) model, and the unbounded parallelism model. We assume that a scheduler can insure good performance in the normal case of a system with a finite number of processors, provided the programmer has insure that the two extreme cases work well [2, 4, 6].

We will use the notation of the Miranda programming language in our examples [16, 17], extended to include annotations to indicate which parameter passing mechanism should be used where. We will take call-by-value to be the default parameter passing mechanism. We will assume that all data constructors leave their components unevaluated. Finally, we will assume that tail recursion can run in constant space and that storage may be reclaimed as soon as it is no longer accessible.

## 2. Sequential Evaluation

Call-by-need (or lazy evaluation) [1, 10-12, 15, 19] has been advocated as an optimal evaluation strategy for functional programs on a sequential processor. Except for some special cases with higher order functions (which are not a problem, because these cases can be avoided without serious inconvenience), and a constant factor overhead, call-by-need is optimal with respect to evaluation time. Unfortunately, call-by-need is not optimal with respect to space.

Consider the tail recursive function *sum* defined by:

```
sum x = s 0 x
        where
        s a [] = a
        s a (b:x) = s (a + b) x
```

With call-by-need, the value of the first argument to *s* is not needed until the bottom level of recursion is reached. Hence, an intermediate result of size O(*n*) will be required to sum a list of length *n*. In like manner, strict call-by-value will cause the list being summed to be fully evaluated when *sum* is first called, again requiring O(*n*) storage. Using call-by-value with lazy data constructors gives the desired performance in this case. The elements of a lazy list can be summed in constant space. In general, call-by-value tends to collapse expressions to single value where possible, while lazy data constructors prevent large structures from being generated.

In the above example, strictness analysis could be used to determine that the first parameter of *s* could be passed by value. If parameters are passed by value only when a function is strict in that parameter, then the above approach can be used while preserving normal order semantics [9]. On the other hand, some cases will be missed. If we wish to compute the product of elements in a list, we might use

```
prod x = p 1 x
         where
         p a [] = a
         p a (0:x) = 0
         p a (b:x) = p (a * b) x
```

where *p* is not strict in its first parameter. Do we want the programmer to have to annotate to get normal order semantics, or to get constant space performance? We choose the former, but do not feel strongly on the matter.

The basic recursive pattern used in *sum* and *prod* can be encapsulated in a higher order function, usually called *reduce*, *insert* of *fold*. There are several minor variations to this function. In order to get the desired performance, it is important to use a tail recursive version such as *fold* defined by

```
fold f a [] =a
fold f a (b:x) = fold f (f a b) x
```

with *sum = fold (+) 0* and *prod = fold (*) 1*.

There are times when one wants strict or partly strict data constructors. It is easy to write pseudo-constructors using the lazy constructors. For example,

```
eager_cons a b = a:b
```

will evaluate its arguments, using the call-by-value default, before placing them in a lazy data structure. To convert a lazy list into a fully evaluated list, one can use

```
force [] = []
force (a:x) = eager_cons a (force x)
```

to recursively evaluate each element. Similarly, partially strict pseudo-constructors may be defined by using annotations to control parameter passing. (We will hang annotations on type specification statements.) A version of *cons* that is strict in its first parameter only is defined below.

```
left_eager_cons :: * -> name [*] -> [*]

left_eager_cons a b = a:b
```

Before proceeding, we should clarify what we mean by lazy data constructors. The arguments of the data constructor are not evaluated, and when they are later selected and expanded, the original components of the constructor remain unevaluated. This is consistent with our use of call-by-name rather than call-by-need. In fact, list operations and other constructors can be defined as higher order functions with parameters passed by name [3].

A second problem with call-by-need is that it is very difficult to throw away results that take up too much space. (In conventional programming, the question of saving a result vs. recomputing it is a common example of a time/space trade-off where there is no single right answer.) For example, suppose that *x* is a very large list with elements that are very cheap to compute. If we compute *f x* where *f x = sum x / prod x* using call-by-need, then the entire list will be generated while the sum is being computed and will be retained for use in computing the product.

We have included call-by-name as a parameter passing mechanism so that results can be thrown away. When an expression is evaluated, the original expression is not overwritten with the resulting value. Hence, *f*, as defined above, will run in constant space but will evaluate the list twice. To prevent reevaluation of the list, at the cost of storing the list, we could use

```
f2 x = sum y / prod y
       where
       y = force x
```

A better solution, which runs in constant space and evaluates the list elements only once is

```
f3 x = s / p
       where
       (s, p) = double_fold (+) 0 (*) 1 x

double_fold f a g b [] = (a, b)
double_fold f a g b (c:x) = double_fold f (f a c) g (g b c) x
```

This transformation has been previously proposed by Wadler [18].

Having disqualified our example of a situation where it might be desirable to throw away results, let us consider another. Using a style of programming recommend by Hughes [13], many combinatorial problems can be solved by searching a lazy tree of values. We can define a binary tree in Miranda by

```
tree * ::= Empty | Node * (tree *) (tree *)
```

and construct a tree of sublists of a list using the function

```
sublists x = make_tree [] x

make_tree y [] = Node y Empty Empty
make_tree y (a:z) = Node y (make_tree (a:y) z) (make_tree y z)
```

so that the leaves contain the distinct subslists of the original lists, and each internal node contains a list which is a sublist of each list in a lower level node. We can use this tree to solve the problem of determining whether a list has a sublist that sums to a particular value. We can transform the above tree to a tree of booleans using the function

```
sumcheck goal Empty = Empty
sumcheck goal (Node x left right)
    = Node True Empty Empty, goal = sum x
    = Node False Empty Empty, goal < sum x
    = Node False (sumcheck goal left) (sumcheck goal right)
```

which indicate for each node whether the node is a solution or not. Pruning occurs when a solution is found or a sublist with a sum that is too big is encountered.

A backtracking algorithm to search a tree of this form is

```
search (Node True Empty Empty) = True
search (Node False Empty Empty) = False
search (Node False left right) = search left ∨ search right
```

which uses the fact that ∨ only evaluates its second argument if its first argument evaluates to *False*. (We will return to this example in the next section.)

The above backtracking algorithm would not work if we were solving a different problem involving searching a tree with infinite paths, unless the leftmost solution was to the left of the left-most infinite path. A breadth-first algorithm can be uses to search such a tree. However, the number of nodes in a tree tends to grow exponentially with the depth of the tree, so the storage required for a breadth-first search may grow exponentially with respect to the storage required for a depth-first search to the same depth. One common solution to the problem is to search, using depth-first, to depth 1, then search again to depth 2, and so forth. This method is not as expensive as it first appears. For example, in a full binary tree the majority of nodes are leaves, so each search inspects twice as many nodes as the previous search. The total number of node visits is only about twice that of a straight breadth-first search in this case. The repeated depth-first searches stop as soon as either a solution is found or it is found that all nodes have been considered in a given depth-first search. (Of course, in an infinite tree with no solutions, no search will terminate.) The search algorithm is

```
choice ::= No | Maybe | Yes

depth_search 0 (Node any left right) = Maybe
depth_search n (Node True Empty Empty) = Yes
depth_search n (Node False Empty Empty) = No
depth_search n (Node False left right)
    = max[depth_search (n-1) left, depth_search (n-1) right]

full_search x n = True, result = Yes
                = False, result = No
                = full_search x (n+1), result = Maybe
                  where
                  result = depth_search n x
```

where *No* < *Maybe* < *Yes* in Miranda.

Now the point of this example is that we still have an exponential growth in storage requirements if call-by-need is use. However, if lazy list constructors are based on call-by-name rather than call-by-need, then we get the desired performance. The tree is reexpanded on each depth-first search, with storage requirements being bounded by the depth of the tree.

## 3. Parallel Evaluation

It is well know that call-by-value can be used to increase parallelism in a functional programming language. If several parameters are passed by value, they can be evaluated in parallel. Sometimes a programmer may want to use call-by-value for a parameter even if the function is not strict in that parameter, in order to increase parallelism.

Recall the backtracking search algorithm of the previous section. We might want to change the last equation to

```
search (Node False left right)
    = strict_or (search left) (search right)

strict_or a b = a ∨ b
```

which could result in an exponential growth in potential parallelism at the cost of doing some work that might be avoided with the sequential algorithm. A better solution to this problem will be given later.

Sometimes it is desirable to inhibit parallel evaluation. Consider a function application of the form

```
f (g a) (update a i x)
```

where *a* is an array, *g* is a function that will return a simple result, or at least a result that does not contain a reference to *a*, and *update* is an array update operation that can modify an array in place if there are no other references to the array. In this case we are likely to want the first argument to be evaluated before the second. We could do this as follows:

```
sequence f (g a) (update a i x)

sequence :: (* -> ** -> ***) -> * -> name ** -> ***
sequence h x y = h x y
```

Notice that *(g a)* is passed to *sequence* by value, and so must be evaluated before *h x y* is computed. On the other hand *(update a i x)* is passed by name, and so can not be evaluated until after the computation of *h x y* has been initiated.

Consumer/producer parallelism can be supported in a limited form by using call-by-value in combination with call by need. Consider the expression

```
fold f a (map g [1..n])
```

where *f* and *g* are functions which are expensive to compute. As usual, *map* is defined by

```
map f [] = []
map f (a:x) = f a : map f x
```

As expressed, the evaluation of *f* and *g* will alternate. That is, the computation will proceed as follows:

```
fold f a (map g [1..n])
    = fold f a (map g (1:[2..n]))
    = fold f a (g 1 : map g [2..n])
    = fold f (f a (g 1)) (map g [2..n])
```

At this point, the evaluations of *(f a (g 1))* and *(map g [2..n])* will proceed in parallel. The important thing to notice is that *map* returns a list with unevaluated components. Therefore, in the evaluation of *(f a (g 1))*, it is necessary to compute *g 1* before starting the application of *f* to its arguments. What we would like is for *map* to evaluate the head of its result, so one application of *g* can take place in parallel with the application of *f*. That is, *g 2* should be evaluated in parallel with the evaluation of *f a x* where *x* is the value of *g 1*. We can accomplish this by computing

```
fold f a (semiforce (map g [1..n]))

semiforce [] = []
semiforce (a:x) = left_eager_cons a (semiforce x)
```

so that

```
fold f a (semiforce (map g [1..n]))
   = fold f a (semiforce (map g (1:[2..n])))
   = fold f a (semiforce (g 1 : map g [2..n]))
   = fold f a (left_eager_cons (g 1) (semiforce map g [2..n]))
   = fold f a (x:semiforce map g [2..n])
   = fold f (f a x) (semiforce map g [2..n])
```

Now *semiforce* will force the evaluation of the next application of *g* in parallel with the application of *f* to *a* and the already evaluated value *x = g 1*.

There is a serious limitation to this approach. Strict synchronization is required. The applications of *f* and *g* must start together and must both terminate before the next application of *fold* can take place. In procedural programming, buffering is used to relax the synchronization requirements. This is probably not really necessary in this case, but in the case of a long multistage pipeline, the cost of keeping the entire pipeline in step could be expensive, involving communication delays proportional the the length of the pipeline on every cycle.

The solution to this problem is to use call-by-speculation. We can define

```
buffer [] = []
buffer (a:x) = buffer_cons a (buffer x)

buffer_cons :: speculation * -> name [*] -> [*]
buffer_cons a b = a:b
```

Now *buffer_cons* is identical to *left_eager_cons* except that the first parameter is passed by speculation rather than value. Therefore *buffer_cons a x* will initiate the evaluation of *a*, but will return a result before the evaluation of *a* has terminated. This lets the producer fill the single buffer, *a*, whenever it wants, and lets the consumer proceed until the value of *a* is actually used, at which point the consumer will be delayed until the producer has filled the buffer.

Call-by-speculation has other uses. It reduces the synchronization requirements and results in an evaluation strategy that is more like data-driven evaluation than demand-driven evaluation. Consider the evaluation of

```
f (g x) (g y) (g z)
```

where

```
f a b c = h a b + h a c + h b c
```

where the applications of *g* and *h* all take a similar large amount of time. If this computation were run with two processors, two of the arguments of *f* could be evaluated during one time period. During a second time period, the final argument would be evaluated while one processor was idle. During the third time period, two applications of *h* could be evaluated, with the final application taking place during a forth time period. If the parameters to *f* were passed by speculation, then one of the applications of *h* could take place in parallel with the evaluation of the final argument to *f*.

Perhaps the most important use of call-by-speculation is with algorithms having a high potential for parallelism if and only if some wasted work will be tolerated. (Every NP-complete problem has solutions with this property-- algorithms that will always run in polynomial time with unbounded parallelism but will sometimes run in polynomial time sequentially.)

Call-by-speculation can be used to accomplish the same thing that or-parallelism is used for in Prolog. Returning to the backtracking algorithm last mentioned at the start of this section, we could replace the *strict_or* function with a *speculative_or* where

```
speculative_or :: bool -> speculation bool -> bool
speculative_or a b = a ∨ b
```

This will permit the desired parallelism without requiring that the speculative work be performed if it is not necessary.

It is useful to be able to associate priorities with speculative work so that work that is more likely to be required can have a higher priority. In the absence of speculative work, it is not desirable to involve the programmer with low level scheduling issues such as this. At best, priorities will make at most a constant factor difference in the time required to evaluate an expression that does not contain speculative work. However, with speculative evaluation, priorities become much more important. If speculative evaluation is performed only by processors that would otherwise be idle, then at least one processor is always doing mandatory work (so aside from overheads, speculative computation can not slow things down). However, with *n* processors, *n - 1* could be doing speculative work which will later be found not to be required, while speculative work which will be required is not being done. In general, with *n* processors, associating good priorities with speculative work can result in an O(*n*) speedup.

We note that with the above backtracking example, a clever compiler could figure out which computations were the most important. In general, this will not be possible. However, it may be possible to adapt some of the work on strictness analysis to determining priorities for speculative evaluation in some simple cases. For example, in

```
speculative_or a (speculative_or b c)
```

the evaluation of *b* will always be required whenever *c* is. While this function is not strict in either *b* or *c*, it is more strict in *b* than *c*.

More information about speculative evaluation may be found in [5].

We will close this section with some comment on the interaction of annotations with nondeterminism.

The problems with referential transparency in a language allowing nondeterminism are well known. For example, using the *amb* construct proposed by McCarthy [14], the value of *amb 1 2 = amb 1 2* may be either *True* or *False* depending on whether *amb* makes the same choice both times. If we define a function *f x = x = x* (where the first "=" is the *is defined to be* symbol in Miranda and the second "=" is the relational equality test) then *f (amb 1 2)* will always be *True* if *x* is passed by value or speculation, but could be *True* or *False* is *x* is passed by name. Our annotations now have significant semantic meaning!

This problem can be solved by using a technique proposed in [8], which also preserves referential transparence. In this paper, we will use the more familiar *amb* operator.

There is a further problem with using a function such as *amb* in the context of a lazy interpreter. If *a* and *b* are two lists, then what is the value of *amb a b*? It can be argued that the list constructor is not strict, so once either list is known to be not completely undefined, it may be returned. (e.g. The list ⊥:⊥ could be returned.) With our annotations, we can control when decisions are made. For example, if we want to return a list only if it is fully defined, then we could write *amb (force a) (force b)*.

Now let us consider the problem of nondeterministicly merging two lists. The question is when do we decide from which list we will select the next element. Usually, we will want to select the next element from a list as soon as the value of the first element of that list is available. We can write a function, *interleave*, to do this type of nondeterministic merging as follows:

```
interleave :: speculation * -> speculation * -> *
interleave a b
    = amb (step a b) (step b a)
      where
      step [] x = x
      step (a:x) y = left_eager_cons a (interleave x y)
```

Notice that *interleave* uses call-by-speculation parameter passing so it can return a value before anything is known about one of the parameters provided enough is known about the other. The use of *left_eager_cons* prevents either of the two alternative computations from terminating before the value of the next element is actually available. (If we used the ordinary lazy cons, ":", then we would be selecting the next element from a particular list as soon as we knew that the list was not *[]*.)

There is one minor problem concerning the efficiency of *interleave* under certain circumstances. Suppose a large amount of work must be done in computing each list element (which would argue in favor of our approach of not selecting as element until its value is available). We might often end up computing the first element of *a* and the first element of *b* in parallel. The work on the element which was not selected would be discarded. We can solve this problem by using

```
nondeterministic_merge a b = interleave (buffer a) (buffer b)
```

In general, annotations make it possible to specify how much computation should be performed before a nondeterministic choice is made.

## 4. Conclusion

We have seen that the use of call-by-value, call-by-name, and call-by-speculation in combination gives a programmer a high degree of control over the time and space requirements of a program and the potential degree of parallelism in the program. In most cases, reasonable defaults make it unnecessary for the programmer to specify the parameter passing mechanism. In most of the remaining cases, standard functions such as *buffer* or *backtrack* may be used to achieve a common behavior.

In some ways, call-by-speculation is similar to a fork operation, in that it allows a computation to be initiated. Call-by-value is similar to a join in that it will not let a thread of control proceed until a particular computation has terminated.

## References

1. Arvind, *Sharing of Computations in Functional Language Implementations*, Laboratory for Computer Science, Massachusetts Institute of Technology, July 1984.
2. F. W. Burton and M. R. Sleep, "Executing functional programs on a virtual tree machine", *Proc. 1981 Conf. Functional Programming Languages and Computer Architecture*, Portsmouth, New Hampshire, Oct. 1981, 187-194.
3. F. W. Burton, "Annotations to control parallelism and reduction order in the distributed evaluation of functional programs", *ACM Trans. Prog. Lang and Systems 6*, 2 (Apr. 1984), 159-174.
4. F. W. Burton and M. M. Huntbach, "Virtual tree machines", *IEEE Trans. on Computers C-33*, 3 (Mar. 1984), 278-280.
5. F. W. Burton, "Speculative computation, parallelism, and functional programming", *IEEE Trans. on Computers C-34*, 12 (Dec. 1985), 1190-1193.
6. F. W. Burton, "Storage management in virtual tree machines", *IEEE Trans. on Computers*, To appear.
7. F. W. Burton, "Functional programming for concurrent and distributed computing", *Comp. J.*, To appear.
8. F. W. Burton, "Nondeterminism with referential transparency in functional programming languages", To appear.
9. F. W. Burton, D. Maurer, H. Oberhauser and R. Wilhelm, "A space efficient optimization of call-by-need", *IEEE Trans. on Software Eng.*, To appear.
10. D. P. Friedman and D. S. Wise, "Cons should not evaluate its arguments", *Third Int. Coll. Automata Languages and Programming*, Edinburgh, Scotland, 1976, 257-284.
11. P. Henderson and J. M. Morris, "A lazy evaluator", *Conf. Rec 3th ACM Symp. on Prin. of Prog. Lang.*, Atlanta, Ga., Jan. 1976, 95-103.

12. R. J. M. Hughes, "Super Combinators: A new implementation method for applicative programs", *Proc. 1982 ACM Symposium on LISP and Functional Programming*, Pittsburgh, Penn., Aug. 1982, 1-10.

13. R. J. M. Hughes, "Why functional programming matters", Programming Methodology Group Memo PMG-40, Dept. of Computer Science, Chambers University of Technology and University of Goteborg, Goteborg, Sweden, 1984.

14. J. McCarthy, "A basic mathematical theory of computation", in *Computer Programming and Formal Systems*, P. Braffort and D. Hirschberg (editor), North Holland, Amsterdam, 1963, 33-70.

15. D. A. Turner, "A new implementation technique for applicative languages", *Software-Practice & Experience 9* (1979), 31-49.

16. D. A. Turner, "Functional programs as executable specifications", in *Mathematical logic and programming languages*, C. A. R. Hoare and J. Shepherdson (editor), Prentice Hall, Englewood Cliffs, NJ, 1985, 29-54.

17. D. A. Turner, "Miranda: A non-strict functional language with polymorphic types", in *Functional Programming Languages and Computer Architecture*, vol. 201 , J. Jouannaud (editor), Springer-Verlag, Berlin- Heidelberg-New York-Tokyo, 1985, 1-16.

18. P. Wadler, "Applicative style programming, program transformation, and list operators", *Proc. 1981 Conf. Functional Programming Languages and Computer Architecture*, Portsmouth, New Hampshire, Oct. 1981.

19. C. P. Wadsworth, *Semantics and Pragmatics of the Lambda-Calculus*, Oxford University, 1971.

# Parallel Garbage Collection for Graph Machines

Ashoke Deb

Memorial University of Newfoundland

Department of Computer Science

St. John's, NFLD, CANADA, A1C 5S7

UUCP address : ashoke@garfield.uucp

EAN address : ashoke@garfield.mun.cdn

## 1 Introduction

The functional and logic programming languages are elegant and powerful. They provide facilities such as nondeterministic programming, infinite data types, polymorphism, strict and nonstrict operations. In these computational environments, either the program manipulates graph structures in order to produce the manifest value, or the program itself is represented as a graph which as a result of successive reductions will arrive at the least fix point, which is the manifest value.

During the course of modifications to the computation graph, a set of nodes may get disconnected. Such unreachable subgraphs are called *garbage*. The garbage collector identifies them and makes them available for eventual use by the computational process.

There are two basic techniques for detecting garbage, *mark-sweep* and *reference counting*. In the mark-sweep technique, a mark bit is maintained with each node. During the marking phase, every node reachable from the root of the computational graph is marked. At the end of the marking phase, all unmarked nodes are garbage, and will be collected. Finally, all marked nodes are reset. In the reference counting technique, each node has a reference count field which maintains the number of references to that node. When the reference count of a node becomes zero, that node is detected as a garbage, the reference counts of each of the offspring-nodes of a garbage node are decremented by one, and so on.

Among the advantages of the mark-sweep technique are that it requires only one bit for marking and that the algorithm is fairly straightforward. A serious disadvantage is that if the memory occupancy rate is high, particularly at the equilibrium state, the overhead of marking the entire computational graph is intolerable. Also, the cost of garbage collection is not proportional to the actual activity in the computation (i.e., the number of edge deletions and additions) but to the amount of resident data [6].

One of the advantages of reference counting is that it is *incremental*; i.e., it is possible to collect some garbage without having to detect all of it. Also, even in the case of high memory occupancy, detection of garbage nodes can be done without overloading either the storage space or the time. One of the disadvantages of reference counting is that it cannot reclaim cyclic structures.

In order to overcome the delay of mark-sweep garbage collection, several parallel algorithms [2,7,12,13] have been proposed. In these algorithms, there is a *collector* running in parallel with the computational process, called the *mutator*, and garbage is collected as it is produced. These algorithms are quite interesting and important in their own right, particularly as an exercise in proving parallel programs correct. Several proposals on distributed multiprocessor garbage collection using some variations of the mark-sweep technique also exist [1,10,11].

Some of the criticisms regarding reference counting have been countered by several authors. In regards to the size of the count field, Clark and Green [4] have shown that in most list processing programs, about 97% of the reference counts are one. That empirical study indicates that one can use the hash linking technique of Bobrow [3] to maintain hash links to only those cells with reference counts greater than 1. In regards to the arithmetic overhead involved in updating the reference counter, Wise [17] argues that such overhead can be minimized by moving the counter fields to separate memory at the same address as the node counted and by performing the increment/decrement operation at the hardware level. The inability to collect cyclic structures can be overcome in some special cases [8]. Others have suggested hybrid techniques, where a reference counting technique is used most of the time and a mark-sweep algorithm is used infrequently to collect the remnant cyclic structures [6].

Parallel/distributed mark-sweep algorithms have additional disadvantages. Primarily, they require very fine-grained synchronization between the mutator and the collector. Hickey [9] showed that the productivity of on-the-fly garbage collection is at most 150% of the productivity of a serial garbage collector, where productivity is the proportion of time during which the mutator is not waiting for garbage collection. Extending mark-sweep algorithms in the multi-collector environment becomes either non-realtime [1] or quite expensive [11]. Also, the collector's effort is still proportional to the resident data, not the amount of computational activity.

This motivates us to present a new garbage collection algorithm. These are some of its highlights:

- The algorithm uses reference counting to identify garbage.
- It collects cyclic structures (without having to use another mark-sweep phase).
- The number of nodes visited is, in most cases, proportional to the amount of garbage collected.
- It is incremental.
- It requires little synchronization between the mutator and the collector(s).
- The algorithm can be applied (with minimum alteration) in an interleaved environment, with the mutator and the collector sharing the same cpu; in a parallel environment with one collector running in parallel with the mutator; or with multiple collectors operating on different memory segments.

We first present the algorithm for the simplest environment, where the collector and the mutator share the same cpu. This is primarily meant to expose the basic idea behind the algorithm.

There is another reason for specially presenting the algorithm for a sequential environment. Many researchers [15,16] are of the opinion that *it is impossible, or at least unthinkable, for reference counting to collect islands of inaccessible garbage containing cyclic structures, and that a marking and gathering policy, independent of reference counting technique, is clearly superior for gathering cells that ordinary reference counting fails to release.* This is simply not so.

For the parallel environments, extensions to the algorithm are small, the basic idea remaining the same; these extensions will be highlighted. There also, we first deal with a single collector operating concurrently with the mutator, showing how little interaction is needed for an on-the-fly collector using reference counting.

Finally, we show how the algorithm can be used in a multi-collector system.

## 2 Interleaved system

In the interleaved system, the mutator is halted during a garbage collection phase and vice versa. Sequential and non-realtime systems fall into this category.

While the mutator is running, it performs a sequence of edge deletion operations, DELETE(Q.x), meaning that an edge from some father node Q to a son node Q.x is deleted. As a result of these operations, garbage nodes may be produced.

There is an important difference between the conventional treatment of the DELETE operation and ours.

In the usual reference counting techniques, DELETE(Q.x) operation performs the following steps:

1. Decrement the reference count of the node Q.x.
2. Set the field Q.x to Nil.

In our system, there is a special structure GC (the Garbage Can), and the DELETE operation is defined as follows:

**DELETE(Q.x): Place the pointer value Q.x in GC;**
**Set the field Q.x to Nil.**

Note that unlike the traditional interpretation of DELETE, the chore of decrementing the reference count is not done during the mutation phase, but transferred to the collector.

The garbage collector may use all or part of the entries in the Garbage Can in order to identify garbage nodes in the memory.

Without loss of generality, we assume that the computational graph is a rooted, possibly cyclic, binary graph. In such a graph, a non-nil node Q is represented by a record consisting of five fields:

| | |
|---|---|
| Q.left | : the left son of Q |
| Q.right | : the right son of Q |
| Q.visit | : a binary digit, initially 1 |
| Q.E | : the in-degree of Q |
| Q.C | : initially 0, to be used by the later algorithms |

At the beginning of each garbage collection phase, for each node Q, Q.visit and Q.C are set to 1 and 0 respectively. The garbage collector consumes the first s entries of GC, denoted by GC[1..s], to find the garbage nodes. GC may be viewed as the root of all garbage, a garbage node being one that is *exclusively reachable* from GC.

We will now describe the algorithm **Detectgarbagefrom**. This algorithm makes use of two algorithms - **Colorgreen** and **Colorred**. At the end of the Colorred phase, a node Q is garbage if and only if Q.C > 0. At the end of detection, the **Reset** algorithm makes the appropriate modifications to the reference count fields. [1]

**Comment**: Given a node Q, the following algorithm will "color" all nodes reachable from Q in such a way that a node P will have the color "green" (encoded as P.C $\geq$ 1), if P is exclusively reachable from Q; otherwise P will have color "red" (encoded as P.C = -1). At the end, all green-colored nodes are garbage nodes.

```
Detectgarbagefrom(GC[1..s]):

  For i := 1 To s Do Colorgreen(GC[i]);
  For i := 1 To s Do Colorred(TRUE, GC[i]);
  For i := 1 To s Do Reset(GC[i]);
```

**Comment**: In the following algorithm, all nodes P reachable from Q are visited, and at the termination, P.C = the total number of input edges of P reachable from Q.

```
Colorgreen(Q):

  If Q ≠ Nil Then
     increment Q.C by 1;
     If Q.visit ≠ 0 Then
       set Q.visit to 0;
       Colorgreen(Q.left);
       Colorgreen(Q.right)
```

**Comment**: In the following algorithm, any node P reachable from Q is visited and at the termination, P.C = -1 (i.e., P is not garbage) if and only if P can be visited from another non-garbage node. The parameter M and the local variable B are boolean.

```
Colorred(M, Q):

   B ≡ M = TRUE and Q.C = Q.E;
   If (Q ≠ Nil) and (Q.C ≠ -1) and (¬ B or Q.visit ≠ 1) Then
       If ¬ B Then set Q.C to -1 ;
       set Q.visit to 1;
       Colorred(B, Q.left);
       Colorred(B, Q.right);
```

For the sake of analysis and appreciation of the algorithm, the following should be pointed out:

a) After the Colorgreen phase, P.C = P.E does not necessarily imply that P is a garbage node.

b) On the other hand, after the Colorgreen phase,P.C < P.E is a sufficient condition that P is not garbage.

[1] Such updating of the reference count does not have to be done at the software level. Wise [17] shows how this overhead can be minimized by transferring it to the hardware level.

c) If Q is detected to be a non-garbage node (because either Q.C < Q.E or Q is visited from another non-garbage node, indicated by M = FALSE), then any node reachable from Q (including Q) will be marked as non-garbage. (Q.C ← -1, and subsequently P.C ← -1.)

d) Any node P will be visited at most once as a possible garbage (i.e., P.C = P.E), and at most once as a garbage node (as explained in (c) above).

**Reset(Q):**

```
If Q ≠ Nil Then
  If Q.C = Q.E ≠ 0 Then
    set Q.E to 0;
    If Q.visit ≠ 0 Then
      set Q.visit to 0;
      Reset (Q.left);
      Reset (Q.right);
  If Q.C = -1 Then
    decrement Q.E by 1;
```

**Comment:**

e) Note that Reset(Q) is called when either

  i) Q is in GC or

  ii) Reset(P) is called and P is a garbage node (i.e., P.C = P.E ≠ 0) and Q = P.left (or P.right).

f) In Reset(Q), if Q is a garbage node (Q.C = Q.E ≠ 0) then

  i) Q.E is set to 0 exactly once, and

  ii) its son nodes are visited exactly once.

g) In Reset(Q), if Q is a nongarbage node, then because of (e) above, Q's reference count field must be decremented.

It is also interesting to recall the traditional reference counting technique again at this point. In the traditional approach, when the reference count of a node becomes zero, it initiates a "cascade" of many decrement operations; i.e., the reference count of the son nodes are decremented, if their reference counts become zero then the reference counts of the grandson nodes are decremented and so on.

In our Reset algorithm, we do not have to do that. Have we eliminated this overhead? In fact, we have absorbed these successive decrement operations within the Colorgreen algorithm in the process of counting up the C field!

## 2.1 Proof of correctness

In this subsection, we outline the proof of some theorems, depicting some properties of the algorithms presented. We omit detailed formalism here.

*Notation*: $N \xrightarrow{*} P$ means that the node P is reachable from the node N by a path of length $\geq$ 1.

**Definition 1** : *A node* P *is exclusively reachable from a set of nodes* R *if and only if* $N \xrightarrow{*} P$ *implies that* $N \in R$.

**Definition 2** : *Given a node* P *and a set of nodes* R, *we define* I(R, P) *as the set of input edges to* P *in paths from* R *to* P.

Also. Let GC = {GC[i],$1 \leq i \leq s$}. Let $G_1$ = {P | GC[i] $\xrightarrow{*}$ P, $1 \leq i \leq s$}.

**Theorem 1** : *For all* $P \in G_1$, *the following holds:*

{P.C = 0 *and* P.visit = 1}

For i := 1 To s Do Colorgreen(GC[i])

{P.C = | I(GC,P) | *and* P.visit = 0}.

Proof: Theorem 1 can be proved directly by induction. Each node P reachable from GC is "visited" exactly once, modifying P.visit from 1 to 0 the first time P is arrived at. Also each edge "reachable" from GC is traversed exactly once; and each time a node P is arrived at via a reachable edge to P, P.C is incremented by 1. Hence, at the termination, for any P, P.C is exactly equal to the number of "reachable" input edges to P from GC.

**Theorem 2** : *For all* $P \in G_1$, *the following holds:*

{P.C = | I(GC,P) | = k *and* P.visit = 0}

For i:= 1 To s Do Colorred(TRUE,GC[i])

{P.C = -1 *iff* P *is not exclusively reachable from* GC,
P.C = k, *otherwise, and*
P.visit = 1}.

Proof: To prove this, note that for a call Colorred(M,Q)

(i) if M = FALSE and Q.C $\neq$ -1, then Q.C and Q.visit are set to -1 and 1, respectively; and all nodes in the left- and right-subtree of Q will be set as a result of Colorred(FALSE, Q.left) and Colorred(FALSE, Q.right)

(ii) if M = FALSE and Q.C = -1, then there exists some node R, R $\xrightarrow{*}$ Q and Colorred(FALSE,R) is called. Therefore, Q $\xrightarrow{*}$ P also implies R $\xrightarrow{*}$ P; and Colorred(FALSE,R) will cause P.C = -1 and P.visit = 1.

(iii) if M = TRUE and Q.C < Q.E, then also Q.C = -1, Q.visit = 1 and the left- and right-subtree of Q are reset as a result of Colorred(FALSE,Q.left) and Colorred(FALSE,Q.right).

(iv) if M = TRUE and Q.C = Q.E, then Q.visit will be set to 1 and the left- and right-subtree will be scanned as a result of Colorred(TRUE,Q.left) and Colorred(TRUE,Q.right), if and only if Q has not been scanned before.

From the above algorithms, the following facts are immediate:

**Fact** : For all Q, $Q \in G_1$ , after the Colorred phase,

Q is garbage iff (i) Q.C > 0
(ii) Q.C = Q.E .
Q is non-garbage iff (i) Q.C = -1
(ii) Q.C < Q.E .

# 3 Parallel system

The obvious way to remove the delay of the computational process imposed by garbage collection is to collect garbage in parallel with the mutator process.

Some of the difficulties in such real-time garbage collection are that (i) links are added and deleted while a collection phase is in progress, and (ii) the necessary communication and synchronization between the mutator and the collector need be kept minimum.

On-the-fly collectors based on mark-sweep are particularly susceptible to these difficulties, where both the mutator and the collector participate in "coloring" the same field and both processes traverse the same set of nodes, the computation graph.

As in the previous section, the only communication between our collector and the mutator is by message passing. Also, additions and deletions of links during a collection phase are used to facilitate, rather than to hinder (as in mark-sweep), the garbage detection.

## 3.1 From interleaved to parallel collector

In moving from the sequential to the parallel environment, the following changes are made:

(i) In the parallel environment, the garbage collector algorithms Colorgreen, Colorred and Reset remain identical to their sequential counterpart. (With only a minor change in the Colorgreen algorithm, as noted in (ii) below.)

(ii) Each node Q now contains an extra bit, Q.color. Q.color is 0 prior to each collection phase. Q.color will be set to 1 if Colorgreen visits the node Q.

(iii) In addition to GC, there is NG, called **New Garbage**, which will hold the links newly disconnected during the current collection phase, as described in (iv) below.

(iv) The modified DELETE is as follows:

**DELETE (Q.x) : If Q.color = 0 Then place Q.x to GC**
**Else place Q.x to NG;**
**Set the field Q.x to Nil.**

In the parallel system, prior to each collection phase, for node Q, Q.visit, Q.C and Q.color are set to 1, 0 and 0, respectively. Also, the elements of NG are added to GC for the new collection phase.

The purpose of having NG is to catch, during a collection phase, the deleted links that participated in counting the C fields. It should be also noted that having both NG and GC in the parallel system requires no more total space than having only GC in the interleaved system.

We now present the algorithm for the parallel system. For the future reference we will label each phase in the algorithm.

**Detectgarbagefrom (GC[1..s]) :**

**step1: For i := 1 to s Do Colorgreen (GC[i]);**
**step2: For i := 1 to s Do Colorred (TRUE, GC[i]);**
**step3: For each element $\alpha$ in NG, Do Colorred (FALSE, $\alpha$);**
**step4: For i := 1 to s Do Reset (GC[i]);**
**step5: For each element $\alpha$ in NG, Do Reset ($\alpha$).**

The changes in the above algorithm from its sequential counterpart are the addition of step3 and step5.

## 3.2 Correctness

In this section, we informally outline the argument that can be used to establish the proof of correctness of the algorithm, although we do not provide any such formal proof.

- Let $G_1$ be the graph visited by the collector during step1. Then for any node $P \in G_1$, P.color = 1.
- During a collection phase, for any node $P \in G_1$, suppose a new edge is *added* from P to Q, or from Q to P. Then both P and Q must be non-garbage prior to (and after the) addition of this edge. Therefore, prior to this addition, $P.E \geq P.C$ and $Q.E \geq Q.C$.

  Since addition of a new edge increments the E-field, the relation between the E and the C fields may only strengthen (from inequality to strong inequality). Therefore, they will still be recognized as non-garbage.
- Suppose, an edge P.x is *deleted* from P to Q (Q = P.x). Then prior to this deletion, both P and Q were non-garbage.
  - If P.x is deleted prior to step3, then P.x will be in NG (according to the DELETE operation). Later in step3, the edge P.x will be visited and Q (= P.x) will be recognized as non-garbage (as a result of Colorred (FALSE, $\alpha$) ).
  - If P.x is *not* deleted prior to step3, then P.x must have been visited by step2. Thus, Q (= P.x) will be recognized as non-garbage.
  - Any edge P.x deleted after step3 of a particular collection phase and before the next collection will remain in NG and will be added to GC for the next collection.

# 4 Multiple collectors

A natural progression in accelerating the garbage collection process is from a single collector to multiple collectors, running in parallel with the mutator. However, the benefit of employing multiple collectors can be substantially diminished if the synchronizations among the collectors are not kept to minimum.

We present a system where each collector operates independently and the only communication is by message passing. More importantly, the only synchronization needed is between steps, as explained below.

## 4.1 The logical structure

We assume that there are N collectors, $C_1, ..., C_N$, and the memory is divided into N segments, $M_1, ..., M_N$. Also, associated with each collector $C_i$, there are structures $GC_i, NG_i, Buddy_i$ and a bit $AV_i$. $C_i$ operates on $M_i$ and collects garbage nodes in $M_i$ (Figure 1).

For each i, $GC_i$ and $NG_i$ have similar purposes to those of their counterparts in the parallel system (see section 3), except that the elements of $GC_i$ and $NG_i$ are in the range of the memory segment $M_i$.

When a pointer p is deleted, it will be placed in the $GC_i$ (or $NG_i$), if $p \in M_i$. The actual implementation of this, though, may be varied. One way to implement this would be to employ a processor (or the supervisor itself) to gather these deleted pointers and *sort* them into appropriate

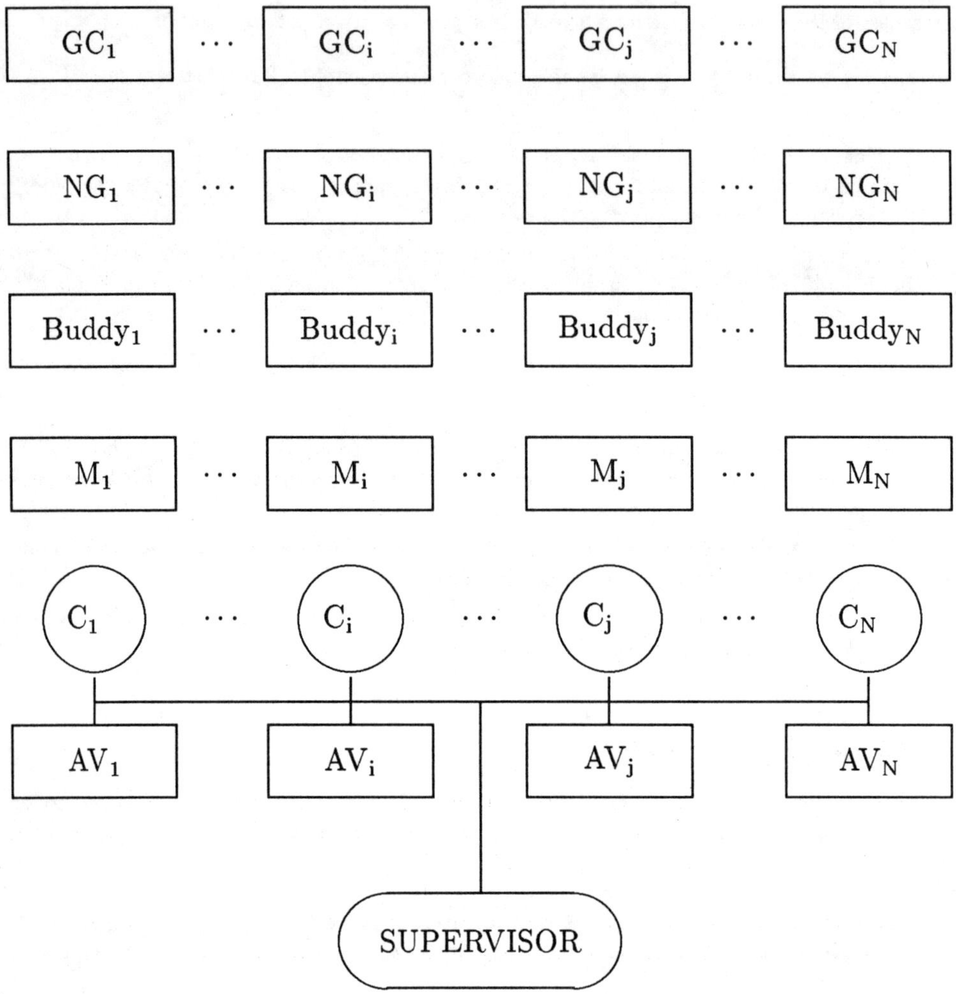


Figure 1: The logical structure of a multi-collector system.

partitions. However, in this paper, we have augmented the DELETE operation, as shown below, to directly pass the deleted pointers to their destination partitions.

**DELETE (Q.x) :** **Let Q.x $\in$ $M_i$.**
**If Q.color = 0then place Q.x in $GC_i$**
**else place Q.x in $NG_i$.**
**Set the field Q.x to Nil.**

A collector $C_j$ may also encounter a distant pointer p $\in$ $M_i$ during its traversal of $M_j$. In this case, the collector $C_j$ sends a packet $\beta$ to the structure $Buddy_i$ of $C_i$.

$\beta$ **= p, if $C_j$ is in Colorgreen or Reset phase.**
**= (TRUE,p), if $C_j$ is in Colored(TRUE,p),**
**= (FALSE,p), if $C_j$ is in Colorred(FALSE,p).**

In this presentation, we also assume that there is no delay in transferring a packet from one collector to the other.

Each garbage collector $C_i$ uses $GC_i$, $NG_i$ and $Buddy_i$ to detect garbage nodes in $M_i$, moving from the Colorgreen phase to the Colorred and finally to the Reset phase. The transition from one phase to the next is initiated by the Supervisor, simultaneously for all the collectors.

**Comment** The SUPERVISOR checks to see that all collectors are currently available and that there are no more distant pointers left, in which case it will send the switching signals to all collectors and set them busy.

**SUPERVISOR : When, for all i, $1 \leq i \leq N$,**
**$AV_i$ = "available", and $Buddy_i$ = empty**
**Then, for each j, $1 \leq j \leq N$, Do**
**Set $switch_j$ to 1;**
**Set $AV_j$ to "busy".**

We now give the garbage detection algorithm, $Detectgarbagefrom_j$, of the collector $C_j$. In this, step1, step2, step3, step4 and step5 are identical to those in the parallel system, with $GC_j$ and $NG_j$ replacing GC and NG, respectively.

**$Detectgarbagefrom_j(GC_j[1..s_j])$:**

**Local variable : $switch_j$ ;**
**Begin**
**$New_j$(step1);**
**$New_j$(step2);**
**$New_j$(step3);**
**$New_j$(step4);**
**$New_j$(step5);**
**End.**

**Comment** Each new step starts if the corresponding switch is set by the Supervisor. After the step is performed, the distant pointers sent by other collectors will be processed in ConsiderBuddy.

**New$_j$(step):**

```
Testlocalandset_j (switch_j);
step;
ConsiderBuddy_j(step).
```

**Comment** In this algorithm, if there are distant pointers in the structure Buddy, then the collector stays in the busy state and processes all those pointers, after which the collector indicates its availability to the Supervisor. At this stage, however, the collector may not switch to the next phase if not all collectors have finished processing their respective Buddies. This is required because while processing a remote pointer, a collector may generate further remote pointers for other collectors.

**ConsiderBuddy$_j$(step):**

```
Case step Do
 step1:          phase := Colorgreen;
 step2 ,step3:   phase := Colorred;
 step4 ,step5:   phase := Reset;
Repeat
 If Buddy_j ≠ empty Then
           set AV_j to "busy";
           For each β ∈ Buddy_j, Do phase(β);
 Else Set AV_j to "available";
Until switch_j = 1.
```

**Comment** This is simply testing and reseting of the local switch, which is set by the Supervisor.

**Testlocalandset$_j$(switch$_j$):**

```
If switch_j = 1 then set switch_j to 0.
```

# 5 Conclusion

Traditional reference counting falls short of being general; it cannot collect cyclic structures. On the other hand, mark-sweep algorithms are expensive, because only the resident data items are scanned, irrespective of the amount of real activity. Parallel mark-sweep algorithms are also unacceptable because they require very fine-grained synchronization with the mutator, therefore slowing down the computational process.

We have presented a novel garbage collection technique that uses reference counters and also collects cyclic structures, with little need for synchronization with the mutator.

Although in this paper we have considered interleaved, parallel and multi-collector environments, there are other computational scenarios which need be investigated.

An apparent disadvantage is the additional storage requirement for the C-field for each node. The C-field is equal to the E-field and several empirical studies show that in most cases the E-field is 1.

The space for the additional fields should be weighed against the simplicity, the efficiency, and the absence of synchronization overhead. We believe that, if the garbage nodes are collected fast and without interruption of the computational process, the total effective space utilization factor will be higher.

# References


[1] Ali, K. A. M. *Object-oriented Storage Management and Garbage Collection in Distributed Processing systems.* Royal Institute of Technology, Stockholm, Ph.D. Thesis, Dec. 1984.

[2] Ben-Ari, M. Algorithms for on-the-fly garbage collection. *ACM Trans. on Prog. Lang. and Sys. 6*, 3 (July 1984), 333-344.

[3] Bobrow, D. A note on hash linking. *Comm. ACM 18*, 7 (July 1975), 413-415.

[4] Clark, D. W., and Green, C. C. An empirical study of list structure in Lisp. *Comm. ACM 20*, 2 (Feb. 1977), 78-87.

[5] Cohen, J. Garbage collection of linked data sturctures. *Comput. Surv. 13*, 3 (Sep. 1981), 341-367.

[6] Deutsch, L. P., and Bobrow, D. G. An efficient incremental automatic garbage collector. *Comm. ACM 19*, 9 (Sept. 1976), 522-526.

[7] Dijkstra, E. W., Lamport, L., Martin, A. J., Scholten, C. S., and Steffens, E. M. F. On-the-fly Garbage Collection: An exercise in cooperation. *Comm. ACM 21*, 11 (Nov. 1978), 966-975.

[8] Friedman, D. P., and Wise, D. S. Reference Counting can manage the circular environments of mutual recursion. *Info. Proc. Letters 8*, 1 (Jan. 1979), 41-45.

[9] Hickey, T., and Cohen, J. Performance analysis of on-the-fly garbage collection, *Comm. ACM 27*, 11 (Nov. 1984), 1143-1154.

[10] Hudak, P., and Keller, R. M. Garbage collection and task deletion in distributed applicative processing systems. *Proc. ACM symposium on LISP and Functional Programming*, Aug. 1982, 168-178.

[11] Hughes, J. A Distributed Garbage Collection Algorithm, *Lecture Notes in Comp. Sci., vol. 201*, Springer-Verlag, Sept. 1985.

[12] Kung, H., and Song, S. An efficient parallel garbage collection system and its correctness proof. Tech. Note., Dept. of Comp. Sci., CMU, Sept. 1977.

[13] Lamport, L. Garbage collection with multiple processes: An exercise in parallelism. *Proc. 1976 Int. Conf. on Parallel Processing*, 1976, 50-54.

[14] Lieberman, H., and Hewitt, C. A real-time garbage collector based on the lifetime of objects. *Comm. ACM 26*, 6 (June 1983), 419-429.

[15] Schorr, H., and Waite, W. M. An efficient machine-independent procedure for garbage collection in various list structures. *Comm. ACM 10*, 8 (Aug. 1967), 501-506.

[16] Standish, T. A. *Data Structure Techniques*, Addison-Wesley, 1980.

[17] Wise, D. S. Design for a Multiprocessing Heap with On-board Reference Counting. *Lecture Notes in Comp. Sci., vol. 201*, Springer-Verlag, Sept. 1985.

# GRAPH REDUCTION IN A PARALLEL VIRTUAL MEMORY ENVIRONMENT

*Ian Watson & Paul Watson*

University of Manchester,
Department of Computer Science,
Oxford Road,
Manchester M13 9PL,
ENGLAND.

*ABSTRACT*

This paper discusses the problems of implementing a parallel machine architecture to support graph reduction. In particular it examines the area of memory organization and the mechanisms which are required to ensure that the full benefits of this implementation strategy are maintained whilst achieving an efficient physical structure. The major areas of interest are the maintenance of locality and the use of storage management techniques which minimize communication.

## 1. INTRODUCTION

The Flagship project aims to produce a declarative programming system which comprises a programming environment through to a highly extensible parallel machine. The machine structures which are being considered have been derived as a result of experience gained from the Imperial College ALICE work [1] and the Manchester University Dataflow Project [2].

Although both these projects have been successful in devising interesting novel computational mechanisms and providing some moderately convincing results in the area of extensible speedup performance, they cannot really be considered as practical parallel machines. This paper concentrates on more practical memory structures which are being considered for a new machine architecture, in particular the issue of virtual memory and the resulting memory management issues which result.

## 2. COMPUTATIONAL MODELS

We will not re-visit the issues of explicit versus implicit parallelism or the need to move away from Von-Neumann architectures in order to achieve sensible use of parallelism. We will assume that the benefits of computational model supporting a functional programming style are accepted. It is worth noting here that the Graph Reduction and Dataflow models are often regarded as competing approaches in this area. We believe that they are in fact very close. Graph Reduction builds the computational graph at run-time and achieves re-entrancy by code copying whereas dataflow builds a single computational graph at compile time and re-uses it by run-time tagging. In essence the underlying principles are identical but, with the benefit of a great deal of practical study, we believe that Graph Reduction approach is more flexible particularly in the handling of data structures and achieving demand driven evaluation.

## 3. REQUIREMENTS OF THE MEMORY STRUCTURE

Graph reduction has a fundamental requirement for a global memory space at the level of an abstract machine architecture. There have been a number of proposals for parallel machine structures which perform graph reduction within a single machine but communicate with other machines by some form of message passing. It can be argued that this is a viable approach to parallel machine architecture either because there are a worthwhile set of problems which can be broken into largely independent sub-computations or that the overheads of recomputation of parts of the graph, which would otherwise be shared, can be recomputed locally at acceptable cost. Whatever the arguments, this cannot be called graph reduction. We believe that, in practice, the set of problems which are amenable to such treatment is small and that any general purpose parallel machine must tackle the issue of global memory space.

Clearly, in such an environment, the major issue is the way in which memory / processor communication takes place. Whatever structure is devised for the physical machine, it is inevitable that much of the physical memory will not be immediately accessible by any single processor. The performance of any parallel machine is critically dependent on the way in which remote access is handled. We hope to show that an acceptable solution is provided by a compromise between previous fine grain approaches and a more conventional closely coupled multi-processor architecture.

## 4. THE COMPUTATIONAL MECHANISM

The ALICE machine was one of the first practical proposals for a parallel machine based on Graph Reduction principles, and provided the starting point for much of the work described here.

ALICE uses an approach where the unit of reduction is a packet of information which is large enough to hold a function call together with its arguments; the computation proceeds by taking any such packet which is complete and replacing it by a set of packets which represent the right hand side of the function definition. The abstract structure of the machine is a central global pool of packets with a number of processors or 'agents' picking out packets to process in this manner.

The physical implementation distributes this central pool in a number of separate stores [3].

The physical processors are connected via a delta-network switch to these stores so that all are equally accessible. The consequence of this structure is that a great deal of communication takes place over the network as each processor performs its task. The situation is worsened if the function arguments are themselves graph structures in that more information must be fetched for each re-write. To alleviate this problem, a special high speed network chip has been designed from ECL technology [3] to support agents built from transputers. This has proved adequate for the prototype implementation but one must question the apparent mismatch in technology and the viability of such a structure for a large scale high performance machine. The original ALICE philosophy, in common with most Dataflow work, was that the computational model had solved the parallel machine store latency problem and as long as all that was required was network bandwidth then viable structures could be devised. Practical experience with both ALICE and the Manchester Dataflow machine have indicated that the communication costs are high.

Having observed these apparent drawbacks, there has been a shift back to 'coarse-grain' parallelism. For example, the GRIP project [4] advocates the use of a relatively conventional bus-based multi-processor structure although the individual processors execute a serial graph reduction model rather than conventional serial code. Each processor is able to operate on a graph in its own local store and as a result, quite large graph manipulation operations may be able to take place without reference to any external memory.

One of the major benefits of the fine grain Graph Reduction / Dataflow approaches is that when an external communication is required, the process size of the machine corresponds to a single packet operation and can hence be suspended with minimal overhead while that communication takes place. In the approach described above the overheads of process suspension are certain to be greater and the effect of a communication, although less frequent, will be more serious.

A second consideration is the division of the computation in to processes and the manner in which they are allocated to physical processors. The fine-grain approach is able to exploit maximum parallelism and the allocation can be performed dynamically as resources become available. The coarse-grain approach requires these decisions to be made much more statically and as a result efficient usage of parallel resources will be more difficult to achieve.

We accept the wisdom of the closely coupled store processor structure in an attempt to minimize external communication, but observe that this does not necessarily imply that such processors must execute a serial computational model. We advocate a machine structure of closely coupled store processor pairs connected by a communication network which will probably be a packet switched delta-network. However, the store of each processor is regarded as having a packet structure similar to that in the ALICE system. The computational graph is then distributed over the machine in such a way that closely interconnected sections of the graph reside in the same store. This is indicated pictorially in Figure 1.

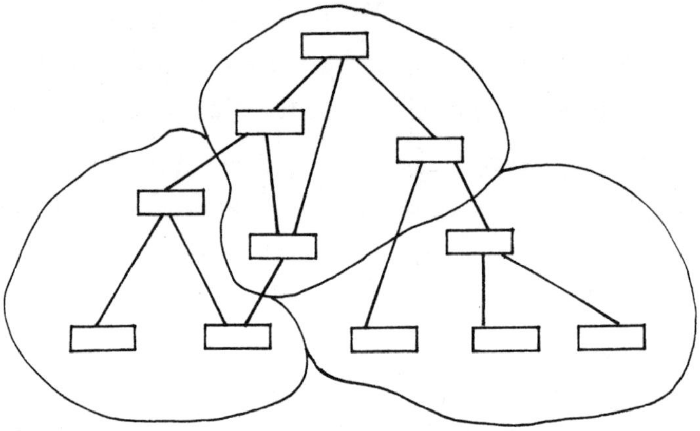

Figure 1. The Graph Partitioning

Each processor can then perform operations on packets in its own store, in many cases without resorting to any external communication. However, each packet is still an independent process and as such can be suspended at minimal cost if necessary. In addition it should be clear that the boundaries of graph to physical processor allocation are arbitrary and as such can be changed dynamically. Sections of the computation can thus be moved between processors as and when required by the dictates of resource utilization.

There are two major questions which arise. One concerns the efficiency of this approach when compared with a serial graph reducer. It might be thought that the overheads of packet manipulations when compared with stack evaluation were considerable. Studies have shown that, if careful decisions are taken about the structure and manipulation of packets that the overheads are not great, particularly as one moves from 'toy' examples towards more realistic computation where large structures are being manipulated.

The second concern is with locality. It is clearly possible to take a 'snapshot' of a computational graph at any point and draw sensible divisions which ensure that closely connected packets are clustered in the same physical store. However, if one allows dynamic shifting of the divisions it must be possible for this sensible division to get severely damaged. In the limit one could picture an allocation where every packet in the graph were physically separate from its logical neighbours. The seriousness or otherwise of this problem is currently being studied via simulation. Although we do not believe that the most pathological cases will occur easily, there is no doubt that some positive strategy is required.

The details of the rewriting mechanisms used are beyond the scope of this paper. In essence they are very similar to those used in the ALICE machine. The major difference is that the code specifying the packet manipulations is held in packets which are treated as normal data-structures in the abstract model. Practically, this means that any mechanisms devised for the handling of data in the physical machine are equally applicable to code. The code takes the form of super-combinators but with internal local computation such as eager expression evaluation taking place without building any computational graph.

## 5. THE IMPLICATIONS FOR THE MEMORY MECHANISMS

If one envisages the computational process as the transformation of a piece of graph held locally in a store which is closely coupled to the processor, then it is possible to see how parallel computation can proceed independently on sections of the graph. However, it will be clear that, in processing a section of the graph close to a physical store division, it may be necessary to refer to parts of the graph which reside in other stores; they may be arguments of a function of which the root resides locally for example.

It is important to remember at this point that we are dealing with graph reduction. That implies that a piece of graph is likely to be shared from several places. If our structure were a simple tree it would make sense, once it were identified that something was required locally, to move required sections from remote storage to the place they were required. However in a situation where sharing occurs such a strategy could easily lead to 'thrashing' with competing computations fighting to move data backwards and forwards. Clearly the sensible strategy is to take copies of remote data for local use. But again in a sharing environment we would like to ensure that each reference to a remote piece of graph, from a single processor, did not result in a separate copy being taken in a single local store. The implication is that everything should always be referred to by its global address and that things which reside in any processors memory should merely be regarded as instances of a global memory space. In this environment several copies of the computational graph may exist in separate stores simultaneously. The situation can be explained pictorially as shown in Figure 2.. The mapping of global address space to physical stores is not one of single boundaries as indicated in Figure 1., but of dynamic boundaries which may overlap arbitrarily.

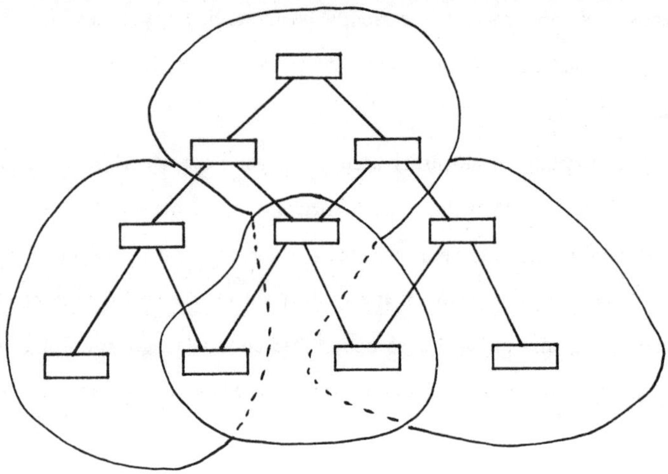

Figure 2. The Overlapped Memory Mapping

There are two ways to view this structure. One may regard the local stores as caches or as real instances of a global virtual memory space. We prefer the latter view as it will later fit in to a more general picture of a system which may include secondary storage although our current experimental structures have more in common with caching. The structure requires that all addresses are global and that each memory access which takes place must first go through a 'translation' phase to see if an instance of that memory resides in local store. If so, it is accessed, otherwise a mechanism is invoked to make a local instance. There is a major issue as to whether the instances should be on an individual packet or a page of packets basis. The former is clearly most general but can probably only be implemented using a cache-like hashing scheme. The latter requires that some locality exists in the use of packets if large sections of memory are not to be wasted; it also raises an issue as to whether complete pages should be fetched or only those parts required immediately. These issues are also currently being studied by simulation.

One of the fundamental problems with cache systems is that of coherence. Are all copies of a particular piece of memory consistent? It is interesting to note that in pure, referentially transparent, graph reduction systems it does not matter in terms of correctness. Because no updates can occur, the only thing we risk is that premature copying may lead to re-evaluation. True graph reduction would impose the restriction that copies could only be made of things which had reached a 'normal form' (of some sort !). We are experimenting with the implications of various strategies. However, we regard the Flagship machines as supporting a wider range of computation, in particular those which can be described by a more general graph rewrite system such as DACTL [5]. In these circumstances the implications of copying and normal forms are more complex.

## 6. MEMORY MANAGEMENT

Graph Reduction requires continuous storage allocation and reclamation. It is inevitable that our distributed global address space will make these tasks more difficult.

First we will consider the issue of free address allocation. Clearly we cannot accept the situation where more than one processor can allocate the same global address. This implies that the memory space must be divided between processors for the purposes of allocation. Any memory location although allowed many distributed instances hence has an owning processor. An unresolved issue is whether that ownership should be static or dynamic. There is in fact a range of options from hard wired division through static 're-boot time' allocation to short term dynamic ownership. The static approach has the benefit of simplicity but the disadvantage of pre-allocating areas of address space to particular processors before overall memory usage is known. This can decrease the usable global address space significantly if memory usage between processors is uneven. The dynamic approach is attractive in some ways but will introduce complex issues of synchronization and central control.

We intend to use reference counting as our first line storage reclamation mechanism. Firstly it is easier to implement in a distributed environment but just as importantly it has properties which are crucial to the maintenance of locality. If we wish to have any form of page structured virtual memory, it is important that closely connected graphs reside on closely related page addresses. During the transformation of a computational graph there are a large number of short lifetime nodes. A reclamation mechanism which re-uses such things immediately is essential to preserve local relations.

We are not wholly convinced that reference counting mechanisms such as those proposed by Brownbridge [6] or Hughes [7] for dealing with cyclic structures are going to prove practical in a distributed environment. We are currently studying the feasibility of occasional mark scan mechanisms such as that proposed by Hudak [8] in order to deal with cyclic structures which may arise.

Conventional reference counting schemes have one disadvantage when implemented in distributed systems. That is the requirement that the ordering of count increments and decrements is maintained otherwise cells can become garbaged incorrectly. In large extensible parallel systems, the need to maintain ordering over communication networks can be a serious restriction. An alternative is to introduce synchronization by acknowledgements but this increases communication and destroys parallelism. A scheme which avoided the ordering restriction would be worthwhile.

The form of our distributed memory system also has implications for reference counting. Clearly if several instances of a particular packet exist, then the decision to reclaim its memory location are not entirely straightforward. We have already indicated that any piece of memory will have a unique owner who is able to allocate it. This owner must also be the major participant in any garbage collection activity; it is he who needs to know its availability. The simplest scheme is to send all increment and decrement messages to the unique owner. As long as the above ordering restrictions are maintained then the owner can simply add the cell to its free list when the count reaches zero. One can rely on some form of cache/page rejection mechanism in the real memory in other processors to remove instances as convenient. The major drawback of this approach is that we introduced the idea of local instances to avoid communication. However, if we communicate reference count adjustment across the network for each packet copy etc. we will have largely negated the advantage aimed for. A more acceptable solution is to maintain the algebraic sum of adjustments on the local instance and resolve these with the owner at the time an instance is rejected from local memory; this greatly reduces the amount of communication required.

Unfortunately there are still problems. In the presence of lazy evaluation, it is possible to generate references to a piece of graph during a rewriting operation without ever having made a local copy. In those circumstances, the only way to ensure that the reference counting is correct, is to send all adjustment messages to the owner. This will clearly involve as much communication as the simple solution suggested above.

Given both this potential communication problem and that of synchronization of updates, we believe that a reference counting system which reduces the number of such messages would be very worthwhile. We therefore propose to use a 'weighted reference count' scheme [9]. When a cell is created, it has a reference count field which is set to a large value. The initial pointer to the cell has a 'weight' associated with it which is a number equal to that of the count on the cell. Whenever a new reference to a cell is created, it must occur as the result of copying an existing pointer. In that case, the weight value of the existing pointer is divided between it and the copy. Whenever a pointer is destroyed, the weight is subtracted from the count held on the cell. Although there is some overhead involved in this scheme, it is not as serious as it might appear. If the original count is a power of two and the weight division is done one a binary basis, then all weight values can be compressed logarithmically.

One minor drawback of this scheme is that subdivisions of the weight value will eventually result in a value of one being reached. In these circumstances it is necessary to make copies. In practice, the frequency of this is small and, in any case, it does not imply copying of the whole of large structures but merely the root cell by which they are referenced.

## CONCLUSIONS

A graph reduction system must have the concept of global memory space, otherwise the benefits of full sharing will be lost. However, a simple real address space distributed statically amongst a number of physical processors is not entirely satisfactory because the sharing of remote parts of a graph can result in many duplicated messages when the correct strategy should be to exploit locality and move data between stores where necessary. This in turn may result in 'thrashing' if different parts of the computation disagree about where the data should reside.

The Flagship solution is to regard the global space as virtual and to provide a mechanism where a number of copies may exist in physical real stores. The full benefits of this require storage management mechanisms which themselves require a minimum of communication. Garbage collection is the most critical area and we believe that a weighted reference count scheme is the most appropriate.

## REFERENCES

[1] J. Darlington & M. Reeve, "ALICE- A Multi-Processor Reduction Machine for The Parallel Evaluation of Applicative Languages", Proc of 1981 ACM Conf on Functional Programming Languages & Computer Architecture.

[2] J.R. Gurd, C.C. Kirkham, I. Watson, "The Manchester Prototype Dataflow Computer", Communications of the ACM, vol.28. no.1., Jan.85, pp.34-52.

[3] A.J. Field, PhD. Thesis, Imperial College, University of London, 1985.

[4] S.L. Peyton-Jones, "GRIP-a parallel graph reduction machine", University College London, Dept. of Computer Science, Internal Note 1665, 1985.

[5] J.R.W. Glauert, N.P. Holt, J.R. Kennaway, M.R. Sleep, "An Active Term Rewrite Model for Parallel Computation", Internal Report, ept. Computing, University of East Anglia, Norwich, England, 1985.

[6] D.R. Brownbridge, "Cyclic Reference Counting for Combinator Machines", Proc. ACM Functional Programming Languages and Computer Architecture, Nancy, France, 1985.

[7] J. Hughes, "A Distributed Garbage Collection Algorithm", Proc. ACM Functional Programming Languages and Computer Architecture, Nancy, France, 1985.

[8] P. Hudak & R.M. Keller, "Garbage Collection and Task Deletion in Distributed Applicative Processing Systems", Proc. Conf. on LISP and Functional Programming, ACM, 1984.

[9] P. Watson, "Parallel Reduction of Lambda Calculus Expressions", PhD. Thesis, Dept. Computer Science, University of Manchester, England, 1986.

# Performance Measurement of a G-machine Implementation

*Richard B. Kieburtz*
Oregon Graduate Center
19600 N.W. von Neumann Dr.
Beaverton, Oregon 97006 USA

## 1. Introduction

What opportunities are there to enhance the performance of a language-directed computer architecture, relative to the performance of a general purpose, register-oriented, random-access machine? Is the potential performance improvement great enough to justify the architecture? These are the questions we seek to answer in the G-machine project.

Limits to the performance of contemporary computer systems are imposed by the time taken to move data within the system. The standard operations of arithmetic add no performance cost, in the sense that an addition, subtraction or multiplication can be performed in the same time taken to move a datum from one machine register to another, or from a memory to a register. Thus arithmetic only contributes to the cost of computation because it requires the movement of data.

Operations are expensive if they perform multiple data movements in sequence. Of particularly noticeable cost are operations that require a complete change of processor state, for then the contents of all active processor registers are moved to or from locations in a memory. Operations can be made cheaper by five principal techniques: (i) shortening data paths, typically by providing additional processor state in the form of register files or by providing high-speed cache memory to hold frequently-referenced data; (ii) overlapping data movements in time by pipelining data paths, (iii) overlapping instruction executions in time by pipelining the instruction fetch and execute cycle; (iv) aggregating related data so that the aggregate is moved along a data path as an atomic datum; (v) concurrently activating independent functional units to perform operations on multiple data in parallel.

In the design of a G-machine processor, each of these techniques is used in some degree. In addition, the G-processor is freed of the obligation to execute an operating system, with the attendant requirement to provide interrupt service for I/O devices and to execute imperative-language programs. It runs as a slave processor to a conventional host, which provides all O/S services required by the G-machine. In consequence, the G-machine need not support context switching for interrupt service or multiprocessing. It communicates with its host by primitive signaling and by dual-port access to its graph store (G-memory) and its program store.

### 1.1. The abstract architecture

The G-machine [Joh83] is an abstract architecture for evaluating applicative expressions by graph reduction. Unlike combinator reducers, in which control is inferred by inspecting the expression graph at each step in a computation, the G-machine is a programmed engine. It interprets the graph to infer control only when control cannot be predicted from a static semantic analysis of the source language program. Data-dependent control is specifically

The research reported here was partially supported by the National Science Foundation under grant No. DCR-8405247.

confined to case (and conditional) discrimination and to initiating evaluation of an application whose value is demanded. Advantages of programmed graph reduction have been discussed in [Kie85, Pey87]. Techniques for compiling applicative language programs into sequential code (G-code) for the G-machine are described in [Aug84, Joh83, Joh85, Joh86].

The G-machine differs from conventional computer architectures in two important respects. Firstly, it is a pure stack machine. Not only does it use an evaluation stack to hold intermediate values during the evaluation of arithmetic expressions, but more importantly, the environment of a computation is represented by a stack of pointers (the P-stack) into the graph undergoing reduction. G-code instructions either obtain their operands from the top of a stack, or else provide stacktop-relative indices to an operand embedded deeper in a stack. These indices are computed (by a compiler) from the current stack depth and the binding height, or De Bruijn index, of each occurrence of a variable in the source-language program. Performance of a G-machine can be enhanced by hardware support to achieve a very fast stack implementation.

A second important difference from conventional computers is that the data comprise a graph and require a tagged, dynamically allocated, list-structure store. Tagged cells are necessary because an expression may have many possible points of reference. The first reference to demand a value causes evaluation of the expression. Subsequent references find the value left in the cell by the prior evaluation. A tag distinguishes an unevaluated application cell from the value to which it may reduce. Without such sharing of evaluations, some algorithms would be subject to a functional increase in computational complexity.

In the graph reduction model, all data are stored in nodes of an expression graph. The rate of allocation of memory cells by graph reducers is substantially greater than that of Lisp evaluators because the Lisp model provides data storage in assignable variables and arrays as well as in an expression graph. The allocation rate may be reduced somewhat by programmed recycling of memory cells. This strategy can be implemented by a compilation scheme that detects opportunities to recycle memory cells.

The one-bit reference count [SCN84] is a runtime scheme that uses tags in pointer values to detect opportunities for recycling storage cells. This strategy is very effective with combinator reducers, which are profligate in their allocation of storage, but harder to use effectively with the G-machine because reclaimed storage is not so frequently reallocated immediately.

In the current design of a hardware-supported G-machine, the memory consists of tagged cells, so that the necessary tags are aggregated with the data held by each cell and fetched simultaneously. Furthermore, an intelligent memory controller performs dynamic allocation and recycling of memory cells without intervention by the G-processor. Thus, ALLOC (allocate storage for a graph node) is a primitive command asserted to the G-memory by the G-processor, just as it would assert read or write to a designated address. The memory controller employs hardware-assisted reference counting as the primary mechanism for detecting recyclable storage.

Another important performance consideration for functional language evaluators is that programs tend to consist of a large number of rather small functions (that is, the function bodies are small expressions). Function calls are executed more frequently than are procedure calls in a typical iterative language program. The efficiency of the function call and return mechanism is critical to performance.

### 1.2. Data formats

The machine data types of a computer architecture are the data formats for which the instructions are specialized. In the present G-machine design, the allocation unit of G-memory is a two-word cell in which each word consists of a 32-bit data field and a pair of tag bits. The words of a cell are individually addressable. When a word is fetched, its tag bits are fetched with the data. The tag bits are called the "is_evaluated" and the "is_pointer" tags, although the "is_evaluated" tag is not interpreted independently of its neighbor. The machine data types presently recognized are:

| | word 0 | | word 1 | |
|---|---|---|---|---|
| **application** | 01 | pointer | 01 | pointer |
| **boxed pair** | 1- | datum | 11 | pointer |
| **unboxed pair** | 10 | basic value | 10 | basic value |

The tags pattern 00 is reserved to distinguish the head of a vector, but is not yet used.

The terms *boxed* and *unboxed* have been used by Luca Cardelli [Car84]. A boxed pair is the head of a linked data structure. In particular, the second word of a boxed pair always contains a pointer, although the first word can contain either a basic value or a pointer. An unboxed pair is used to represent a basic value, or a pair of such values. A basic value is an individual signed or unsigned 32-bit datum.

The G memory actually represents more tags per cell than those shown here, in order to accommodate garbage collection. However, the G-processor is not required to participate in garbage collection, and sees only the two tag bits per word illustrated in the diagram above.

### 1.3. The internal architecture

Figure 1 shows the internal organization of the G-processor. Its major functional units are the P-stack, which holds the representations of environments for function calls being evaluated, the ALU in which arithmetic and logical expressions are evaluated, the instruction fetch and translation unit (IFTU), and the processor control unit (PCU) responsible for the dispatch of micro-instructions.

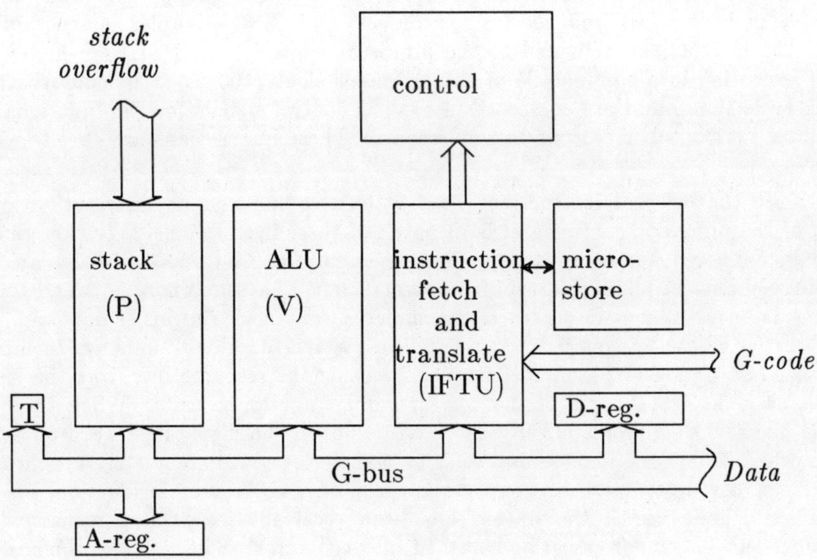


Figure 1 -- Functional units of the G-processor

In the compilation strategy for the G-machine [Joh85], a compiled function body refers only to locally bound variables. The bindings are represented by a sequence of pointers into the G-memory (or directly, by basic values), held in an initial sequence of the P-stack. References to variables are implemented by indexing into the P-stack to obtain an address used to fetch data from G-memory. The "is_pointer" tag accompanies each word in the P-stack, enabling a conditional test to determine whether a word at the top of the stack is a pointer or is the direct representation of a basic value. The T-register holds the tag values corresponding to the datum at the top of the P-stack.

Only a finite initial segment of the P-stack can be accessed by indexed stack instructions. The environment of an individual function body must be contained in this interval, which is implemented by a register file 128 words long in the current design. When a function call occurs to evaluate an application nested in the function body currently undergoing evaluation, the P-stack grows, and may overflow this fixed set of active stack registers. To accommodate this without the costly action of dumping the stack contents to memory, the P-stack register file is connected to a high-speed static RAM memory. The register file can shift its entire contents one unit upward or downward on every stack operation. When a value is pushed into the P-stack, the contents shift upward, and the value in the last register is written out to the RAM memory. When a value is popped, the value most recently written to the RAM is read back into the stack in the 128th register.

The ALU has an independent evaluation stack (the V-stack) of bounded size, but with no automatic overflow mechanism. The top two elements of this stack are dual-ported so that the arithmetic unit can receive two operands simultaneously. The presence of an independent stack serving the ALU means that ALU operations can be overlapped in time with P-stack operations or with fetches (or stores) from the G-memory to the P-stack.

Data are moved among the P- and V-stacks, the instruction fetch unit and the G-memory on an internal processor bus (G-bus). Addresses for memory operations are latched in the A-register, which is also connected to the G-bus.

### 1.3.1. The instruction pipeline

This G-machine implementation uses a novel instruction pipeline that combines many benefits of both RISC (reduced-instruction set) and CISC (complex instruction set) computers. The instructions delivered to the processor control unit (PCU) are highly synchronous, most executing in a single cycle of the processor clock (100 ns is the conservatively designed clock cycle time), and are composed of explicit control signals for the functional units. They resemble horizontal microcode in this respect. These instructions are 48 bits wide, in a fixed format.

Since the internal instructions need little decoding and no distribution logic, the main task of the processor control unit is dispatch of these instructions. For this purpose it uses a technique known as scoreboarding, first used on the CDC 6600. When an instruction is dispatched that requires the use of functional unit A, a semaphore is raised to indicate that unit A is busy. When the instruction completes, unit A clears this semaphore. Each instruction carries a set of resource bits that indicate which functional units are required for its execution. Instruction dispatch is controlled by using the resource bits from the instruction as a mask over the semaphores of the corresponding functional units. When the conjunction of these masked semaphores is zero, the instruction can be dispatched. Scoreboarding is useful even when some functional units, such as the P-stack, are completely synchronous and can execute a new instruction in every clock cycle. A READ microinstruction, for instance, can hold the P-stack until the datum has been received from the G-memory (which is not presumed to be synchronous) and pushed into the stack. The units scoreboarded are the P-stack, G-memory, ALU, and the literals queue of the instruction fetch unit.

The G-code instructions produced by a compiler are translated into micro-instruction sequences by an instruction translation unit (ITU). Most G-code instructions are not very complex, and many translate into single microinstructions. Since the G-code formats are 16 bits for normal instructions, 32 bits for jumps and 48 bits only for long literals, the compression of instruction fetch bandwidth is at least 4:1. The instruction fetch bandwidth has not been found to be a performance limiting parameter of this design. In addition to affording some compression in the bandwidth required of the control memory, the IFTU allows buffering of the instruction pipeline, which is extremely important to good performance.

## 2. G-machine instructions

The instructions executed by the G-machine, which we often refer to as G-code, are conveniently grouped into eight classes.

### 2.1. The CALL group: function calls and returns

| |
|---|
| EVAL |
| CALL |
| CALLGLOBFUN |
| TCALL |
| RETURN |
| RET_INT |

There are several ways that the reduction of a function application can be initiated. Most general is the EVAL instruction, which is invoked on an expression known to be an application, or which may possibly be an application, in order to produce its value. EVAL tests the evaluation tag on a cell of graph memory. If the tag is set, indicating that the cell has already been evaluated, then the EVAL instruction completes without further action. On the other hand, if the cell is the root of an application graph, as shown in Figure 2(a), then EVAL traverses the graph, pushing pointers to the arguments into the P-stack, and inspects the function descriptor that it finds at the end of the normal-order spine of the graph (Figure 2(b)). A function descriptor contains the number of arguments expected by the function and the entry point address of the code compiled for the function body. The EVAL microcode determines whether there are sufficient arguments in the stack to proceed with reduction of the application, and if so, reconfigures the P-stack so that it contains a pointer to the redex and pointers to each of the arguments of the application (Figure 2(c)). In the stack below the redex pointer are stored (i) the return address, and (ii) the excess argument count, which is the difference between the number of arguments stacked for this application and the number expected by the function. EVAL then invokes the function body referred to by the descriptor. In case the stack contains insufficient arguments, the expression represents a partial application and is not evaluated further. EVAL is the most complex instruction of the G-machine.

If an applicative expression is to be suspended (lazy evaluation) the compiler emits code to construct an application graph of the form shown in Figure 2. Very often, however, a compiler can determine that an applicative expression must be evaluated when it is first encountered in an expression, rather than suspended. This is the case when following a strict, or call-by-value computation rule, or with lazy evaluation whenever an expression is the argument of a strict function or operator, or when it represents a value to be returned by a function body (i.e., a tail-call).

When an applicative expression is to be evaluated immediately, the function is invoked by a CALLGLOBFUN instruction if it is represented by function constant, and by a CALL instruction if it is represented by a variable or is the value of a conditional expression. In either case, the compiler does not emit code to build a graph representing the applicative expression. Instead, it compiles code to push pointers to the argument expressions into the P-stack, setting up the environment for the call. Use of the CALLGLOBFUN instruction permits some additional economies in the calling sequence because the compiler has determined that the number of stacked arguments is correct, and can locate the code of the

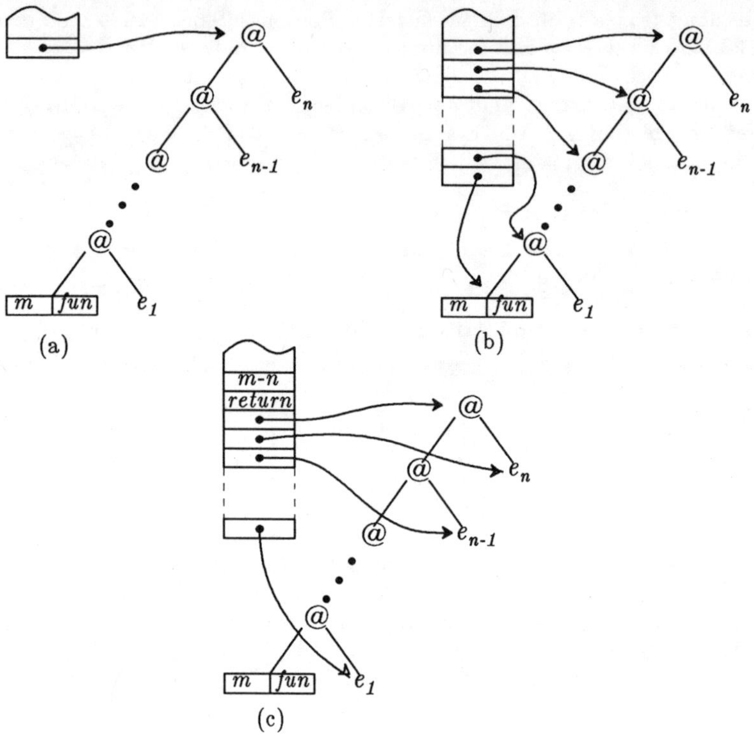


Figure 2 -- snapshots of the EVAL sequence

function body directly by address.

When an applicative expression is a tail-call, pointers to the arguments (or the argument values themselves, for call-by-value computation) are pushed into the P-stack, and the call to the function body is implemented by a control jump to the entry point, without saving a return address. When the function to be applied is represented by a variable, the TCALL instruction is used.

The RETURN instruction is complicated by the need to determine whether the returned value represents a function whose arguments are present in the P-stack. This involves checking whether the excess argument count saved in the P-stack at the point of call is greater than zero. In case it is, the returned value should be applied to the arguments stacked in the environment of the previous call, and so RETURN reenters the EVAL sequence. Otherwise, the return address is simply retrieved from the P-stack.

### 2.2. The ALLOC group: storage allocating instructions

| ALLOC |
|---|
| MK_VAL |
| MK_PR |
| MK_V1_PR |
| MK_V_PR |
| MK_APP |

The basic storage allocation instruction is ALLOC, which demands of the memory a new cell whose tags have been initialized, but whose data content is indeterminate. Other storage-allocating instructions also initialize the contents of the cell from values or pointers stored in the stacks of the G-machine.

MK_VAL writes a basic value into the first word of a newly allocated cell, MK_PR writes pointers from the P-stack into both words of a cell. MK_V1_PR writes a basic value in the first word and a pointer in the second. It is emitted to construct representations of values of a disjoint union type. MK_V_PR writes a pair of basic values to represent, for instance, an extended precision arithmetic value. MK_APP creates an application node with pointers taken from the P-stack.

### 2.3. The UPDATE group

| UPDATE |
|---|
| UPDATE_V1 |
| UPDATE_PR |

An UPDATE instruction overwrites the contents of a cell addressed from the P-stack with new values, to implement the basic graph reduction operation that replaces an application node by its value. UPDATE copies the new values (or pointers) from another cell pointed to from the top of the P-stack. UPDATE_PR takes both the new values directly from the P-stack. UPDATE_V1 takes one value from the V-stack and one from the P-stack.

### 2.4. The ALU group: arithmetic and logical instructions

| ADD | CIRC |
|---|---|
| ADDC | SHLA |
| SUB | SHRA |
| SUBB | SHRL |
| MUL | AND |
| DIV | OR |
| MOD | NEG |
| INCR | NOT |
| DECR | |

Instructions of the arithmetic and logical group obtain operands from the V-stack and return results there.

### 2.5. The READ group: data fetch instructions

| FST |
|---|
| SND |
| GET_FST |
| GET_SND |

Data fetch instructions extract the contents of the first or second word of a cell, bringing them to the P-stack (FST, SND) or to the V-stack (GET_FST, GET_SND).

### 2.6. The STACK group

| COPYP | COPYV |
|---|---|
| MOVEP | MOVEV |
| POPP | POPV |
| ROTP | ROTV |
| NROTP | NROTV |
| VROTP | |

This group constitutes the "pure" stack instructions; no other functional unit of the G-machine is affected. The instruction codes ending in the letter "P" refer to the P-stack, those

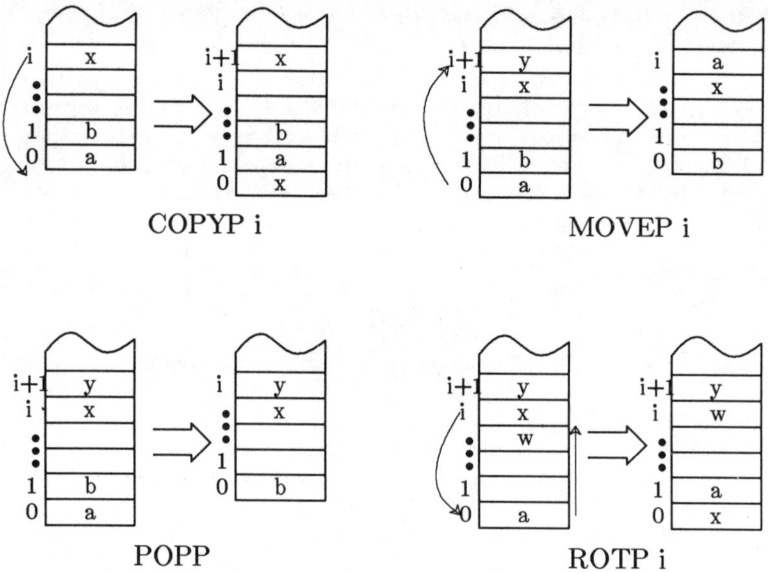


Figure 3 -- P-stack transitions induced by the stack instructions

ending in "V" to the V-stack. The meanings of these instructions are illustrated most conveniently by the diagrams in Figure 3. NROTP is the inverse of ROTP. The instruction VROTP has exactly the action of ROTP, but takes its argument from the contents of the top of the V-stack. The stack rotation instructions, ROTP, NROTP and VROTP are used to reconfigure the P-stack for function calls and returns.

### 2.7. The JMP group: control transfer instructions

| | |
|---|---|
| JZERO | JNOT_ZERO |
| JNEG | JNOT_NEG |
| JCARRY | J_OFLO |
| J_IF_PTR | JNOT_PTR |
| JMP | |

The first six conditional jump instructions test condition codes set by a preceding ALU operation. The instructions J_IF_PTR and JNOT_PTR test the "is_pointer" tag of the datum at the top of the P-stack. JMP is an unconditional jump.

### 2.8. The LITERALS group

| |
|---|
| GET_BYTE |
| PUSHLITERAL |
| PUSHCONST |

These are pseudo-instructions that extract literal data from the instruction stream. GET_BYTE and PUSHLITERAL extract one- and four-byte literal data, respectively, and push it into the V-stack. PUSHCONST extracts a pointer (to a compiler-initialized cell) and pushes it into the P-stack.

### 2.9. SIGNAL

SIGNAL

SIGNAL interrupts the host processor to request service. The G-processor remains in a wait state until restarted by a signal from the host processor.

## 3. Performance Measurement

Performance of the G-machine design has been evaluated by executing code compiled from a number of functional language (LML) source programs and executed in three different ways.

### 3.1. Microsimulation

Small examples (typically executing several tens of G-code instructions) have been run on a microsimulator (Microsim). This simulator accurately represents the hardware design of the processor to the level of individual register transfers. Timing is accurate to a 25-nanosecond clock phase at this level of simulation. Microsim is written using the N.2[1] architecture simulation system, in which a processor is described in terms of a register transfer language, whose codes are interpreted by the simulation system in each test run.

Microsim was used to debug and validate the detailed design, both with respect to its logic and to its timing, but it is too slow for use as the performance measurement tool. Instead, Microsim was used to determine average execution times for individual G-code instructions.

Since most of the G-code instructions expand into straight-line sequences of microcode, it might appear that the execution time for an instruction is always the same, and could be determined by simply counting microinstructions and multiplying by the clock cycle time. However, not all microinstructions can execute in a single clock cycle. In particular, to read from G-memory requires a minimum of 3 clock cycles (300 ns) even if no delay is caused by contention for memory cycles with the concurrent collector. This is a timing constraint of the memory technology we have assumed (1986 silicon CMOS dynamic RAM). An arithmetic or logical operation by the ALU also takes at least three cycles to complete, including depositing the result back into the top cell of the V-stack.

Because of the instruction dispatch strategy followed in this design, it is sometimes the case that the next microinstruction can be dispatched even though a preceding microinstruction has not completed execution. If the two instructions use no common functional unit of the processor, then dispatch of the next instruction is not inhibited. Therefore, when the last microinstruction in the expansion sequence of a G-code instruction requires multiple clock cycles, the full time for its completion may or may not be attributed to that particular G-code instruction, depending upon whether dispatch of the first microinstruction of the following G-code instruction is delayed.

By measuring the time attributed to each execution of a G-code instruction in a test program, the average execution time for each species of G-code instruction is obtained for that program. By averaging these single-program averages over a number of different test programs, stable average execution times are obtained for all species of G-code instructions. These stable averages have been used as a basis for all further performance measurements of the design.

[1] N.2 is a trademark of Endot, Inc.

### 3.2. Macrosimulation

The macrosimulator (Macrosim) is an instrumented interpreter for G-code. It is coded in Modula-II, and runs approximately as fast as a Lisp interpreter, exclusive of the extra code inserted for instrumentation purposes. It includes a garbage collector, and is therefore quite satisfactory for simulating the execution of good-sized test programs.

Even though it does not simulate the execution of each G-machine microinstruction, Macrosim can provide data on microinstruction execution. It uses a table that summarizes the number of microinstructions that would be executed by the G-processor in the expansion of each microinstruction. When the microsequence is data dependent, as are EVAL, RETURN and UPDATE, for instance, Macrosim's table gives the number of microinstructions that would be executed and the average execution time corresponding to each data-dependent option. Thus Macrosim furnishes not only accurate counts of G-code instructions by species, but also the time attributed to them and the total number of microinstructions that would be executed by the actual G-processor.

Since the garbage collection implemented in Macrosim does not simulate the reference counted garbage collection of the G-machine design, time for garbage collection has not been counted in the performance statistics reported here.

### 3.3. G-code translation

In order to obtain a comparison with the performance of an implementation on a conventional computer, the G-code output by our compiler has been expanded into native machine code for a DEC VAX-11, an architecture with which there is wide familiarity. Furthermore, the code generator has been instrumented so that in the VAX code sequence emitted for each G-code instruction, there is included code to increment two counters, one which counts the number of executions of that G-code instruction species, and another that counts the number of VAX instructions executed in its expansion (exclusive of instrumentation instructions). We thus obtain the number of VAX instructions executed when the abstract G-machine architecture is implemented by compiling G-code into VAX code.

Of course, there are some basic differences between the VAX implementation and the G-machine design. Since the VAX is not a tagged-memory architecture, The tags required by the G-machine are implemented with an extra 16-bit word extending the storage allocated for each graph node. Also, the EVAL instruction is implemented as a parameterless, assembler-coded procedure, rather than by in-line code expansion. However, the G-code sequences executed by Macrosim and those translated into VAX code are identical.

Code generation is attribute-driven and reasonably sophisticated. It takes advantage of opportunities to use the auto-increment addressing modes of the VAX. Operands are held in registers when possible, with the consequence that in the examples we have run, the contents of the G-machine V-stack are always held in registers, never in memory.

### 3.4. A Comparative Architecture Measure

Since we can measure the simulated time require for the G-machine design to execute test programs and measure the number of VAX instructions executed to run the same test programs, compiled by the same compiler to the level of G-code, we have a way to directly compare the designed G-machine implementation to that of a VAX. The measure is

$$\textbf{equivalent VAX Mips} = \frac{\text{number of VAX instructions executed}}{\text{time for G—processor execution (microseconds)}}$$

This measure is more meaningful than raw benchmark timings because it is not contaminated by variables that are uncontrollable or hard to predict, such as time spent in performing operating system services, cache performance, or different garbage collection algorithms. It could be made more universal if the technology-dependent performance assumptions made in the design of the G-processor could be factored out. Primarily this means the designed clock frequency. Unfortunately, not all of the technology assumptions can be expected to scale uniformly. A faster processor does not necessarily mean that a faster G-memory will be available to accompany it.

Nevertheless, this relative architectural performance figure is more meaningful than figures such as the number of function evaluations per second, which is wildly unstable across a mix of different test programs, or the number of graph reductions per second, which depends upon compilation strategy at least as much as upon machine architecture and implementation.

## 4. Results of Simulation Tests

The G-machine design has been evaluated by simulated execution of a variety of test programs and using several different computation rules, realized by variants of the compiler. The invariants of the performance measurements are these:

The **clock cycle time** of the G-processor is 100 nanoseconds. Times are given in units of these cycles.

The **IFTU efficiency** is the ratio

$$1 - \frac{\text{number of cycles in which the IFTU inserts a NOP}}{\text{total number of cycles}}$$

The IFTU inserts a NOP into the dynamic microinstruction sequence whenever the execution unit demands a new microinstruction but the IFTU is unable to produce one because of delay in the instruction pipeline. These delays can occur when the pipeline is perturbed by a conditional jump or by a return from a function call. Detailed results of IFTU performance will be reported in a separate paper, but measurements have shown the IFTU efficiency to lie between .90 and .95 on several test programs constructed to produce worst-case performance from the IFTU. In the data presented here, we have not accounted for IFTU efficiency, which has been assumed to be 1.0.

The **G-memory contention** figure is the ratio

$$\frac{\text{average memory response time in the presence of collection}}{\text{average memory response time with no collection}} - 1$$

expressed as a percentage. It accounts for delay expected due to contention for G-memory cycles by the concurrent garbage collector. This figure has been measured at 4 per cent when the garbage collector uses hardware-assisted reference counting to detect collectable nodes [Ran86], and at 8 per cent when simple cyclic structures (all cycles pass through the root of the subgraph to be collected) are collected. In the data presented here, degradation of G-memory response due to concurrent collection has not been accounted for.

### 4.1. Computation rules

Several different computation rules have been followed in compiling code from functional language source programs. The source language is LML [Joh84], a purely functional language with lazy evaluation rules. However, it is obviously not necessary to adhere to a lazy evaluation rule in compiling programs, and we have created several versions of our compiler in order to compare the performance of different computation rules.

**Lazy evaluation**, is the rule of the original source language. It means that functions evaluate their arguments only by need, and that the constructors of all compound data types are lazy as well. N-tuples are compiled as pairs, nested to the right, and each pair construction is lazy. Values of defined constants (the definitions appearing in let expressions) are also evaluated only by need. Nested abstractions are transformed into global function declarations by lambda-lifting [Joh85] rather than by m.f.e. abstraction [Hug82] and so do not achieve "full-laziness". No attempt is made by the compiler to identify textually separate occurrences of equivalent subexpressions.

**Moderately-lazy evaluation** is the rule in which the arguments of all functions are evaluated but constructors may suspend evaluation of their arguments if declared to be lazy. LML has no strictness declarations for constructors, and we have chosen to make pair construction and the constructors of all programmer-defined types strict. List construction is lazy, however.

**Strict evaluation** means that the list constructor also evaluates its arguments. Strict evaluation still fails to achieve the full economy of call-by-value, however. Values are uniformly represented by nodes in the data graph, and so arguments are bound in a function call by passing pointers to the graph nodes that represent their values, even when these values are unboxed.

In **call-by-value** computation, basic values are passed directly rather than indirectly. In the G-machine, this means that basic values are passed in the P-stack. In the presently operational version of our compiler, only arithmetic and conditional expressions occurring as operands are actually passed by value. Applicative expressions are passed by reference, because the result of a function application produces a pointer to a graph node that represents the value. The G-machine distinguishes between basic values and pointers in the P-stack by testing the "is-pointer" tag that accompanies each element of the stack.

### 4.2. The test suite

The test suite consists of eight short, but interesting programs. The LML code is listed in the Appendix. *Ackermann* and *Tak* are functions of two and three arguments, respectively, which generate a large number of recursive function calls to exercise the global function call mechanism. *Linfib* computes the first 100 numbers of the Fibonacci sequence, but uses a tail-recursive function with two accumulator arguments. All versions of the compiler make some optimizations for tail-recursion, but only the call-by-value version eliminates use of the expression graph to communicate arguments. It produces iterative code almost like Fortran or "C" code for this function.

*Towers_of_Hanoi* is a list-processing example that performs little arithmetic but generates many function calls. *Isort* and *Treesort* are also list-processing examples that do no numeric computation other than to generate test data. *Treesort* makes use of a programmer-defined data type for order trees.

*Digits_of_e* and *Primes* both compute finite prefixes of infinite sequences of integer values. The algorithms would generate infinite streams of values, incrementally, but generation is curtailed by the function *cnthd*, which terminates the computation when a specified number of values have been produced. Both algorithms can be executed using the moderately-lazy computation rule as well as by lazy evaluation, but would not terminate under a computation rule in which the list constructor was also strict.

### 4.3. Statistics

Statistics are given for each program executed by Macrosim, subject to the assumed invariants stated above. The statistics we present are:

- total execution time, in seconds (cycles $\times 10^{-7}$)
- microinstruction dispatch rate, (cycles/microinstruction)
- microcode expansion ratio (microinstructions/G-code executed)
- rate of G-memory allocations, ($\times 10^5$)
- total number of EVAL's executed
- G-code execution profile, by instruction group
- relative performance, in equivalent VAX Mips

The microinstruction dispatch rate is a measure of the internal efficiency of the processor. At a rate of 1.0 the processor would execute one microinstruction per clock cycle. The microcode expansion ratio indicates the degree of stress imposed upon the fetch of G-code from control memory. The product of microinstruction dispatch rate and microcode expansion ratio determines the average number of clock cycles in which a G-code instruction must be fetched in order not to delay the processor. This product gives the inverse of the required bandwidth to the control memory, in units of instructions per 100 ns.

The G-memory allocation rate has been presented to quantify the requirement imposed upon the G-memory controller. The total number of EVAL instructions executed has been included to help compare alternate computation rules. The G-code execution profile indicates how the processor is spending its time, and in particular, helps to determine where effort should be directed to obtain further performance improvements.

| **Ackermann** | | | |
|---|---|---|---|
| | *Lazy* | *Strict* | *by-Value* |
| time | 0.220 | 0.075 | 0.064 |
| dispatch rate | 1.29 | 1.50 | 1.57 |
| expansion ratio | 5.78 | 2.36 | 1.92 |
| allocation rate | 0.93 | 2.06 | 0.80 |
| equivalent VAX Mips | 3.45 | 3.84 | |
| number of EVAL's | 20615 | 1 | 1[2] |
| G-code profile: | | | |
| CALL | .630 | .181 | .213 |
| ALLOC | .128 | .182 | .025 |
| UPDATE | .000 | .000 | .000 |
| ALU | .077 | .228 | .267 |
| READ | .072 | .211 | .247 |
| STACK | .045 | .076 | .105 |
| JMP | .017 | .048 | .057 |
| LIT | .025 | .053 | .062 |

[2] **The single EVAL instruction initiates computation on an initial application graph in the G-memory.**

For *Ackermann*, the strict, call-by-value computation ran 3.4 times faster than when the lazy computation rule was used. The G-code for this example contained no explicit UPDATE instruction. The update was accomplished by a RET_INT instruction and the cost is charged to the CALL group.

| Tak | *Lazy* | *Strict* | *by-Value* |
|---|---|---|---|
| time | 1.691 | 0.542 | 0.491 |
| dispatch rate | 1.31 | 1.49 | 1.53 |
| expansion ratio | 7.23 | 2.98 | 2.62 |
| allocation rate | 1.13 | 1.76 | 0.97 |
| equivalent VAX Mips | 3.23 | 3.43 | |
| number of EVAL's | 190828 | 1 | 1 |
| G-code profile: | | | |
| CALL | .643 | .264 | .291 |
| ALLOC | .155 | .130 | .030 |
| UPDATE | .053 | .166 | .184 |
| ALU | .039 | .123 | .136 |
| READ | .053 | .165 | .182 |
| STACK | .043 | .094 | .113 |
| JMP | .008 | .026 | .029 |
| LIT | .010 | .019 | .021 |

| Linfib | *Lazy* | *Strict* | *by-Value* |
|---|---|---|---|
| time | .00292 | .00074 | .00069 |
| dispatch rate | 1.27 | 1.51 | 1.72 |
| expansion ratio | 5.93 | 2.23 | 1.43 |
| allocation rate | 1.01 | 2.67 | 0.0 |
| equivalent VAX Mips | 3.61 | 4.43 | |
| number of EVAL's | 300 | 1 | 1 |
| G-code profile: | | | |
| CALL | .649 | .018 | .019 |
| ALLOC | .139 | .313 | .000 |
| UPDATE | .001 | .003 | .003 |
| ALU | .051 | .202 | .303 |
| READ | .069 | .274 | .367[2] |
| STACK | .061 | .108 | .159 |
| JMP | .010 | .040 | .072 |
| LIT | .016 | .030 | .048 |

[2] Accounts for reading values held in the P-stack, including test of the "is_pointer" tag.

*Linfib* shows the most dramatic speedup from lazy to strict evaluation, nearly four times. This is entirely attributable to suspension of the arguments in repeated calls to the function under lazy evaluation, even though these are tail calls. Notice that when the the computation rule is strict evaluation, the G-memory allocation rate has become 267,000 nodes/second, and all of these allocations are for graph nodes in which to pass an integer-valued argument (by reference) in a tail-recursive function call. The call-by-value model eliminates all graph node allocations.

| **Towers of Hanoi** | | | |
|---|---|---|---|
| | *Lazy* | *Mod. lazy* | *Strict* |
| time | 0.233 | 0.164 | 0.048 |
| dispatch rate | 1.27 | 1.26 | 1.40 |
| expansion ratio | 9.32 | 7.52 | 2.67 |
| allocation rate | 1.03 | 1.12 | 1.38 |
| number of EVAL's | 26622 | 14337 | 1024[2] |
| G-code profile: | | | |
| CALL | .661 | .617 | .390 |
| ALLOC | .141 | .149 | .054 |
| UPDATE | .103 | .098 | .132 |
| ALU | .008 | .011 | .038 |
| READ | .040 | .057 | .195 |
| STACK | .026 | .039 | .116 |
| JMP | .008 | .015 | .041 |
| LIT | .010 | .011 | .023 |

[2] The EVAL instructions occurred in library routines for the *hd* and *tl* functions. The library was not specialized to the call-by-value computation model.

Because list construction is lazy in the moderately-lazy computation rule we have used, the majority of function calls remain suspended under this rule. *Towers_of_Hanoi* executes 4.85 times faster under the fully strict computation rule than with lazy evaluation. This also includes call-by-value passage of numeric arguments.

| **Treesort** | | | |
|---|---|---|---|
| | *Lazy* | *Mod. lazy* | *Strict* |
| time | 0.051 | 0.033 | 0.29 |
| dispatch rate | 1.36 | 1.54 | 1.48 |
| expansion ratio | 4.15 | 2.88 | 2.48 |
| allocation rate | 1.40 | 1.75 | 1.85 |
| number of EVAL's | 8030 | 1589 | 1086[2] |
| G-code profile: | | | |
| CALL | .449 | .260 | .192 |
| ALLOC | .184 | .189 | .183 |
| UPDATE | .065 | .089 | .101 |
| ALU | .049 | .077 | .088 |
| READ | .130 | .204 | .232 |
| STACK | .075 | .105 | .118 |
| JMP | .025 | .040 | .044 |
| LIT | .014 | .021 | .024 |

[2] The EVAL's counted in call-by-value evaluation were from the code of library functions *hd* and *tl*. They did not produce function calls.

| **Isort** | | | |
|---|---|---|---|
| | *Lazy* | *Mod. lazy* | *Strict* |
| time | 0.111 | 0.078 | 0.026 |
| dispatch rate | 1.28 | 1.28 | 1.46 |
| expansion ratio | 7.25 | 5.84 | 2.22 |
| allocation rate | 0.84 | 0.88 | 0.95 |
| number of EVAL's | 16058 | 8837 | 1 |
| G-code profile: | | | |
| CALL | .625 | .575 | .281 |
| ALLOC | .115 | .119 | .029 |
| UPDATE | .111 | .097 | .106 |
| ALU | .014 | .020 | .060 |
| READ | .073 | .103 | .316 |
| STACK | .035 | .050 | .134 |
| JMP | .014 | .023 | .053 |
| LIT | .008 | .008 | .012 |

| **Primes** | | |
|---|---|---|
| | *Lazy* | *Mod. lazy* |
| time | 0.798 | 1.527 |
| dispatch rate | 1.39 | 1.41 |
| expansion ratio | 4.84 | 4.93 |
| allocation rate | 0.88 | 0.92 |
| equivalent VAX Mips | 3.86 | 3.65 |
| number of EVAL's | 104984 | 140157 |
| G-code profile: | | |
| CALL | .537 | .518 |
| ALLOC | .120 | .126 |
| UPDATE | .052 | .054 |
| ALU | .086 | .092 |
| READ | .100 | .107 |
| STACK | .077 | .074 |
| JMP | .013 | .014 |
| LIT | .011 | .011 |

In both *Primes* and *Digits_of_e*, execution time is increased by changing from lazy evaluation to the moderately lazy model. The slowdown factor is four times when evaluating the first 15 digits of $e$!

In the moderately lazy model, evaluation is suspended at list constructions, rather than at function applications. Consequently, the function that terminates the computation in these two example programs,

```
counthd lis n = if n=0 then nil
                else (hd lis) . counthd (tl lis) (n-1)
```

evaluates its first argument to a constructed list in its last call when its second argument is zero, even though it does not return the head of the list. Thus the program always evaluates an extra element of the stream of values it is printing. This same phenomenon of somewhat over-eager evaluation also occurs for some other functions in these two programs when the moderately lazy model is used. The number of instructions executed in the moderately lazy model may actually exceed that in the lazy model.

For the *Digits_of_e* example, there is no uniform *a priori* bound on the amount of lookahead computation necessary to resolve precisely the value of an arbitrarily chosen digit in the sequence. The sixteenth digit requires a particularly large amount of lookahead computation

| Digits_of_e (15 digits) | | |
|---|---|---|
| | *Lazy* | *Mod. lazy* |
| time | 0.066 | 0.270 |
| dispatch rate | 1.30 | 1.31 |
| expansion ratio | 5.55 | 5.98 |
| allocation rate | 1.13 | 1.23 |
| equivalent VAX Mips | 3.91 | 3.83 |
| number of EVAL's | 7859 | 21856 |
| G-code profile: | | |
| CALL | .596 | .553 |
| ALLOC | .155 | .166 |
| UPDATE | .027 | .038 |
| ALU | .028 | .028 |
| READ | .102 | .116 |
| STACK | .071 | .075 |
| JMP | .008 | .011 |
| LIT | .011 | .013 |

to resolve its value. In the moderately lazy model, the value of the sixteenth digit is eagerly computed in order that *counthd* will produce the fifteenth, adding a very significant computational burden. We ran *Digits_of_e* to produce 12, 15, 18, 20, 22, 25, and 30 digits, and noted an increase in computation time in the moderately lazy model only for 15 and 30 digits.

## 5. Conclusions

Simulation of the hardware G-machine demonstrates the performance improvement achievable through architecture. The relative performance, of the G-processor, expressed in terms of a well-known architecture for general-purpose computation, is that of a 3.6 Mips VAX, or roughly a VAX 8600. The design of the G-processor does not rely on any particular device technology, although there are a few components that may be difficult to realize without a custom VLSI implementation. The technology assumptions made in the design have been conservative. Commercial, silicon VLSI processors of much greater design complexity have achieved basic clock rates in excess of twice that for which we have designed.

The G-processor achieves its performance through a few basic mechanisms and design strategies:

a) Its internal architecture is a RISC, most of whose instructions execute synchronously and can be dispatched on nearly every clock cycle. The measured instruction dispatch efficiency (1.26 to 1.72) indicates the effectiveness of the pipelined data path.

b) The context for a function call is maintained by the processor in the top segment of the P-stack, which is held in a register file within the processor. The overhead of a context switch is minimal, and consists of saving (or restoring) a return address and an excess argument count in the P-stack.

c) Data are tagged to distinguish pointers from basic values and applications from reduced data structures. Tags are fetched from memory along with data and can be tested in the following cycle. Pointer tags are saved in the P-stack, which allows basic values, as well as pointers, to reside in the P-stack. This allows true call-by-value optimizations to be superimposed upon a graph-reduction model of computation.

d) A very low-latency instruction pipeline provides buffering of the instruction flow, and actually enhances the performance of the G-processor relative to a RISC design that fetches instructions directly from control memory to the execution unit.

e) Concurrent management of the G-memory supports a high rate of graph node allocations without interrupting the activity of the G-processor.

The design is notable for its versatility. It supports several models of computation, ranging from lazy evaluation of recursive functions through iteration with call-by-value. Several mechanisms are available to invoke function calls, providing efficient evaluation of function applications whether suspended or not, and whether or not the compiler can identify the function being applied.

These measurements also show, more directly than any previous results, the wide range of performance discrepancies between the different models of computation. For algorithms that are not structured to require lazy evaluation, the measured cost of laziness is a factor of 1.8 to 4.8 times in execution speed. However, the two algorithms designed to use lazy streams were seen to save in execution time by factors of 2 and 4 times by evaluating applicative expressions in normal order as well as suspending the stream constructor.

### 5.1. Extensions of the design

Further extensions of the design can make the G-machine even more flexible and powerful. Some of these extensions will come in the realm of compiler optimizations, some as hardware extensions.

1) **Vector nodes** in the expression graph have been suggested (and implemented in the Chalmers University LML compiler) by Augustsson and Johnsson [Aug86, Joh86]. The use of vectors in memory follows the principle of data aggregation. A vector can be used to represent the suspended application of a function to multiple arguments, to represent an n-tuple by contiguous storage locations in preference to a linked list of nested pairs, and to implement a functional array [Hud86, Hug85].

2) **Result-by-value.** Presently, functions always embed the result of an application in a graph node. Sometimes, when all uses of the result can be located within a single expression context by a compiler, it would be possible to avoid the graph representation entirely, by leaving the result in the P-stack. This is done only with the results of arithmetic expressions by the current call-by-value version of the compiler. More generally, it could be done for the values of applicative expressions, particularly for applications denoted by **let** definitions. This strategy should reduce the memory allocation rate and the execution time for functions able to take advantage of it.

3) **Large-grain parallel evaluation** is also a possibility, with the G-machine design described here serving as a node processor. Since it has been designed to run as a slave to a host processor, it is a good choice to use as a graph reducer to run in conjunction with a host that operates as a network communications controller in a distributed-memory multiprocessor, or with a host that schedules tasks in a shared-memory multiprocessor. Neither the communications control, the synchronization, nor the scheduling of tasks is the responsibility of the G-processor, which is specialized to perform sequential graph reduction rapidly.

## 6. Acknowledgements

This research would not have been possible without the dedicated efforts of Linda Rankin, who designed the concurrent memory manager and the control unit of the G-processor, Shyue-Ling Kuo, who designed and simulated the instruction fetch and translation unit, Siroos Farahmand-nia, who integrated Microsim, the microsimulator for the G-processor, Boris Agapiev, who implemented Macrosim and gathered most of the data reported here, and Raman Tenneti, who implemented the retargetable G-code translator.

**References**

[Aug84] Augustsson, L., "A Compiler for Lazy ML," *Proc. of 1984 ACM Conf. on Lisp and Functional Programming*, Austin, Texas, 1984, pp. 218-227.

[Aug86] Augustsson, L., "The revised G-machine," PMG Memo 45, Dept. of Computer Science, Chalmers University of Technology, Gothenburg, 1986.

[Car84] Cardelli, L., "Compiling a functional language," *Proc. of 1984 ACM Conf. on Lisp and Functional Programming*, 1984, pp. 208-217.

[Hud86] Hudak, P., "Programming with and implementing functional arrays," *Proceedings of the 1986 Graph Reduction Workshop*, Santa Fe, New Mexico, 1986.

[Hug82] Hughes, R. J. M., "Super-combinators," *Proc. of 1982 ACM Conf. on Lisp and Functional Programming*, Pittsburgh, 1982, pp. 1-10.

[Hug85] Hughes, J., "Lazy memo-functions," in *Functional Programming Languages and Computer Architecture*, vol. 201, J. Jouannaud (ed.), Springer-Verlag, Nancy, 1985, pp. 129-146.

[Joh83] Johnsson, T., *The G-machine -- an abstract architecture for graph-reduction*, Dept. of Computer Sciences, Chalmers Univ. of Technology, Gothenburg, 1983.

[Joh84] Johnsson, T., "Efficient compilation of lazy evaluation," *Proc. of 1984 ACM/SIGPLAN Notices Conf. on Compiler Construction*, Montreal, 1984.

[Joh85] Johnsson, T., "Lambda lifting: transforming programs to recursive equations," in *Functional Programming Languages and Computer Architecture*, vol. 201, J. Jouannaud (ed.), Springer-Verlag, Nancy, 1985, pp. 190-203.

[Joh86] Johnsson, T., "Generating target machine code from G-machine code," *Proceedings of the 1986 Graph Reduction Workshop*, Santa Fe, New Mexico, 1986.

[Kie85] Kieburtz, R. B., "The G-machine: a fast, graph-reduction evaluator," *Proc. of IFIP Conf. on Functional Prog. Lang. and Computer Arch.*, Nancy, 1985.

[Pey87] Peyton-Jones, S. L., *The implementation of functional programming languages*, Prentice-Hall International, London, 1987.

[Ran86] Rankin, L. J., *A dual-ported real memory architecture for the G-machine*, M.S. thesis, Oregon Graduate Center, Beaverton, Oregon, 1986.

[SCN84] Stoye, W. R., Clarke, T. J. W. and Norman, A. C., "Some practical methods for rapid combinator reduction," *Proc. 1984 ACM Sympos. on Lisp and Functional Programming*, 1984, pp. 159-166.

# Appendix -- Test Programs

*****************************

**Ackermann**

*****************************

```
letrec
   A x z = if x=0 then z+1
           else if z=0 then A (x-1) 1
           else A (x-1) (A x (z-1))
in A 3 4
```

*****************************

**Tak**

*****************************

```
letrec
   tak x y z =
         if ~(y < x) then z
         else   tak (tak (x-1) y z)
                    (tak (y-1) z x)
                    (tak (z-1) x y)
in
   tak 18 12 6
```

*****************************

**Linfib**

*****************************

```
letrec
   fib x y n =
         if n=0 then y
         else fib y (x+y) (n-1)
in fib 0 1 100
```

*****************************

**Towers_of_Hanoi**

*****************************

```
letrec append x y =
         if null x then y
         else (hd x) . append (tl x) y
and
   move x y = 10 * x + y
in
letrec hanoi f s t n =
         if n = 1 then [move f s]
         else
            append (hanoi f t s (n - 1))
                   (move f s . hanoi t s f (n - 1))
in  hanoi 1 2 3 10
```

*****************************

**Treesort**

*****************************

```
import type Ordertree =
   empty + tip(int) + node(Ordertree # int # Ordertree);
letrec
   map f lis =
      if null lis then []
      else f (hd lis) . map f (tl lis)
and
   reduce f lis z =
      case lis in
         [] : z
      || (a.l) : f a (reduce f l z)
      end
and
   insert n t =
```

```
      case t in
         empty : tip n
      || tip m : if n<m then node(tip n, m, empty)
                 else node(empty, m, tip n)
      || node(t1, m, t2) :
                 if n<m then node(insert n t1, m, t2)
                 else node(t1, m, insert n t2)
      end in
letrec
   flatten t =
      case t in
         empty : []
      || tip n : [n]
      || node(t1, n, t2) : flatten t1 @ n.flatten t2
      end
and count x y =
       if x > y then []
       else x.count (x+1) y in
let sort lis = flatten(reduce insert lis empty)
and rand r = 56789 * r % 12345 in
sort (map rand (count 1 100))
```

****************************

**Isort**

****************************

```
letrec map f l =
   if null l then
      nil
   else
      f (hd l) . map f (tl l)
in
letrec insert x l =
   if null l then
      [x]
   else
      if x < hd l then
         x . l
      else
         hd l . insert x (tl l)
in
letrec sort l =
   if null l then
      nil
   else
      insert (hd l) (sort (tl l))
in
let rand r = (r * 56789) % 12345
in
letrec count a b =
   if a > b then
      nil
   else
      a . count (a+1) b
in
   sort (map rand (count 1 100))
```

****************************

**Primes**

****************************

```
   letrec
      from x = x . from (x+1)
   and
      counthd lis n =
         if n = 0 then []
```

```
            else hd l. counthd (tl l) (n-1)
      and
         filter p seq =
            case seq in
               (a.rest) :
                  if a%p ~= 0 then a.filter p rest
                  else filter p rest
            end
      in
         sieve seq =
            case seq in
               (p.rest) :
                  p.sieve (filter p rest)
            end
      in counthd (sieve (from 2)) 250
```

****************************

**Digits_of_e**

****************************

```
      letrec
         carry c lis =
            case lis in
               (h1.t1): case t1 in
                           (h2.t2): (h1+h2/c).(h2 % c).t2
                        end
            end
      in
      letrec
         mult lis =
            case lis in
               (h.t) : 10*h.mult t
            end
      and
         norm c lis =
            case lis in
               (h1.t1):
                  case t1 in
                     (h2.t2):
                        if (h2+9) < c then h1.norm (c+1) t1
                        else carry c (h1.norm (c+1) t1)
                  end
            end
      in
      letrec
         convert lis =
            case lis in
               (h.t) : h . convert (norm 2 (0.mult t))
            end
      and
         counthd lis n =
            if n = 0 then []
            else hd lis. counthd (tl lis) (n-1)
      and
         seq = x.seq x in
   counthd (convert (2.seq 1)) 15
```

# A Flexible Architectural Study Methodology

Steven Tighe, Ken Zink, and Richard Brice
Parallel Processing Program, MCC

William Alexander
Database Program, MCC

**Abstract**

An efficient emulation/simulation system for evaluating architectures and scheduling strategies for reduction systems is described. Execution traces of example programs are generated by the emulator. The execution method of the emulator exercises all possible parallelism available in the execution model under study. The trace of each program execution is then reduced to an "architecturally neutral" precedence graph. The precedence graph can then be used repeatedly in simulations to study the effects of changes in architecture or scheduling strategy.

## 1.0 Introduction

A simulation system has been developed to allow study of reduction as an execution model on a wide range of parallel computer architectures. The challenge was to design a simulator efficient and flexible enough that a very wide range of architectures could be investigated. The system has been used in the analysis of combinator graph reduction as an execution model.

Combinator graph reduction as a basis for the implementation of computer languages is an idea introduced by David Turner [Tur79] to support his purely functional language SASL [Tur76]. Because combinator graph reduction includes no operations with side effects, it is an interesting candidate for a parallel processing model of execution.

The problem of simulating graph reduction architectures has been addressed a number of times in the literature. Previous studies include Keller's Rediflow simulator[Kel84], Hudak's work with DAPS [Hud84] and Darlington's Alice simulator [Dar81]. These projects have generally not included parallelism that involves eager evaluation. Previous work at MCC has investigated the potential of eager evaluation under ideal conditions [Tig87].

The project described in this paper was begun with three goals in mind. The first was to extend previous work to include the parallelism generated by eager evaluation; the second was to design a simulator which would allow the study of a very wide range of hardware and software architectures to support reduction; the third was to make the simulator run as fast as possible.

### 1.1 Eager Parallelism

Most previous studies have been restricted to parallelism that is often called "strict" parallelism. This type of parallelism occurs when an operator is known to need two or more arguments in order to complete. In this situation, both arguments can often be evaluated in parallel. As an example of this type of parallelism, consider the simple problem:

(a+b) + (c+d)

If additions are the unit of both work and time, a parallel system taking advantage of strict parallelism could add a to b and c to d during the same time and add the two results on the next time step. In this way, three units of work could be completed in two time units.

Eager parallelism allows work to be performed before it is known to be needed. While this approach may provide opportunity for computational speed-up, it may result in work being done that would not have been done in a strict parallel system. For example, take a simple conditional:

IF (a=b) THEN (a+c) ELSE (a+b)

One way to speed up this evaluation is to perform the two additions (a+c) and (a+b) simultaneously with the test for equality. Then when the test is complete, the proper answer will be available. This allows the evaluation of the conditional in one unit of time, but requires an extra unit of work. Work that is done because it *might* be needed at some future time is called *eager*.

As the first example illustrates, speed-up is available from clever scheduling of operations in order to take advantage of strict parallelism. However, as the second example illustrates, another type of speed-up is available at the cost of evaluating work that might *not* be used. In this study, both kinds of parallelism were investigated.

### 1.2 Simulation vs. Emulation

One common way to model systems is to use an event driven simulator with a statistically generated model workload. Unfortunately, because the run-time performance properties of the graph reduction model have only recently begun to receive significant attention, that model is not well enough understood to reduce it to its statistical properties. Furthermore, the applications have not matured to the level required to extract statistical workloads. Therefore, a statistically driven simulation approach was ruled out and a method for efficient simulation based on event trace data was developed.

The distinction between *emulation* and *simulation* is a key to achieving the goals of flexibility and efficiency. As separate systems, the emulator and simulator could each be designed and optimized for the particular part of the problem they were to address. The *emulator* faithfully represents the behavior of a parallel graph reduction engine. A SASL source program is compiled into a combinator graph. The emulator will then perform the reduction steps as required by the program. In addition, the emulator will produce a detailed trace of its internal behavior. The *simulator* only models the resource usage represented by these computations; it does not have to recreate them. The trace information gathered from the emulator is used to govern the simulation resource usage patterns. This division of labor has enabled the design of an efficient system in which the computation required by the compilation and graph reduction stages are done only once and produce an execution trace which will support numerous simulation studies.

## 2.0 Description of System

The details of SK graph reduction itself are exclusively handled by the emulator. The challenge of the emulator's design was to make it unconstrained and "architecturally neutral". Architectural neutrality is achieved by emulating a parallel processor with an unlimited number of processors and an unlimited global memory. All important architectural characteristics, like communication and process spawning costs are not given values in the emulator but are left as parameters to be given values in the simulator. This means that the *trace file* produced by the emulator is free of any architecturally imposed constraints. (The *statistics* produced by the emulator are *not* architecturally neutral, for reasons that will be discussed in section 3.0.)

In the graph reduction model, the program code and data are both directly represented in the graph; therefore, after a sub-graph has been evaluated, its result is available for any other function which needs it. To support this type of data sharing requires that the trace include not only a record of the activity related to the reduction of a sub-graph, but all accesses to the value of the reduced sub-graph as well. One obvious reason for maintaining information about accesses is the need to model memory accesses in the simulator. However, a more important reason results from the way in which the architectural description and scheduling policies may affect the sequence of events during a simulation. It is very difficult to determine, until simulation run-time, the order in which reduction will occur.

All of the information needed by the simulator is captured by giving each reduction process a unique identifier in the emulator. A process retains its unique ID and each record in the trace, of an access to the graph, contains the ID. Using this information, a precedence graph is constructed from the trace, which includes the computational cost (number of reduction steps) required by each process and the relationships which exist among the different processes.

Details specific to the architecture being studied are provided as input to the simulator. These details include such parameters as the number of processors, the amount of memory and the cost of communication among the processors. Other input to the simulator describes the scheduling and resource allocation policies.

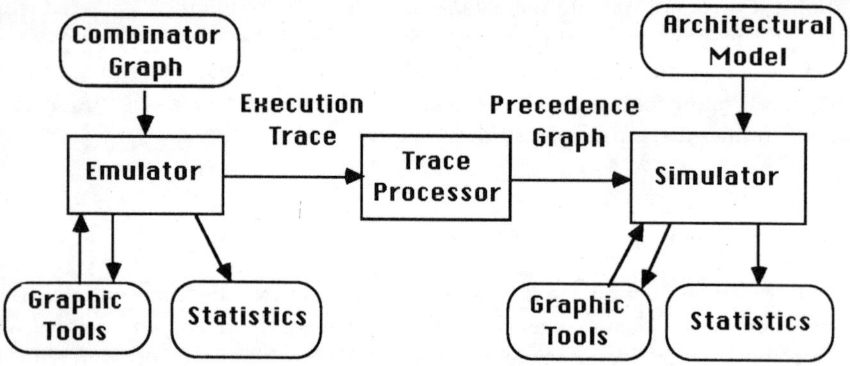


Figure 1. Overall System Design.

Figure 1 illustrates the interaction of the three components of the system. The emulator produces an execution trace which captures the behavior of a neutral graph reduction engine. The trace processor converts the trace into a precedence graph. Many different simulations may then be run on a single precedence graph to study the effects of architectural and scheduling decisions.

### 3.0 Emulator

The six component parts of the emulator are illustrated in Figure 2. Evaluation is initiated by the printer in response to a request to print the value of the program. Management of the actual reduction activity is handled by the process manager. It loads the context of a process, calls the reduction engine to perform one reduction, and then continues to the next process. The program graph is represented as nodes and pointers in a large heap space; each process is modeled by a separate structure in a large array. The graphics tools provide an observation and debugging window into all parts of the emulator.

The input to the emulator is an SK combinator graph which is constructed by a SASL compiler. The graph is made up of nodes, each containing two cells, a tag field, a statistics field (which is further divided) and a pointer for handling process suspension. The sizes of the graphs range from very small (on the order of 5-10 nodes) to very large for complex applications.

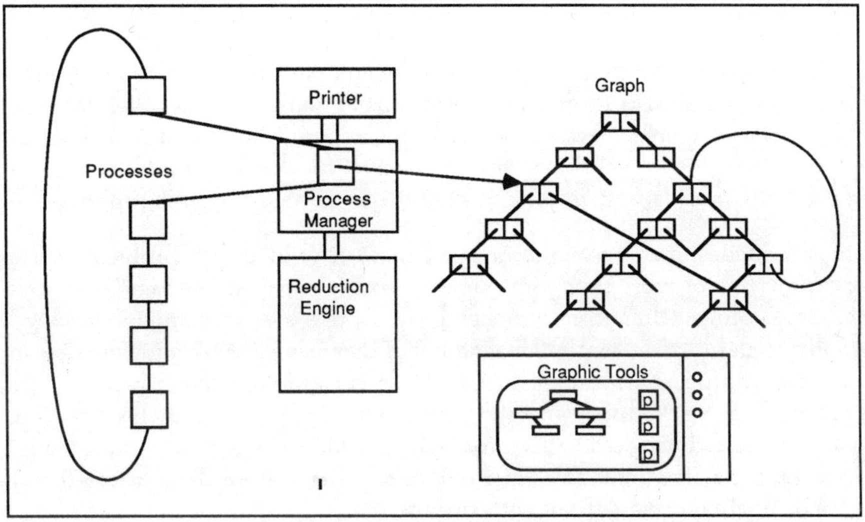


Figure 2. Emulator Design.

Each process in the system is represented as a structure in an array. The process structure contains a stack, a pointer into the combinator graph and a set of about ten registers for gathering statistics. Emulating parallelism among the processes is the responsibility of the process manager. The process manager activates each process in the active queue, round-robin, until the queue is completed; then the next cycle through the queue is made. During its activation, each process is allowed to perform the equivalent of one reduction. If a trace of the program execution has been selected, relevant events are recorded in the trace file by each process during its activation.

One of the characteristics of an "architecturally neutral" emulator is that it has no temporal dimension. However, the emulator, with its graphics interface, is very convenient for conducting preliminary studies or architecture independent studies (e.g., the effects of higher order strictness analysis). To be able to quantify the results of these studies requires some minimal architectural specification to impose a temporal domain. (As soon as the property of time is introduced into an emulation, some architectural properties have been specified, whether intentional or not.) For these reasons, a very simple mechanism of time accounting is included in the emulator. The fundamental unit of time in the emulator is one reduction, and all reductions are assumed to have roughly the same complexity. (Burroughs has, in fact, built a machine that performs SK combinator reduction as its basic machine step [Ric85].) Process spawning is assumed to take some number of reductions, generally one for simple experiments. The temporal properties of the emulator have been introduced very carefully so that they do not compromise the architectural neutrality of the trace file.

### 3.1 Managing Parallel Reduction

Each individual process performs normal order reduction which is explained in [Tur79]. The algorithms necessary to control and synchronize a number of parallel reduction processes are explained here. They are similar to those found in [Cla86]. These include techniques for managing synchronization of access to the graph, initiating and suspending parallel processes, and keeping track of nodes traversed by each process.

Although in a well designed reduction system there should be no problems except wasted work from two or more processors simultaneously reducing the same sub-graph, we have chosen to prevent this from happening by using a system of node marking to lock sub-graphs. Nodes can be "virgin", meaning that they have not been touched by any reduction process; "reducing", meaning that they are currently on the stack of the first process to reduce them; or "reduced", which means that they have already been reduced. When a node is in the reducing state it is effectively locked, and any process needing it is suspended. Nodes that have already been reduced, or have been demonstrated to be "pure code" and so will not change, are not locked.

The act of initiating a parallel reduction process is called spawning. Spawning of parallel tasks in this system can be done on the arguments of any primitive operator or any user-defined function. The simplest spawning strategy is to spawn both arguments of the strict dyadic operators in parallel. More involved strategies allow spawning on the strict arguments of user-defined functions (to the extent that strictness analysis can determine which are strict) and speculative spawning of arguments that are not known to be strict.

Suspension of reduction processes can occur for two reasons, the first involves the locking of nodes. If a sub-graph required by a process is locked (currently being reduced by some other process), the process is suspended until the sub-graph becomes available. The second possible reason for suspending a process involves a strict operator with unreduced arguments. When a process encounters an operator which requires one or more arguments which are not yet reduced, it will spawn processes to reduce them if they are still "virgin". The process initiating the spawns is then suspended until completion of the first argument. If another required argument that was spawned at the same time has not completed, the process is immediately suspended again.

Each node contains a pointer for suspended processes. Multiple suspended processes waiting on the same node can be chained through a pointer in the process structure. Whenever the state of a node changes from reducing to reduced the pointer is checked, if there are suspended processes, these are awakened.

During reduction, some record of a process's path along the graph must be kept. This can be done by stacking pointers to the nodes already traversed or reversing pointers in the head of each node to point to its father. Reversing pointers is obviously more space efficient, but in a parallel system it also automatically locks nodes. This is a problem when the nodes that are locked have been previously reduced, or when the nodes represent pure code (although in the case of pure code, the penalty paid for reversing pointers is only one reduction). Because of the disadvantages of automatic node locking this system was designed to stack pointers.

One surprising and fortunate observation has been that the stacks in the processors do not have to be very large. A stack size of twenty is sufficient for most of the programs that have been run so far. It is probably simple to construct a case that requires more stack space, but in practice this seems to happen rarely.

### 3.2 Graphic Tools

The emulator graphic tools are a set of programs that allow the state of the emulator to be examined at any point. Any process or node in the combinator graph can be examined. The entire active graph, and the location of each process on that graph, can be displayed.

Most system displays present the processors as being stable and display the state as changing in each processor. The primary display for this tool presents the code (which is the combinator graph) as stable and shows the processes moving through the graph. An auxiliary display shows the processes, with their internal state, in a more traditional manner. Other features include single and multi-step reduction, dynamic debug facilities and a running display of evaluation statistics.

These tools have been invaluable for learning about the behavior of combinator graphs. They have also been enormously helpful in the tedious process of debugging the emulator and verifying it with the simulator described below.

### 3.3 Control of Parallelism

The graphics interface is also used to provide control of the three major means of generating and controlling parallelism in graph reduction models of execution: eager evaluation, strictness analysis and run-time clipping. The control of eager evaluation allows selecting an evaluation policy for combinator arguments via the graphics interface. The evaluation of each argument of each combinator may be selected as (1) lazy–do not evaluate this argument until it is needed, (2) eager–begin evaluation of this argument and continue without waiting for its result or (3) strict–begin evaluating this argument and wait until it is completed.

The strictness analysis was done manually, so the graphics tools provide a window into the library that allows setting the calling method of each function to evaluate its arguments as either (1) lazy–do not evaluate this argument until it is needed or (2) eager–begin evaluation of this argument immediately upon function invocation but do not wait for its completion.

Recursive eager evaluation can create "black hole" or "cancer" calculations which swamp the resources of the system. Although the emulator is intended to be unconstrained, it does have real limits. Some of the programs under study can exceed the real limits of the emulator if permitted to generate eager work without constraints. For those programs, some run-time clipping is invoked to keep the execution within the emulator limits.

When an eagerly spawned process disappears (is "clipped"), it leaves the sub-graph reduced or partially reduced. The preferred mode of operation is to generate as much eager work as possible in the emulator and then use a combination of scheduling policies, including clipping, to constrain wasted work in the simulator.

## 4.0 Precedence Graph

Information about the execution of the program on the emulator has to be expressed in a form that will allow it to be simulated on an almost arbitrary parallel architecture under a variety of execution policy assumptions. It was, therefore, very important to design the simulation to be as independent as possible of specific features of the emulation, and, in fact, to be independent of specifics of the graph reduction execution model. The primary design consideration was this independence, but program execution efficiency and program manageability were also important.

The simulation of architectures for support of reduction based computations requires the transformation of measurements taken from running systems into a form suitable as input to a discrete event-driven simulator. Simple chronological traces tagged with the time of occurrence are inadequate for this purpose. The trace must make explicit the precedence which exists among individual events. A simple chronology only describes the time at which some event occurred in the system which produced the trace; it does not capture the information which is necessary to know the earliest moment at which the event might have occurred in some other system, i.e. the precedence information which enabled the event to occur. This information is important because the order in which events occur in the simulator may be very different from the order in which they occur in the trace.

In fact, the time at which events occur in the trace is of absolutely no interest. Only two things matter: the precedence which relates the reduction of individual sub-graphs, and the computational resources (e.g., counts of reduction steps) required to reduce a sub-graph. The precedence graph is necessary in order to be able to "rearrange time" during simulation.

### 4.1 Form of the Precedence Graph

One outgrowth of the attempts to define an appropriate data structure to represent precedence in a form suitable as input to the simulator was a very general graph language for describing parallel computations. The language has two primary data structures, nodes and arcs, and a small number of control primitives which are applied to the arcs and nodes based on the execution strategy being used by the simulator.

The nodes in the precedence graph correspond to the reduction processes which occurred in the trace. The arcs describe the dependency relationship between the nodes in the precedence graph. An arc from A to B means that the computation in A might need the data represented by B. A process to evaluate B may be spawned via the arc from A to B. If the computation in A requires the actual data value of B, the relationship or process spawn along that arc is strict and A cannot continue until B is evaluated. If the computation in A merely manipulates a pointer to B, the relationship is non-strict and a process spawn via that arc would be eager. An eager spawn initiates evaluation of the sub-process, B, without suspension of the initiating process, A.

In general, nodes may have many outgoing arcs and also many incoming arcs. Both nodes and arcs must include a rich set of information to support our intended experiments. Each node consists of some amount of initial computing followed by an arbitrary number of tuples $\{D_i, C_i\}$. Each tuple consists of a reference arc to some data, $D_i$, followed by some additional computing, $C_i$. Thus, a node may be arbitrarily large in size, depending on the granularity of computation, the number of computations and the number of tuples included in the node. The tuples within a node are processed sequentially, according to the scheduling parameters selected for a given simulation. (Note that each of the tuples could be a separate node in the precedence graph, but the trace processor, which builds the precedence graph, insures that no potential for parallelism is lost by combining them into a single node for storage efficiency.)

Pictorially, some process k might be represented as in Figure 3 where $D_1$ represents a strict reference arc to a single data item, $D_2$ represents the strict reference arcs to two data items and $D_j$ represents a non-strict reference arc to a data item. The eager evaluation of $D_j$ would imply spawning a process to perform the evaluation in parallel with $C_j$, and any subsequent data references and computations.

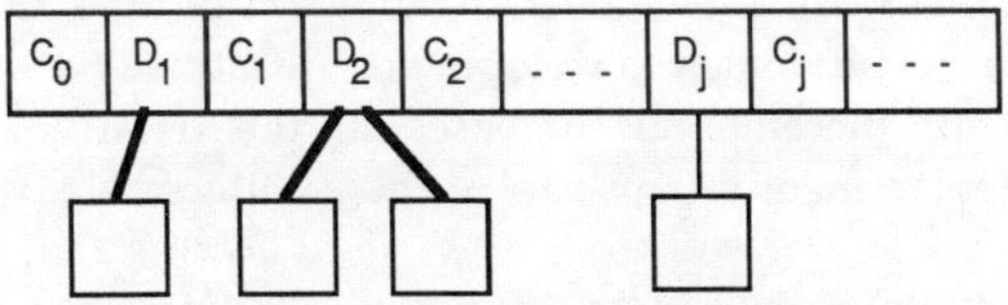


Figure 3. Partial Precedence Graph.

Generally, the control primitives are represented by data fields within the nodes and arcs. For example, one field within a node can record why it was initiated and how many processes are awaiting its completion. Control primitives include:

spawn sub-task and wait (don't wait) for reply,
spawn parallel sub-task set, wait for all/first to complete,
change sub-task evaluation type from eager to strict,
kill remaining active processes in a parallel sub-task set, and
resume all processes awaiting completion of some sub-task.

As the recursive processing of nested sub-graph evaluation unwinds and child processes restart those which are awaiting their completion, control information can be recorded in local program variables. Some of the control information can be subtle and dynamic because of the non-deterministic character of the evaluation. Sub-graphs, started by one procedure eagerly, can be strictly required by other procedures before their completion.

### 4.2 Precedence Graph Builder

The graph builder program takes the sequential event trace information produced by the execution of a program in the emulator and produces the precedence graph of the execution used by the simulator. The emulator implementation ensures that the first reference to a data value results in a unique process being spawned to perform the evaluation of that data value. The spawn event in the trace file identifies the new process, the parent process and the combinator node at the head of the sub-graph representing the data value. In order to determine the precedence relations, the graph builder creates a process node for each spawn event and an arc from its parent. Any subsequent references to that data value in the trace file result in the creation of an arc from the process of reference to the process node representing the data value. Then, during simulation, the evaluation of the data value can be initiated via any one of the arcs pointing to it. Along with the spawn/data reference information, the graph builder also records the computational statistics in the process node tuples.

## 5.0 Simulator

The simulator is intended to be used to evaluate system architectures and program execution strategies. Because it is expected that many architectures and execution strategies will be of interest and that these will evolve over some period of time, the simulator design must support a hierarchical, evolutionary development. Additionally, the implementation must be modular to permit incremental increases in the detail of the architectural specification and/or easy replacement of any part of the architectural specification, e. g. the memory system.

The simulator uses an underlying system of simulation support routines collectively called CSIM [Sch85]. CSIM is a process oriented simulation language and run-time environment based on the C programming language. It is patterned after the ASPOL language by MacDougall [Mac72]. The CSIM run-time library provides:

spawning and management of pseudo-parallel processes,
a simulated parallel execution of all concurrently active processes,
management of shared and exclusive use resources,
synchronization primitives,
simulated time synchronization across processes,
automatic statistics accumulation, and
extensible statistics gathering.

### 5.1 Program Simulation

A program execution to be simulated is represented by a precedence graph. The simulated program execution begins and terminates at a single process node which represents the program result(s). Each of the data references in that process node points to the data used or manipulated by that process. A data node in the precedence graph represents either the actual data value or a process that must be evaluated to produce the data value.

The simulator must keep track of why each process node was initiated (eager/strict) as well as its current state of processing, e.g. unstarted, in progress, blocked. Numerous arcs might point to a given process node, and each arc might represent a different reason for initiating the process. For example, reduction of a sub-graph might have begun eagerly if one path through the computation graph were followed and might have begun as a strict operation if some other path were followed. The scheduling and allocation policies used by the simulator will determine which arc is used, no matter which path initiated the process in the emulator.

As an example of a simulation run, consider the following functional program:

```
add (add 3 4) (mult 5 6)
   WHERE
   add n m = n + m
   mult s t = s * t
?
```

The execution of this program would be represented in a precedence graph with the seven nodes shown in Figure 4. C1 and C2 actually represent the add operations and C3 represents the multiply.

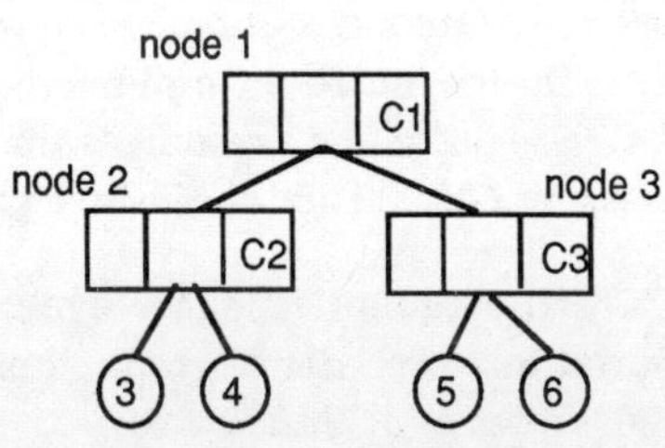


Figure 4. Representation of a Small Program

The simulator first initiates a process representing node 1. The resource utilization and time requirements for process initiation are simulated within that routine and determined by the parameterization of the architecture under study. Once initiated, process A would examine the data structures for node 1 and determine that it has two potentially parallel data references to perform. Because these are strict data references, the C1 computation cannot be performed until the evaluation of nodes 2 and 3 complete. Within the simulation, depending on the execution strategy, a number of choices are available: processes to evaluate nodes 2 and 3 could be spawned to execute in parallel; or rather than spawn additional processes, process A could evaluate nodes 2 and 3 serially, itself; or A could choose to evaluate one of the nodes itself and spawn another process to evaluate the other one in parallel. These choices are dependent on the selected execution strategy, and the simulation outcome will vary with the strategy and the architectural parameters chosen. Similarly, the execution strategy determines whether to spawn processes to perform eager evaluation of non-strict data references.

Assume process A elects to evaluate node 2 itself and spawns a parallel process B to evaluate node 3. Process A may complete its evaluation before process B is completed. If so, it sets a flag for node 3, indicating that there is at least one process waiting for its evaluation, and enters a wait state. Upon the completion of the evaluation of node 3, process B will post an event to all waiting processes that node 3 is now evaluated. Once nodes 2 and 3 are evaluated, process A can now proceed to the computation C1. This computation is simulated by causing simulated time to pass. The simulation is now complete and the simulation statistics are printed out for evaluation of the architecture and execution strategy.

Because of the sharing of sub-graphs, several processes may wait on the completion of a particular node. Note, however, that if a node is referenced from a number of other nodes, different execution strategies may well change when and by whom the evaluation of that node is initiated.

## 6.0 Problems

Like most emulator or simulator systems, real computing resources limit both the size of problems which can be used in experiments and the detail of the architecture being simulated. Allocating a potentially unlimited number of emulated processors in the most eager possible way tends to use large amounts of real resources and generate an enormous amount of trace information. In the current implementation on a SUN 3 with 32 megabytes of UNIX swap space, the "unlimited" resources that the emulator can provide are up to 60,000 concurrent processes and 600,000 nodes of active graph space.

The 4 queens problem with virtually unlimited eager evaluation produces a trace file of over 20 megabytes. When reduced to a precedence graph, that graph approaches the current simulator limits of about 100,000 nodes and 300,000 arcs.

One of the most difficult steps in the development of the simulation system turned out to be the verification of the emulator and the simulator. When the simulator is parameterized to simulate the architecture represented in the emulator, then the performance results produced by the simulator should match those produced by the emulator. Initially, the numbers simply didn't match. A number of assumptions made on the basis of logic and convenience in the design of both the emulator and the simulator seemed unimportant but turned out to be incompatible in some very subtle ways. As an example, in the emulator it is possible for process spawns to occur at the beginning or end of a reduction cycle, depending on the type of spawn. In the simulator, all spawning is done at the end of each cycle. This difference in spawning caused an incremental divergence of the two systems with every early spawn. After several thousand cycles, this small difference could snowball in such a way that the results could vary significantly. Therefore, in order to be able to validate the simulator against emulator results, and vise versa, this architectural convenience had to be faithfully represented in the simulator.

## 7.0 Conclusion

The system described above has been used to perform a number of experiments on parallel graph reduction. The preliminary results are promising and include the study of a range of applications from simple toy problems, such as 8 queens, to reasonably complex applications, including an E-MYCIN kernel and an image processing application.

The major problem of studying eager evaluation is simply keeping track of all the extra work. The decision to separate the simulation system into two parts proved to be valuable. Each part alone is smaller and uses fewer computational resources than they would combined; consequently, the total system can handle significantly larger problems than it could have had the two parts been combined. The emulator can currently handle up to 600,000 nodes and 60,000 processes active at one time (although there is no limit to the number of processes that can occur during a run).

Even small programs can produce traces with one million events and a precedence graph with one hundred thousand nodes. The construction of this graph represents a substantial investment of time and computing resources. Because of the simulator and emulator are independent, each of these graphs can be saved and used many times. One instance of a graph represents the interaction of an application and execution model. The simulator represents the behavior of some architecture designed to support the combined application and execution model. Applications and execution models evolve relatively slowly in comparison to the designs of execution strategies and architectures. Thus, building the graph separately results in a valuable savings in compute time.

Equally important, the simulator is much simpler because it does not need to create precedence information from raw trace data at the moment it is required. Debugging parallel simulations is hard work. The traces and graphs are large enough, and the simulation complex enough, that the correct answer is almost impossible to verify by hand. Separating the total correctness problem into two subsets has greatly simplified the task of producing correct simulations.

The system is currently implemented on a SUN 3/160 with 8 Megabytes of memory; the overall speed of the system is satisfactory. The emulator performs about 1000 reductions a second. The fastest single SK reduction engine in the world, Burroughs' NORMA machine, does about 300,000 reductions a second (which Burroughs claims makes SASL run at about the speed of C on a VAX), which means that the emulator is simulating real-time at a rate of about 300 to 1. The speed of the simulator varies with the size of the problem. Small problems run as fast as the emulator. Large problems (100,000 nodes) run at around 100 reductions per second. The process of building the intermediate graph is the slowest part of the system: a 25 Megabyte trace file took several hours for the precedence graph builder to process into an eight Megabyte execution graph; however, because the graph can be reused, this only has to be done once for each application.

Future work will include continued work on the simulator as more architectures are studied in more detail. The system will also be used to investigate varying grain sizes of reduction. One extremely interesting question is the extent to which the intermediate representation and simulator developed for graph reduction can serve as a common foundation for studying other parallel execution models such as those for logic languages and LISP.

Detailed traces have been produced from two other parallel execution models. One is patterned after MULTI-LISP and consists of the addition of the FUTURES construct to Common Lisp. The second is an extension of the Warren Abstract Machine to include AND parallelism in a PROLOG-like language. The overwhelming similarity in the precedence information required by parallel simulations of architectures to support these different execution models is encouraging. However, some important differences remain to be sorted out.

## Acknowledgements

We are grateful to the members of the Parallel Processing Program of MCC for the environment that supported this work, particularly to Steve Lundstrom, who heads the program, and Scott Danforth for his role in getting the study started. We are also indebted to Dr. David Turner for providing the original SK-combinator reducer, which was extended to build our current emulator.



## References

[Cla86] C. Clack, S. Peyton Jones, "The Four-Stroke Reduction Machine", *Proceedings of the 1986 ACM Conference on Lisp and Functional Programming,* Cambridge, Mass., August 4-6, 1986.

[Dar81] J. Darlington, M. Reeve, "ALICE: A Multi-processor Reduction Machine for the Parallel Evaluation of Applicative Languages", *ACM Functional Programming and Computer Architecture Conference,* Portsmouth, New Hampshire, October 1981.

[Hud84] P. Hudak, B. Goldberg, "Experiments in Diffused Combinator Reduction", *1984 ACM Symposium on LISP and Functional Programming,* Austin, Texas, 1984.

[Hud85] P. Hudak, B. Goldberg, "Serial Combinators: 'Optimal' Grains of Parallelism", *Functional Programming and Computer Architecture Conference,* Nancy, France, September, 1985.

[Kel84] B. Keller, F. Lin, "Simulated Performance of a Reduction Based Multi-processor", *IEEE Computer*, July, 1984.

[Mac72] M. MacDougall, "ASPOL Reference Manual", Control Data Corporation, Sunnyvale, CA, 1972 (latest revision).

[Ric85] H. Richards, Jr., "An Overview of the Burroughs NORMA", Burroughs Austin Research Center, January, 1985.

[Sch85] H. Schwetman, "CSIM: A C-Based Process Oriented Simulation Language,"

MCC Technical Report PP-080-85, 1985 (latest revision).

[Tig87] S. Tighe, "A Study of the Parallelism Inherent in Combinator Reduction", *Proceedings of the 1987 Hawaii International Conference on System Sciences*, Kona, Hawaii, Jan. 6-9, 1987.

[Tur76] D. A. Turner, "SASL Language Manual", Computing Laboratory, University of Kent, Canterbury, England.

[Tur79] D. A. Turner, "A New Implementation Technique for Applicative Languages", *Software—Practice & Experience, Vol. 9,* 31-49, 1979.

# Arrays, Non-Determinism, Side-Effects, and Parallelism: A Functional Perspective (Extended Abstract)

Paul Hudak
Yale University
Department of Computer Science

November 1986

**Abstract**[1]

Incremental, functional updates to arrays, executed in a non-deterministic manner, are shown to achieve the same effect (in both efficiency and functionality) as parallel assignment to imperative arrays. The strategy depends critically on the ability of a compiler to recognize not only that the incremental updates can be done destructively, but also that the updates may be done in any order. Special functional syntax is introduced that "captures" typical array-assignment patterns. The syntax falls somewhere in between the purely functional and (im)purely imperative, and makes the inferencing problem fairly easy. If nothing else, the work represents an interesting intellectual excercise in the relationship between non-determinism, side-effects, and parallelism.

## 1 Introduction

By now most language theorists/designers/programmers know how to express imperative programs in a functional style. For example, loops are replaced with tail-recursions (in which actual parameters in the recursive call replace assignment to loop variables); multiple assignments to a single variable are replaced with creation of new variables in let expressions (or subsumed completely within larger expressions); and assignments to aggregate data structures such as arrays are replaced with some equivalent expression using "functional" (i.e., "immutable") aggregates.[2] It is this last example, in particular assignment to arrays, and more particularly *parallel* assignment to arrays, that I wish to investigate in depth in this paper. My goal is to achieve with functional arrays the same *observable behavior* and *run-time efficiency* that one gets using arrays in an imperative language. Such efforts for sequential implementations have in fact already been fairly successful, and thus in this paper I will concentrate on parallel implementations.

[1]This research was supported in part by the US Government under DOE grant FG02-86ER25012.

[2]Functional programming advocates have argued endlessly why the resulting programs are better than imperative ones, but I do not wish to pursue that issue here.

The reason that assignment to cells in an array (as opposed to individual variables) is being singled-out for investigation is that such assignments are a common source of parallelism in conventional imperative languages; i.e., it is possible to perform concurrent operations to different parts of the same array, and to do so in a deterministic manner (usually by ensuring that no parallel assignments are made to the same location). Unfortunately, the "standard" translation of a *sequential* imperative program with mutable arrays into a functional program with immutable arrays results in a sequentialization, via data dependencies, of the immutable arrays – after all, that is what the semantics of the sequential imperative language dictates. So the question is now this: when one extends the semantics of an imperative language to accommodate *parallel* operations to arrays, how can one achieve the same behavior in a functional language?

One approach to solving this problem is to invent special functional arrays whose elements are defined all at once (and presumably in parallel) at the time the array is first created – I refer to these as *monolithic arrays*. Another approach, and the primary one to be investigated in this paper, is to use what I call *incremental arrays*, in which an "empty" array is incrementally "modified" until the desired result is achieved.[3] Although the incremental approach depends critically on clever optimization strategies to make it efficient, I argue that it is the most general of approaches, and is in fact suitable for modeling both sequential and parallel assignments to mutable arrays.

In the context of incremental arrays, the primary message that I wish to convey is this: Parallel assignment to arrays in an imperative language introduces a form of *non-determinism* which is manifested as different intermediate states of the array, whether or not the final array is deterministic. For example, in:

$$\begin{array}{l} \text{parbegin} \\ \quad a[i] := e_1; \\ \quad a[j] := e_2 \\ \text{parend} \end{array}$$

consider the value (i.e., state) of the array $a$ at the point when the second assignment is about to take place; the array may have one of two values, depending on whether or not the first assignment occurred already. If $i = j$ a non-deterministic result is possible, but even if $i \neq j$, making the final result deterministic, there is an implicit occurrence of non-determinism in the program. It is probably not surprising then that the solutions presented in this paper involve combining explicit non-determinism with incremental arrays. Doing so raises some rather interesting questions about the relationship between parallelism and non-determinism; indeed, in this context I may raise more questions than answers.

I will proceed by first summarizing briefly the incremental and monolithic approaches to arrays, including a summary of research on copy-avoidance strategies to allow the incremental updates to be done efficiently, and making an argument for why the incremental approach is most general. Next, "functional non-determinism" is introduced, followed by ways to combine it with incremental arrays to achieve the same behavior as parallel assignment to arrays. Syntax for common parallel array-assignment patterns is then described, which is (uncomfortably) similar to imperative language syntax. Finally, I will discuss strategies for implementing the resulting programs as efficiently as their imperative counterparts.

[3] The terms *monolithic* and *incremental* were adopted in [16], where the reader may find an excellent summary of a variety of monolithic arrays.

# 2 Functional Arrays

The *list* is the most popular data structure in Lisp and functional programming languages, due in a large part to its natural integration with recursive algorithms, but also, I believe, due to historical biases against other, less general data structures. One of these other data structures, central to most conventional languages, is the *array.* Arrays are traditionally used because of certain efficiencies that make them superior to lists in many applications:

1. A read or an update to any element can be done in *constant time.* In other words, one can index into the array directly without any kind of searching.

2. An update costs *zero space.* In other words, an update does not result in the allocation of any extra space to store the new value.

3. The elements can be stored in *contiguous memory locations*, thus saving the space required by links (pointers) in more general data structures.

In general I will concentrate first and foremost on achieving in a functional way the same *observable behavior* (i.e., the same *value*) as a particular imperative statement. Only then will I consider efficiency. On the other hand, the properties mentioned above are what I will use to define an acceptable realization of an array data structure.

## 2.1 Incremental Arrays

The simplest, and perhaps most naive, version of a functional array is the *incremental array*, having three operators:

1. $newarray(n)$ creates a new array of length $n$ whose elements are initialized to some default value, say *nil*.

2. $select(a, i)$ returns the $i$th element in the array $a$.

3. $update(a, i, x)$ returns a new array $a'$ such that $select(a', i) = x$, and $select(a', j) = select(a, j)$ if $i \neq j$.

Henceforth I will use the notation "$a[i]$" as syntactic sugar for "$select(a, i)$," and "$a[i] := x$" as syntax for "$update(a, i, x)$." Thus the notation "$a[i] := x$" will be "overloaded," in that if it appears in an imperative setting it denotes *assignment*, whereas in a functional setting it denotes an *expression* whose value is an array.[4]

The main difficulty with the incremental approach is that it suggests that a *copy* of the entire array is made upon each update. However, this does not have to be, and there exist several strategies for avoiding the copying. The most fundamental strategy is based on the simple observation that if the array being updated has *no other references* to it at the time of the update, then the update could be done *destructively*, effectively reusing the old array, and thus achieving the same efficiency as an imperative assignment to the array. An array having this property is said to be *single-threaded* [13].

[4]Although the use of ":=" may seem strange in a functional setting, it is technically no different from the use of any other binary infix operator, such as in "$a + b$" (denoting a numeric value) or "$a = b$" (denoting a boolean value).

Determining whether or not an array is single-threaded can be done either at run-time via conventional reference counting, or at compile-time via static analysis. Examples of the former include Goldberg's implementation of a graph reducer on commercial multiprocessors [2], and Wise's proposal for automatic reference counting via hardware support [17]; related to Wise's work is O'Donnell's proposal for hardware support of "associative aggregates" [12]. Examples of compile-time analyses include the efforts of Mycroft [10], Schwarz [14], Schmidt [13], Hudak [5], and Hudak and Bloss [6].

If in fact an array is *not* single-threaded, one can still avoid copying in many cases; techniques for doing so are described in [6]. Alternatively, one could use the strategies proposed in [12].

As a simple example, consider the canonical problem of initializing every element in some array to 0. In an imperative language I might write:

$$\begin{array}{l} \text{for } i = 1, \ldots, n \\ \text{do } a[i] := 0 \end{array}$$

The obvious solution using an incremental array is:

$$\begin{array}{ll} \text{let } \mathit{init}(lo, hi, acc) = & \text{if } lo > hi \text{ then } acc \\ & \text{else } \mathit{init}(lo+1, hi, a[i] := 0) \\ \text{in } \mathit{init}(1, n, a) & \end{array}$$

which, using syntactic sugar for tail-recursion, can be rewritten as:

$$\begin{array}{l} \text{for } i = 1, \ldots, n \\ \text{next } a \text{ is } a[i] := 0 \\ \text{return } a \end{array}$$

Although this looks quite similar to the imperative program, it is completely functional, and essentially captures in a functional way a "disciplined" use of imperative assignment in a loop.[5] More importantly, it is quite easy, at least for the analysis described in [5], to determine that each of the incremental updates could be done destructively, and thus the "compiled code" for these two programs would essentially be the same.

It is beyond the scope of this paper to discuss these optimization techniques any further, other than to make the following intuitive claim: If a sequential imperative program with mutable arrays is translated "faithfully" into a functional program with incremental updates to immutable arrays, then it is always possible to do the incremental updates destructively. This is because if the translation is indeed faithful, then it is impossible for there to be some other reference to the old array when the update is about to be performed.[6] For the remainder of this paper I will assume that the copy avoidance problem has been completely solved. I view the eventual acceptance of this fact by programmers to be philosophically no different from the acceptance of efficient procedure calls, and in particular the acceptance of a compiler's ability to turn tail-recursions into loops (note that tail-recursion optimization can be viewed as the reuse of activation records, much like reusing an array during an incremental update – the activation record in effect has a reference count of zero!).

[5] We will return to more of these syntactic games in Section 5.

[6] An attempt at formalizing this notion may be found in [6].

## 2.2 Monolithic Arrays

The main idea behind the monolithic approach can be captured by the function *mka* (for "make-array"), defined as follows: $mka(n, f)$ returns an n-element array, say $a$, such that $a[i] = f(i)$.[7] Versions of *mka* can be found in several functional languages, including FEL [8], Alfl [3], and VAL [9].

In addition, many other variations of *mka* are possible, such as the function *mkga* (for "make-general-array"), defined such that $mkga(n, f)$ returns an n-element array, say $a$, such that $a[i] = x$ if there exists $j$ in the range $1, ..., n$ where $f(j) = [i, x]$; otherwise $a[i] = nil$.[8] This function can be conveniently used, for example, to solve the "matrix inversion problem" described by Wadler in [16]. Wadler also describes four other kinds of monolithic arrays, the most interesting being variations of *mkga* which are defined to do useful things when there exists more than one $j$ such that $f(j) = [i, x]$. Steele uses a very similar kind of array called a *xector* in Connection Machine Lisp [15].

Another interesting kind of monolithic array is Arvind, Pingali, and Nikhil's *I-structures* in Id-Nouveau [11]. I-structures are in some sense a *blend* of imperative and functional arrays, in that assignments are made to elements of an I-structure using a syntax not unlike assignment to an imperative array, yet it is an error to assign to the same location more than once, thus preserving the "single-assignment" semantics of a functional language. The primary deficiency of I-structures is that they are not referentially transparent. As an aside, the parallelism of I-structures (which Arvind calls "parallel-call-by-value" [1]) can be achieved using a *lazy* version of *mka* with an operational semantics that specifies the eager, parallel evaluation of the elements (such an approach is used, for example, in ParAlfl [4,7]).

## 2.3 The Generality of Incremental Arrays

The modularity and high degree of parallelism are nice features of monolithic arrays, and I have experimented with several versions of them in Alfl [3] and ParAlfl [4,7]. It is perhaps not surprising, however, that incremental arrays are in some sense more general. After all, incremental updates seem to induce the smallest possible change in behavior to an array. They can be used to define any of the monolithic arrays described in the previous section, as well as the ones described in [16]. To highlight this generality, imagine for a moment a situation in which we wish to update every element in some array of length $n$. Using the monolithic approach this will consume space proportional to $n$, but using the incremental approach (with copy avoidance optimization) it consumes *no* space. Now suppose we do this operation $k$ times (fairly common in scientific computing, for example). The monolithic approach consumes $O(kn)$ space, but the incremental approach + optimization still consumes *no* space. So the monolithic approach appears to be *less efficient* in this case than the incremental approach + optimization.

On the other hand, the monolithic approach is attractive since it seems to be better suited to parallelism. In particular, in $mka(n, f)$ one could evaluate every $f(i)$ in parallel, and store the results into the new array in parallel as well (meaning in any order). But

[7] *mka* is the same as *array1* used in [16].
[8] *mkga* is the same as *array2* used in [16].

note that *mka* can be defined easily in terms of incremental updates:

$$\begin{aligned} mka(n, f) \;\; = \;\; & \text{let } a = newarray(n) \\ & \text{in for } i = 1, ..., n \\ & \quad \text{next } a \text{ is } a[i] := f(i) \\ & \quad \text{return } a \end{aligned}$$

Furthermore, most parallel implementations of functional languages would dynamically "unwind" the loop, and all of the $f(i)$ would be evaluated in parallel, as discussed above. On the other hand, the actual array *updates* cannot be done in parallel, since the recursion specifies a particular order for them. Although this might not seem too severe, it might be for very large arrays. Fortunately, there is a way around even this problem, to be discussed later.

To conclude this section, I will describe a simple solution to the "histogram problem," which Arvind posed [1] as a problem not well-suited to I-structures.[9] My solution demonstrates both of the points made in the previous two paragraphs: incremental updates can solve the problem with minimal space consumption and with parallelism limited only by the sequential array updates imposed by recursion.

The histogram problem can be described as any situation in which a large amount of data is being accumulated into $n$ "bins," where the final heights of the bins represent the histogram of some observed phenomenon. We can represent the bins collectively as an array $a$ of length $n$, and we can describe the updates that we wish to induce as $a[f(i)] := g(a[f(i)])$ for $i$ from 1 to $k$, $k$ being much larger than $n$, and where the range of $f$ is $1, ..., n$. Thus, for example, if we're just tallying events of some sort, $g$ might simply add one to its first argument. A sequential solution using incremental arrays is very simple:

$$\begin{aligned} & \text{let } a = newarray(n) \\ & \text{in for } i = 1, ..., k \\ & \quad \text{next } a \text{ is } a[f(i)] := g(a[f(i)]) \\ & \quad \text{return } a \end{aligned}$$

It is easy to determine in this example that $a$ is single-threaded (assuming $g$ is single-threaded), and thus this solution consumes $O(n)$ space. Furthermore, there is still potential for a large amount of parallelism, in that each of the $f(i)$ could be evaluated concurrently (in fact, let's assume that the execution of $f(i)$ corresponds to the "event" being counted). On the other hand, note as in the previous example that the updates are done in the particular order specified by the recursion, thus limiting to some extent the degree of parallelism, or at least the degree of asynchrony. This latter deficiency will be removed in a later section.

# 3 An Introduction to "Functional Non-Determinism"

Most programmers (including the very idealistic among us) admit the need for non-determinism, despite the semantic difficulties that it introduces. It seems to be an essential ingredient of "real" systems, such as operating systems and real-time controllers. In imperative languages non-determinism is typically manifested by running in parallel several

[9] It is an interesting social phenomenon whereby researchers jump on an "open problem" like vultures on an open carcass – Wadler provides a different solution to the histogram problem in [16].

processes that are side-effecting some global state – the non-determinism is thus *implicit* in the semantics of the language. In functional languages non-determinism is manifested through the use of primitive operators such as *amb* or *merge* – the non-determinism is thus made *explicit.*[10] Quite a few papers have been published on the use of such primitives in functional programming, and it appears quite reasonable to program conventional non-deterministic applications using them. In this section a few non-deterministic operators are introduced that are siufficient for our purposes. I will refer to them as *pseudo-functions*, since I wish to use them in the same way that normal functions are used, but their non-deterministic behavior makes the word "function" a misnomer.

By way of introduction, McCarthy defined a binary pseudo-function called *amb* having the following behavior:

$$\begin{aligned} amb(e_1, \bot) &= e_1 \\ amb(\bot, e_2) &= e_2 \\ amb(e_1, e_2) &= \text{either } e_1 \text{ or } e_2, \text{ chosen non-deterministically} \end{aligned}$$

The "operational reading" of $amb(e_1, e_2)$ is that $e_1$ and $e_2$ are evaluated in parallel, and the one that completes first is returned as the value of the expression. Using *amb* one can define things such as *merge* that non-deterministically merge two lists, or *streams.* The pseudo-function *merge* is useful, for example, in combining streams of characters from different computer terminals.

Instead of using *amb*, however, I will define a slightly different primitive, called *choose_one*, that takes a non-empty finite list and returns a pair consisting of a non-deterministically chosen element of the list, and the list with the one element removed:

$$\begin{aligned} choose\_one([e]) &= [e, []] \\ choose\_one([e_1, ..., e_n]) &= [e_i, [e_1, ..., e_{i-1}, e_{i+1}, ..., e_n]] \\ &\quad \text{where } e_i \neq \bot \text{ is chosen non-deterministically} \end{aligned}$$

As with *amb*, *choose_one* only chooses proper (i.e., non-$\bot$) elements of the list to return as the first element in the pair. In fact, *amb* can be implemented in terms of *choose_one* very easily:

$$amb(e_1, e_2) = hd(choose\_one([e_1, e_2]))$$

However, it is *not* so easy to implement *choose_one* in terms of *amb*, for two reasons: First, doing so requires *nesting* of *amb*, thus raising questions of fairness (which I would like not to deal with in this paper). Second, and perhaps more serious, it is not clear how to specify the removal of the chosen element from the list, since doing so requires testing the elements, some of which may be $\bot$.

Now let us define a pseudo-function called *apply_all* whose behavior is captured by:

$$apply\_all(a, [f_1, ..., f_n]) = f_{k_1}(f_{k_2}(...(f_{k_n}(a))))$$
$$\text{where } k_1, ..., k_n \text{ is an arbitrary permutation of } 1, ..., n.$$

Although *apply_all* can be defined quite simply in terms of *choose_one*:

$$\begin{aligned} apply\_all(a, []) &= a \\ apply\_all(a, fs) &= \text{let } [f, rest] = choose\_one(fs) \\ &\quad \text{in } apply\_all(f(a), rest) \end{aligned}$$

I prefer to think of it as primitive, for reasons to be disclosed later.

---

[10]To me it is *advantageous* to make the non-determinism (and its associated insidious properties) explicit, but others may disagree. However, I will not argue this "religious" point in this otherwise objective paper.

# 4 Parallel Array Operations

In this section a series of typical parallel array operations are considered, and "pseudo-functional" solutions are given for each.

## 4.1 Simple Parallel Updates

Recall the canonical problem of updating every element in an array just once. Both an imperative and a functional program for doing this were given in Section 2.1, and I argued that the functional program, with a suitable optimization using either run-time or compile-time reference counting, could be made to run as efficiently as the imperative program. If these two programs were run on a parallel system, pretty much the same behavior would result, since the inherent structure of the algorithm is strictly sequential.

But now suppose we wish to run this program in such a way that every update is done in parallel. The use of a monolithic array is one solution, although this creates an entirely new array. On the other hand, in an imperative language we might write something like:

$$\text{forall } i = 1, ..., n$$
$$a[i] := 0;$$

where I assume that "forall" is a parallel construct that creates, in this case, $n$ copies of its body and executes them in parallel. This program does not allocate new space, and thus is more efficient than the monolithic approach. How can we achieve such behavior in a functional program?

To answer this question we must ask what it is this program is really "saying" in a declarative sense. The answer is that it is defining a "new" array whose elements are created *in any order*; i.e., non-deterministically. This theme often repeats itself: parallel behavior involving side-effects in an imperative language can often only be expressed using non-determinism in a functional language. A variation of this theme, mentioned in the Introduction, is that *implicit* non-determinsm in an imperative language must be made *explicit* in a functional language.

With this hint, and the previous discussion about non-determinism, the most naive solution would be to non-deterministically choose one execution order from all possible execution orders in the above program. For example, for $n = 3$, we could write:

$$\begin{aligned} \textit{choose\_one}(\ [\ &((a[1] := 0)[2] := 0)[3] := 0,\\ &((a[1] := 0)[3] := 0)[2] := 0,\\ &((a[2] := 0)[1] := 0)[3] := 0,\\ &((a[2] := 0)[3] := 0)[1] := 0,\\ &((a[3] := 0)[1] := 0)[2] := 0,\\ &((a[3] := 0)[2] := 0)[1] := 0\ ]\ ) \end{aligned}$$

It should be clear that this style of translation results in a program of size $n*n!$, whereas the original program's size is independent of $n$. This is clearly unsatisfactory. The observant reader will notice, however, that one could define a recursive function to generate, for any $n$, the above list of expressions given to *choose_one*, thus making the program size independent of $n$. Yet this is still unsatisfactory, since, for example, the list is still of length $n * n!$. A better solution is to use *apply_all*:

$$\begin{aligned} apply\_all(\ a,\ [\ & \lambda a.a[1] := 0, \\ & \lambda a.a[2] := 0, \\ & \lambda a.a[3] := 0\ ]\ ) \end{aligned}$$

Although this program grows linearly with $n$ (since the list is of length $n$), one could again write a recursive function to generate the list, which in all fairness simply corresponds to the tail-recursive interpretation of "forall" (in other words, if we allow ourselves functional syntax corresponding to "forall," then the programs would be the same size – such syntax in fact will be introduced in Section 5).

## 4.2 The Closure of *apply_all*

The use of *apply_all* results in a more intuitive translation of a set of parallel imperative statements than *choose_one*. Indeed, in general it would seem that we could replace something like:

$$\begin{aligned} &\text{parbegin} \\ &\quad a[i_1] := e_1; \\ &\quad a[i_2] := e_2; \\ &\quad \ldots \\ &\quad a[i_k] := e_k; \\ &\text{parend} \end{aligned}$$

in an imperative language, where $e_1$ through $e_k$ are arbitrary expressions, with:

$$\begin{aligned} apply\_all(\ a,\ [\ & \lambda a.a[i_1] := e_1, \\ & \lambda a.a[i_2] := e_2, \\ & \ldots \\ & \lambda a.a[i_k] := e_k\ ]\ ) \end{aligned}$$

in a functional language; note that the program size is identical in either case. Is this in fact a valid translation? The answer depends on $e_1$ through $e_k$: If each of these expressions makes no references to the array $a$, then yes, it is a valid translation. But what if each of them makes a single reference to some other element in $a$? By "parameterizing" this reference we can write such a program imperatively as:

$$\begin{aligned} &\text{parbegin} \\ &\quad a[i_1] := f_1(a[j_1]); \\ &\quad a[i_2] := f_2(a[j_2]); \\ &\quad \ldots \\ &\quad a[i_k] := f_k(a[j_k]); \\ &\text{parend} \end{aligned}$$

The conventional imperative interpretation of this program would allow arbitrary interleavings of the array references with the array updates, and thus the non-deterministic possibilities increase considerably! Unfortunately, the functional translation given earlier would ensure "atomic" behavior of each update, which is not what is desired. (We will return to this atomic kind of behavior shortly). How can we express this yet-higher degree of non-determinism in a functional language?

Here is a solution. First define *apply_all** by:

$$
\begin{array}{rcl}
apply\_all*(a,[\,]) & = & a \\
apply\_all*(a,lst) & = & \text{let } [f, rest] = choose\_one(lst), \\
 & & \qquad val = f(a) \\
 & & \text{in if } array?(val) \\
 & & \qquad \text{then } apply\_all*(val, rest) \\
 & & \qquad \text{else } apply\_all*(a, cons(val, rest))
\end{array}
$$

*apply_all** can be thought of as the "closure" of *apply_all* (thus the "*"), since it keeps applying the functions in the list until an array is yielded.

Returning to the example, the correct translation is then:

$$
\begin{array}{ll}
apply\_all*(\ a, & [\ \lambda a.\lambda a'.a'[i_1] := f_1(a[j_1]), \\
 & \quad \lambda a.\lambda a'.a'[i_2] := f_2(a[j_2]), \\
 & \quad \ldots \\
 & \quad \lambda a.\lambda a'.a'[i_k] := f_k(a[j_k])\ ]\ )
\end{array}
$$

In other words, each of the functions $\lambda a.\lambda a'.a'[i] := f(a[j])$ is "executed" twice; once to capture the value $a[j]$, and once to do the update to $a[i]$. After the first execution it is "thrown back into the pool" of things to choose from non-deterministically, and thus by the time the update happens the array may not be the same as it was when $a[j]$ was read. This is just the kind of non-determinism exhibited by the original program.

## 4.3 Parallel Atomic Updates

Now let us return briefly to the "atomic" behavior that was achieved "accidentally" in the original functional translation. To be fair, we should ask ourselves how one would achieve *that* kind of behavior in an *imperative* language (it is easy to conjure up situations where such behavior would be desired – indeed, the histrogram problem is one such example). Here is a typical solution using a semaphore:

```
sem := semaphore(k)
parbegin
   begin P(sem);
         a[i1] := f1(a[j1]);
         V(sem);
   end;
   begin P(sem);
         a[i2] := f2(a[j2]);
         V(sem);
   end;
   ...
   begin P(sem);
         a[ik] := fk(a[jk]);
         V(sem);
   end;
parend
```

Whereas in a functional language one would write:

$$\begin{aligned} apply\_all(\ a,\ [\ & \lambda a.a[i_1] := f_1(a[j_1]), \\ & \lambda a.a[i_2] := f_2(a[j_2]), \\ & \ldots \\ & \lambda a.a[i_k] := f_k(a[j_k])\ ]\ ) \end{aligned}$$

What can we conclude from all this? Simply put, the profundity-of-the-day: "In imperative languages one must work hard to *ensure* mutual exclusion, i.e., to *reduce* the degree of non-determinism; and in a functional language one must work hard to *defeat* mutual exclusion, i.e., to *increase* the degree of non-determinism."

## 4.4 Recursive Nesting of Parallel Array Operations

Consider this definition of a "divide-and-conquer" procedure in an imperative language for updating an entire array (where we assume the array $a$ is "global" to the definition):

$$\begin{aligned} init(lo, hi) = \ & \text{if } lo \geq hi \\ & \text{then } a[lo] := 0 \\ & \text{else begin } mid := (lo + hi)/2; \\ & \qquad\quad \text{parbegin} \\ & \qquad\quad\ \ init(lo, mid); \\ & \qquad\quad\ \ init(mid + 1, hi) \\ & \qquad\quad \text{parend} \\ & \quad\ \text{end}; \end{aligned}$$

which would then be used in a context such as: "$a := \ldots;\ init(1, n)$." Here is an equivalent solution in a functional language:

$$\begin{aligned} init(lo, hi) = \ & \text{if } lo \geq hi \\ & \text{then } [\lambda a.a[lo] := 0] \\ & \text{else let } mid = (lo + hi)/2 \\ & \qquad \text{in } append(\ init(lo, mid), \\ & \qquad\qquad\qquad\quad init(mid + 1, hi)) \end{aligned}$$

which would then be called via $apply\_all(a, init(1, n))$. With such a technique it is possible to define parallel, functional, "destructive" versions of things like Hoare's original formulation of Quicksort.

# 5 It's All Syntax

Now for the fun of it let's play the time-honored game of *syntactic sugaring.*[11] The philosophy behind this game-playing is that if there is a particular pattern of use of non-determinism and arrays that we like, why not provide special syntax for it? For sake of argument let us continue the trend set in the previous sections, and assume that the traditional imperative style of programming with arrays is one such desirable pattern.[12]

[11] We have already done this to some extent with syntax for *update* and tail-recursion.

[12] This may be a rather strong assumption, but it makes for good exercise in any case.

Before proceeding, note that if $apply\_all(a, lst)$ is a proper use of $apply\_all$ (i.e., the elements of $lst$ are all of form $\lambda a.exp$, where $exp$ evaluates to an array), then $apply\_all(a, lst)$ is equivalent to $apply\_all*(a, lst)$. Thus we will concentrate only on syntax for $apply\_all*$. But before doing that, we need syntax for lambda expressions.

Recall that we defined "$a[i] := exp$" as syntax for "$update(a, i, exp)$." Another popular syntax for assignment is "$a[i] \leftarrow exp$," and thus we will use that as syntax for "$\lambda a.\lambda a'.a'[i] := exp$." (Note that the original assignment to $a$ is changed to $a'$, but that $exp$ remains the same, and may or may not have references to $a$.) Furthermore, let "$\langle a[i] \leftarrow exp\rangle$" be syntax for "$\lambda a.a[i] := exp$." This will be used to achieve the "atomic" behavior discussed in Section 4.3; in this sense the brackets "$\langle ...\rangle$" can be viewed as syntax for a lock, and can be read "do this atomically."

Next define the following three syntactic sugarcubes:

1. Let "f_parbegin $s_1$; $s_2$; ...; $s_n$ f_parend" be syntax for:

$$\begin{array}{rcl} & & append1(s_1, append1(s_2, ...append1(s_{n-1}, s_n))) \\ \text{where } append1(s_1, s_2) & = & append(listify(s_1), listify(s_2)) \\ listify(s) & = & \text{if } list?(s) \text{ then } s \text{ else } [s] \end{array}$$

2. Let "with $a$ do $s$" be syntax for "$apply\_all*(a, s)$."

3. Just as in Section 2.1 syntax was defined for a for-loop, I will now define syntax for a forall-loop. Let:

$$\begin{array}{l} \text{forall } i = 1, ..., n \\ \text{do } exp(i) \end{array}$$

be syntax for:

$$\begin{array}{l} \text{let } loop(lo, hi, acc) = \text{ if } lo > hi \text{ then } acc \\ \qquad\qquad\qquad\qquad\qquad\quad \text{else } loop(lo+1, hi, cons(exp(i), acc)) \\ \text{in } loop(1, n, []) \end{array}$$

The reader should be cautioned that this syntax is (intentionally) very misleading. In particular, "f_parbegin ... f_parend" is really not "executing" anything – it is just packaging together a lot of lambda expressions into a list.

We are now in a position to try on our new clothes. Consider first the histogram problem from Section 2.3; a sugared, pseudo-functional solution is:

$$\begin{array}{l} \text{with } a \text{ do} \\ \text{forall } i = 1, ..., n \\ \text{do } \langle\ a[f(i)] \leftarrow g(a[f(i)])\ \rangle \end{array}$$

This is an improvement over the functional solution given in Section 2.3, in that the updates can now be done in any order (and therefore presumably in parallel). Note that the updates are still done *atomically*, which is necessary in order to achieve the correct histogram.

Now consider the sample programs taken from the previous section; the reader is urged to compare the following sugared, pseudo-functional solutions to the imperative ones given previously. From Section 4.1:

$$\begin{array}{l} \text{with } a \text{ do} \\ \text{f\_parbegin } a[1] \leftarrow 0; \\ \qquad\qquad\quad a[2] \leftarrow 0; \\ \qquad\qquad\quad a[3] \leftarrow 0 \\ \text{f\_parend} \end{array}$$

or, better yet:

$$\begin{array}{l} \text{with } a \text{ do} \\ \text{forall } i = 1, \ldots, 3 \\ \text{do } a[i] \leftarrow 0 \end{array}$$

From Section 4.2:

$$\begin{array}{l} \text{with } a \text{ do} \\ \text{f\_parbegin } a[i_1] \leftarrow f_1(a[j_1]); \\ \qquad\qquad\quad a[i_2] \leftarrow f_2(a[j_2]); \\ \qquad\qquad\quad \ldots \\ \qquad\qquad\quad a[i_k] \leftarrow f_k(a[j_k]) \\ \text{f\_parend} \end{array}$$

From Section 4.3:

$$\begin{array}{l} \text{with } a \text{ do} \\ \text{f\_parbegin } \langle\ a[i_1] \leftarrow f_1(a[j_1])\ \rangle; \\ \qquad\qquad\quad \langle\ a[i_2] \leftarrow f_2(a[j_2])\ \rangle; \\ \qquad\qquad\quad \ldots \\ \qquad\qquad\quad \langle\ a[i_k] \leftarrow f_k(a[j_k])\ \rangle \\ \text{f\_parend} \end{array}$$

Finally, from Section 4.4:

$$\begin{array}{ll} \text{with } a \text{ do } init(1, n) & \\ \text{where } init(lo, hi) = & \text{if } lo \geq hi \\ & \text{then } a[lo] \leftarrow 0 \\ & \text{else let } mid = (lo + hi)/2 \\ & \qquad \text{in f\_parbegin} \\ & \qquad\qquad init(lo, mid); \\ & \qquad\qquad init(mid + 1, hi) \\ & \qquad\quad \text{f\_parend} \end{array}$$

What can be said about all of these syntactic games? At least two things:

- "Imperative folk" might like the results, since it looks more like conventional assignment statements and explicit parallelism in an imperative language.
- "Functional folk" might like the results, since it imposes constraints on, or encourages a discipline for, the use of assignment-like operations.

How far can one go with these games? Note, for example, that I have given no translation strategy that accounts for an *arbitrary* number of references to an *arbitrary* number of different arrays. To do this, and to deal with other insidious behaviors of imperative languages (such as side-effects in the presence of aliasing), one would have to resort to machinery such as used in denotational semantics to capture the notion of store, continuation, etc. I am certain that this can be done, and may be useful in capturing a "pseudo-functional semantics" for a parallel imperative language, but that is not my purpose here – I am just trying to make apparent the range of syntactic possibilities. On the other hand, although I don't claim to know what the *right* choice of syntax is, I *am* suggesting that syntax such as presented here may be a better alternative than either the completely unconstrained and undisciplined possibilities in an imperative language, or the somewhat over-constrained and over-disciplined equivalances in a functional language.

# 6 Efficiency

In this section I wish to argue that all of the pseudo-functional examples given in the last section can be done as efficiently as their imperative counterparts. My argument will only be "partial" in the sense that the above claim is only for pseudo-functional programs that are designed to "mimic" a particular parallel imperative program – a more general claim is difficult to make. This is analogous to arguing that tail-recursion can be implemented as efficiently as an imperative loop, in that such a claim cannot be made about recursion in general.

For the remainder of the discussion I will make the reasonable assumption that assignment to an array location is atomic.

**Proposition 1:** The expression:

$$\begin{aligned} apply\_all(\ a,\ [\ & \lambda a.a[i_1] := e_1, \\ & \lambda a.a[i_2] := e_2, \\ & \ldots \\ & \lambda a.a[i_k] := e_k\ ]\ ) \end{aligned}$$

with the constraint that each $e_j$ and $i_j$, $j = 1, ..., k$, makes no reference to $a$, may be evaluated by executing all of the updates destructively and in parallel.

**Proof:** (Informal) It should be clear that each lambda expression is single-threaded. Further, since each $e_j$ and $i_j$ makes no reference to $a$, then an atomic assignment to an array location is sufficient to ensure the correct non-deterministic behavior.

**Proposition 2:** The expression:

$$\begin{aligned} apply\_all{*}(\ a,\ [\ & \lambda a.\lambda a'.a'[i_1] := e_1, \\ & \lambda a.\lambda a'.a'[i_2] := e_2, \\ & \ldots \\ & \lambda a.\lambda a'.a'[i_k] := e_k\ ]\ ) \end{aligned}$$

with the constraint that each $i_j$ makes no references to $a$ and each $e_j$ makes at most *one* reference to $a$, may be evaluated by executing all of the updates destructively and in parallel. In particular, each single-reference may be implemented by reading the array element asynchronously at the time it is needed during the evaluation of $e_i$.

**Proof:** Essentially the same as for Proposition 1.

Note that the "firing" of either an array reference or array update at the time it is needed is analogous to the operational reading of $amb(e_1, e_2)$ that says the expression that completes first is the one that is chosen.

Implementing any of the "atomic" kinds of updates requires either a locking mechanism or, interestingly, just a way to make the lambda expressions *compose* themselves (pardon the pun) as they are non-deterministically chosen. Thus conventional graph-reduction techniques can be used to ensure the mutual exclusion.

The observant reader will note that there is a hidden source of inefficiency in the pseudo-functional programs as given: namely, the space required of the *lists* that hold the lambda expressions in the argument to *apply_all* or *apply_all**. In general this is a real cost (especially considering the amount of list appending that might occur in a recursive application). However, if the "disciplined" use of the pseudo-functions as captured by the syntax in Section 5 is used, then it should be clear that the lists could be eliminated altogether, since they are being replaced by whatever it is that coordinates the parallel execution of the updates (which must exist in any case).

# 7 Conclusions

I have presented an argument for the use of incremental arrays plus non-determinism to achieve the effect of parallel assignment to imperative arrays. Furthermore, I have suggested a possible syntax for expressing such behavior that falls somewhere in between the purely functional and (im)purely imperative. The strategy depends crucially on the ability of a compiler to recognize not only that the incremental updates can be done destructively, but also that the updates may be done in any order.

Much work remains. The syntax presented in Section 5 is interesting in its similarity to imperative language syntax, but far too similar for me to feel comfortable with. For example, the syntax for lambda expressions completely destroys referential transparency, exactly the criticism made earlier about I-structures. Of course, non-determinism in general causes this problem, but surely there are better alternatives to the syntax given here. Finally, none of the described techniques have been implemented. Without doing so, and gaining experience writing and debugging programs in this way, we can never be sure how good the ideas are.

# 8 Acknowledgements

I wish to thank Los Alamos National Laboratory and MCC for their generous support of the Santa Fe Graph Reduction Workshop, which inspired many of the ideas presented here. A special thanks is extended to Joe Fasel and Bob Keller for their organizational efforts. Also thanks to Adrienne Bloss for commenting on an earlier draft of this manuscript, and to the rest of my Wrestling Team at Yale for the inspiration induced by their critical scrutiny of every new idea.

# References

[1] Arvind. Data structures for parallel computing. October 1986.

[2] B. Goldberg. *Multiprocessor Execution of Functional Programs.* PhD thesis, Yale University, expected Spring 1987.

[3] P. Hudak. *ALFL Reference Manual and Programmer's Guide.* Research Report YALEU/DCS/RR-322, Second Edition, Yale University, October 1984.

[4] P. Hudak. Para-functional programming. *Computer*, 19(8):60–71, August 1986.

[5] P. Hudak. A semantic model of reference counting and its abstraction (detailed summary). In *Proc. 1986 ACM Conf. on LISP and Functional Prog.*, pages 351–363, ACM, August 1986.

[6] P. Hudak and A. Bloss. The aggregate update problem in functional programming systems. In *12th ACM Sym. on Prin. of Prog. Lang.*, pages 300–314, ACM, 1985.

[7] P. Hudak and L. Smith. Para-functional programming: a paradigm for programming multiprocessor systems. In *12th ACM Sym. on Prin. of Prog. Lang.*, pages 243–254, January 1986.

[8] R.M. Keller. *FEL programmer's guide.* AMPS TR 7, University of Utah, March 1982.

[9] J.R. Mcgraw. The val language: description and analysis. *TOPLAS*, 4(1):44–82, January 1982.

[10] A. Mycroft. *Abstract Interpretation and Optimizing Transformations for Applicative Programs.* PhD thesis, Univ. of Edinburgh, 1981.

[11] R.S. Nikhil, K. Pingali, and Arvind. *Id Nouveau.* Computation Structures Group Memo 265, Massachusetts Institute of Technology, Laboratory for Computer Science, July 1986.

[12] J.T. O'Donnell. An architecture that efficiently updates associative aggregates in applicative programming language. In *Functional Programming Languages and Computer Architecture*, pages 164–189, Springer-Verlag LNCS 201, September 1985.

[13] D.A. Schmidt. Detecting global variables in denotational specifications. *ACM Trans. on Prog. Lang. and Systems*, 7(2):299–310, 1985.

[14] J. Schwarz. Verifying the safe use of destructive operations in applicative programs. In B. Robinet, editor, *Program Transformations – Proc. of the 3rd Int'l Sym. on Programming*, pages 395–411, Dunod Informatique, 1978.

[15] Jr. Steele, Guy L. and W. Daniel Hillis. Connection machine lisp: fine-grained parallel symbolic processing. In *Proc. 1986 ACM Conference on Lisp and Functional Programming*, pages 279–297, ACM SIGPLAN/SIGACT/SIGART, Cambridge, Massachusetts, August 1986.

[16] P. Wadler. A new array operation. In *Workshop on Graph Reduction Techniques*, Springer-Verlag LNCS, this volume, October 1986.

[17] D.S. Wise. Design for a multiprocessing heap with on-board reference counting. In *Functional Programming Languages and Computer Architecture*, pages 289–304, Springer-Verlag LNCS 201, September 1985.

# A New Array Operation

Philip Wadler

Programming Research Group, Oxford University, UK

and Programming Methodology Group, Chalmers University, Göteborg, Sweden

Providing efficient operations on arrays in a functional language is problematic. No completely satisfactory solution has yet been found. This paper proposes a new solution, which is a variant on the "monolothic" approach to array operations. The new solution is also not completely satisfactory, but does have advantages complementary to other proposals. It makes a class of programs easy to express, notably those involving the construction of histograms. It also allows for parallel implementations without the need to introduce non-determinism.

The work reported here was motivated by discussions at the Workshop on Graph Reduction, of which this is the proceedings. In particular, it was motivated by the contributions of Arvind and Paul Hudak (see this volume), and especially Arvind's observation that the histogram problem is difficult to solve in a satisfactory way.

After the first draft of this paper was written, I discovered that Simon Peyton-Jones independently suggested the same idea, also prompted by one of Arvind's talks. After the second draft was written, I discovered that Guy Steele and Daniel Hillis use a very similar notion in Connection Machine Lisp [SH86]. Apparently the simple innovation described in this paper is an idea whose time has come.

This paper is organized as follows. Section 1 briefly surveys previously proposed array operations. Section 2 introduces the new operation. Section 3 gives a small example of its use. Section 4 discusses two variations on the array operation. Section 5 concludes.

## 1 Background

Most of the proposed methods for incorporating arrays into functional languages fall into one of two groups. These might be called the *incremental* and *monolithic* approaches.

In the incremental approach, the basic operation has the form

$$update\ A\ i\ x$$

and returns a new array that is identical to $A$ except that the position with index $i$ has value $x$. (Paul Hudak and John Hughes, independently, have suggested the alternate notation $A[i] := x$ for the term above.) A naive implementation of *update* would copy the entire array, but this is unacceptably expensive. Usually, we expect that each *update* operation is passed the only existing pointer to the array $A$, so that the array may be updated destructively. David Schmidt calls programs satisfying this condition *single threaded* [Sch85].

Several ways exist to take advantage of single-threading at run-time. If reference counts are used, $A$ may be destructively updated whenever its reference count is one. Sören Holmström has proposed a simple but ingenious run-time technique that does not require reference counts [Hol83]. John Hughes has suggested a variation of Holmström's method, which may be valuable in some cases where single-threading does not apply [Hug85]. Difficulties with adapting Holmström's and Hughes's methods to a lazy language are discussed by William Stoye [Sto85].

Methods of detecting single-threading at compile-time have also been suggested. Schmidt has given syntactic constraints that guarantee it [Sch85]; Hudak has described analysis methods that discover it [HB85,Hud86]; and Holmström has proposed a type-system that enforces it [Hol86]. The compile-time methods proposed to date (so far as I know) all assume a strict language, while the run-time methods work with either a lazy or a strict language.

In the monolithic approach, the basic operation has the form

$$array1\ n\ f$$

and returns a new array $A$ such that $A[i] = f\ i$ for each $i$ in the range from 1 to $n$. This may be generalized slightly, to give the operation

$$array2\ n\ g$$

that returns a new array $A$ such that $A[i] = x$ where $(i, x) = g\ j$ for some $j$ in the range 1 to $n$. It is assumed that each possible value of $i$ will appear in a pair exactly once; that is, if $j_1 \neq j_2$ and $(i_1, x_1) = g\ j_1$ and $(i_2, x_2) = g\ j_2$ then $i_1 \neq i_2$.

For example, the *permutation inversion* problem is as follows. Given an array $A$ that defines a permutation on 1 to $n$, we wish to find an array $B$ such that $B[A[i]] = i$ for each $i$ from 1 to $n$. Setting

$$\begin{aligned} B &= array2\ n\ g \\ \textbf{where}\ g\ j &= (A[j], j) \end{aligned}$$

solves the problem.

The incremental and monolithic approaches have complementary advantages. It is easy to see how to define the *array1* and *array2* operations in terms of *update*. Thus, the monolithic approach is less general than the incremental approach. On the other hand, the incremental approach requires that the programmer be careful to maintain single threading and that the implementation be careful to exploit it. The monolithic approach is therefore easier to implement and use, where it is applicable. More importantly, the monolithic approach is also well suited for use in parallel computation, since all $n$ applications of $f$ or $g$ can be computed in parallel. Using parallelism with the incremental approach poses more problems, as analysis is required to discover which updates can be performed in parallel.

In a call of *array2 n g*, the only purpose of $g$ is to yield the $n$ pairs $(i_1, x_1), \ldots, (i_n, x_n)$, where $(i_j, x_j) = g\ j$. It would be just as easy to pass an association list of $(i, x)$ pairs. Thus, in place of *array2* we can use the operation

$$array3\ n\ ixs$$

where $ixs$ is a list of length $n$ containing pairs of the form $(i, x)$. For example, the permutation inverse problem can be solved by setting

$$B = array3\ n\ [(A[j], j) \mid j \leftarrow [1\,..\,n]]$$

(this uses list comprehension notation, as found in KRC [Tur81] and Miranda [Tur85]). Recall that $g$ was restricted so that each index appeared in exactly one pair; we can express an equivalent restriction on $ixs$ by insisting that if $(i_1, x_1)$ and $(i_2, x_2)$ are in $ixs$ then $i_1 \neq i_2$.

Does this new formulation hamper parallelism? Not necessarily. The function form obviously has sufficient parallelism, since each of $g\ 1, \ldots, g\ n$ can be evaluated in parallel. In the list form, we need to form the list $[g\ 1, \ldots, g\ n]$. If the computation of $g$ is relatively expensive compared to the cost of counting from 1 to $n$, then this should provide sufficient parallelism. Usually, this will be the case. If it is not, then to achieve sufficient parallelism we would have to use trees in place of lists; this is described in Section 4.

The operations described here have been independently proposed by several people. Bob Keller is one of the people who suggested *array1* [Kel80,Kel82], Paul Hudak is one of the people whose suggested *array2* [Hud84], and David Turner is one of the people who suggested *array3* [Tur86].

Another important approach to array operations uses "I-structures", developed by Arvind and others [ANP86]. The I-structure approach resembles the monolothic approach in that it is more restricted than the incremental approach, but better suited for parallelism.

A completely different approach to the array problem is to forego arrays altogether, on the grounds that tree structures may be more suitable for parallel architectures. This theme has been developed by David Wise [Wis85,Wis86].

The above survey is brief and incomplete. Many proposals for operations on arrays have circulated for a long time without attribution, simply as part of the folklore. I apologize to those whose names and work I have unintentionally omitted.

## 2 A new array operation

This section presents a new array operation, belonging to the monolithic class. The operation will be derived in several steps by modifying the *array3* operation.

In a lazy functional language, one may choose whether array operations should be strict or lazy. For this paper, we will choose to make the operations strict. The main reason for this choice is that it makes it easier to reason about the time and space complexity of programs that use arrays. Some difficulties in the interactions between arrays and lazy evaluation are discussed in William Stoye's dissertation [Sto85].

We begin by relaxing the assumption in *array3* that each possible index is specified exactly once. That is, we will allow several elements with the same index to appear in the association list that pairs indexes with values. For example, the pairs

$$(i, x_1), \ldots, (i, x_m)$$

which all have the same index $i$, might appear in the association list $ixs$ (although not necessarily adjacent to each other). The question now arises, what value should we give $A[i]$? An obvious choice is to use the last value with index $i$, in this case $x_m$. A more general choice would be to combine all of the values with index $i$ to get the value of $A[i]$. Say the combining function is called $h$, then we should set $A[i] = h\ [x_1, \ldots, x_m]$.

More precisely, we define the operation

$$\mathit{array4}\ h\ n\ ixs$$

to return an array $A$ such that

$$A[i] \quad = \quad h\ [x \mid (i, x) \leftarrow ixs]$$

for each $i$ from 1 to $n$. (Pairs in $ixs$ with an index outside the range 1 to $n$ are ignored. Alternatively, once could require that if any such pairs are present then an error value is returned instead of an array.)

A particularly interesting case arises if $h$ is defined using the familiar *fold* function. This function, also sometimes called *accumulate* or *reduce*, is defined as follows:

$$\begin{aligned} \mathit{fold}\ f\ a\ [\,] &= a \\ \mathit{fold}\ f\ a\ (x : xs) &= \mathit{fold}\ f\ (f\ a\ x)\ xs \end{aligned}$$

For example,

$$sum \;=\; fold\,(+)\,0$$

defines the function *sum* that computes the sum of a list.

Because we have chosen to use strict operations on arrays, it is convenient to use a version of *fold* that is strict in its second argument:

$$\begin{aligned} strictfold\ f\ a \;&=\; strict\ g\ a \\ &\quad \textbf{where} \\ &\quad g\,b\,[\,] \;=\; b \\ &\quad g\,b\,(x : xs) \;=\; strict\ g\,(f\ b\ x)\ xs \end{aligned}$$

Here *strict* is defined so that *strict f x* first evaluates $x$ and then applies $f$. It is easy to prove that if $f$ is strict in its first argument (that is, $f \perp x = \perp$), then $f = strict\,f$ and $fold\,f\ a = strictfold\,f\ a$. Note that *strictfold* can be computed quite efficiently, since $f\,b\,x$ can be computed directly without forming a closure.

So we define

$$array5\ f\ a\ n\ ixs$$

to be equivalent to

$$array4\ (strictfold\ f\ a)\ n\ ixs$$

This function can be computed in an efficient manner, as follows. First, an array $A$ of size $n$ is allocated, and each element of $A$ is initialized to the value $a$. Then, each element of the list *ixs* is processed in turn. If the element is $(i, x)$, then if $A[i]$ currently contains the value $b$ it is replaced by $f\ b\ x$. (In imperative terms: $A[i] := f\ (A[i])\ x$.)

Note that the operation of setting $A[i]$ to the last value in *ixs* with index $i$ can be derived as a special case. We simply compute

$$\begin{aligned} &array5\ right\ a\ n\ ixs \\ &\textbf{where}\ right\ x\ y \;=\; y \end{aligned}$$

Elements of the array that are unspecified by *ixs* will be set to the initial value $a$.

However, it is the general case that is of more interest. For example,

$$array5\ (+)\ 0\ n\ ixs$$

is useful for accumulating the list of pairs *ixs* into a histogram. Finding histograms of this sort is quite an important operation. However, neither the monolithic operations suggested by Hudak nor the I-structures suggested by Arvind can compute such histograms efficiently. Indeed, as mentioned previously, this paper was motivated by Arvind's observation that computing histograms is an important problem not addressed by I-structures.

Further, there is even more scope for parallelism when $f$ is a an associative and commutative operator, as in the case of $(+)$. It is easy to see that for such operators, the order in which the operations occur is irrelevant. Hence, in such situations all elements of the list *ixs* may be evaluated in parallel. We therefore define our final operation

$$array6\ f\ a\ n\ ixs$$

which is identical to *array5* except that $f$ is assumed to be associative and commutative. This new operation yields wide scope for parallelism. Indeed, this method yields the maximum amount of parallelism possible for this sort of computation. (However, computation of the list *ixs* may introduce unnecessary sequencing; ways to avoid this are discussed in Section 4.)

# 3 The plotting problem

As a more detailed example, somewhat similar to histograms, consider the *plotting problem.* We are given a point $(x_{max}, y_{max})$ and a list of points $xys$, such that every point $(x, y)$ in $xys$ has $0 \leq x < x_{max}$ and $0 \leq y < y_{max}$. We wish to return a two-dimensional array $A$, such that $A[x, y]$ contains *True* if $(x, y)$ is in $xys$ and *False* otherwise.

This requires that we choose a representation for two-dimensional arrays. We will pick the traditional row-major representation. The two-dimensional array $A$ with dimensions $x_{max}$ by $y_{max}$ will be represented by a one-dimensional array $B$ with dimension $x_{max} \times y_{max}$, and the value of $A[x, y]$ will be found in $B[(x_{max} \times y) + x + 1]$.

We first define an analogue of *array6* for two-dimensional arrays:

$$\begin{aligned}&\mathit{array6in2d}\ f\ a\ (x_{max}, y_{max})\ xyzs\\ &= \ \mathit{array6}\ f\ a\ (x_{max} \times y_{max})[((x_{max} \times y) + x + 1, z) \mid ((x, y), z) \leftarrow xyzs]\end{aligned}$$

The solution is then given by

$$\begin{aligned}&\mathit{plot}\ (x_{max}, y_{max})\ xys\\ &= \ \mathit{array6in2d}\ \mathit{or}\ \mathit{False}\ (x_{max}, y_{max})\ [((x, y), \mathit{True}) \mid (x, y) \leftarrow xys]\end{aligned}$$

Note that *or* is associative and commutative, and thus satisfies the requirements of an argument to *array6*. Also note that passing *False* as the second argument neatly achieves the desired effect of setting an array element to *False* if it is not specified by the list $xys$.

Another way to represent a two-dimensional array is by an array of arrays. The bad news is that such a representation does not work well with the array operations suggested here. The good news is that this is not a severe problem, as the row-major representation is quite adequate.

# 4 Alternative representations

In the above, finite collections of $(i, x)$ pairs are represented by lists. Lists have the advantage that they are conceptually simple, and that most functional languages contain features that make them easy to manipulate. Further, if the list we are forming is, say,

$$[g\ j \mid j \leftarrow [1\,..\,n]]$$

and the function $g$ is relatively expensive to compute, then little potential parallelism is lost by representing collections of pairs in this way. On the other hand, if $g$ is relatively inexpensive, then it would be better to use a different representation for the collection.

One alternative representation would be to represent the list above by $g$ and $n$, thus returning to the functional representation used by *array1* and *array2*. Since $g$ can be applied in parallel to all the values from 1 to $n$, this offers maximum parallelism. Although adequate for most purposes, this representation may be cumbersome for some applications.

Another possibility is to use a representation based on trees. Several researchers have suggested that lists be represented using a tree data type [Kel80,SH82,Mee84]. This type could be declared

$$\mathit{tree}\ \alpha \ ::= \ \mathit{Empty} \mid \mathit{Unit}\ \alpha \mid \mathit{Join}\ (\mathit{tree}\ \alpha)\ (\mathit{tree}\ \alpha)$$

(this uses the notation of Miranda). Here *Empty* denotes the empty list, *Unit* $x$ denotes the list containing the single element $x$, and *Join* $xs$ $ys$ denotes the result of appending the lists $xs$ and $ys$. Clearly, such trees can be generated in a divide-and-conquer manner, and thus provide sufficient parallelism.

The *fold* operation is defined over such trees as follows:

$$\begin{array}{lcl} \mathit{fold}\ f\ a\ \mathit{Empty} & = & a \\ \mathit{fold}\ f\ a\ (\mathit{Unit}\ x) & = & x \\ \mathit{fold}\ f\ a\ (\mathit{Join}\ xs\ ys) & = & f\ (\mathit{fold}\ f\ a\ xs)\ (\mathit{fold}\ f\ a\ ys) \end{array}$$

Meertens points out that this definition is only sensible if $f$ is an associative operation with unit $a$, and that in this case *fold* corresponds to a homomorphism from lists onto the range of the fold. Meertens also notes that if in addition $f$ is commutative, then this is equivalent to making the *Join* operation commutative, so that the trees represent bags rather than lists. Note that this fits neatly with the use of *fold* in ***array5*** and ***array6***. For more on folding and homomorphisms, see Meertens [Mee84] or Bird [Bir86].

# 5 Conclusions

A new monolithic array operation has been proposed. The operation is similar to earlier proposals, except that it uses the *fold* operation to combine multiple values with the same index. The new operation conveniently handles certain important kinds of computation, such as histogram, and provides parallelism in a convenient way.

This method solves the important problem of calculating histograms in a fashion suitable for parallel computation. I-structures, devised by Arvind and his colleagues, sacrifice referential transparency in order to increase the opportunities for parallelism, but do not provide a good solution to the histogram problem. I am grateful to Arvind for pointing out this defect of I-structures, which in turn inspired this paper.

Although the method presented here is useful for an important class of problems, it is not the best solution for all problems involving arrays. Further experience is needed to assess its advantages and shortcomings.

**Acknowledgements.** I am grateful to Joseph Fasel (of Los Alamos National Laboratory). His special attention made it possible for me to attend the Workshop on Graph Reduction, and his hard work was largely responsible for the workshop's success.

I thank Paul Hudak, Bob Keller, Simon Peyton-Jones, David Turner, and David Wise for commenting on an earlier draft of this paper and providing background information for Section 1.

This work was performed in part while on a research fellowship sponsored by ICL.



# References

[ANP86] Arvind, R. S. Nikhil, and K. Pengali. *Id Nouveau Manual.* Computation Structures Group Memo 265, MIT, 1986.

[Bir86] R. S. Bird. An introduction to the theory of lists. In *Marktoberdorf Workshop on Logics of Programming*, August 1986.

[HB85] P. Hudak and A. Bloss. The aggregate update problem in functional programming systems. In *12th ACM Symposium on Principles of Programming Languages*, pages 300–314, 1985.

[Hol83] Sören Holmström. A simple and efficient way to handle large data structures in applicative languges. In *Proceedings SERC/Chalmers Workshop on Declarative Programming*, University College London, 1983.

[Hol86] Sören Holmström. Types for unshared objects in functional programming. Chalmers University of Technology, Göteborg, Sweden, 1986. To appear.

[Hud84] P. Hudak. *ALFL Reference Manual and Programmer's Guide.* Research Report YALEU/DCS/RR-322, Second Edition, Yale University, October 1984.

[Hud86] P. Hudak. A semantic model of reference counting and its abstraction (detailed summary). In *ACM Symposium on Lisp and Functional Programming*, pages 351–363, August 1986.

[Hug85] R. J. M. Hughes. *An efficient implementation of purely functional arrays.* Technical Report, Programming Methodology Group, Chalmers University of Technology, Göteborg, Sweden, 1985.

[Kel80] R. Keller. Divide and CONCer. In *ACM Symposium on Lisp*, pages 196–202, Palo Alto, California, August 1980.

[Kel82] R.M. Keller. *FEL programmer's guide.* Technical Report AMPS TR 7, University of Utah, March 1982.

[Mee84] L. Meertens. Algorithmics: towards programming as a mathematical activity. In J. W. de Bakker, *et. al.*, editor, *Mathematics and Computer Science*, North-Holland, 1984.

[Sch85] D. A. Schmidt. Detecting global variables in denotational specifications. *ACM Transactions on Programming Languages and Systems*, 7:299–310, 1985.

[SH82] M. R. Sleep and S. Holmström. A short note concerning lazy reduction rules of append. *Software Practice and Experience*, 12(11):1082–4, November 1982.

[SH86] G. L. Steele Jr. and W. D. Hillis. Connection Machine Lisp: fine-grained parallel symbolic processing. In *ACM Symposium on Lisp and Functional Programming*, pages 279–297, Cambridge, Massachusetts, August 1986.

[Sto85] W. Stoye. *The Implementation of Functional Languages using Custom Hardware.* PhD thesis, Cambridge University, December 1985. Computer Laboratory Technical Report No. 81.

[Tur81] D. A. Turner. Recursion equations as a programming language. In J. Darlington, P. Henderson, and D. Turner, editors, *Functional Programming and Its Applications*, Cambridge University Press, 1981.

[Tur85] D. A. Turner. Miranda: A non-strict functional language with polymorphic types. In *Proceedings of the Conference on Functional Programming Languages and Computer Architecture*, Springer-Verlag, September 1985.

[Tur86] D. A. Turner. *Applicative arrays—an alternative proposal.* Technical Memo, Burroughs Austin Research Center, April 1986.

[Wis85] D. S. Wise. Representing matrices as quadtrees for parallel processors. *Information Processing Letters*, 20:195–199, May 1985.

[Wis86] D. S. Wise. Parallel decomposition of matrix inversion using quadtrees. In *International Conference on Parallel Processing*, pages 92–99, 1986. (IEEE Cat. No. 86CH2355-6).

# I-Structures: Data Structures for Parallel Computing

Arvind† (MIT)
Rishiyur S. Nikhil† (MIT)
Keshav K. Pingali‡ (Cornell University)

**Abstract**

It is difficult simultaneously to achieve elegance, efficiency and parallelism in functional programs that manipulate large data structures. We demonstrate this through careful analysis of program examples using three common functional data-structuring approaches— lists using `Cons` and arrays using `Update` (both fine-grained operators), and arrays using `make-array` (a "bulk" operator). We then present I-structures as an alternative, defining precisely the parallel operational semantics of Id, a language with I-structures. We show elegant, efficient and parallel solutions for the program examples in Id. I-structures make the language non-functional, but do not raise determinacy issues. Finally, we show that even in the context of purely functional languages, I-structures are invaluable for implementing functional data abstractions.

# 1 Introduction

There is widespread agreement that only parallelism can bring about significant improvements in computing speed (several orders of magnitude faster than today's supercomputers). Unfortunately, most of the computing models that we are comfortable and familiar with are based on current von Neumann, sequential architectures, and do not seem extensible in any straightforward way to parallel machines.

Functional languages have received much attention as appropriate vehicles for programming parallel machines, for several reasons. They are high-level, "declarative" languages, insulating the programmer from architectural details. Their operational semantics in terms of rewrite rules offers plenty of exploitable parallelism, freeing the programmer from having to "identify" parallelism explicitly. They are *determinate*, freeing the programmer from details of scheduling and synchronization of parallel activities.

---

*Authors' addresses:*
† MIT Lab. for Computer Science, 545 Technology Square, Cambridge, MA 02139, USA
‡ Dept. of Computer Science, 303A Upson Hall, Cornell University, Ithaca, NY 14850, USA

This research was done at the MIT Laboratory for Computer Science. Funding for this project is provided in part by the Advanced Research Projects Agency of the Department of Defense under the Office of Naval Research contract N00014-84-K-0099.

Keshav Pingali was also supported by an IBM Faculty Development Award.

In this paper, we focus on the issue of *data structures*. We first demonstrate some difficulties in the treatment of data structures in functional languages, and then propose an alternative, called "I-structures". Our method will be to take some test applications, and compare their solutions using functional data structures, and using I-structures. We study the solutions from the point of view of

- efficiency (amount of unnecessary copying, speed of access, number of reads and writes, overheads in construction, *etc.* ),
- parallelism (amount of unnecessary sequentialization), and
- ease of coding.

We hope to show that it is very difficult to achieve all three objectives using functional data structures.

Since our ideas about I-structures evolved in the context of scientific computing, most of the discussion will be couched in terms of *arrays*.[1] All our program examples are written in Id, which is a functional language augmented with I-structures. It is the language we use in our research on parallel architectures. Of course, the efficiency and parallelism of a program depend on the underlying implementation model. Our findings are based on our own extensive experience with dataflow architectures— in particular the MIT Tagged-Token Dataflow Architecture, the centerpiece of our research [3,15]. We have also carefully studied other published implementations of functional languages. However, it is beyond the scope of this paper to delve into such levels of implementation detail, and so we conduct our analyses at a level which does not require any knowledge of dataflow on the part of the reader. In Section 5, we present an abbreviated version of the rewrite-rule semantics of Id, which captures precisely the parallelism of the dataflow machine; we leave it to the intuition of the reader to follow the purely functional examples prior to that section.

While the addition of I-structures takes us beyond functional languages, Id does not lose any of the properties that make functional languages attractive for parallel machines. In particular, Id remains a higher-order, determinate language, *i.e.,* its rewrite-rule semantics remains confluent. In the final section of the paper, we discuss the implications of such an extension to a functional language. We also show that I-structures are not enough— there are some applications that are not solved efficiently whether we use functional data structures or I-structures. This class of applications is a subject of current research.

## 2 The Test Problems

In this section we describe four small example applications which we use to study functional data structures and I-structures.

[1] However, it would be erroneous to infer that our conclusions are relevant only to programs with arrays.

## 2.1 Example A

Build a matrix with

$$A[i,j] = i + j$$

Note that the computation for each element is independent of all the others.

## 2.2 Example B (Wavefront)

Build a matrix with:

$$A[1,j] = 1$$
$$A[i,1] = 1$$
$$A[i,j] = A[i-1,j] + A[i-1,j-1] + A[i,j-1]$$

The left and top edges of the matrix are all 1. The computation of each remaining element depends on its neighbors to the left and above. In a parallel implementation one can thus imagine the computation proceeding as a "wavefront" from the top and left edges to the bottom-right corner of the matrix:

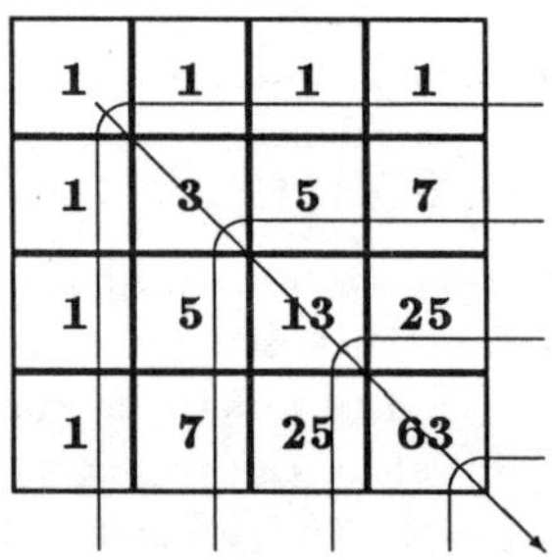


## 2.3 Example C (Inverse Permutation)

This problem was posed to one of us (Arvind) by Barendregt. Given a vector $B$ of size $n$ containing a permutation of integers $1..n$, build a new vector $A$ of size $n$ such that:

$$A[B[i]] = i$$

The computation for each of $A$'s components is independent of the others. (This is called an inverse permutation because the result $A$ also contains a permutation of $1..n$, and when the operation is repeated with $A$ as argument, the original permutation is returned.)

### 2.4 Example D (Shared Computation)

Build two arrays $A$ and $B$ of size $n$ such that

$$A[i] = f\ (h\ i)$$
$$B[i] = g\ (h\ i)$$

such that the $h$ part of the computation for every $i$'th element of the two arrays is shared.[2]

This example illustrates shared computation across arrays. Sharing could also occur across indices in a single array— for example, the computations for $A[2i]$ and $A[2i+1]$ may have a common sub-computation. And of course, in other applications the two types of sharing may be combined.

## 3 Fine-Grained Functional Data Structure Operations

We begin by looking at two data-structuring operations traditionally found in functional languages. In Section 3.1, we look at "Cons", a pairing operation, and in Section 3.2, we look at "Update", an operation that specifies a single, incremental change in an array. We call them "fine-grained" operations because more useful operations such as a vector sum, matrix multiplication, *etc.* must be programmed in terms of a number of uses of these primitives.

### 3.1 Cons: Simulating Large Data Structures Using Lists

Functional languages have traditionally had a two-place "Cons" constructor as a basic data-structuring mechanism. Given Cons, one can of course write suitable array abstractions as a first step towards solving our examples. In this section we quickly reject this as a serious solution.

A typical representation for arrays using Cons would be to maintain an array as a list of elements, a matrix as a list of arrays, and so on. An abstraction for a general access to an array component may be defined as follows:

```
Def select A i = If (i == 0) Then hd A
                 Else select A (i-1) ;
```

Because of the list traversal, selection takes $O(n)$ reads, where $n$ is the length of the array.

Now consider a vector sum, programmed in terms of select:

```
Def vector_sum A B i = If (i > n) Then nil
                       Else cons ((select A i) + (select B i))
                                 vector_sum A B (i+1) ;
```

[2] Here we use juxtaposition to indicate function application— notation that is common in functional languages. Application associates to the left, so that "f x y" stands for "(f x) y".

This function performs $O(n^2)$ reads, where a corresponding FORTRAN program would perform only $O(n)$ reads.

This problem can be mitigated at the expense of ignoring the `select` abstraction and taking advantage of the underlying list representation so that the list-traversing overhead is not cumulative:

```
Def vector_sum_2 A B = If (null A) Then nil
                       Else cons ((hd A) + (hd B))
                                 vector_sum_2 (tl A) (tl B) ;
```

This solution performs $O(n)$ reads (though it is still inefficient because it is not tail-recursive).

Unfortunately, *every* new abstraction must be carefully recoded like this because combinations of given abstractions are not efficient. For example,

```
vector_sum_2 A (vector-sum_2 B C)
```

creates and traverses an intermediate list unnecessarily.

Coding new abstractions efficiently is difficult because the list representation dictates a preferred order in which arrays should be constructed and traversed, an order that is extremely difficult to circumvent. Consider one of the most basic array operations: multiplication of two matrices `A` and `B` as described in a mathematics textbook:

$$C[i,j] = A[i,*] \circ B[*,j]$$

where $\circ$ is the "inner-product". But this requires a traversal of `B` by column, which is very inefficient in our list representation. One may propose first to transpose `B`; but even a transpose is not easy to code efficiently (we invite the reader to attempt it!), and even if it were, we still pay the overhead of making an intermediate copy of the matrix.

Finally, the use of a "fine-grained" data-structuring primitive such as `Cons` places an enormous burden on the storage allocator because of the large number and frequency of requests. Note also that in many typical implementations where a `Cons` cell occupies twice the storage of a number (for two pointers), the storage requirements for the list representation of a vector of numbers can be more than twice the storage for the numbers alone.

For the rest of the paper, we will assume primitives that allocate contiguous storage for each array, so that there is not much storage overhead, and so that array accesses take constant time.

## 3.2 Update: A Functional Array Operator

Instead of simulating arrays using lists, one could provide array operators directly. We now describe one such set of operators.

An array is allocated initially using the expression

```
array (l,u)
```

which returns an array whose index bounds are (l,u), and all of whose locations contain some standard initial value (call it "nil").[3]

The expression

```
update A i v
```

returns an array that is identical to "A", except at index "i", where it contains the value "v". Despite its imperative-sounding name, this is a *functional* operation— it returns a *new* array and does not disturb A.

A component of an array is selected using the expression

```
A[i]
```

which returns the value at index "i" from array "A".

For multi-dimensional arrays, we could nest 1-dimensional arrays, or we could introduce new primitives such as

```
matrix ((li,ui),(lj,uj))

update A (i,j) v

A[i,j]
```

These operations leave a lot of room for choosing the internal representation of arrays. In order to achieve constant time access, at the expense of $O(n)$ allocation and update, we will only look at representations that allocate arrays as contiguous chunks of memory. Other researchers have looked at implementations based on trees, where selection and update are both $O(log\ n)$, and where it is possible to have extensible arrays: Ackerman [1] studied implementations based on binary trees, and Thomas [7] studied implementations based on 2-3 trees.

But none of these implementations are adequate in of themselves— they all involve far too much unnecessary copying and unnecessary sequentialization, as we will demonstrate in the next section. Thus, they are always considered along with some major compile-time and/or run-time optimizations to recoup efficiency and parallelism, and these are discussed in subsequent sections.

[3] In Id, the comma is an infix tupling operation, so that the expression "e1,...,en" denotes a $n$-tuple whose components are the values of e1, ..., en respectively.

### 3.2.1 Copying and Sequentialization of Update

A direct implementation of the "update A i v" operator would be:

- allocate an array with the same index bounds as "A",
- copy all elements from "A" to the result array, except at location "i",
- store value "v" in location "i" of the result array,
- return the pointer to the result array.

The array selection operation would simply read a memory location at an appropriate offset from the pointer to the array argument.

Example A will suffice to demonstrate that such a direct implementation is grossly inefficient. Here is a solution that allocates an array, and then uses (tail-) recursion to traverse and fill it with the appropriate contents:

```
Def A = {  A = matrix ((1,m),(1,n))
          In
            traverse A 1 1 } ;

Def traverse A i j =
          {  next_A = update A (i,j) (i+j) ;
           In
             If       (j < n) Then traverse next_A i (j+1)
             Else If (i < m) Then traverse next_A (i+1) 1
             Else next_A } ;
```

We use the syntax

```
{ BINDING ... BINDING In EXPRESSION }
```

for blocks, which are like "letrec" blocks in other functional languages, and follow the usual static scoping rules.

We prefer to use the following loop syntax to express the tail-recursions:

```
{  A = matrix ((1,m),(1,n))
 In
   {For i From 1 To m Do
      Next A = {For j From 1 To n Do
                   Next A = update A (i,j) (i+j)
                 Finally A}
    Finally A}
```

In the first iteration of the inner loop body, the "A" on the right-hand side refers to its value in the surrounding scope (in this case, the matrix of "nil"s allocated at the top of the block). In each iteration of the loop, the phrase "Next A" binds the value of A for the next iteration. The phrase "Finally A" specifies the ultimate value to be returned at the end of the iteration.

There are two major difficulties in such a program. The first is its profligate use of storage. It is clear that, using a direct implementation of update, we would create $(mn+1)$ arrays, of which only one– the final one– is of interest. Each intermediate array carries only incrementally more information than the previous intermediate array.

The second criticism of this program is that it *over-specifies the order of the updates.* In the problem specification, each element can be computed independently of the others. However, because of the nature of the update primitive, it is necessary for us to chain all the updates involved in producing the final value into a linear sequence.

The necessity to sequentialize the updates also affects program clarity adversely— it is an extra (and unnecessary) bit of detail to be considered by the programmer and reader. Consider a solution for the wavefront problem (Example B):

```
{  A = matrix ((1,m),(1,n))
 In
   {For i From 1 To m Do
       Next A = {For j From 1 To n Do
                      v = If (i == 1) or (j == 1) Then 1
                          Else    A[i-1, j ]
                                + A[i-1,j-1]
                                + A[ i ,j-1] ) ;
                   Next A = update A (i,j) v
                  Finally A}
    Finally A}
```

It takes some careful thought to convince oneself that the above program is correct— that the array selections in computing "v" actually read previously computed values and not "nil", the original contents of A. For example, if the recurrence had been specified instead as $A_{i-1,j}+A_{i-1,j+1}+A_{i,j+1}$ (with appropriate boundary conditions), the programmer would have to realize that the j iteration would have to be reversed to count down from n to 1. This is a great departure from the "declarative" nature of the original recurrence specification.

### 3.2.2 Using Reference Counts to Reduce Storage Requirements

Several researchers have recognized that we can use reference counts to improve the efficiency of the update operation. The idea is very simple: assume that associated with each data structure is a number, called its "reference count" (RC), which counts the number of pointers to it that are currently outstanding. The RC of a structure is incremented every time a copy of its pointer is made and decremented every time its pointer is discarded.

If the RC of the argument array is 1 when the update operation executes, there can be no other references to the array. The update operation can thus safely be performed *in situ*, by destructively writing the value into the existing array and returning a pointer to the existing array. This is of course *much* cheaper than allocating and filling a new array. This solution has been studied carefully in [1]. Unfortunately, except where the program is written with an artificial sequentialization of array accesses and updates, opportunities for this optimization occur but rarely in a parallel machine.

We must also consider that *every* update operation now pays the overhead of checking the RC. Further, the space and time behavior of the program becomes very unpredictable, because whether or not the RC is 1 depends on the particular schedule for processes chosen by the operating system. This can depend, for example, on the current load and configuration of the machine.[4]

In [8], Hudak has proposed a technique called "abstract reference counting", in which a program is analyzed statically to predict the reference counts of arrays at various program points (see also [6]). When the analysis predicts that the reference count of the array argument to an update operation will be one, the compiler generates code to perform an update *in situ*.

Hudak's analysis was performed with respect to a *sequential* operational semantics, and relies on the sequential chaining of the collection of update operations. In this regard, Hudak reports great success in his experiments. We believe that it will be possible to predict that in our program for Example A, the reference count for each update will indeed be one; thus exactly one array will be allocated, and *all* the updates will be done destructively, resulting in a program as efficient (and as sequential) as its FORTRAN counterpart!

Another problem is that the analysis can be sensitive to the order in which the programmer writes his program. Consider a program to compute an array that is identical to a given array A except that the i'th and j'th elements are exchanged:

```
1)    { vj = A[j]
2)     In { vi = A[i]
3)         In { B = update A i vj
4)              In update B j vi }}}
```

Consider a sequential operational semantics that specifies that the bindings of a block are executed before the body of the block. Static analysis may then predict that lines 1 and 2 have been completed before executing line 3, and so the reference count of A in line 3 should be 1. Thus the update can be done in place. Similarly, the update in line 4 can also be done in place. But the programmer could easily have written the program with lines 2 and 3 exchanged:

```
1)    { vj = A[j]
```

[4]Maintaining RCs at run time also raises other issues which are beyond the scope of this paper, such as how much additional code/network-traffic there is to maintain RCs; how much contention there is at the RC field of an array amongst all operations on that array; how atomically to increment/decrement the RC field; how to avoid races between increment and decrement operations, *etc.*

```
3)      In {  B = update A i vj
2)          In {  vi = A[i]
4)              In update B j vi }}}
```

The reference count of A in line 3 is no longer 1 because of the outstanding reference in line 2, and so the update in line 3 *cannot* be done in place. The update in line 4 can still be done in place.

Now consider a *parallel* operational semantics for the language. A precise example of such a semantics is given in Section 5, but, for now, imagine that the bindings of a block can be executed in parallel with the body, with sequencing, if any, based only on data dependencies. All four lines of the program are now initiated in parallel. Since there are no data dependencies between lines 2 and 3, their order of execution is unpredictable. Thus, static analysis cannot draw any definite conclusions about the reference count of A in line 3.

### 3.2.3 Using Subscript Analysis to Increase Parallelism

We have seen that the nature of the `update` primitive requires the programmer to sequentialize the sequence of updates in computing an array. Reference count analysis sometimes determines that these updates may be done in place.

If static analysis could further predict that the subscripts in the sequence of updates were disjoint, then the updates would commute— they could then all be done in parallel. Using such analysis on our program for Example A in Section 3.2.1, the compiler could generate code to perform all the $mn$ updates in parallel.

Subscript analysis has been studied extensively, most notably by Kuck *et al.* at the University of Illinois [11,14] and Kennedy at Rice University [2]. Most of this work was done in the context of vectorizing compilers for FORTRAN. In general, this is an intractable problem, but in the commonly occuring case where the subscripts are of the form $ai + b$ ($a$ and $b$ are constants, $i$ is a loop index), subscript analysis can reveal parallelism. However, there is a significant cost to this analysis, both in terms of compilation speed and in terms of the effort to develop a compiler.

Compared to FORTRAN, subscript analysis is on the one hand easier in functional languages due to referential transparency, but on the other hand more difficult because of dynamic storage allocation.

An example of a program where subscript analysis cannot extract any useful information is a solution to Example C, the Inverse Permutation problem:

```
Def B = {  B = array (1,u)
         In
           For i From 1 To n Do
             Next B = update B A[i] i
           Finally B } ;
```

In order to parallelize the loop, the compiler needs to know about the contents of A, such as the fact that it contains a permutation of $1..n$. This is in general too much to ask of compile-time analysis. This situation is not artificial or unusual— it occurs all the time in practical codes, such as in sorting algorithms that avoid copying large elements of arrays by manipulating their indices instead, and in Monte Carlo techniques and Random Walks.

## 3.3 Discussion

We hope we have convinced the reader of the inadequacy of "fine-grained" functional data structuring mechanisms such as `Cons` and `Update`, especially in a parallel environment. (Some of these problems are solved using the "`make-array`" primitive discussed in the next section.)

Writing programs directly in terms of these primitives does not result in very perspicuous programs— `Cons` requires the programmer continuously to keep in mind the list representation, and `update` requires the programmer to devise a sequential chaining of more abstract operations. In both cases, it is advisable first to program some higher-level abstractions and subsequently to use those abstractions.

Both operators normally involve substantial unnecessary copying of intermediate data structures and substantial unnecessary sequentialization. It was possible to avoid these overheads only when the compiler could be assured that a) reference counts were one, and that b) the subscripts in a chain of updates were disjoint.[5] Automatic detection of these properties does not seem tractable in general.

There is a disquieting analogy with FORTRAN here. Our functional operators force *over-specification* of a problem solution, and static analysis attempts to relax unnecessary constraints. Parallelizing FORTRAN compilers face the same problem, albeit for a different reason (side effects).

# 4 Make-Array: A Bulk Functional Data Structure Operation

Many researchers (notably Keller) have proposed a "bulk" array-definition primitive that treats an array as a "cache" for a function over a rectangular subset of its domain [10,5]. For example, the expression

```
make-array (l,u) f
```

where `(l,u)` is a pair (2-tuple) of integers and `f` is a function, returns an array whose index bounds are `(l,u)`, and whose i'th component contains `(f i)`. We will often refer to `f` as the "filling function".

One can think of the array *as a cache* for `f` because for `i` within bounds, `A[i]` returns the same value as `(f i)`, but (hopefully) at significantly lower cost.

[5]Originally, I-structures were functional data structures with these two properties [4].

Higher dimensional arrays may be constructed either by nesting arrays, or by generalizing the primitive. Thus,

```
make-matrix ((li,ui),(lj,uj)) f
```

produces a matrix where the `(i,j)`'th component contains `f (i,j)`. The first argument is a pair of pairs of integers, and specifies the index bounds of the matrix.

## 4.1 Example A

We can now readily see the solution for Example A:

```
Def f (i,j) = i + j;

Def A = make-matrix ((1,m),(1,n)) f
```

which is concise and elegant and does not pose any serious problem for an efficient, parallel implementation.

## 4.2 Strictness of `make-array`

Before moving on to the remaining examples, it is worth noting that `make-array` need not be strict, *i.e.*, the array may be "returned" before any of the component values have been filled in.

An eager implementation (such as a dataflow implementation) may behave as follows: the bounds expression is first evaluated, and storage of the appropriate size allocated. Then, $n$ independent processes are initiated, computing `(f 1)`, ..., `(f n)` respectively—each process, on completion, writes into the appropriate location in the array. Meanwhile, a pointer to the array is returned immediately as the result of the `make-array` expression. Some synchronization mechanism is necessary at each array location, so that a consumer that tried to read some `A[i]` while it is still empty is made to wait until the corresponding `(f i)` has completed. (Later, we shall see how I-structures provide this synchronization.)

A lazy implementation of `make-array` may behave as follows: the bounds expression is first evaluated, and storage of the appropriate size allocated. Each location `A[i]` is then loaded with the suspension for `(f i)`, and the pointer to the array is returned. A subsequent attempt to read `A[i]` will trigger the evaluation of the suspension, and the value will overwrite the suspension.

This kind of non-strictness permits a "pipelined" parallelism in that the consumer of an array can begin work on parts of the array while the producer of the array is still working on other parts. Of course, even the `Cons` and `Update` operators of Section 3 could benefit from this type of non-strictness.

## 4.3 Example B (Wavefront)

A straightforward solution that comes to mind immediately for the wavefront example is:

```
Def f (i,j) = If (i == 1) or (j == 1) Then 1
              Else    f (i-1,  j )
                    + f (i-1, j-1)
                    + f ( i , j-1) ;

Def A = make-matrix ((1,m),(1,n)) f ;
```

But this is extremely inefficient because "`f (i,j)`" is evaluated repeatedly for each `(i,j)`, not only to compute the `(i,j)`'th component, but also during the computation of every component to its right and below. (This is the typical exponential behavior of a recursively defined Fibonacci function.)

The trick is to recognize that the array is a "cache" or "memo" for the function, and to use the array itself to access already-computed values. This can be done with a recursive definition for `A`:

```
Def f X (i,j) = If (i == 1) or (j == 1) Then 1
                Else    X[i-1,  j ]
                      + X[i-1, j-1]
                      + X[ i , j-1] ;

Def g = f A ;
Def A = make-matrix ((1,m),(1,n)) g;
```

Here, the function `f` is a curried function of two arguments, a matrix and a pair of integers. By applying it to `A`, `g` becomes a function on a pair of integers, which is a suitable argument for `make-matrix`. The function `g`, in defining `A`, carries a reference to `A` itself, so that the computation of a component of `A` has access to other components of `A`.

In order for this to achieve the desired caching behavior, the language implementation must handle this correctly, *i.e.*, the `A` used in `g` must be the *same* `A` produced by `make-matrix` and *not* a new copy of the definition of `A`.

Assuming the language implementation handles this correctly, the main inefficiency that remains is that the `If-Then-Else` is executed at every location.

Note that in a general recursive definition, it will be impossible to predict statically in what order the components must be filled to satisfy the dependencies, and so a compiler cannot "pre-schedule" the computation of the components of an array. Thus, any implementation necessarily must use some of the dynamic synchronization techniques mentioned in Section 4.2.

## 4.4 Example C (Inverse Permutation)

Unfortunately, `make-array` does not do so well on Example C (Inverse Permutation):

```
Def find i j = If (B[j] == i) Then j
               Else find i (j+1) ;

Def f i = find i 1 ;

Def A = make-array (1,n) f ;
```

The problem is that each (f i) that is responsible for filling the i'th location of A needs to search B for the location that contains i, and this search must be linear. Thus, the cost of the program is $O(n^2)$.

It is possible to come up with a slightly different array primitive that addresses this problem. Consider

```
make-array-1 (l,u) f
```

where each (f i) returns (j,v), so that A[j] = v, *i.e.*, the "filling function" f is now responsible for computing not only a component value, but also its index.[6] Example C may now be written:

```
Def f i = B[i],i ;

Def A = make-array-1 (1,n) f ;
```

Of course, if B does not contain a permutation of $1..n$, a run-time error must be detected— either two (f i)'s will attempt to write the same location, or some (f i) will attempt to write out of bounds.

Note that this new primitive, make-array-1, no longer has the simple and elegant characterization of make-array as being a "cache" for the filling function— the relation between the array and the filling function is no longer straightforward. In supplying a filling function for make-array-1, the programmer must now take care that the indices computed by f form a permutation of $1..n$.

## 4.5 Example D (Shared Computation)

A straightforward solution to the shared computation problem may be written as follows:

```
Def fh i = f (h i) ;
Def gh i = g (h i) ;

Def A = make-array (1,n) fh ;
Def B = make-array (1,n) gh ;
```

[6] We first heard this solution independently from David Turner and Simon Peyton-Jones, in a slightly different form— instead of having a filling function f, they proposed an association-list of index-and-value pairs. This solution is also mentioned by Wadler in [17].

This solution, of course, does not share any computation— `(h i)` is repeated for each `i` for `A` and for `B`.

One possible way out is first to cache the values of `(h i)` in an array `C`:

```
Def C = make-array (1,n) h ;

Def fh i = f C[i] ;
Def gh i = g C[i] ;

Def A = make-array (1,n) fh ;
Def B = make-array (1,n) gh ;
```

The drawback is the overhead of allocating, writing, reading and deallocating the intermediate array `C`.

To regain the sharing, one could imagine the following scenario performed by an automatic program transformer. The two `make-arrays` are expanded into, say, two loops. Recognizing that the loops have the same index bounds, they are fused into a single loop. Within the resulting loop, there will be two occurrences of `(h i)`; this common sub-expression can then be eliminated.

We believe that this scenario is overly optimistic. It is very easy to modify the example very slightly and come up with something for which an automatic program transformer would have no chance at all— for example, by changing or displacing the index bounds of one array, or by having a sharing relationship that is not one-to-one, *etc.*

## 4.6 Discussion

Any functional data-structuring primitive must specify two things: the storage to be allocated and the contents of that storage. For example, `Cons e1 e2` specifies a two-cell storage allocation, together with the contents of the two cells.

For a large data structure such as an array, it is obviously not feasible to enumerate expressions specifying the contents of all the components. Thus, functional primitives for large data structures must specify a regular way to *generate* the contents. Thus, `make-array` takes the "filling" parameter `f`, and sets up $n$ independent processes, with the $i$'th process responsible for computing and filling the $i$'th location.

We saw two problems with this process structure. First, this early binding of the $i$'th process to the $i$'th location could not handle things like the inverse permutation. Second, there was no convenient way to express shared computation between the filling processes.

The variant `make-array-1` solved the first problem, by leaving it up to each of the $i$ processes to decide which index $j$ it was responsible for; but it still did not address the issue of shared computation, which could only be performed with the overhead of constructing intermediate arrays or lists. In recent correspondence with us, Phil Wadler has conjectured that, using the version of `make-array-1` that uses association lists of index-and-value pairs together with his "list-less transformer" [16], these problems may indeed be solved without any overhead of intermediate lists. We have yet to investigate the viability of this approach.

All the examples we have seen are quite small and simple; even so, we saw that the first, straightforward solution that came to mind was in many cases quite unacceptable, and that the programmer would have to think twice to achieve any efficiency at all. The complications that were introduced to regain efficiency had nothing to do with improving the algorithms— they were introduced to get around the language limitations.

We are thus pessimistic about relying on a fixed set of functional data structuring primitives. We have encountered situations where the problems illustrated above do not occur in isolation— recursive definitions are combined with shared computations across indices and across arrays. In these situations, writing efficient programs using functional array primitives has proven to be very difficult, and is almost invariably at the expense of program clarity. Perhaps, with so many researchers currently looking at this problem, new functional data-structuring primitives will emerge that will allow us to revise our opinion.

# 5 I-Structures

In the preceding discussion of functional data structures, we saw that the source of inefficiency is the fact that the various primitives impose too rigid a structure on the processes responsible for filling in the components of the data structure. Imperative languages do not suffer from this drawback, because the allocation of a data structure (variable declaration) is decoupled from the filling-in of that data structure (assignment). But imperative languages, with unrestricted assignments, complicate parallelism because of timing and determinacy issues. I-structures are an attempt to regain that flexibility without losing determinacy.

In the Section 5.1 we present the operations that define I-structures with an informal and intuitive explanation.

In Sections 5.2 and 5.3, respectively, we present a precise operational semantics for the functional subset of Id, and for Id with I-structures. This operational semantics is defined by a confluent set of rewrite rules and a reduction strategy which may be loosely described as "Parallel Call-by-Value" or *the dataflow rule of computation*. It provides a basis against which to evaluate the efficiency and parallelism of I-structures.

The rewrite rules and the associated reduction strategy of Id are unusual compared to what experts in functional language may be familiar with. To begin with, they describe exactly what computations are shared, an issue that can (and usually is) left unspecified for other functional languages. This allows us then, even for the functional subset of Id, to describe the dataflow rule of computation, which is a parallel normalizing rule. Further, this specification of sharing (or exactly how many times an expression is evaluated) is necessary to preserve confluency in a language with I-structures.

Because of these subtleties, we invite the reader to study Sections 5.2 and 5.3 carefully, even though they may be skipped at a first reading.

## 5.1 I-structure operations

One can think of an I-structure as a special kind of array, each of whose components may be written no more than once. To augment a language with I-structures, we introduce three new constructs.

An I-structure is allocated by the expression

```
array (l,u)
```

which allocates and returns an "empty" array whose index bounds are `(l,u)`. I-structures are first-class values, and they can contain other I-structures, functions, *etc.* We can simulate multi-dimensional arrays by nesting I-structures, but for efficiency reasons Id also provides primitives for directly constructing multi-dimensional I-structures:

```
matrix ((li,ui),(lj,uj))
```

and so on.

A given component of an I-structure `A` may be assigned (written) no more than once, using a "constraint statement":

```
A[i] = v
```

Operationally, one thinks of this as assigning, or storing the value `v` into the `i`'th location of array `A`. It is a run-time error to write more than once into any I-structure location— the entire program is considered to be in error.

Syntactically, the assignment statement appears intermixed with the bindings in a block.

A component of an I-structure `A` may be selected (read) using the expression

```
A[i]
```

This expression returns a value only after the location becomes non-empty, *i.e.*, after some other part of the program has assigned the value.

There is no test for "emptiness" of an array location. These restrictions— write-once, deferred reads, and no test for emptiness— ensure that the language remains *determinate*; there are no read-write races. Thus the programmer need not be concerned with the timing of a read relative to a write. All reads of a location return a single, consistent value, albeit after an arbitrary delay.

Semantically, one can think of each location in an I-structure as containing a *logical term*. Initially, the term is just a logic variable— it is completely unconstrained. Assignment to that location can be viewed as a refinement of, or constraint on the term at that location. This is what motivates our calling it a "constraint statement". The single-assignment rule is sufficient to preclude inconsistent instantiations of the initial logic variable. Of course, the single-assignment rule is not a *necessary* condition to avoid inconsistent instantiations. We could take the view that assignment is really *unification*, and then multiple writes would be safe so long as the values unify. Id does not currently take this view, for efficiency reasons.

Some machine-level intuition: Conceptually, I-structures reside in an "I-structure memory" unit. When an allocation request arrives, an array is allocated in free space, and a pointer to this array is returned. Every location in the array has an extra bit that designates it as being "empty".

Every request to read a location of an I-structure (*i.e.*, a selection) is accompanied by a "tag", which can be viewed as the name of the continuation that expects the result. The I-structure controller checks the "empty" bit at that location. If it is not empty, the value is read and sent to the continuation. If the location is still empty, the controller simply queues the tag at that location.

When a request to store a value in a location of an I-structure arrives (*i.e.*, an assignment), The I-structure controller checks the "empty" bit at that location. If it is empty, the value is stored there, the bit is toggled to "non-empty", and if any tags are queued at that location, the value is also sent to all those continuations. If the location is not empty, the controller generates a run time error.

## 5.2 Id and its Operational Semantics: the Functional Subset

In this section we give the reader a feel for the parallelism in dataflow execution of Id. Previously, this has only been described in terms of dataflow graphs, the "machine language" of the dataflow machine. Here we describe it in terms of a confluent set of rewrite rules that capture precisely the behaviour of dataflow graphs. We consider this description more abstract and probably easier to understand for those already familiar with rewrite rules. Our description here is necessarily brief— the reader is referred to [13,12,15] for more comprehensive treatments.

The reader is invited to read this section carefully, as the rewrite rules incorporate many subtle differences from other execution models for functional languages that we are aware of.

### 5.2.1 Syntax

Here is an abbreviated syntax for Id:

```
PROGRAM ::= FUN-DEF ";" ... ";" FUN-DEF ";" EXPRESSION

FUN-DEF ::= "Def" IDENTIFIER ARG ... ARG "=" EXPRESSION

ARG     ::= IDENTIFIER

EXPRESSION ::=
            CONSTANT | IDENTIFIER
```

```
         "If" EXPRESSION "Then" EXPRESSION "Else" EXPRESSION
         EXPRESSION EXPRESSION
         BLOCK

BLOCK   ::= "{" BINDING ";" ... ";" BINDING "In" EXPRESSION "}"

BINDING ::= IDENTIFIER "=" EXPRESSION
```

*i.e.*, a program is a list of function definitions, and a main expression. The arguments in function definitions are curried, and the definitions may be recursive and mutually recursive. Blocks are analogous to the "letrec" construct in other functional languages; the bindings may be recursive and mutually recursive, they follow the usual static scoping rules, and the order of bindings in a block is not significant.

For simplicity of exposition here, the bindings in blocks do not include syntactic function definitions (there are no arguments), so we do not have to worry about dynamically handling free variables. This is not a serious restriction— we use the technique of lambda-lifting [9] to "compile out" internal function definitions, leaving programs in the above standard form.

Since all function definitions occur only at the top (program) level, one can think of the set of function definitions as a set of combinator definitions, or rewrite rules.

We define the "arity" of a function identifier as the number of arguments in its definition. This is a *syntactic*, not semantic attribute. Thus the following two definitions

```
Def f a   = plus (sqr a)   ;
Def g a b = plus (sqr a) b ;
```

define semantically equivalent functions of two arguments, but have arities 1 and 2 respectively, by virtue of their different syntactic definitions.

Though arity is a syntactic attribute, it determines which expressions are redexes (rewritable), which in turn affects the definition of "normal forms", or answers. Thus `(f 5)` is a redex, and so is not in normal form, whereas `(g 5)` is not a redex, and *is* in normal form. This "pattern matching" based on the syntactic definition is fairly standard in rewrite-rule systems.

We have the usual conventions that

- Applications associate to the left
- Infix expressions "`x op y`" are syntactic sugar for "`opfn x y`", for some known set of infix operators such as `+`.
- Parentheses are used for grouping

### 5.2.2 Values

The purpose of a program is to produce a *value* for the main expression. A value is an expression in "normal form". We will use the generic symbol "`V`" to denote a value.

Values include all constants, such as numbers, and the booleans `true` and `false`. Values also include partial applications of the form

```
f x1 ... xj
```

where the `xi`'s are identifiers or values (not general expressions), and the arity of `f` is greater than `j`. Thus, for example, if `g` is function of arity 2, then `g` and `(g x)` are values, but `(g x y)` is not. Similarly, if there is a binding

```
a = EXPRESSION
```

then the expression "a" is not a value— its arity (0) is satisfied, and so it is not a partial application.

These technical definitions of values may seem an over-specification for a functional language; but they are necessary in a language with I-structures, where we must specify, for each expression, things like its sharing behavior, how many times it is evaluated, *etc.*

These values have a one-to-one correspondence with "tokens" in the dataflow machine.

### 5.2.3 Machine States and Termination

A machine state is described as follows:

```
E ; B ; ... ; B
```

where `E` is an expression and `B`'s are bindings. `E` is called the "main" expression. There may be zero or more bindings, and their order is not significant.

The initial state of the machine is just

```
E
```

*i.e.*, an expression and no bindings, where `E` is the main expression of the program.

The machine is in its final state when no rewrite rules can be applied *anywhere* in the machine, *i.e.*, it is of the form

```
V ; IDENTIFIER = V ; ... ; IDENTIFIER = V
```

(all the `B`'s are quiescent). Note that the main expression may reduce to a value *before* all the bindings become quiescent.[7]

[7]On our dataflow machine this phenomenon is reflected in the sometimes unnerving behavior that an answer may by printed long before termination is reported!

### 5.2.4 Rewrite Rules

Each rewrite rule identifies a sub-expression somewhere in the machine state and shows what that sub-expression reduces to. The sub-expression may occur in E, or in the right-hand sides of any of the B's; thus one should imagine all the components of the machine state executing in parallel.

We use the notation

```
    (e)
    E ; B ; ... ; B
```

to focus our attention on a sub-expression "e" that occurs somewhere in E or somewhere in one of the right-hand sides of the B's, *but not within the* Then *or* Else *arms of a conditional expression.* Such an "e" is called a *redex.*

The dataflow rule requires *all* such redexes to be reduced ultimately. Of course, in a real implementation only a subset of redexes can be rewritten at any step, and so some form of fair scheduling is required in order to produce answers.

**1)** Identifiers may be substituted by *values* (not expressions!) that they are bound to:

```
    (x)
    E ; B ; ... ; x = V ; ... ; B
==>
    (V)
    E ; B ; ... ; x = V ; ... ; B
```

The fact that substitution must wait for the expressions to reduce to values motivates our characterizing it as a "call-by-value" semantics.

**2)** Conditionals:

```
    (If true Then E1 Else E2)
    E ; B ; ... ; B
==>
    (E1)
    E ; B ; ... ; B
```

and similarly for the `false` case.

**3)** Partial Applications (arity not satisfied):

Suppose the program has a definition

```
    Def f x1 ... xn = Ebody ;
```

Then

```
    (f Earg1 ... Eargj)
    E ; B ; ... ; B
==>
    (f x1' ... xj')
    E ; B ; ... ; B ; x1' = Earg1 ; ... ; xj' = Eargj
```

where j < n, and x1' through xj' are new identifiers.

As a result of this transformation, the partial application now conforms to the definition of a value, and so may be passed around and shared (for example, applied to different remaining arguments). However, by Rule **1**, the shared arguments x1 through xj are evaluated exactly once.

**4**) Full Applications (arity satisfied):

Suppose the program has a definition

```
    Def f x1 ... xn = Ebody ;
```

Then

```
    (f Earg1 ... Eargn)
    E ; B ; ... ; B
==>
    (Ebody')
    E ; B ; ... ; B ; x1' = Earg1 ; ... ; xn' = Eargn
```

where x1' through xn' are new identifiers and Ebody' is an appropriately alpha-renamed version of Ebody. Note the parallel behavior: the body of the function is evaluated in parallel with the arguments, but by Rule **1**, the arguments are not substituted in the body until they have reduced to values. This is what motivates the name "Parallel Call-By-Value", and demonstrates how non-strict functions can return values before receiving argument values.

**5**) Blocks:

```
    ({ B1 ; ... ; Bn IN E0 })
    E ; B ; ... ; B
==>
    (E0')
    E ; B ; ... ; B ; B1' ; ... ; Bn'
```

where E0' and B1' through Bn' use new identifiers for the identifiers bound in the block.

### 5.2.5 An Example

Consider the following program:

```
Def twice f x = f (f x) ;

twice (plus 5) (2 * 3)
```

Here is a possible rewrite sequence (in each step we have underlined redex(es) that we have chosen to rewrite):

```
    twice (plus 5) (2 * 3)
    ^^^^^^^^^^^^^^^^^^^^^^
==>
    f1 (f1 x1) ; f1 = plus 5 ; x1 = 2 * 3
    ^^  ^^                          ^^^^^
==>
    plus 5 (plus 5 x1) ; f1 = plus 5 ; x1 = 6
                   ^^
==>
    plus 5 (plus 5 6) ; f1 = plus 5 ; x1 = 6
            ^^^^^^^^
==>
    plus 5 11 ; f1 = plus 5 ; x1 = 6
    ^^^^^^^^^
==>
    16 ; f1 = plus 5 ; x1 = 6
```

### 5.2.6 Discussion

It is important to note that Id, with these operational semantics, supports non-strict functions, *but it is eager, not lazy.* Given an application (f x), the body of f is invoked in parallel with the evaluation of x. If f does not need x, it may return a result immediately. However, x is still computed; it is not left up to f to "demand" x. Unfortunately, in the literature, non-strictness is often equated with laziness.

The effect of eager evaluation in producing answers becomes an issue only when an argument to a function is not used, because resources are used in evaluating it. An extreme case of this occurs when such an argument computation diverges. Consider the following program:

```
Def diverge x = diverge x ;

Def f y = 5 ;

f (diverge 1)
```

A correct implementation of Id should print the value 5 even though the program does not terminate. This issue does not arise in any of the other examples discussed in this paper.

Two aspects of the rewrite rules deserve close scrutiny: A) the rewrite rules have been carefully constructed to ensure that arguments are evaluated exactly once, and B) there is a qualification that the redex not be within the arms of an `If-Then-Else`. In the functional subset of Id, these affect only the efficiency of the reduction— they avoid wasting machine resources— and do not affect the outcome of the program. When Id is augmented with I-structures, these restrictions become semantically necessary.

## 5.3 Operational Semantics of Id with I-Structures

### 5.3.1 I-structure Values

We first introduce a notation for I-structure values.

```
    <X1, ..., Xu>
```

denotes an I-structure, where the `Xi`'s are identifiers. One can think of an I-structure value as a pointer to an array in I-structure memory, with `X1...Xu` the names of the locations in the array.

### 5.3.2 Rewrite Rules for I-structure Operations

There are two new rewrite rules for I-structures:

```
    array (v1,v2)
=>
    <Xv1, ..., Xv2>
```

where `v1` and `v2` are integer values, and `Xv1, ..., Xv2` are new identifiers. Note that this rewrite rule has a characteristic never found in functional languages, but common in logic programming languages— new variables are introduced on the right-hand side.

The rule

```
    <X1, ..., Xu>[V]
=>
    XV
```

specifies the selection of one of the locations of the I-structure value. For a minor technical reason having to do with termination, the identifiers inside an I-structure value are never substituted by their values.

### 5.3.3 An Example

Consider the following program:

```
Def pair x y = {  a = array (0,1) ;
                  a[0] = x ;
                  a[1] = y
                In
                  a } ;

pair 5 (pair 6 7)
```

Here is a possible rewrite sequence (some of the machine state descriptions take two lines, and we have omitted introducing identifiers for function arguments that are already values):

```
    pair 5 (pair 6 7)
    ^^^^^^^^^^^^^^^^^

==>
    {a=array(0,1); a[0]=5 ; a[1]=y1 IN a} ; y1 = pair 6 7
                                                 ^^^^^^^^

==>
    {a=array(0,1); a[0]=5 ; a[1]=y1 IN a} ;
    ^^^^^^^^^^^^^^^^^^^^^^^^^^^^^^^^^^^^^^
        y1={a=array(0,1); a[0]=6 ; a[1]=7 IN a}
           ^^^^^^^^^^^^^^^^^^^^^^^^^^^^^^^^^^^^

==>
    a1 ; a1=array(0,1) ; a1[0]=5 ; a1[1]=y1 ;
            ^^^^^^^^^^
        y1=a2; a2=array(0,1); a2[0]=6 ; a2[1]=7
                  ^^^^^^^^^^

==>
    a1 ; a1=<X0,X1} ; a1[0]=5 ; a1[1]=y1 ;
    ^^                ^^        ^^
        y1=a2; a2=<Y0,Y1> ; a2[0]=6 ; a2[1]=7
           ^^               ^^        ^^

==>
    <X0,X1> ; a1=<X0,X1> ; <X0, X1>[0]=5 ; <X0,X1>[1]=y1 ;
                           ^^^^^^^^^^^     ^^^^^^^^^^ ^^
        y1=<Y0,Y1>; a2=<Y0,Y1> ; <Y0,Y1>[0]=6 ; <Y0,Y1>[1]=7
                                 ^^^^^^^^^^     ^^^^^^^^^^

==>
    <X0,X1> ; a1=<X0,X1> ; X0=5 ; X1=<Y0,Y1> ;
        y1=<Y0,Y1>; a2=<Y0,Y1> ; Y0=6 ; Y1=7
```

Note that the resulting I-structure value <X0,X1> appears before termination. On the dataflow machine, an I-structure descriptor (pointer) is printed before termination. If the user subsequently queries the zero'th element (say) of the I-structure, the component value will be fetched (corresponding to dereferencing X0 to the value 5).

The astute reader may ask: since computation may go on after a value is printed, what is to be made of a run-time error that then occurs? There is no problem of determinacy here— a program with a run-time error is considered to have the value "$\top$", the "inconsistent" element at the top of the semantic domain. A value printed out before termination is to be considered as an approximation to the final value; a run-time error only refines this approximation to $\top$.

## 5.4 The Programming Examples

The assignment statement is the simplest form of a constraint statement, and this can be generalized into compound constraint statements.

By embedding assignments in loops, we can now also have loops that behave purely as constraint statements, by omitting the "`Finally e`" clause.

A function "`f`" may not return any value of interest— it may contain only constraint statements. A call to such a function is itself a constraint statement, using the syntax

```
Call f x
```

Constraint statements— assignments, constraint loops, constraint calls— appear intermixed with the bindings in a block or loop-body.

Let us now see how our programming examples would be expressed using Id with I-structures.

### 5.4.1 Example A

The first example is straightforward:

```
{ A = matrix ((1,m),(1,n)) ;
  {For i From 1 To m Do
     {For j From 1 To n Do
        A[i,j] = i + j }}
In
  A }
```

Recall that the loop is a *parallel* construct, so in the above program, the loop bodies can be executed in any order— sequentially forwards, as in FORTRAN, or all in parallel, or even sequentially backwards!

The matrix A may be returned as the value of the block as soon as it is allocated. Meanwhile, $m \times n$ loop bodies execute in parallel, each filling in one location in A. Any consumer that tries to read A[i,j] will get its value as soon as the corresponding loop body completes.

### 5.4.2 Example B (Wavefront)

```
{  A = matrix ((1,m),(1,n)) ;
   {For i From 1 To m Do
      A[i,1] = 1 } ;
   {For j From 2 To n Do
      A[1,j] = 1 } ;
   {For i From 2 To m Do
      {For j From 2 To n Do
         A[i,j] = A[i-1,j] + A[i-1,j-1] + A[i,j-1] }}
In
   A }
```

The matrix A may be returned as the value of the expression as soon as it is allocated. Meanwhile, *all* the loop bodies are initiated in parallel, but some will be delayed until the loop bodies to their left and top complete. Thus a "wavefront" of processes fills the matrix.

Note that we do not pay the overhead of executing an `If-Then-Else` expression at each index, as in the functional solution.

It is worth emphasizing again that loops are *parallel* constructs. In the above example, it makes no difference if we reverse the index sequences:

```
{For i From m Downto 2 Do
   {For i From n Downto 2 Do ...}}
```

The data dependencies being the same, the order of execution would be the same. This is certainly not the case in imperative languages such as FORTRAN.

### 5.4.3 Example C (Inverse Permutation)

```
{  A = array (1,n) ;
   {For i From 1 To n Do
      A[B[i]] = i }
In
   A }
```

The array A may be returned as the value of the expression as soon as it is allocated. Meanwhile, all the loop bodies execute in parallel, each filling in one location. If B does not contain a permutation of $1..n$, then a run-time error will arise, either because two processes tried to assign to the same location or because some process tried to write out of bounds.

### 5.4.4 Example D (Shared Computation)

```
{ A = array (1,n) ;
  B = array (1,n) ;
  {For i From 1 To n Do
     z = h i ;
     A[i] = f z;
     B[i] = g z }
In
  A,B }
```

The arrays A and B may be returned as the value of the expression as soon as they are allocated. Meanwhile, all the loop bodies execute in parallel, each filling in two locations, one in A and the other in B. In each loop body, the computation of (h i) is performed only once.

# 6 Using I-structures to Implement Array Abstractions

From the point of view of programming methodology, it is usually desirable for the programmer first to implement higher-level array abstractions and subsequently to use those abstractions.

## 6.1 Functional Array Abstractions

As a first example, we can implement the functional make-array primitive:

```
Def make-array (l,u) f = { A = array (l,u) ;
                           {For i From l To u Do
                              A[i] = f i }
                         In
                           A } ;
```

Note that there is all the parallelism we need in this implementation. The array A can be returned as soon as it is allocated. Meanwhile, all the loop bodies execute in parallel, each filling in one component. Any consumer that attempts to read a component will get the value as soon as it is filled. It is likely that even if we stick to a purely functional language and supply make-array as a primitive, some such I-structure-like implementation mechanism will be necessary to achieve this parallelism.

Similarly, here is an efficient, parallel implementation for make-matrix:

```
Def make-matrix ((li,ui),(lj,uj)) f =
                       { A = matrix ((li,ui),(lj,uj)) ;
```

```
                    {For i From li To ui Do
                       {For j From lj To uj Do
                          A[i,j] = f (i,j) }}
                 In
                    A } ;
```

A functional vector sum:

```
Def vector_sum A B = {  l,u = bounds A ;
                        C = array (l,u) ;
                        {For i From l To u Do
                               C[i] = A[i] + B[i] }
                     In
                        C } ;
```

Again, the solution has all the parallelism we need. The array C is returned as soon as it is allocated. Meanwhile, independent processes execute in parallel, each computing one sum and storing it in one location in C.

The functional make-array-1 primitive:

```
Def make-array-1 (l,u) f =
                     {  A = array (l,u) ;
                        {For i From l To u Do
                           j,v = f i ;
                           A[j] = v }
                      In
                        A } ;
```

A primitive to make two arrays in parallel:

```
Def make-two-arrays (l,u) f =
                     {  A = array (l,u) ;
                        B = array (l,u) ;
                        {For i From l To u Do
                           va,vb = f i ;
                           A[i] = va ;
                           B[i] = vb }
                      In
                        A,B } ;
```

We leave it as an exercise for the reader to use make-two-arrays to produce an elegant solution to the shared computation problem (Example D).

It is clear that, with I-structures in the language, it is straightforward for the programmer to implement any desired functional array abstractions— the solutions are perspicuous, efficient, and there is no loss of parallelism.

## 6.2 Non-Functional Array Abstractions

It has been our experience that functional abstractions are not the only ones that lead to compact, elegant programs. Consider the following (non-functional) "array-filling" abstraction:

```
Def fill A ((li,ui),(lj,uj)) f =
                        {For i From li To ui Do
                           {For j From lj To uj Do
                                 A[i,j] = f (i,j)}} ;
```

which fills a rectangular region of the given matrix A. Our wavefront program can then be written as follows:

```
{  A = matrix ((1,m),(1,n)) ;
   border   (i,j) = 1 ;
   interior (i,j) = A[i-1,j] + A[i-1,j-1] + A[i,j-1] ;
   Call fill A ((1,m),(1,1)) border ;
   Call fill A ((1,1),(2,n)) border ;
   Call fill A ((2,m),(2,n)) interior
In
   A }
```

Of course, for more efficiency, we could define special abstractions for filling in horizontal or vertical regions:

```
Def fill_col  A ((li,ui),j) f = {For i From li To ui Do
                                      A[i,j] = f (i,j)} ;

Def fill_row  A (i,(lj,uj)) f = {For j From lj To uj Do
                                      A[i,j] = f (i,j)} ;
```

and use them to fill the borders of our matrix.

# 7 Limitations of I-structures

While we believe that I-structures solve some of the problems that arise with functional data structures, we have frequently encountered another class of problems for which they still do not lead to efficient solutions.

Consider the following problem: we are given a very large collection of generators (say a million of them), each producing a number. We wish to compute a frequency distribution (histogram) of these values in, say, 10 intervals. An efficient parallel solution should allocate an array of 10 "accumulators" initialized to 0, and execute as many generators as it can in parallel. As each generator completes, its result should be classified into an interval j, and the j'th accumulator should be incremented. It does not matter in what order the accumulations are performed, so there are no serious determinacy issues, except the following synchronization requirement: there is a single instant when the resulting histogram is ready (*i.e.*, available to consumers)— it is ready when all the generators have completed. To avoid indeterminacy, no consumer should be allowed to read any location of the histogram until this instant.

A second example: In a system that performs symbolic algebra computations, consider the part that multiplies polynomials. A possible representation for the polynomial

$$a_0 + a_1x + a_2x^2 + a_3x^3 ... + a_nx^n$$

would be an array of size $n + 1$ containing the coefficients $a_0$, ..., $a_n$. To multiply two polynomials $A$ and $B$ of degree $n$ together, we need first to allocate an array of size $2n$, with each location containing an "accumulator" initialized to 0; then, for each $j$, initiate $(j+1)$ processes to compute $a_0 \times b_j$, $a_1 \times b_{j-1}$, ..., $a_j \times b_0$; as each of these processes completes, its result should be added into the $j$'th accumulator. The order of the accumulation at any index does not matter.

The synchronization requirement here is more complex. A consumer for a location in the result array may read it as soon as the $j + 1$ processes attached to it have completed; this may occur before other locations are ready. Contrast this with the histogram example where the entire array became available to consumers at a single instant.

These problems cannot be solved efficiently either with any of the functional data structures that we have seen so far, or with I-structures. There are two fundamental problems to be addressed:

1. How to model the accumulators. With I-structures and functional data structures, once a location in an array has a value, it cannot be updated at all, even though the update occurs in a safe, structured manner.

2. How to express the termination of the accumulation. In the histogram example, the termination was a global condition. In the polynomial example, termination is tied to each location.

We are currently studying some proposed solutions to this problem.[8]

# 8 Conclusion

In this paper, we have studied the issue of data structures for parallel computing. We saw that with functional data structures, it can be difficult simultaneously to achieve efficiency, parallelism, and program clarity. We showed that I-structures go a long way towards solving this problem.

[8] In [17], Wadler has proposed yet another functional array operation to handle such "accumulation" problems. This construct combines an association-list of index-and-value pairs, together with a reduction operator to specify the array. We do not yet know what are the implementation issues for this construct.

Further, it is clear that whether or not I-structures are available in the programming language, something like I-structures are needed at the machine level to *implement* functional primitives such as make-array, in order that they have maximum parallelism— the read/write synchronization mechanism seems fundamental.

The introduction of any non-functional feature (such as I-structures) into a functional language is not without cost— the language loses referential transparency, with all the attendant implications on the ability to reason about programs, do program transformations for optimization, *etc.* In the case of I-structures, the loss of referential transparency is evident— this program

```
{ a = array (1,10)
In
  a, a }
```

is not semantically equivalent to this program

```
array (1,10), array (1,10)
```

Even so, it is still much easier to reason about programs with I-structures than it is to reason about programs in unconstrained imperative languages, because of the absence of timing issues.

A functional language with I-structures can be made referentially transparent by adopting a "relational" syntax (like logic programming languages) rather than a functional one. Referential transparency is lost in Id because the "array" construct allocates an array without naming it. To fix this, we first replace it with a new construct called "array_bounds". Array allocation is then achieved by the constraint statement:

```
array_bounds (x) = (l,u)
```

which instantiates "x" to be an array with bounds "(l,u)". The array is thus not allocated anonymously.

But this is not enough; functional abstraction still allows us to produce anonymous arrays:

```
Def alloc (l,u) = {  array_bounds (x) = (l,u)
                    In x } ;
```

To prevent this, abstraction (indeed all constructs) must be converted to a relational form. For example, a procedure cannot return a value explicitly; rather, it must take an additional argument which it instantiates to the returned value. The alloc procedure would be thus be written as follows:

```
Def rel_alloc (l,u) x = { array_bounds (a) = l,u ;
                          x = a }
```

For example, the invocation "`rel_alloc (1,10) a`" will instantiate "a" to an array of size 10. Further, to specify that "a" is a "place-holder" argument rather than a value passed to `rel_alloc`, we must annotate it appropriately, say with "`^`". The invocation must therefore be written as "`rel_alloc (1,10) ^a`".

By adopting this annotated relational syntax, we believe that we *could* achieve referential transparency, at the cost of complicating the syntax considerably. We are unconvinced that this is a useful thing to do.

Because I-structure operations compromise referential transparency, as a matter of programming style we strongly encourage the programmer to use only functional abstractions wherever possible. A good Id programmer will separate the program into two parts— a part that defines convenient functional data-structure abstractions in terms of I-structures, and the rest of the program that uses only those abstractions and does not explicitly use I-structures. The latter part of the program is then purely functional, and amenable to all the tools available to manipulate functional languages.

However, to permit efficient parallel implementation of those functional data structure abstractions, I-structures are indispensible, and must be available in the language. As we have argued above, it does not seem possible to define new abstractions entirely in terms of functional primitives without loss of efficiency and parallelism; but as shown in Section 6, I-structures make this absolutely straightforward.

**Acknowledgements:** Vinod Kathail, who wrote the first Id compiler, brought to light many of the subtle issues on copying and parallelism, and was instrumental in clarifying our understanding of rewrite rules. The current Id compiler, written by Ken Traub, has been invaluable in testing our ideas. K. Ekanadham of IBM first explored the use of I-structures to program high-level abstractions.

## References

[1] William B. Ackerman. *A Structure Memory for Data Flow Computers.* Master's thesis, Technical Report TR-186, MIT Lab. for Computer Science, Cambridge, MA 02139, 1978.

[2] J.R. Allen and K. Kennedy. *PFC: A Program to convert FORTRAN to Parallel Form.* Technical Report MASC-TR82-6, Rice University, Houston, TX, March 1982.

[3] Arvind and David Ethan Culler. Dataflow architectures. In *Annual Reviews in Computer Science*, pages 225–253, Annual Reviews Inc., Palo Alto, CA, 1986.

[4] Arvind and Robert E. Thomas. *I-structures: An Efficient Data Type for Parallel Machines.* Technical Report TM 178, Computation Structures Group, MIT Lab. for Computer Science, Cambridge, MA 02139, September 1980.

[5] Henk Barendregt and Marc van Leeuwen. *Functional Programming and the Language TALE.* Technical Report TR 412, Mathematical Institute, Budapestlaan 6, 3508 TA Utrecht, The Netherlands, 1985.

[6] J. Driscoll, N. Sarnak, D. Sleator, and R. Tarjan. Making data structures persistent. In *Proc. 18th Annual ACM Symposium on Theory of Computing, Berkeley, CA*, pages 109–121, May 1986.

[7] Kim P. Gostelow and Robert E. Thomas. A view of dataflow. *AFIPS Conference Proceedings*, 48:629–636, June 1979.

[8] Paul Hudak. A semantic model of reference counting and its abstraction. In *Proc. 1986 ACM Conf. on Lisp and Functional Programming, MIT, Cambridge, MA*, pages 351–363, August 1986.

[9] Thomas Johnsson. Lambda lifting: transforming programs to recursive equations. In *Springer-Verlag LNCS 201 (Proc. Functional Programming Languages and Computer Architecture, Nancy, France)*, September 1985.

[10] Robert M. Keller. *FEL (Function Equation Language) Programmer's Guide.* Technical Report AMPS Technical Memorandum No. 7, University of Utah, Department of Computer Science, April 1983.

[11] David J. Kuck, R.H. Kuhn, D.A. Padua, B. Leasure, and M. Wolfe. Dependence graphs and compiler optimizations. In *Proc. 8th ACM Symp. on Principles of Programming Languages*, pages 207–218, January 1981.

[12] Rishiyur S. Nikhil. *Id Nouveau Reference Manual.* Technical Report (Forthcoming), Computation Structures Group, MIT Lab. for Computer Science, Cambridge, MA 02139, 1987.

[13] Rishiyur S. Nikhil, Keshav Pingali, and Arvind. *Id Nouveau.* Technical Report CSG Memo 265, Computation Structures Group, MIT Lab. for Computer Science, Cambridge, MA 02139, July 1986.

[14] David A. Padua and Michael J. Wolfe. Advanced compiler optimizations for supercomputers. *Communications of the ACM*, 29(12), December 1986.

[15] Kenneth R. Traub. *A Compiler for the MIT Tagged Token Dataflow Architecture.* Master's thesis, Technical Report TR-370, MIT Lab. for Computer Science, Cambridge, MA 02139, August 1986.

[16] Philip Wadler. Listlessness is better than laziness: lazy evaluation and garbage collection at compile time. In *Proc. 1984 ACM Conf. on Lisp and Functional Programming, Austin, TX*, pages 45–52, August 1984.

[17] Philip Wadler. A new array operation for functional languages. In *Proceedings of the Graph Reduction Workshop, Santa Fe, New Mexico, October 1986*, 1987.

# Parallel Execution of an Equational Language†
*(Summary)*

*Bharat Jayaraman*
*Gopal Gupta*

*Department of Computer Science*
*University of North Carolina at Chapel Hill*
*Chapel Hill, NC 27514*

## 1. Introduction

Amongst a number of recent proposals for integrating functional and logic programming [RS82, D83, F84, GM84, SY84, DP85, YS86, JS86], one promising approach is *equational programming* [F84, DP85, YS86, JS86]. In our recent formulation of equational programming [J84, JS86, JSG86], a program is a collection of *conditional rewrite rules*, where the conditions are expressed as a set of *equations*, and the top-level goal to be solved is also a set of equations. We have shown that this framework is powerful enough for first-order functional and Horn logic programming [JS86], and has the following properties:

- incremental, rule-based programming with pattern-directed rules;
- functional, rather than relational, notation;
- formal semantics based on *complete set of solutions*;
- operational semantics based on *object refinement*;
- potential for parallel execution.

The computational paradigm underlying our class of equational languages is *equation solution*, and this forms the main subject of our paper. This paradigm subsumes expression evaluation in a functional language because an expression $e$ to be evaluated is simply viewed as an equation $v = e$, where $v$ is some distinct variable. It also subsumes goal solution in logic languages because a goal $g$ to be solved can also be viewed as an equation, namely, $g = true$. Because solutions are not necessarily unique, our formal semantics for a set of equations is expressed as the *complete set of solutions*, similar to complete set of unifiers in a equational theory [H80]. *Object refinement* is a *sound* and *complete* operational strategy for computing solutions that are at least as general as the complete set of solutions [JSG86].

An important difference in our approach from other equational languages is that the formal semantics of our language is *domain-theoretic*, rather than based on *equational logic* [F84, O85, YS86]. Thus constructors and function symbols are sharply distinguished in our language, and the denotation of an expression is a set of *terms* over the constructors. Operationally, *object refinement* differs from related approaches such as *narrowing* [H80, GM84, YS86] in that it does not use *unification* as a primitive operation (object refinement is a generalization of unification in that equations are not restricted to terms), and reductions are restricted to the outermost term. These semantic differences effectively allow a more efficient search for solutions than a narrowing-based language: An equation of two expressions, $e_1 = e_2$, can be solved by reducing its constituent expressions until a constructor is produced at the outermost level. A narrowing interpreter based on equational logic, on the hand, may have to cross-compare expressions emanating from $e_1$ and $e_2$ unless its programs may be assumed to be canonical, that is, confluent and terminating.

It is generally believed that preserving completeness in the implementation of any logic programming language would lead to unacceptable computational requirements. The basis for

† This research is supported by the National Science Foundation under grant DCR-8603609.

this belief is that preserving completeness requires a breadth-first search for solutions, which leads to explosive storage requirements if implemented naively. Our research aims to minimize the storage requirements of a breadth-first search by combining the ***outermost reduction*** strategy found in functional languages along with a ***shared environment*** for representing multiple solutions. This environment sharing technique exhibits some interesting storage economies: For example, in the goal equation

$$append(x,\ y) = [1\ 2 \ldots \mathrm{n}]$$

where ***append*** is the usual LISP function that concatenates two lists, the storage needed to represent the n+1 solutions is O(n) rather than O($n^2$), as might happen without environment sharing. Similar improvements are possible in other cases too.

Our execution strategy exploits parallelism in solving equations, especially or-parallelism which arises in producing multiple solutions. This parallel equation solver orders equations judiciously to minimize the amount of computation performed, and also performs a number of space-saving optimizations. The net result is a space-efficient, parallel equation solver.

The rest of this paper divides into the following sections: section 2 gives a brief description of an equational language called EqL (which serves as a concrete setting for our work), and also its operational semantics (which serves as a basis for the parallel equations solver); section 3 describes the parallel equation solver; and section 4 presents conclusions and further comparisons with related work.

## 2. EqL: An Equational Language

### 2.1. Language Features

An EqL program is a collection of conditional rewrite rules, where each rule is of the form

$$f(p_1, \ldots, p_n) \Rightarrow exp$$

or of the form

$$f(p_1, \ldots, p_n) \Rightarrow exp \ \textbf{where}\ \mathcal{S},$$

where $p_i$ is a pattern, *exp* is an expression, and $\mathcal{S}$ is a set of equations. The top-level goal is a set of equations to be solved.

A couple of syntactic points should be noted: First, the right-hand sides of operation definitions may contain variables not present on the left-hand sides. These newly introduced variables are referred to as *logical variables*, and they derive their values as a result of solving the associated equations. Note that there are no global variables in this language. Second, the set of equations forms a *conjunctive* set, i.e., they are simultaneously true or false. Although *disjunctive* sets are useful and can be supported in the language, we do not consider them in this paper. However, at execution time multiple sets of equations need to be processed because rules may be nondeterministic.

The use of conditional rewrite rules for functional programming should be clear. We therefore give a simple example to show how any Horn logic program can be mechanically converted to an EqL program. The following program is written using DEC-10 Prolog syntax [CM81]:

$rev([\,], [\,]).$
$rev([H|T], Y) \ :- \ rev(T, Z), apd(Z, [H], Y).$
?- $rev(X,$ [1,2,3,4]).

The mechanically derived EqL program would be:

$$rev([\,],[\,]) \Rightarrow true$$
$$rev(cons(H,T),Y) \Rightarrow true \ \ \textbf{where}\ rev(T,Z) = true$$
$$apd(Z,[H],Y) = true$$

$?\text{-}\ rev(X, [1\ 2\ 3\ 4]) = true$

It should be noted that the above program is not the best way to define the 'reverse' function in EqL. For example, the program below is a clearer definition of 'reverse':

$$rev([\,]) \Rightarrow [\,]$$
$$rev(cons(H,T)) \Rightarrow apd(rev(T),[H])$$
$$?\text{-}\ rev(X) = [1\ 2\ 3\ 4]$$

More examples of EqL are given in other papers [JS86, JSG86]. The language we consider here for implementation is a restriction of an earlier version [JS86] in that set expressions are not included.

## 2.2. Operational Semantics

We review here the operational semantics of EqL because it serves as the basis for the parallel equation solver. In order to simplify the presentation of the semantics we restrict the left-hand sides of operation definitions to be of the form $f(v_1, \ldots, v_n)$ where each $v_i$ is a variable. Definitions with patterns on the left-hand sides, such as $f(p_1, \ldots, p_n) \Rightarrow exp$, where each $p_i$ is a pattern, can be effectively replaced by

$$f(v_1, \ldots, v_n) \Rightarrow exp \textbf{ where } v_1 = p_1 \ldots v_n = p_n.$$

If the right-hand side of an operation definition were of the form $exp$ **where** $\mathcal{S}$ then the condition $\mathcal{S}$ simply augments the set of equations, $v_1 = p_1 \ldots v_n = p_n$.

In order to express the solution of equations using reduction rules, we represent an equation $e_1 = e_2$ as a single expression $\mathcal{E}(e_1, e_2)$, where $\mathcal{E}$ is a unique constructor. In general, a set of equations $e_{11} = e_{12}, e_{21} = e_{22}, \ldots e_{n1} = e_{n2}$ is represented by a tree of nested *and*'s as follows:

$$and(\mathcal{E}(e_{11}, e_{12}), \ldots, and(\mathcal{E}(e_{n-1\ 1}, e_{n-1\ 2}), \mathcal{E}(e_{n1}, e_{n2}))).$$

We define, similar to Hullot [H80], an *occurrence*, as a string of 'l' and 'r' (for left and right respectively), for uniquely identifying an equation within an equation tree.

### Reduction rules

Below we express the operational semantics of an EqL program in terms of two relations: the $\rightarrow$ relation for performing pure *reductions*, and the $\rightsquigarrow$ relation for performing variable bindings. These are one-to-many relations over equations, and are mutually exclusive and collectively exhaustive for computing solutions to a set of equations. We use the special symbol *sat* to denote the satisfaction of two equations.

1. *Unit Identity:*

   $\mathcal{E}(a, a) \rightarrow sat$ IF atom($a$)

   $\mathcal{E}(v, v) \rightarrow sat$ IF variable($v$)

2. *And Elimination:*

   $and(sat, sat) \rightarrow sat$

3. *Decomposition:*

   $\mathcal{E}(cons(e_1, e_2), cons(e_3, e_4)) \rightarrow and(\mathcal{E}(e_1, e_3), \mathcal{E}(e_2, e_4))$

4. *Function Application:*

   $\mathcal{E}(f(e_1, \ldots, e_n), e) \rightarrow and(\mathcal{E}(exp\ \delta, e), \mathcal{S}\delta)$

   $\mathcal{E}(e, f(e_1, \ldots, e_n)) \rightarrow and(\mathcal{E}(exp\ \delta, e), \mathcal{S}\delta)$

if there is an operation definition $f(v_1, \ldots, v_n) \Rightarrow exp$ **where** $\mathcal{S}$, and $\delta = \{v_1 \leftarrow e_1, \ldots, v_n \leftarrow e_n\}$. An operation definition of the form $f(v_1, \ldots, v_n) \Rightarrow exp$ is just syntactic sugar for $f(v_1, \ldots, v_n) \Rightarrow exp$ **where** *sat*.

5. *Logical Variable Binding:*

There are basically two sets of cases: (a) the binding of a variable or an atom to a variable, and (b) the binding of a cons-term to a variable. The rules for case (a) are:

$\mathcal{E}(v, a) \leadsto_\sigma sat$ where $\sigma = \{v \leftarrow a\}$

$\mathcal{E}(a, v) \leadsto_\sigma sat$ where $\sigma = \{v \leftarrow a\}$

$\mathcal{E}(v_1, v_2) \leadsto_\sigma sat$ where $\sigma = \{v_1 \leftarrow z, v_2 \leftarrow z\}$

where $v, v_1$ and $v_2$ are distinct unbound variables, $a$ is an atom or [ ], and $z$ is some distinct new variable. The rules for case (b) are:

$\mathcal{E}(v, cons(e_1, e_2)) \leadsto_\sigma and(\mathcal{E}(z_1, e_1), \mathcal{E}(z_2, e_2))$

$\mathcal{E}(cons(e_1, e_2), v) \leadsto_\sigma and(\mathcal{E}(z_1, e_1), \mathcal{E}(z_2, e_2))$

where $v$ is an unbound variable, $\sigma = \{v \leftarrow cons(z_1, z_2)\}$, and $z_1$ and $z_2$ are two distinct variables. Note that we do not permit the binding of an expression to a variable because an expression might not denote any value.

**Solution derivation**

Given a set of equations, $\mathcal{S}$, if an equation $eqn_1$ at occurence $o$ rewrites to $eqn_2$, i.e., $eqn_1 \rightarrow eqn_2$, then we write

$\mathcal{S} \rightarrow \mathcal{S}[o \leftarrow eqn_2]$.

The reader may note that we have used the same notation $\rightarrow$ to denote two relations: one between two equations, and the other between two sets of equations. However, from context it will be easy to distinguish which of the two relations is intended.

A derivation starting from $\mathcal{S}$ and using rules 1-4 is called a **$\rightarrow$-derivation**, and is denoted by $\rightarrow^*$.

We now sketch the solution of a set of equations $\mathcal{S}$. We associate **solution bindings** $\sigma$ with a set of equations S, and denote this pair by (S **with** $\sigma$). These bindings associate a term with each logical variable in the equations. Initially these bindings are empty, thus initially we have (S **with** $\phi$). As the set of equations get reduced, the associated solution bindings get "refined." The following two rules express this refinement:

1. ($\mathcal{S}$ **with** $\rho_1$) $\leadsto$ ($\mathcal{S}[o \leftarrow eqn]$ **with** $\rho_1$) IF $\mathcal{S}[o] \rightarrow eqn$

2. ($\mathcal{S}$ **with** $\rho_1$) $\leadsto$ (($\mathcal{S}[o \leftarrow eqn]$)$\rho_2$ **with** $\rho_1 \cup \rho_2$) IF $\mathcal{S}[o] \leadsto_{\rho_2} eqn$

A derivation starting from ($\mathcal{S}$ **with** $\phi$) and using rules 1-5 is called a **$\leadsto$-derivation**, and is denoted by $\leadsto^*$.

**Definition:** $\sigma \downarrow V(\mathcal{S})$ *is a* **computed solution** *of a set of equations* $\mathcal{S}$ *if there is a* $\leadsto$-*derivation of (sat* **with** $\sigma$*) starting from (*$\mathcal{S}$ **with** $\phi$*), i.e.,*

($\mathcal{S}$ **with** $\phi$) $\leadsto^*$ (*sat* **with** $\sigma$).

where $\sigma \downarrow V(\mathcal{S})$ refers to the subsitution $\sigma$ restricted to the variables in $\mathcal{S}$.

Note that we have altogether introduced four relations: the relation $\rightarrow$ between two equations, the relation $\rightarrow$ between two *sets* of equations, the relation $\leadsto_\rho$ between two equations, and the the relation $\leadsto$ between two *sets* of equations.

We illustrate the operational semantics by a simple example. Consider the following program:

$null(x) \Rightarrow true$ **where** $x = [\,]$
$null(x) \Rightarrow false$ **where** $x = cons(x_1, x_2)$

and a top-level equation

?- $null(y) = true$.

The top-level equation is expressed using the constructor $\mathcal{E}$ as $\mathcal{E}(null(y), true)$. A $\leadsto$-derivation of *sat* for this equation is shown below:

$\mathcal{E}(null(y), true)$ **with** $\phi$

$\leadsto$ $and(\mathcal{E}(true, true), \mathcal{E}(y, [\,]))$ **with** $\phi$ (rule 4)

$\leadsto$ $and(sat, \mathcal{E}(y, [\,]))$ **with** $\phi$ (rule 1)

$\leadsto$ $and(sat, sat)$ **with** $\{y \leftarrow [\,]\}$ (rule 5)

$\leadsto$ $sat$ **with** $\{y \leftarrow [\,]\}$ (rule 2)

The above example uses the first definition of *null* in the application of rule 4. Another $\leadsto$-derivation is possible using the second rule for *null*, but this derivation will not terminate with *sat*.

### 2.3 Discussion

The solution of a set of equations is thus a process of gradually "refining" the values of its logical variables until *sat* is derived. This process is therefore referred to as **object refinement**. It is a generalization of the outermost reduction rule in functional languages because refinement includes reduction. It is also a generalization of unification (without the "occurs check") because equations are not restricted to terms.

The $\leadsto$ relation can serve as the basis for a breadth-first procedure for finding the complete set of solutions. At each step of the rewriting process, any equation may selected for rewriting. If a function application appears at the outermost level in one of its constituent expressions it would be rewritten to all of its different possible right-hand sides, thereby resulting in multiple sets of equations to be solved for multiple equations. Note that multiple solutions can arise only because of the presence of multiple definitions for some operation.

In order to prune unproductive computational paths, it is necessary to recognize the following forms of equations in an equation set, where the $e_i$'s are arbitrary expressions.

$\mathcal{E}(a_1, a_2)$ IF $a_1, a_2$ are atoms and $a_1 \not\equiv a_2$
$\mathcal{E}(a, cons(e_1, e_2))$ IF $a$ is an atom
$\mathcal{E}(cons(e_1, e_2), a)$ IF $a$ is an atom
$\mathcal{E}(e, f(e_1, \ldots e_n))$ IF $f$ has no definition.
$\mathcal{E}(f(e_1, \ldots e_n), e)$ IF $f$ has no definition.

An equation-set containing any of the above forms of equations can never reduce to *sat*, and therefore may be abandoned. Indeed, an interpreter should be "on the look out" for these cases and cause early pruning of such branches. This strategy thus gives rise to a form of **lazy equation solution**. Deferring function application until no other kinds of rules apply can also avoid much over-computation.

The *soundness* and *completeness* of the above rules for computing solutions using a breadth-first search was proved in [JS86]. Our completeness is similar to the 'recursively enumerable completeness' defined by Goguen and Meseguer [GM84].

## 3. Parallel Equation Solution

As should be evident from the operational semantics, computation in EqL basically consists of solving equations. There are two obvious sources of parallelism in EqL: (1) and-parallelism arises in solving a single set of equations in parallel, and (2) or-parallelism arises in solving multiple sets of equations in parallel. These are analogous to the forms of parallelism in logic languages [CK81]. A further source of and-parallelism in EqL is the splitting of an equation into two equations as a result of the decomposition rule (rule 3).

In this paper, we present our preliminary ideas on solving a set of equations in parallel. We sketch an or-parallel, and-sequential execution model; that is, we do not attempt to obtain any and-parallelism. We eventually would like to also obtain and-parallelism, but our initial model does not support it. The main difficulty in obtaining and-parallelism is in determining which equation in a conjunctive set should produce the binding for an unbound variable that occurs in function calls in several equations. Exploiting and-parallelism without this knowledge can lead to much over-computation; however, see [CK85] for an approach to this problem.

### 3.1. Frames

The run-time environment is organized as a *tree of frames*, where each frame has two main parts:

(1) an entry for each of its **relevant** variables (which will be defined later); and

(2) a set of equations to be solved.

There are three important facets to our execution strategy:

(1) how the relevant variables of a frame are determined;

(2) how multiple solutions are represented; and

(3) which equation is selected next for processing.

We summarize these issues in the subsequent subsections, but first we give a brief overview of the execution of an EqL program:

Initially, a frame is created for the top-level goal. The relevant variables in the case of the top-level goal are all the variables appearing in the goal. One equation from this set is chosen to be solved (by a selection process to be described later). If the chosen equation has a function call at the outermost level and this function is defined by $k$ rules, $k$ new frames are created. The equations associated with these new frames are then determined (using rule 4 of the operational semantics). Because equations are solved by *reduction*, i.e., a set of equations is continually rewritten to another set, the equations for the new frames are constructed in such a way that execution in the 'calling' frame need not be resumed. Thus, only leaf frames serve as execution sites, and the degree of or-parallelism at any point may be measured by the number of leaf frames corresponding to partially solved equations.

Because all variables possess the *single assignment* property, a *bound* variable in a non-leaf frame may be shared by the entire subtree rooted at this frame. Indeed, during equation solution we will try to select those equations first that cause variables to become bound, so the effect of environment sharing is enhanced. Non-leaf frames therefore need to be retained to record these variable bindings. Unbound variables, however, could assume different values in different or-parallel paths, and are therefore not shared.

A computation terminates when all equations in all leaf frames are fully solved or are detected to be trivially unsolvable (these cases are mentioned at the end of the previous section). Computations, however, can be nonterminating in this language.

### 3.2. Relevant Variables

We say that a call $f(e_1, \ldots, e_n)$ is deterministic if $f$ has exactly one definition; otherwise, we say the call is nondeterministic.

We denote by $\mathcal{V}(f)$ the set of variables appearing in some operation definition $f(p_1, \ldots, p_n)$ $\Rightarrow$ *exp* **where** $\mathcal{S}$.

For a deterministic call $f(e_1, \ldots, e_n)$, the set of relevant variables for the created frame is $\mathcal{V}(f)$.

For a nondeterministic call $f(e_1, \ldots, e_n)$, the set of relevant variables is the union of $\mathcal{V}(f)$ and the set of **affected** variables. The set of affected variables are those *unbound* variables in the environment of the (nondeterministic) call that *could possibly become* bound as a result of the call.

When a frame is created, space is allocated for each of its relevant variables. (We later mention an optimization for reducing the space allocated for affected variables.) When a non-deterministic call occurs, the number of bindings for an affected variable initially is equal to the number of different definitions for the operation being called. A count field is therefore maintained with each variable indicating the initial number of bindings. The actual bindings for an affected variable $x$ are pointers to the respective locations for $x$ in the created frames. The number of bindings for an affected variable $x$ can ultimately be greater than the initial count. This happens when the location for $x$ in one of the nondeterministic branches itself becomes an affected variable in some subsequent call.

### 3.3. Equation Selection

Ordering equations judiciously can sometimes dramatically reduce the amount of computation needed. Our ordering strategy is fairly simple and performs well: equations with function (operation) calls at the outermost level are deferred as much as possible. This ensures that a function call has as many variables bound as possible, thereby narrowing-down nondeterministic choices as early as possible. Thus the order in which equations are solved is as follows:

(1) Examine whether there are any trivially unsolvable equations, and if so abandon the set of equations and delete the associated frame.

(2) Select equations for which equation-solution rules 1, 3, and 5 are applicable.

(3) Select equations for which rule 4 (function application) is applicable.

Case 3 may be refined further into a number of cases. We mention below some useful heuristics for selecting an equation in this set of equations: (a) Function calls which do not result in any further calls are preferred over those that do. (b) Deterministic calls are preferred over non-deterministic calls. (c) Function calls with more fully defined arguments are preferred over those with less defined arguments. It is important to note that the above orderings do not affect completeness, i.e., they do not prevent any solution from being found.

### 3.4. Example

We illustrate the ideas in the preceding subsections by means of an example. The following program defines the function *append* in LISP:

$append([\,], y) \Rightarrow y$
$append(cons(h,t), y) \Rightarrow cons(h, append(t, y))$
?- $append(x, y) = [1\ 2]$

First, the above program is converted (at compilation time) into the following semantically equivalent program without any patterns on the arguments:

$append(x, y) \Rightarrow y$ **where** $x = [\,]$
$append(x, y) \Rightarrow cons(h, append(t, y))$ **where** $x = cons(h, t)$
?- $append(x, y) = [1\ 2]$

Figure 1 at the end of this paper sketches the sequence of steps during the execution of the above program. Each frame has a numeric label (shown in the right-hand top corner of the frame) indicating when it was created relative to other frames. Although we show the name of the variable, such as $x^2$, $x^3$, $x^4$, etc., in the implementation these fields would contain pointers to the storage locations for the respective affected variables.

This example shows how environment sharing can reduce the amount of storage needed to represent multiple solutions. Note that only a linear amount of space is needed to represent all n+1 solutions to the goal ***append***$(x, y)$ = [1 2 ... n]. The key to achieving this is to share the common portions of the n+1 solutions, as shown in figure 3.1. (Note that the bindings for $y^2$, $y^4$, and $y^6$ actually share portions of the input list [1 2].)

Solutions to the top-level goal can be reported as soon they are found or when the program terminates. We wish to support both modes of solution-reporting and allow the user to specify which is desired. Reporting solutions as soon as they are found could take longer than if reported at the end of the program (because common solution paths must be re-traversed) but offers the advantage that pointers can be de-referenced early. However, relying on reporting solutions at the end of the program is not safe because of the possibility of nontermination.

### 3.5. Optimizations

In the above example we introduced a simple optimization for reducing the number of affected variables to be copied when frames are created for nondeterministic calls: when an affected variable appears as an actual parameter, it need not be allocated space in the frames created. Thus the variables $x_1$ and $y_1$ are not copied into frames 2 and 3 in figure 3.1. We sketch two other optimizations applicable to nondeterministic calls in this section.

The first optimization is to defer the copying of equations where possible. Often only one of the branches of a nondetermistic call might be successful based on the patterns on the arguments. In such cases, it is desirable to avoid copying with each frame all other equations in the environment of the call. Thus, deferring the copying of other equations until equations corresponding to the patterns are successful can reduce the amount of space and time needed for or-parallel execution.

It is possible to further save space and time in solving independent and-parallel equations such as

***append***$(x,\ y)$ = [1 2 ... m]

***append***$(p,\ q)$ = [1 2 ... n].

Here, the (m+1).(n+1) solutions can be computed in O(m+n) time and represented using O(m+n) space because the solutions for $p$ and $q$ need not be not recomputed for each solution of $x$ and $y$. (Note that Prolog implementations would in fact recompute the solutions for $p$ and $q$.) In general, partitioning a set of equations into the finest possible groups such that no two groups have any common variable can reduce the space and time needed for solving such and-parallel equations. However, some extensions to the basic frame model described in this section are needed to accomodate this optimization—frames will no longer form a tree, but instead a acyclic graph.

## 4. Conclusions and Related Work

We believe that efficient equation solution can serve as a unifying computational model for functional and logic languages. This research aims at elucidating the details of such a computational model, relying on parallelism to obtain performance. The main focus is on a space-efficient or-parallel equation solver that does not sacrifice completeness. The salient features of our approach are: (i) environment sharing across or-parallel equations to minimize space; (ii) equation ordering to minimize execution time; and (iii) optimizations to further reduce space and time.

The differences in our approach from Prolog implementations [WPP77] is worth noting: Firstly, the state of the computation cannot be stack-based, as in Prolog implementations, because we are attempting to produce all solutions. Because Prolog is sequential and produces one solution at a time (by depth-first search and backtracking), it is unnecessary for it to determine relevant variables; instead, a *trail stack* is used to record those variables that were bound since the last nondeterministic call and need to be unbound upon backtracking. In contrast, because EqL must produce all solutions (by breadth-first search), it is necessary to determine which variables can be affected during a nondeterministic call so that storage can be allocated for each such variable in each nondeterministic branch.

Most Prolog implementations also employ structure-sharing, a technique whereby a new data object is created using a *molecule*, which consists of a pointer to the 'skeleton' of a term and an environment-pointer. This effectively allows the invariant part of terms from being replicated at run-time, at the expense of longer access time to substructures. Because our computational model is based on reduction, which involves code-substitution at run-time, new data objects are created by copying—this is closer to the non-structure-sharing technique [M82].

The problem of ordering equations is similar to the selection of goals in a Horn-logic program. The use of *mode declarations* by the programmer to explicitly guide and-parallel execution is proposed by Clark and Gregory [CG86]. In our approach, however, we try to order equations without explicit advice from the programmer. This is closer in spirit to the work by Conery and Kibler [CK85], except that they also support and-parallelism.

The parallel implementation of functional and logic languages has been a topic of considerable interest because these languages are free from side-effects and explicit sequencing. Among the many proposed approaches in the literature, the ones most closely related to our work are the Or-parallel models of Haridi [HC83] and Lindstrom [L84], and the reduction model of Onai et al [O85b]. All of these are designed to execute Horn clauses rather than equations. Another major difference is that our execution model employs a combination of reduction and an explicit environment. This way, we seek to combine the parallelism provided by reduction with the storage economies of a shared environment.

## References


[CK81] J. S. Conery and D. F. Kibler, "Parallel Interpretation of Logic Programs," In *Conf. Functional Prog. Lang. and Comp. Arch., ACM*, 1981, pp. 163–170.

[CK85] J. S. Conery and D. F. Kibler, "AND Parallelism and Nondeterminism in Logic Programs," *New Generation Computing*, vol. 3, 1985, pp. 43-70.

[CM81] W. F. Clocksin and C. S. Mellish, *Programming in Prolog,* Springer-Verlag, New York, 1981.

[CG86] K. Clark and S. Gregory, "PARLOG: Parallel Programming in Logic," *TOPLAS*, vol. 8, no. 1, pp. 1-49.

[D83] J. Darlington, "Unifying Functional and Logic Languages," Internal Report, Imperial College, London, 1983.

[DP85] N. Dershowitz and D. A. Plaisted, "Applicative Programming *cum* Logic Programming," In *1985 Symp. on Logic Programming*, Boston, pp. 54–66.

[F84] L. Fribourg, "Oriented Equational Clauses as a Programming Language," *J. Logic Prog.* **2** (1984) pp. 165–177.

[GM84] J. A. Goguen and J. Meseguer, "Equality, Types, Modules, and (Why Not?) Generics for Logic Programming," *J. Logic Prog.* **2** (1984) pp. 179–210.

[H80] J-M. Hullot, "Canonical Forms and Unification," In *Proc. 5th Workshop on Automated Deduction*, Springer Lecture Notes, 1980, pp. 318–334.

[HC83] S. Haridi and A. Ciepielewski, "An OR-Parallel Token Machine," Tech. Report No. TRITA-CS-8303, Royal Inst. of Tech., May 1983.

[J84] B. Jayaraman, "An Equational Language and its Evaluator," TR-CS-84-11, Department of Computer Science, Univ. of N. Carolina, December 1984.

[JS86] B. Jayaraman and F.S.K. Silbermann, "Equations, Sets, and Reduction Semantics for Functional and Logic Programming," In *1986 ACM Conf. on LISP and Functional Programming*, pp. 320-331, Boston, August 1986.

[JSG86] B. Jayaraman, F.S.K. Silbermann, G. Gupta, "Equational Programming: A Unifying Approach to Functional and Logic Programming," In *Int'l Conf. on Computer Languages*, pp. 47-57, Miami Beach, October 1986.

[L84] G. Lindstrom, "OR-Parallelism on Applicative Architectures," In *Second Intl. Conf. on Logic Programming*, Uppsala, July 1984.

[M82] C. S. Mellish, "An Alternative to Structure Sharing in the Implementation of a PROLOG Interpreter," In *Logic Programming*, Ed. K. L. Clark and S.-A. Tärnlund, Academic Press, 1982, pp. 99-106.

[O85] M. J. O'Donnell, "Equational Logic as a Programming Language," M.I.T. Press, 1985.

[O85b] R. Onai, et al, "Architecture of a Reduction-based Parallel Inference Engine: PIM-R," In *New Generation Computing*, vol. 3, 1985, pp. 197-228.

[R85] U. S. Reddy, "Narrowing as the Operational Semantics of Functional Languages," In *1985 Symp. on Logic Programming*, Boston, 1985, pp. 138–151.

[RS82] J. A. Robinson and E. E. Sibert, "LOGLISP: Motivation, Design, and Implementation," In *Logic Programming*, Ed. K. L. Clark and S.-A. Tärnlund, Academic Press, 1982, pp. 299–313.

[SY84] P. A. Subrahmanyam and J-H. You, "Pattern-driven Lazy Reduction: A Unifying Evaluation Mechanism for Functional and Logical Programs," In *11th ACM POPL*, Salt Lake City, 1984, pp. 228-234.

[WPP77] D. H. D. Warren, F. Pereira, and L. M. Pereira, "Prolog: the Language and Its Implementation Compared with LISP," *SIGPLAN Notices* **12**, No. 8 (1977) pp. 109–115.

[YS86] J-H. You and P. A. Subrahmanyam, "Equational Logic Programming: an Extension to Equational Programming," In *13th ACM POPL*, St. Petersburg, 1986, pp. 209-218.

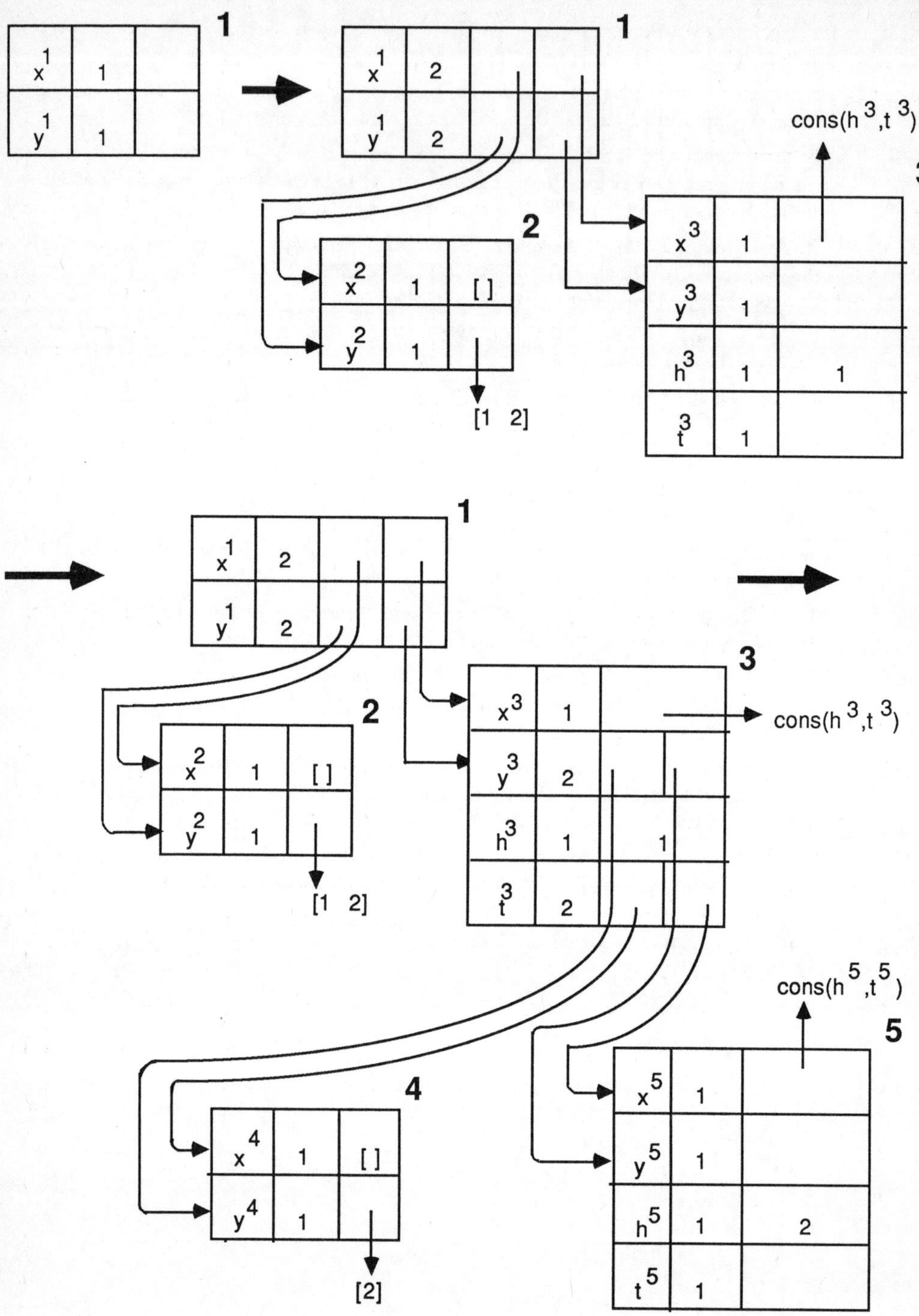


Figure 1: Snapshot execution of append (x, y) = [ 1 2 ]

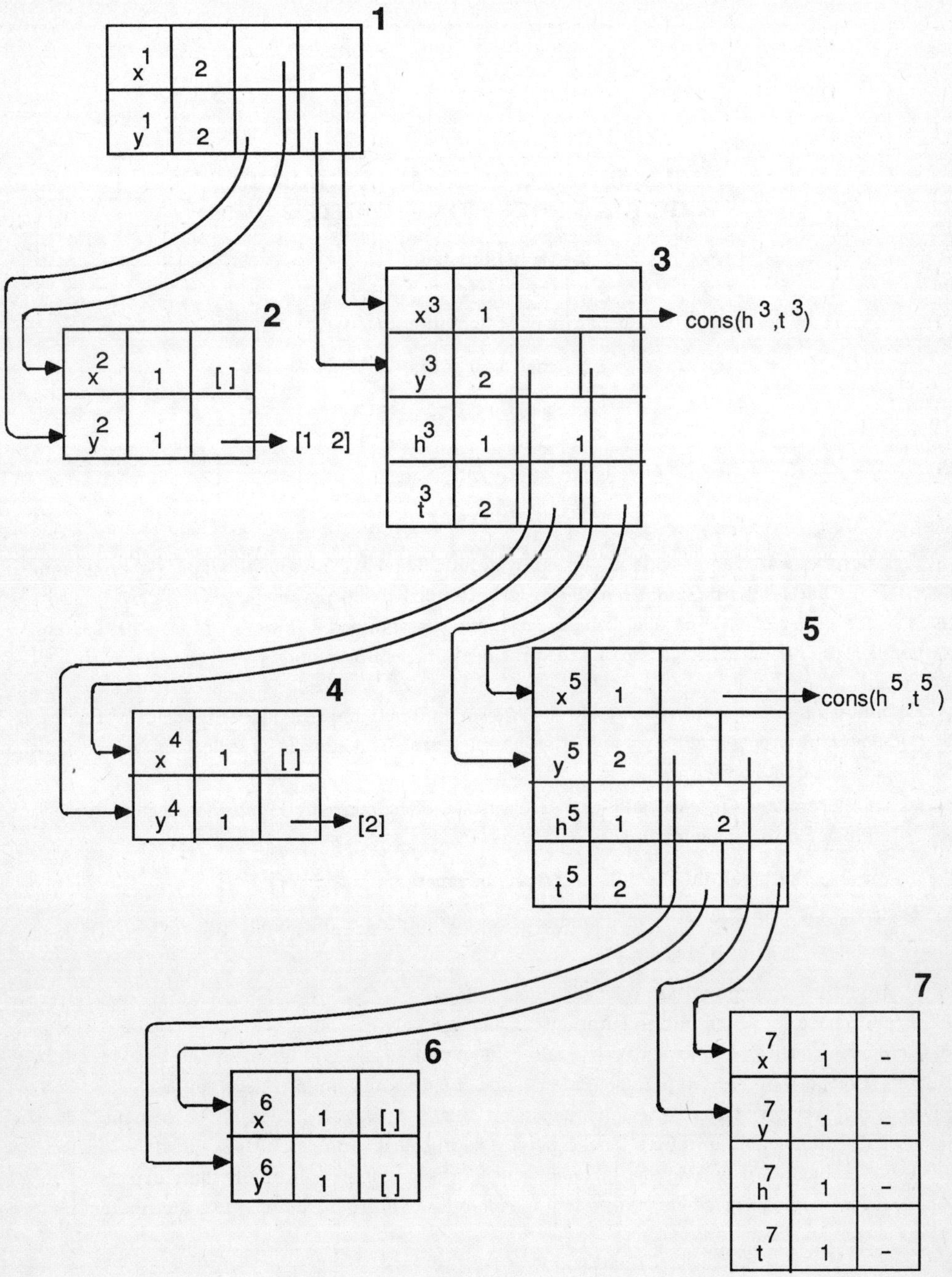


Figure 1 (Continued from previous page)

# IMPLEMENTING LOGICAL VARIABLES ON A GRAPH REDUCTION ARCHITECTURE[1]

**Gary Lindstrom**

Department of Computer Science

University of Utah

Salt Lake City, Utah 84112

## Abstract

Logical variables offer a semantic meeting ground between functional and logic programming languages. That is, if functional languages are extended to do parameter passage by unification, much of the power of AND-parallel logic programming is obtained. However, multi-path search, by OR-parallelism or backtracking, remains the province of logic programming.

We outline an implementation strategy for FGL+LV, a functional language with logical variables, on the Rediflow multiprocessing graph reduction architecture. The aspects of logical variables receiving special consideration include:

a. parallel unification, especially proper treatment of indeterminate behavior, e.g. mutual exclusion on variable binding;

b. variable binding through emulated graph node merging;

c. exploitation of two levels of demand: *assertive* (during unification), and *non-assertive* (ordinary "read-only" usage), and

d. avoidance of meaningless cyclic variable bindings.

The existing base language implementation can be smoothly extended, with a word size increase of only one bit (needed to implement two levels of demand). Lazy evaluation in the base language is retained, except that actual parameters are now made strict to one level of evaluation. This requirement, overlooked in a previous paper on this subject, is argued to be semantically and operationally inescapable in a functional language with logical variables. It also provides insight into the vexing problem of how to apply the occur check in a language with infinite data objects. A novel technique for merging cyclic lists is used to implement logical variable binding in a distributed manner without locking or busy waiting.

---

[1]**This material is based upon work supported by NSF Grant DCR-8506000, and a Shared University Research Grant from the IBM Corporation.**

# 1. Incorporating Logical Variables into Functional Programming

## 1.1. Essential Characteristics

From the perspective of functional programming, logical variables present a radically different basis for information flow. In particular,

- logical variables are not directly bound by uniquely determined expressions, in contrast to the customary dataflow orientation of functional languages;
- rather, they denote values determined by the cumulative application of successive constraints.

## 1.2. Advantages

In a functional programming language with logical variables, this difference can permit novel effects otherwise quite difficult to obtain. These include:

- an *action at a distance* effect, whereby certain kinds of "side-effects" can be achieved within an applicative setting (the Prolog "difference list" trick for constant time list concatenation is a familiar example);
- a stronger *object orientation*, under which logical variables play the role of mutable shared entities, and
- the use of this object orientation as a basis for *isotropic* process communication and synchronization, as in Concurrent Prolog [17, 18].

We investigate here the adoption of logical variables in a deterministic setting, i.e. without support for backtracking or OR-parallel search. This single extension brings into the realm of functional programming such elegant techniques as the Milner algorithm for polymorphic type checking [14]. And, while the absence of OR-parallelism may be lamentable, in partial compensation we retain functional programming's execution *directionality*, which offers a clean solution to the AND-parallel control problem which plagues pure logic programming.

## 1.3. Challenges

If one seeks to add logical variables to a normal order functional programming language, the following challenges quickly become apparent:

a. semantic upward compatibility, including lazy evaluation;

b. preservation of concurrency opportunities;

c. assurance of overall determinacy despite aggressive exploitation of these concurrency opportunities, and

d. efficient implementation on distributed architectures.

A previous paper [13] addresses all but the last of these challenges, which is the subject of this paper.

## 1.4. Related Work

The task of amalgamating functional and logic programming is a highly active research area. Among the efforts most allied to the work reported here are:

- HASL [1], which adds conditional unification to a sequential form of SASL;

- Funlog [19], which defines a concept of *semantic unification* in functional programming, and
- an extension to HOPE involving absolute set expressions [4].

Also relevant are languages based on equational programming, e.g. EQLOG [5], and logic programming languages with mode annotations, e.g. PARLOG [2]. The close connection of this work to Concurrent Prolog [17] is also apparent, particularly in its distributed implementations [21].

Other researchers who have investigated the augmentation of functional programming languages to include logical variables as a separable aspect of logic programming include Danforth [3] and Nikhil, Pingali and Arvind [15].

## 2. A Function Graph Language

### 2.1. FEL and FGL

Our base language is graphical in form, but also possesses an equivalent textual notation. An equational style is used; hence the language is called FEL for Function Equation Language [9]. The essential features of FEL and their notations are as follows:

- *Block notation:* The "main program" of a FEL computation is a block. Blocks have the form

```
{       equations
        result expression
        equations
}
```

  where the `equations` define locally bound names, generally used within the **result** `expression` and recursively among their definitions. Non-local identifiers may be "imported" into nested blocks in observance of static scoping.

- *Function definition and application:* Functions in FEL take single arguments, with colons denoting function application, e.g. `f:p`. Functions are defined by equations of the form

  **function** `fname:parameter = expression`

- *Tuples and lists:* FEL supports tuples (vectors) of any fixed length up to an implementation defined limit. For simplicity, we will deal here only with tuples of length two, i.e. *pairs*. Lisp-like lists may be constructed in the usual manner, using pairs as building blocks. The notations `[a, b]` and `cons:[a, b]` are equivalent in FEL (hence `cons` is the identity function, restricted here to pairs). The operators `head` and `tail` play the role of `car` and `cdr`.

- *Quoting convention:* `'a` evaluates to `a` itself.

- *Comments:* brackets of the form `(* ... *)` surround comments.

FEL in its entirety contains a wide variety of built-in operators and primitive data types. For our purposes here we will restrict our attention to the elementary subset described above.

Programs in FEL can be mapped into an underlying graphical form in a quite straightforward manner. The result is a "data flow" digraph, where the nodes are operators and the arcs express input-output dependencies among them. Referential transparency is exploited through exhaustive common subexpression elimination within function bodies. Such graphs are programs in the

Function Graph Language FGL. Henceforth, we will consider FEL and FGL to be synonymous except for their visual renderings.

## 3. Graph Reduction

### 3.1. Executing FGL by Graph Reduction

FGL has been designed as the native language of a graph reduction architecture, Rediflow [10]. Its reduction rules are quite simple:

- Whenever the operands to an operator are sufficiently evaluated, that operator may be applied to those operands, with the operator's node being replaced by a new node containing the resulting value.
- An application of a programmer defined function may at any time be replaced by a copy of the function's graph, with the actual parameter arc connected to the formal parameter arc in the substituted graph.

Of course, the second rule must be cautiously applied, lest recursion lead to unnecessary divergence. This caution is provided by demand evaluation, as delineated in [7].

### 3.2. Reduction Architectures

Graph reduction offers many advantages as a strategy for executing functional programs. These include:

- very flexible patterns of concurrency and load distribution on parallel architectures;
- clean support for infinite and cyclic data structures through demand evaluation [8];
- good program and data locality effects;
- the support of recursion and iteration (via tail recursion) without the need for explicit token objects or "color" tags and associative memories for their assembly into operator "firing sets", and
- the ability to incorporate a form of data-drive as a submode [20].

The principal prices to be paid for these advantages are potentially costly storage management, and the added overhead of demand propagation. A good survey of reduction architectures may be found in [23]. A recent effort to abstractly describe several graph reduction approaches may be found in [22].

### 3.3. Graph Reduction on Rediflow

Rediflow executes FGL programs through a quite direct emulation of the demand-sensitive node level graph transformation rules sketched in section 3.1. In simplified form, its principal features include the following.

a. Function bodies are encoded as *blocks* of contiguous words [6], with arcs represented as *local* addresses relative to the block beginning address.

b. Individual *word representations* are indicated in figure 3-1. Words usually are created in an *instruction* format as a result of function code block copying, and are transformed into the *data* format as the function executes. This "destructive" style of execution is characteristic of reduction machines.

c. There is space in each format 1 instruction for a fixed number of argument links (currently three). These links are stored in the *anti-dataflow*, or *demand propagation* direction.

d. The arrival of demand at a format 1 instruction causes a *notifier* link (also local) to be set to point to the source of the demand. If the demand bit at the word is zero, that bit is set to one and demand is propagated to appropriate arguments, depending on the strictness characteristics of the operation. At compile time we ensure, by insertion of **ident** operations if necessary, that the result value fan-out of each node obeys a fixed limit (also currently three). Hence adequate space for all possible local notifiers can be pre-allocated in each word.

e. Arrival of demand at a data word causes an answering value notification activity. A similar effect occurs when an operator is applied and its word is converted from instruction to data format. Before the instruction is overwritten with the resulting value, the notifiers are used to propagate value notification.

f. Local addresses work well within function blocks, but global linkage is clearly necessary as well. Since global addresses are likely to be considerably larger than local addresses, special instructions are used where global addresses can arise. At most one global address is stored in each word.

g. The most important of the instructions involving global addresses are a special pair of instructions that serve as "impedance matchers" between local and global addressing modes:

   - **forward** instructions (format 2), which take a *local argument* and *global notifiers*, and

   - **fetch** instructions (format 3), which take a *global argument* and *local notifiers*.

   The instructions **fetch** and **forward** are used for several purposes, including function value reporting and tuple component accessing. In the latter case the tuple component is represented as a **forward**, and all demanders of that component are **fetch**es chained together through their global notifier fields.

### 3.4. Task Based Execution

At any moment in the execution of an FGL program on Rediflow there is a frontier of nodes in the graph where reduction activity is ready to occur. These sites are referenced by a list of *tasks*, which can be distributed over the set of available processors. Tasks are Rediflow's analog of program counters. Typical tasks include:

a. *Demand propagation:* the instruction at location `[b, k]` (offset `k` in the block at global address `b`) is propagating demand to the word at location `[b', k']`.

b. *Value notification:* the word at location `[b, k]` now holds a value and the word at location `[b', k']` should be so notified.

c. *Function application:* an instruction at location `[b, k]` seeks to **apply** a function whose code is of length `l` and referenced in a closure tuple at location `t`.

Other tasks of less interest here support such housekeeping activities as storage management and instrumentation and monitoring activities. Exportation of function tasks forms the basis for Rediflow's dynamic load balancing capability [11].

**Instructions:**

format 1:

```
                ---------------------------------------------
                |0| |   operation   | arguments | notifiers |
                ---------------------------------------------
                 | |
       evaluated |                    (local)     (local)
                demanded
```

format 2:

```
                ---------------------------------------------
                |0| |    forward    | arg | notifier chain  |
                ---------------------------------------------
                 | |
       evaluated |                 (local)     (global)
                demanded
```

format 3:

```
                ---------------------------------------------
                |0| |    fetch     |  argument | notifiers |
                ---------------------------------------------
                 | |
       evaluated |                    (global)    (local)
                demanded
```

**Data:**

```
                ---------------------------------------------
                |1| tag |               value               |
                ---------------------------------------------
                 |
         evaluated

                  tag ∈ {atom, tuple_ref}
```

**Figure 3-1:** Rediflow word formats.

## 4. Extending FGL to FGL+LV

We now consider the language design issues special to accommodating logical variables in FGL. The design of FGL+LV, as we call our extended language, is discussed in detail in [13]. However, we briefly review that design here for the orientation of new readers.

### 4.1. General Characteristics of FGL+LV

We extend FGL to incorporate logical variables as follows:

- *Syntactically*, FGL+LV functions have general *expressions* as formal parameters, rather than distinct *variables* (possibly tupled). Hence constants and function invocations can appear within *formal* as well as *actual* parameters, and variables can have multiple occurrences in either.

- *Semantically*, variables are given values by constraint application, rather than by being bound to particular value expressions.
- *Operationally*, parameter passage by unification is necessary, and information flow between formal and actual parameter computations is bidirectional [16].
- *Functionally*, the language loses local determinacy under parallel evaluation, since individual function invocations may succeed or fail depending on the extent to which constraints have already been applied to variables accessible by the function. However, global determinacy is retained by making function applications strict on unification success.

Thus we interpret *all* symbols not predefined or explicitly bound in equations to be logical variables. A sample FGL+LV function, expressing constant time concatenation of "difference lists," is shown in figure 4-1. Notice:

- `b` denotes the *body* or result arc of the function;
- `p` denotes its *formal parameter* arc, and
- logical variables are expressed as input-less `ubv` nodes.

---

`nconc` function text:

```
function nconc:[[x, y], [y, z]] = [x, z]
```

`nconc` function graph:

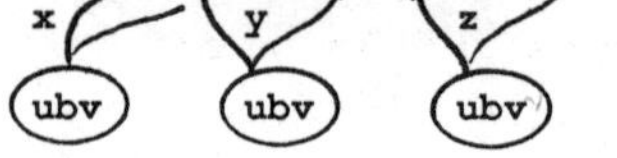


**Figure 4-1:** FGL+LV code and graph for `nconc`.

---

## 4.2. Intuitive Semantics of FGL+LV

Informally, the semantics of FGL+LV are as follows. Suppose `f:a` is an application of function `f` to actual parameter `a`. Let the body of `f` be `b`, and its formal parameter (expression, recall) be `p`. Then:

a. `a` and `p` are evaluated sufficiently to determine whether their denoted values can be unified (in the usual Herbrand "syntactic" sense).

b. If so, `b` is evaluated observing the constraints on variables in `p` that resulted from the unification of **a** and **p**, and **b**'s value becomes the result of **f:a**.

c. If not, a function invocation error has occurred, and the result is undefined.

Notice that, in contrast to Prolog, we view unification as a generalized parameter passage mechanism, rather than as a precondition for function invocation. A failing unification attempt is considered to be a function invocation error. (This is a limitation, toward the elimination of which further research is currently being applied.)

One may observe that function invocation and unification are closely intertwined in FGL+LV: unification can occur during function invocation (obviously), and vice-versa, since function invocations can appear in formal and actual parameter expressions. Moreover, the binding of a logical variable by parameter unification in one function invocation can enable the evaluation of another function invocation to resume, if the latter has been awaiting a binding for that variable. The design of FGL+LV smoothly integrates these two activities, while facilitating parallel execution but preserving overall determinacy.

## 5. FGL+LV Reduction Semantics

### 5.1. Extensions to FGL Reduction Rules

The reduction rules for FGL+LV consist of the existing rules for FGL, with the following changes made to accommodate logical variables:

a. The rule for function invocation in FGL is replaced by a new rule which accomplishes parameter linkage by unification, and makes the invocation result strict on the success of that unification.

b. New rules must be added to accomplish parallel unification, including variable binding by graph node merging.

The special reduction rules for FGL+LV are given in figures 5-1 and 5-2.

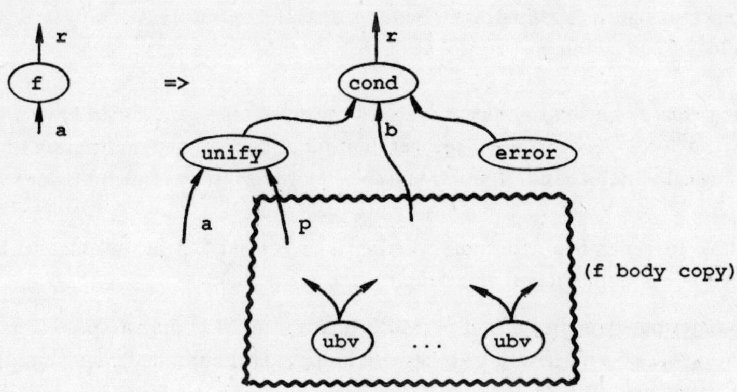


**Figure 5-1:** FGL+LV reduction rules: function invocation.

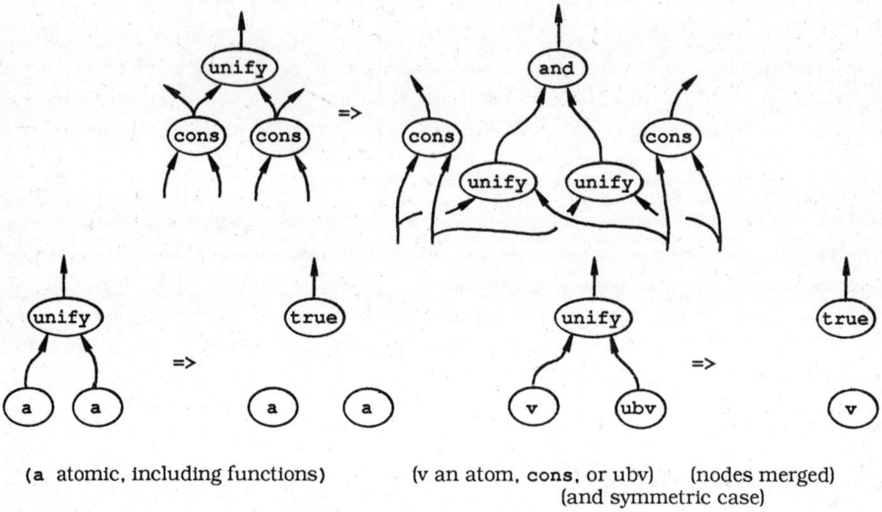


(plus obvious rules for **unify** failure)

**Figure 5-2:** FGL+LV reduction rules: `unify`.

Figure 5-2 corrects a significant error in the rules for `unify` as presented in [13]. The original rules permitted unification by node merging of any node pair where *at least one* of the nodes is an unbound logical variable. Thus variables could be bound to *unevaluated* expressions, as might be desired for lazy evaluation of actual parameters in the spirit of FGL compatibility.

However, this otherwise commendable lenience leads to semantic anomalies in FGL+LV. Consider unifying `[x, ident:x]` with `[y, y]`. The correct result is to bind `x` to `y`, or vice-versa. However, as the reduction sequence in figure 5-3 illustrates, binding unevaluated expressions to logical variables can lead to the binding of *both* `x` and `y`.

Thus it must be realized that *any* unification algorithm must take care not to bind a variable to itself. Consequently *each* argument to a `unify` operation must be analyzed sufficiently to determine whether it yields a logical variable, and, if so, which one. Moreover, note that in any `unify:[a, p]` operation:

a. It is impossible in general to determine whether `a` or `p` yield logical variables without doing one level of evaluation (to an atom, tuple, or logical variable).

b. The need to determine whether `a` and `p` yield logical variables is only a *special case* of the occur check, which we deem to be otherwise inappropriate in a language permitting infinite data structures. Thus:

   i. The unification of `[x, cons:[1, x]]` with `[y, y]` *should* (and *will* here) succeed with `x` and `y` becoming bound to an infinite stream of ones.

   ii. Furthermore, single level parameter evaluation avoids another semantic

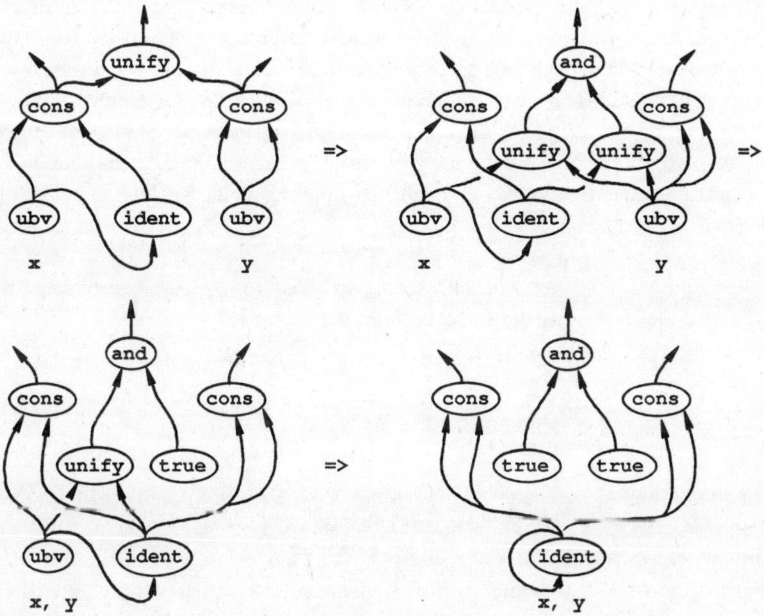


**Figure 5-3:** Unification anomaly with lazy evaluation.

anomaly: successful unification on ⊥-denoting values. For example, consider the unification of `[x, x+1]` and `[y, y]`. This will not complete until a value is externally supplied for **x** or **y**, at which time the unification will fail. Under the scheme in [13], the unification could succeed with **x** bound to **x+1**, thus denoting ⊥.

iii. In short, we argue that *exactly the right level of occur check protection* in a lazily-evaluated language is provided by this evaluation policy.

c. The one level of strictness being proposed does not increase risk of divergence in programs not containing parameter expressions denoting ⊥. This is a much weaker stipulation than that of call-by-value, the usual alternative to lazy evaluation, which precludes infinite values.

d. Although danger of evaluation deadlock may appear to be increased through this added strictness, the increase is illusory. In the example above of **x** undergoing unification with **x+1**, a deadlock will occur if **x** is involved in no other unifications. However, this corresponds directly to a demand deadlock as would occur in FGL if one accessed **x** under the equation **x = x+1**. Hence the only added risk of deadlock is that attributable to one-level evaluation eagerness.

## 5.2. The Role of Demand in FGL+LV

Demand plays an even more central role in FGL+LV than it does in FGL. In FGL, demand is an important implementation technique that guards against divergence, enhances parallelism, and minimizes effort. However, demand is not instrumental in the denotational semantics for FGL. In FGL+LV, on the other hand, demand plays a fundamental *semantic* role. Essentially, this is because sensible semantics for logical variables dictate that careful specification must be made of the conditions under which function applications are performed. If this semantic precaution is not taken, "gratuitous" function applications may cause conflicting bindings to be applied to individual logical variables, thereby causing erroneous results.

Thus, as noted in [15], the presence of logical variables in FGL+LV diminishes *referential transparency*. This is evident when we consider the following example:

```
{       function f:x = y
        a = f:1
        b = f:1
        function g:[2,3] = true
        result g:[a,b]
}
```

In this example, if referential transparency is invoked to combine `a` and `b`, then they both are equated to the *same instance* of the logical variable `y`, and the overall program result is **error**. Otherwise, *two instances* of `y` are generated, and the overall result is **true**. This visibility of common subexpression elimination is lamentable from a semantic purity viewpoint. However, it is part and parcel of the evolution of the language toward object oriented behavior, in which visibility of object identity plays such a substantial role.

The rules of figures 5-1 and 5-2 do not explicitly convey the demand propagation effects necessary in FGL+LV. However, the essentials of these effects (discussed in detail in [13]), may be summarized as follows. *Two* levels of demand are required for a correct semantic rendering of FGL+LV, as indicated in figure 5-4.

Note that assertive demand is *stronger* than non-assertive demand, because:

- assertive demand can elicit a wider class of results, viz. unbound logical variables as well as tuples and atoms, whereas
- an access of a logical variable in an ordinary (non-unification) context can only elicit the variable's eventual binding as a tuple or atom.

An arc to an operator `op` is said to be *transparent* if it delivers a value which will become the value of `op`. Examples of transparent arcs are the `then` and `else` arcs of conditionals, result arcs of function bodies, `ident` instruction inputs, etc. The *predicate* arc of a conditional is an example of a non-transparent arc; this reflects the fact that an assertive demand presented to a conditional cannot sensibly subject the unbound variables occurring in the conditional's predicate expression to a constraint to equal any particular value.

## 6. Executing FGL+LV on Rediflow

We now turn to extending Rediflow's execution mechanism to deal with the special needs of logical variables in FGL+LV.

**Demand varieties:**

*Non-assertive:* indicates conventional demand, and is not satisfied at a **ubv** until that variable becomes bound.

*Assertive:* a stronger demand, indicating "demand in unification context", which is satisfied by a **ubv** as well as by an atom or by a tuple.

**Demand transitions:**

*Non-assertive to assertive:* along input arcs to **unify** applications.

*Assertive to non-assertive:* along non-transparent arcs.

All other arcs propagate demand unchanged.

**External demands**, i.e. those applied to the main program result expression, are always *non-assertive*.

**Figure 5-4:** Demand varieties and transitions.

## 6.1. Requirements

Graph reduction as emulated on Rediflow must be augmented to include the following additional effects:

a. variable binding via graph node merging;

b. parallel unification, including proper treatment of indeterminate behavior, e.g. mutual exclusion on variable binding;

c. emulation of *assertive* and *non-assertive* demand, and

d. avoidance of cyclic variable bindings, as per section 5.1.

## 6.2. Word Formats

In order to support logical variables, we extend the word formats of Rediflow in the following respects.

a. We extend the demand field of each Rediflow word include two bits, indicating the strongest level of demand that has been received. These levels are represented as `0` (undemanded), `N` (non-assertively demanded), and `A` (assertively demanded). As argued in section 5.2, we order these indicators `0 < N < A`.

b. Two new instructions are introduced:

- **ubv**, whose role has been previously indicated, and
- **bind:[v, w]**, which is described below.

c. A new data format, **lv_ref**, is introduced to hold a pointer to a word previously seen to hold a **ubv** instruction. These reference values can be freely copied; clearly, we cannot copy actual words containing **ubv**'s, lest their unique identities be lost.

### 6.3. Unification Pseudo-Functions

Parallel unification on Rediflow is accomplished through use of a `unify` library routine, as well as the `bind` instruction mentioned above (see figure 6-1). These two activities are both pseudo-functions, in the sense that their results are generally indeterminate.

```
function unify:[a, p] =
        if (* a or p assertively evaluate to a ubv and binding is accomplished *)
                bind:[a, p]
        then    true
        else    (* both are evaluated to an atom or pair *)
                if atom:a
                then  if atom:p
                      then  eq:[a, p]
                      else  false
                else  (* a is a pair *)
                      if (* p is also a pair *)
                            tuple:p
                      then  and:[unify:[head:a, head:p],
                                 unify:[tail:a, tail:p]]
                      else  false

function bind:[v, w] =
        (* pseudo-operation which atomically:
                i)    assertively evaluates v and w;
                ii)   if v and w both yield either an atom or a pair, returns false;
                iii)  otherwise at least one of v or w must have yielded a logical
                            variable; if v and w are the same logical variable, true
                            is returned; otherwise, one of the logical variables v or w
                            becomes bound to the other without causing any further
                            evaluation, and true is returned in this case*)
```

**Figure 6-1:** Recursive definition of `unify` pseudo-function.

### 6.4. Logical Variable Access Cycles

Before we present the new task types necessary to support logical variables, we introduce an implementation technique central to our method. Recall in that in section 3.3 it was mentioned that non-local demanders of `forward`'s are chained together to permit global value notification. We now generalize that technique in two respects (see figure 6-2):

- the bases of such chains can be `ubv` instructions as well as `forward` instructions, and
- we close these chains into cycles by having the final `fetch` refer back to the base of the chain.

This technique offers several important advantages:

a. The normal notification mechanism still applies when base words become evaluated. Final notification of the cycle base word, already evaluated, is at worst a `no-op` and can be suppressed if the value notification task carries along with it the address of the base word from which the notification originated.

b. An access path is preserved from any demand entry point (`fetch`) to the base word of the cycle. This is crucial to permit the "demand upgrading" effect whereby a non-assertive demand becomes strengthened to an assertive demand, and must be communicated to the `ubv` or `forward` at the cycle base.

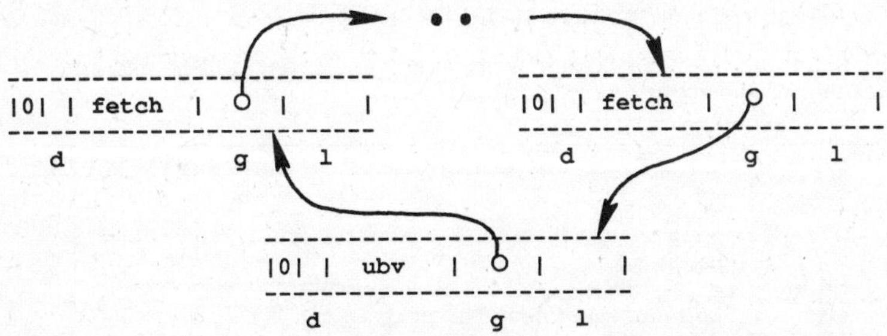


**Figure 6-2:** `fetch` cycle chaining.

c. An efficient technique exists for equating two **ubv**'s in this representation (see section 7.2).

## 6.5. Task Extensions

The new task varieties necessary to support logical variables are relatively few. We outline them now, under the assumption that each is performed atomically. The implications of distributed control on this assumption are discussed in section 7.1.

- *propagate assertive demand:* Such demand will necessarily be local in origin.
  - a. Set a notifier, if the demander is new.
  - b. If the target of the demand is not a **fetch** or **ubv**, (*) set **d** = **A** and, if **d** is thereby changed, propagate assertive demand to the target's arguments, according to its strictness policy; this task is then completed. Otherwise:
  - c. If **d** < **A**, set **d** = **A** and walk the cycle until either (i) an instruction with **d** = **A**, or (ii) the cycle base is found.
  - d. If an instruction with **d** = **A** is found, this task is completed.
  - e. If a **ubv** cycle base is found, perform a *lv_ref notification* task from the base.
  - f. Otherwise, the base must be a **forward**. Do step (*) above.
- *propagate non-assertive demand:* this is the conventional demand propagation task on Rediflow. Set a notifier, and set **d** = **max(d, N)**. If **d** is changed, propagate non-assertive demand to the instruction's arguments, according to its strictness policy.
- *lv_ref notification* from a **ubv** instruction at address **w**: walk the cycle based at **w** and notify all assertive demanders with the value **lv_ref to w**. Store the **lv_ref to w** value only in arguments to **bind** instructions.
- *bind operation:* when both of the arguments to a **bind** instruction are evaluated (to an atom, a **tuple_ref**, or a **lv_ref**), following cases apply:
  - a. If *neither* is **lv_ref**, return **false**.
  - b. If *both* are **lv_ref**'s, check to see if they are identical. If so, simply report **true**. Otherwise, do cycle merging as shown in figure 6-3, and report **true**.
  - c. The remaining case is where one argument evaluated to a value **lv_ref to w**,

and the other to some atom or tuple value **v**. In this case,

i. store **v** at **w** (after taking saving the notifier link);

ii. initiate value notification from **w** using the saved notifier link, and

iii. return **true**.

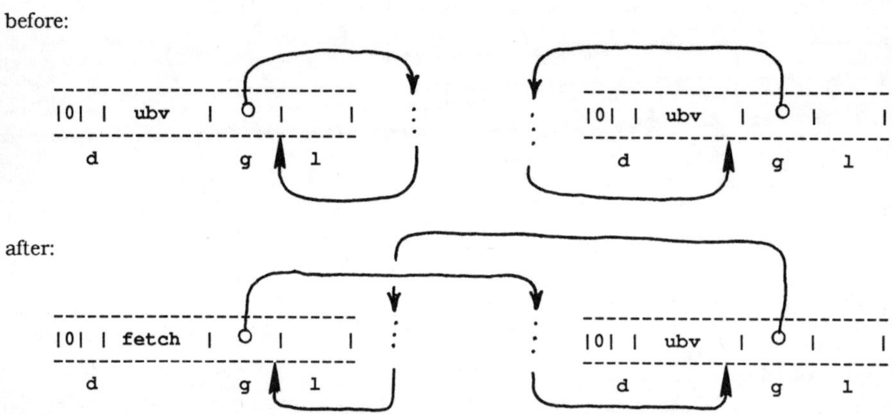


**Figure 6-3:** Logical variable equating by cycle merging.

## 7. Effects of Distributed Control

### 7.1. Requirements

In the previous section, we assumed tasks were accomplished by atomic actions with instantaneous remote memory access. We now consider the effects of distributed control on this idealized model.

a. Simplest to deal with are the decomposition of tasks into pairs of subtasks due to remote memory service delays. This effect is implicit on Rediflow, and does not require special treatment here.

b. The problems of parallel unification [12] are more critical. Under a distributed system such as Rediflow, logical variables may become bound after a **lv_ref** has been returned in response to an assertive demand. Hence **lv_ref**'s to non-local addresses can never be fully "trusted".

c. When a **bind** instruction is executed, a **true** result must be returned if and only if at least one argument evaluated to an **lv_ref** *and* the logical variable referenced became bound as a result of *this* **bind** instruction. The most pressing aspect of this mutual exclusion requirement is the need for a distributed cycle merging algorithm.

d. Finally, even while a **bind** instruction is being performed, new traversals around a cycle involved in the **bind** operation (commencing, for example, as the result of a new *propagate assertive demand* task) must be guaranteed to have connectivity to the cycle base.

## 7.2. Distributed Cycle Merging

We now consider a distributed, message based version of the cycle merging operation described as an atomic action in section 6.4. For simplicity, we initially assume that no two such operations will be in progress at any moment involving the same `ubv`. Figure 7-1 gives such a message processing sequence arising from a `bind` instruction at location `b` with evaluated arguments referencing `fetch` cycle bases at `v1` and `v2`, which must necessarily have been evaluated to *different* logical variable references. We order `v1` and `v2` so that `address(v1) < address(v2)`, the reason for which will become evident shortly.

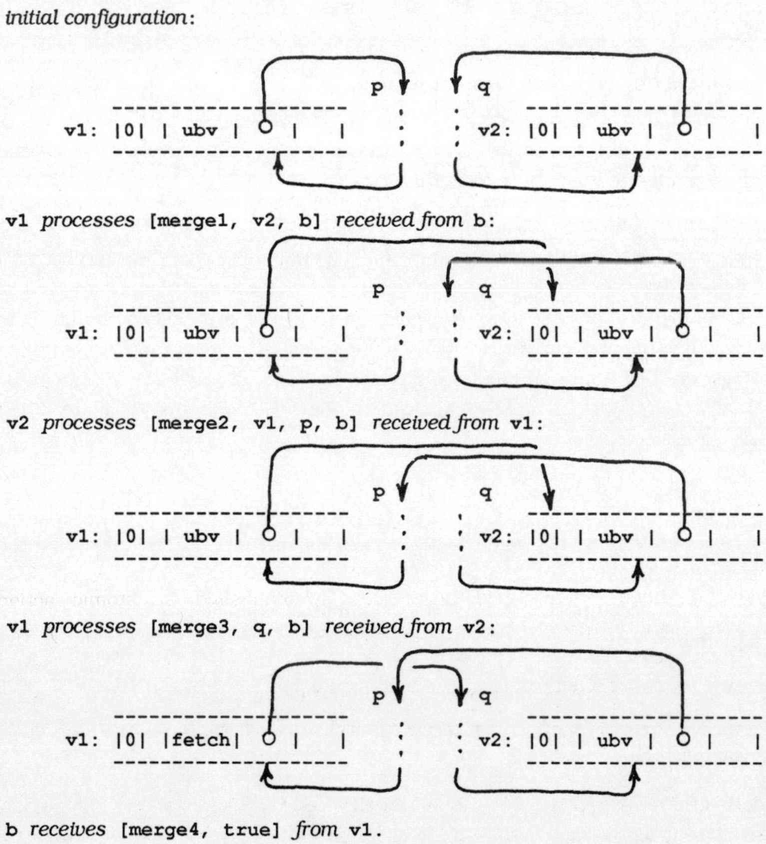


**Figure 7-1:** Message-based cycle merging

Now consider the problems that can arise if two or more such cycle merging activities concurrently involve a given `fetch` cycle.

a. Observe that there is only one message processing event at `v2`, and so the atomic processing of `merge2` messages is sufficient serialization for the second partner in a cycle merging operation.

b. However, a critical period is in effect at `v1` between the time it has processed a `merge1` message and the time it receives the mating **merge3** message.

i. We assume there is a bit available in **ubv** instruction formats to indicate this state (set on **merge1**, cleared on **merge3**).

ii. Now assume another **merge1** message, or a **merge2** message, arrives at **v1** during this period. Such messages are simply *rerouted unchanged* to **v2**, using the global link at **v1**.

iii. Note that such message deflection is not simply busy waiting, for the messages are being deflected to what is to become the base of the enlarged **v1** - **v2** cycle. Moreover, this deflection cannot go on indefinitely, since deflection is always *upward* in address order.

Two other correctness concerns need to be addressed:

a. Recall our stipulation that any "cycle walkers" (propagating assertive demand) must have connectivity at all times to the base (**forward** or **ubv**) of the cycle. The message processing steps enumerated above meet this test.

b. Finally, we must consider the situation when an instruction originally determined to be a **ubv** is in fact already bound when inspected during cycle merging. This manifests as a **fetch** or **forward** instruction (or value) where an **ubv** was expected. In this we simply abort the cycle merging process since the value (in residence or being computed by the **forward**) will be independently notified to **b**, and the **bind** operation automatically retried. However, if the non-**ubv** happens to be the higher-addressed **ubv** of the pair, i.e. the target of a **[merge2, v1, p, b]** operation, then we must first reset **v1** with a **[merge5, p]** message.

## 8. Summary

We have described an extension adding logical variables to the FGL language as executed by emulated graph reduction on Rediflow.

a. This extension is accomplished with relatively minor extension of Rediflow's basic execution method.

b. The necessary word size has been extended by only one bit.

c. A slight added degree of function parameter strictness has been introduced, on sound semantic and operational grounds.

d. This added strictness appears to offer a much needed reconciliation between lazy evaluation and the logical motivations of the occur check.

e. A distributed cycle merging algorithm is presented supporting mutual exclusion of logical variable binding.

f. Many of the techniques employed appear to have application to distributed implementation methods other than graph reduction. However, their relevance to pure logic programming languages is limited by the absence of a means for the environment separation necessary for nondeterminism and OR-parallelism.

## References


[1] H. Abramson.
A prological definition of HASL, a purely functional language with unification based conditional binding expressions.
In D. DeGroot and G. Lindstrom (editors), *Logic Programming: Functions, Relations, and Equations.* Prentice Hall, 1986.

[2] K. L. Clark and S. Gregory.
*A relational language for parallel programming.*
Technical Report DOC 81/16, Imperial College, London, 1981.

[3] S. H. Danforth.
*Logical variables for a functional language.*
Technical Report PP-120-85, Microelectronics and Computer Technology Corp., 1985.

[4] J. Darlington, A. J.Field, and H. Pull.
The unification of functional and logic languages.
In D. DeGroot and G. Lindstrom (editors), *Logic Programming: Functions, Relations, and Equations.* Prentice Hall, 1986.

[5] J. Goguen and J. Meseguer.
EQLOG: equality, types, and generic modules for logic programming.
In D. DeGroot and G. Lindstrom (editors), *Logic Programming: Functions, Relations, and Equations.* Prentice Hall, 1986.

[6] R.M. Keller, G. Lindstrom, and S. Patil.
A loosely-coupled applicative multi-processing system.
In *AFIPS Conference Proceedings*, pages 613-622. June, 1979.

[7] R.M. Keller and G. Lindstrom.
Hierarchical analysis of a distributed evaluator.
In *Proc. International Conference on Parallel Processing*, pages 299-310. August, 1980.

[8] R.M. Keller and G. Lindstrom.
Applications of feedback in functional programming.
In *Conference on functional languages and computer architecture*, pages 123-130. October, 1981.

[9] R.M. Keller.
*FEL (Function Equation Language) programmer's guide.*
Technical Report 7, University of Utah, Department of Computer Science, AMPS Technical Memorandum, 1982.

[10] R.M. Keller, F.C.H. Lin, and J. Tanaka.
Rediflow multiprocessing.
In *IEEE Compcon '84*, pages 410-417. Feb., 1984.

[11] R. M. Keller and F. C. H. Lin.
Simulated performance of a reduction-based multiprocessor.
*IEEE Computer* 17(7):70-82, July, 1984.

[12] G. Lindstrom.
OR-parallelism on applicative architectures.
In Sten-Ake Tarnlund (editor), *Proc. Second International Logic Programming Conference*, pages 159-170. Uppsala University, July, 1984.

[13] G. Lindstrom.
Functional programming and the logical variable.
In *Symposium on Principles of Programming Languages*, pages 266-280. ACM, January, 1985.

[14] R. Milner.
A theory of type polymorphism.
*J. of Comp. and Sys. Sci.* 17(3):348-375, 1978.

[15] R. S. Nikhil, K. Pingali, and Arvind.
*Id Nouveau.*
Technical Report Computation Structures Group Memo 265, MIT Laboratory for Computer Science, July, 1986.

[16] U.S. Reddy.
On the relationship between functional and logic languages.
In D. DeGroot and G. Lindstrom (editors), *Logic Programming: Functions, Relations, and Equations.* Prentice Hall, 1986.

[17] E.Y. Shapiro.
*A subset of Concurrent Prolog and its interpreter.*
Technical Report TR-003, Institute for New Generation Computer Technology, January, 1983.

[18] Shapiro, E., and A. Takeuchi.
Object oriented programming in Concurrent Prolog.
*New Generation Computing* 1(1):25-48, 1983.

[19] P. A. Subrahmanyam and J.-H. You.
FUNLOG: a computational model integrating logic programming and functional programming.
In D. DeGroot and G. Lindstrom (editors), *Logic Programming: Functions, Relations, and Equations.* Prentice Hall, 1986.

[20] J. Tanaka.
*Optimized execution of an applicative language.*
PhD thesis, Univ. of Utah, 1984.

[21] S. Taylor, S. Safra and E. Shapiro.
A parallel implementation of Flat Concurrent Prolog.
October, 1986.
Dept. of Computer Science, Weizmann Inst. of Science; to appear in *International Journal of Parallel Programming.*

[22] K. R. Traub.
An abstract parallel graph reduction machine.
In *Proc. 12th Ann. Symp. on Computer Architecture*, pages 333-341. IEEE Computer Society, 1985.

[23] P.C. Treleaven, D.R. Brownbridge, and R.P. Hopkins.
Data-driven and demand-driven computer architecture.
*Computing Surveys* 14(1):93-143, March, 1982.

# Functional Logic Languages
# Part I
(Preliminary Report)

*Uday S. Reddy*
University of Illinois at Urbana-Champaign

**Abstract**

Functional logic languages are extensions of functional languages with principles derived from logic programming. While syntactically they look similar to conventional functional languages, their operational semantics is based on narrowing, an evaluation mechanism that uses unification for parameter passing. We present here a small formal language based on lambda calculus with existential quantification and set abstraction, and define its denotational and narrowing semantics.

## 1 Introduction

Logic programming languages, typified by Prolog, have come into widespread use over the recent years [CKPR73,Kow79,CM81]. Even though programs written in logic languages exhibit a disarming similarity to functional programs, a deeper analysis shows that logic languages have more expressive power than functional languages [Kow83,Red86a]. The additional expressive power arises from the use of *unification* for parameter passing in logic languages. Subsequently, a number of efforts were made to incorporate this expressive power in functional languages by using unification in their parameter passing mechanisms as well [Lin85,SS84,DFP86,Smo86]. We have shown in [Red85,Red86b] that the result of extending the operational semantics of functional languages with unification amounts to using a mechanism called *narrowing* originally introduced for theorem proving in classical logic with equality [Sla74,Fay79,Hul80]. The languages obtained by such an extension are termed *functional logic languages.*

Let us introduce the concepts of functional logic languages through a simple conventional example: a function for appending lists.

$$\begin{aligned} append\ Nil\ y &\equiv y \\ append\ a.x\ y &\equiv a.(append\ x\ y) \end{aligned}$$

A definition such as this can be written in a number of functional languages such as HOPE [BMS80], Standard ML [Mil84], and Miranda [Tur85]. The use of such a definition in these languages is restricted to the evaluation of ground (variable-free) expressions involving *append*, e.g.,

$$append\ 1.Nil\ 2.Nil \Longrightarrow 1.2.Nil$$

The evaluation method can easily be extended to allow free variables in the input expression in some contexts, e.g.,

$$append\ a.Nil\ 2.Nil \Longrightarrow a.2.Nil$$

But, when we want to evaluate an expression such as

$$append\ t\ 2.Nil$$

no evaluation is possible. The expression is in normal form. However, we would be reluctant to treat the expression as a "literal value", since it has a function application.

When evaluating the expression (*append* 1.*Nil* 2.*Nil*) a conventional functional language *matches* the arguments against the formal parameter patterns. This yields a substitution for the formal parameter variables (e.g. $a = 1$, $x = Nil$, $y = 2.Nil$). A functional logic language, in contrast *unifies* the arguments with functional parameter patterns. This yields a substitution for both the formal parameter variables and the argument variables. The expression (*append* $t$ 2.*Nil*) can be unified with the left hand sides of both the equations for *append*, yielding the unifier substitutions:

$$\begin{array}{l} (1).\ t = Nil,\ y = 2.Nil \\ (2).\ t = a.x,\ y = 2.Nil \end{array}$$

The unifiers can be divided into two parts: an argument substitution $\sigma$ and a formal parameter substitution $\theta$. The latter applied to the right hand side of the equation becomes the result of the evaluation step. The former is a condition under which the original expression is reducible. So, there are two possible evaluation paths for the expression (*append* $t$ 2.*Nil*)

$$\begin{array}{lll} (1)\ append\ t\ 2.Nil & \Longrightarrow & t = Nil \rightarrow 2.Nil \\ (2)\ append\ t\ 2.Nil & \Longrightarrow & t = a.x \rightarrow a.(append\ x\ 2.Nil) \end{array}$$

Such evaluation steps are called *narrowing steps.* The substitutions to the left of $\rightarrow$ in the result expressions are called *narrowing substitutions.* Whenever

$$e \Longrightarrow \sigma \rightarrow e'$$

we say $e$ narrows to $e'$ via the narrowing substitution $\sigma$. As is evident, narrowing is a nondeterministic evaluation process. The *append* expression cited above has an infinite number of nondeterministic narrowings.

Semantically, a nonground expression such as (*append* $t$ 2.*Nil*) denotes a *parametric value* in the domain $Env \rightarrow Value$

$$[\![append\ t\ 2.Nil]\!]\eta = \begin{cases} 2.Nil & \text{if } \eta t = Nil \\ v_1.2.Nil & \text{if } \eta t = v_1.Nil \\ v_1.v_2.2.Nil & \text{if } \eta t = v_1.v_2.Nil \\ \vdots & \\ \bot & \text{otherwise} \end{cases}$$

Conditional terms of the kind $t = Nil \rightarrow 2.Nil$, and $t = a.x \rightarrow x$ similarly denote parametric values.

$$[\![t = Nil \rightarrow 2.Nil]\!]\eta = \begin{cases} 2.Nil & \text{if } \eta t = Nil \\ \bot & \text{otherwise} \end{cases}$$

$$[\![t = a.x \rightarrow x]\!]\eta = \begin{cases} v_2 & \text{if } \eta t = v_1.v_2 \\ \bot & \text{otherwise} \end{cases}$$

The narrowing process produces a compact (not necessarily finite) representation of the parametric value of a nonground expression in terms of such conditional terms. If the set of all nondeterministic narrowings of an expression $e$ is $\{e \Longrightarrow t_i\}$ then

$$[\![e]\!] = \bigsqcup_i [\![t_i]\!]$$

Thus the nondeterminism involved in narrowing is quite meaningful and does not imply any kind of referential opacity.

Let us demonstrate the usefulness of narrowing through a less trivial example. Consider the boolean-valued expression

$$(append\, t\; 2.Nil) = 1.2.Nil$$

The narrowing evaluation should produce a solution for $t$. We show below two nondeterministic narrowing sequences of the expression:

1. $(append\, t\; 2.Nil) = 1.2.Nil$
   $\Longrightarrow (t = Nil \rightarrow 2.Nil) = 1.2.Nil$
   $\Longrightarrow t = Nil \rightarrow (2.Nil = 1.2.Nil)$
   failure in evaluating $2.Nil = 1.2.Nil$

2. $(append\, t\; 2.Nil) = 1.2.Nil$
   $\Longrightarrow (t = a.x \rightarrow a.(append\; x\; 2.Nil)) = 1.2.Nil$
   $\Longrightarrow t = a.x \rightarrow a.(append\; x\; 2.Nil) = 1.2.Nil$
   $\Longrightarrow t = a.x \rightarrow (a = 1) \wedge (append\; x\; 2.Nil) = 2.Nil$
   $\Longrightarrow t = a.x \rightarrow (a = 1 \rightarrow true) \wedge (x = Nil \rightarrow 2.Nil = 2.Nil)$
   $\Longrightarrow t = a.x \rightarrow a = 1 \wedge x = Nil \rightarrow true$
   $\Longrightarrow t = 1.Nil \rightarrow true$

The remaining narrowing sequences all lead to failure. In general, narrowing gives the programming language a capability to *solve* arbitrary constraints for instantiations of the variables.

In this paper, we shall present a formal framework for functional logic languages using an abstract language based on $\lambda$-calculus. This is a natural sequel to [Red85]. Many aspects of functional logic languages that were left tacit in [Red85] are made explicit and formal. In addition, adequate attention is paid to higher order functions and set abstraction.

# 2 Framework

We first define a small abstract language based on $\lambda$-calculus. Its main difference from $\lambda$-calculus in that we use terms as formal parameters in abstractions. Syntactically, the language is very similar to HOPE-like functional languages. But, its operational semantics uses narrowing.

## 2.1 Syntax

Syntactic domains:

*Variable* ranged over by $x$ and $y$
*Atom* ranged over by $a$ and $b$
*Term* ranged over by $t$
*Expression* ranged over by $e$, $d$ and $f$.

To distinguish atoms from variables, we use names beginning with capital letters for atoms. A distinguished atom *true* is assumed to be present in *Atom*. The abstract syntax of the language is

$$\begin{aligned} t \; ::= \;& x \mid a \\ & \mid (t_1, t_2) \quad \text{- pairs} \end{aligned}$$

$$\begin{array}{lll}
e & ::= & x \mid a \\
 & & \mid (e_1, e_2) \qquad\qquad \text{– pair} \\
 & & \mid (\lambda t.\, e) \qquad\qquad \text{– abstraction} \\
 & & \qquad \text{where } t \text{ is linear} \\
 & & \mid (e_1; e_2) \qquad\qquad \text{– union or lub} \\
 & & \mid (e_1\ e_2) \qquad\qquad \text{– application} \\
 & & \mid (e_1 \rightarrow e_2) \qquad\qquad \text{– conditional} \\
 & & \mid (e_1 = e_2) \qquad\qquad \text{– equality} \\
 & & \mid (e_1 \wedge e_2) \qquad\qquad \text{– conjunction}
\end{array}$$

The restriction that $t$ should be *linear* in abstraction means that no variable should appear twice in $t$. This restriction is necessary to ensure that the functions denoted by abstraction expressions are continuous. To avoid the proliferation of parentheses, we use the following precedence rule:

$$\lambda\ \leq ;\ \leq \rightarrow \leq\ \wedge\ \leq =$$

In addition, we assume that ",", $\lambda$, $\rightarrow$ and $\wedge$ associate to the right, and ";" and application associate to the left.

As an example of the syntax, the *append* function of section 1 is written in the abstract language as follows:

$$\begin{array}{l}
(fix\ (\lambda append.\ (\lambda Nil.\ \lambda y.\ y); \\
\qquad\qquad\qquad (\lambda(a, x).\ \lambda y.\ (a,\ (append\ x\ y)))))
\end{array}$$

The functions defined by the various equations are combined using the ";" operator. Conventionally, equational definitions such as that of *append* are translated into the standard $\lambda$ calculus using conditional, e.g.

$$\begin{array}{l}
(fix\ (\lambda append.\ \lambda l.\ \lambda y. \\
\quad (l = Nil \rightarrow y); \\
\quad ((pair?\ l) \rightarrow \textbf{let}\ a \equiv (fst\ l),\ x \equiv (snd\ l)\ \ \textbf{in}\ (a,\ (append\ x\ y))) \\
))
\end{array}$$

Using terms directly as parameters in abstraction expressions provides a convenient notation, and makes it narrowing to define narrowing as we will see later.

## 2.2 Denotational semantics

The denotational semantics uses complete lattices rather than cpo's. There is a top element $\top$ to denote inconsistent values that may be produced by the ";" operator. Let $B$ be the *set of basic values.*

$$B = A + B \times B$$

Let $B_{\perp\top}$ be the flat lattice obtained by augmenting $B$ with $\perp$ and $\top$. The semantic domain $D$ is the complete lattice defined by

$$D = B_{\perp\top} + [D \rightarrow D]$$

$$[\![ - ]\!] \quad : \quad Env \rightarrow D$$

$$\begin{array}{lll}
[\![x]\!]\eta & = & \eta x \\
[\![a]\!]\eta & = & a
\end{array}$$

$$
\begin{aligned}
[\![(e_1, e_2)]\!]\eta &= \begin{cases} ([\![e_1]\!]\eta,\ [\![e_2]\!]\eta) & \text{if } [\![e_1]\!]\eta \in B, [\![e_2]\!]\eta \in B \\ \top & \text{if } [\![e_1]\!]\eta = \top \text{ or } [\![e_2]\!]\eta = \top \\ \bot & \text{otherwise} \end{cases} \\
[\![(\lambda t.\ e)]\!]\eta &= \lambda v.\ \text{if } \exists \rho \in FV(t) \to D \text{ such that } [\![t]\!]\rho = v \text{ then } [\![e]\!](\eta;\rho) \text{ else } \bot \\
[\![(e_1; e_2)]\!]\eta &= [\![e_1]\!]\eta \sqcup [\![e_2]\!]\eta \\
[\![e_1 \to e_2]\!]\eta &= \begin{cases} [\![e_2]\!]\eta & \text{if } [\![e_1]\!]\eta = \mathit{true} \\ \top & \text{if } [\![e_1]\!]\eta = \top \\ \bot & \text{otherwise} \end{cases} \\
[\![e_1 = e_2]\!]\eta &= \begin{cases} \mathit{true} & \text{if } [\![e_1]\!]\eta = [\![e_2]\!]\eta \in B \\ \top & \text{if } [\![e_1]\!]\eta = \top \text{ or } [\![e_2]\!]\eta = \top \\ \bot & \text{otherwise} \end{cases} \\
[\![e_1 \wedge e_2]\!]\eta &= \begin{cases} \mathit{true} & \text{if } [\![e_1]\!]\eta = \mathit{true} \text{ and } [\![e_2]\!]\eta = \mathit{true} \\ \top & \text{if } [\![e_1]\!]\eta = \top \text{ or } [\![e_2]\!]\eta = \top \\ \bot & \text{otherwise} \end{cases}
\end{aligned}
$$

The environment $(\eta; \rho)$ in the semantics of abstraction expression means the environment obtained by updating $\eta$ with the bindings in $\rho$. $FV(t)$ is the set of free variables of $t$.

Note that the pairs constructed by "," are *strict*. It will be seen later that strict data structures are necessary to make good use of the capabilities offered by narrowing. Nonstrict data structures are still available through the use of functions, just as in conventional $\lambda$ calculus.

$$
\begin{aligned}
\mathit{pair} &= \lambda x.\ \lambda y.\ ((\lambda 1.\ x);\ (\lambda 2.\ y)) \\
\mathit{fst} &= \lambda p.\ (p\ 1) \\
\mathit{snd} &= \lambda p.\ (p\ 2)
\end{aligned}
$$

## 2.3 Reduction semantics

For comparison, we first present the conventional reduction semantics for the language.

$$
\begin{array}{llcl}
\alpha: & (\lambda t.\ e) & \Longrightarrow & \lambda t'.\theta[e] \\
 & & & \text{where } t' = \theta[t] \text{ is a renaming of } t \\
\text{APP:} & ((\lambda t.\ d)\ e) & \Longrightarrow & \theta[d] \\
 & & & \text{if there are no common free variables in } t \text{ and } e \\
 & & & \text{and } \theta \text{ is the minimal substitution such that } \theta[t] = e \\
\text{CHOICE1:} & (e_1; e_2) & \Longrightarrow & e_1 \\
\text{CHOICE2:} & (e_1; e_2) & \Longrightarrow & e_2 \\
\text{COND:} & (\mathit{true} \to e) & \Longrightarrow & e \\
\text{EQ1:} & (a = a) & \Longrightarrow & \mathit{true} \\
\text{EQ2:} & ((e_1, e_2) = (e_3, e_4)) & \Longrightarrow & (e_1 = e_3) \wedge (e_2 = e_4) \\
\text{CONJ:} & (\mathit{true} \wedge \mathit{true}) & \Longrightarrow & \mathit{true}
\end{array}
$$

The complex rule for function application can be seen to be just a short hand for the following simpler rules:

$$
\begin{array}{lrcl}
\text{APP1:} & ((\lambda x.\ d)\ e) & \Longrightarrow & [e/x][d] \\
\text{APP2:} & ((\lambda a.\ d)\ a) & \Longrightarrow & d \\
\text{APP3:} & ((\lambda (t_1, t_2).\ d)\ (e_1, e_2)) & \Longrightarrow & ((\lambda t_1.\ \lambda t_2.\ d)\ e_1\ e_2)
\end{array}
$$

Evaluation using these rules does not necessarily preserve the semantics of expressions because of the nondeterminism involved in the CHOICE rules. However, if $(e \Longrightarrow e')$, then $[\![e]\!] \sqsupseteq [\![e']\!]$. We call this *soundness* of the reduction relation. Furthermore, for any expression $e$, there is a set of normalizing reductions $\{e \Longrightarrow e_i\}_i$ such that $[\![e]\!] = \bigsqcup_i [\![e_i]\!]$. We call this *completeness* of the reduction relation.

## 2.4 Narrowing semantics

We can now extend the operational semantics of the language to narrowing. A syntactic (meta-level) substitution is written as

$$[t_1/x_1, \ldots, t_n/x_n]$$

The variables $x_1, \ldots, x_n$ are called the *domain variables (DV)* of the substitution. For narrowing, we need to represent substitutions as parts of expressions. We use the following subclass of expressions for this purpose.

$$\begin{array}{lll} \sigma & \in & \textit{Substitution-expression} \\ \sigma & ::= & (x = t) \quad \text{where } x \text{ is not free in } t \\ & & \mid (\sigma_1 \wedge \sigma_2) \quad \text{where } DV(\sigma_1) \cap FV(\sigma_2) = \emptyset,\ \ DV(\sigma_2) \cap FV(\sigma_1) = \emptyset \end{array}$$

A syntactic substitution $[t/x]$ is represented by the equality $(x = t)$, and compound substitutions with several bindings are represented by conjunctions. Since the correspondence is straightforward, we shall not distinguish between substitutions and their expression representations.

First of all, the rules for application are extended to handle variable arguments:

$$\begin{array}{lrcl} \text{APP2}': & ((\lambda a.\, d)\ x) & \Longrightarrow & x = a \rightarrow [a/x][d] \\ \text{APP3}': & ((\lambda(t_1, t_2)\ d)\ x) & \Longrightarrow & x = (x_1, x_2) \rightarrow ((\lambda t_1.\, \lambda t_2.\, [(x_1, x_2)/x][d])\ x_1\ x_2)) \\ & & & \text{where } x_1 \text{ and } x_2 \text{ are new variables, and} \\ & & & x \text{ is not free in } t_1 \text{ or } t_2 \end{array}$$

All these rules for application can be combined into a single complex rule that uses unification:

$$\begin{array}{lrcl} \text{APP:} & ((\lambda t.\, d)\ e) & \Longrightarrow & (\sigma \rightarrow \theta[\sigma[d]]) \\ & & & \text{where } t \text{ and } e \text{ have no common free variables} \\ & & & \text{and } \sigma \wedge \theta \text{ is a most general unifier such that } \sigma[e] = \theta[t]. \end{array}$$

The other forms of expressions such as conditional are disguised function applications. So, they too need to be extended for solving for variables.

$$\begin{array}{lrcl} \text{COND}': & (x \rightarrow e) & \Longrightarrow & (x = \textit{true} \rightarrow e) \\ \text{EQ1}': & (x = a) & \Longrightarrow & (x = a) \rightarrow \textit{true} \\ \text{EQ2}': & (x = (t_1, t_2)) & \Longrightarrow & (x = (x_1, x_2) \rightarrow (x_1 = t_1 \wedge x_2 = t_2)) \\ & & & \text{where } x_1 \text{ and } x_2 \text{ are new variables, and} \\ & & & x \text{ is not free in } (t_1, t_2) \\ \text{EQ3}': & (t = x) & \Longrightarrow & (x = t) \\ \text{CONJ1}': & (x \wedge e) & \Longrightarrow & (x = \textit{true} \rightarrow [\textit{true}/x][e]) \\ \text{CONJ2}': & (e \wedge x) & \Longrightarrow & (x = \textit{true} \rightarrow [\textit{true}/x][e]) \end{array}$$

By using narrowing, we now have a new kind of irreducible normal forms: conditional terms of the form $(\sigma \rightarrow t)$. The substitutions $\sigma$ in such conditional terms (and conditional expressions) need to be propagated out from all possible reduction positions. This is handled by the following rules:

$$\begin{array}{lrcl} \text{PROP-PAIR1:} & ((\sigma \rightarrow e_1), e_2) & \Longrightarrow & (\sigma \rightarrow (e_1, \sigma[e_2])) \\ \text{PROP-PAIR2:} & (e_1, (\sigma \rightarrow e_2)) & \Longrightarrow & (\sigma \rightarrow (\sigma[e_1], e_2)) \\ \text{PROP-APP1:} & ((\sigma \rightarrow e_1)\ e_2) & \Longrightarrow & (\sigma \rightarrow (e_1\ \sigma[e_2])) \\ \text{PROP-APP2:} & ((\lambda t.\, e_1); (\sigma \rightarrow e_2)) & \Longrightarrow & (\sigma \rightarrow (\sigma[\lambda t.\, e_1]\ e_2)) \\ & & & \text{if } t \text{ is not a variable} \\ \text{PROP-COND:} & ((\sigma \rightarrow e_1) \rightarrow e_2) & \Longrightarrow & (\sigma \rightarrow (e_1 \rightarrow \sigma[e_2])) \\ \text{PROP-EQ1:} & ((\sigma \rightarrow e_1) = e_2) & \Longrightarrow & (\sigma \rightarrow (e_1 = \sigma[e_2])) \\ \text{PROP-EQ2:} & (e_1 = (\sigma \rightarrow e_2)) & \Longrightarrow & (\sigma \rightarrow (\sigma[e_1] = e_2)) \\ \text{PROP-CONJ1:} & ((\sigma \rightarrow e_1) \wedge e_2) & \Longrightarrow & (\sigma \rightarrow (e_1 \wedge \sigma[e_2])) \\ \text{PROP-CONJ2:} & (e_1 \wedge (\sigma \rightarrow e_2)) & \Longrightarrow & (\sigma \rightarrow (\sigma[e_1] \wedge e_2)) \end{array}$$

Note that there is only one propagation rule for conditionals. Because we reduce the condition of a conditional before reducing its body, we do not need to propagate a substitution out from the body. But, there is a special case. If the condition of the conditional is itself a substitution, then we need to take the composition of the two substitutions.

$$\text{COMP:}\quad (\sigma_1 \to (\sigma_2 \to e)) \implies (\sigma_1 \circ \sigma_2 \to e)$$

The propagation of substitutions shows why we need strict pairs. Consider, for example,

$$(x = A \to 0,\ x)$$

Its first component is defined only if $x = A$ in the current environment, but its second component is whatever is the binding of $x$ in the environment. So, in an environment in which $x$ is bound to 0, it would denote $(\perp, 0)$. But, typically, we would like the substitutions that come up in the evaluation of one part of a data structure to be used for instantiating its other parts. In other words, we would be interested in those environments in which all the components of the data structure are defined. The propagation rules PROP-PAIR1 and PROP-PAIR2 which achieve this are valid only for strict pairs. Through these rules, the above expression reduces to

$$x = A \to (0, A)$$

which simply denotes $\perp$ in an environment in which $x$ is bound to 0.

Similar propagation does not happen for nonstrict pairs constructed using the *pair* function. The expression

$$(\mathit{pair}\ (\sigma \to e_1)\ e_2)$$

does not reduce to

$$\sigma \to (\mathit{pair}\ e_1\ e_2)$$

because the rule PROP-APP2 does not apply when the formal parameter of the function is a variable.

The narrowing relation given by these rules is sound. It is also complete except for a minor technical problem, *viz.*, that narrowing can introduce new variables which are not present in the original expression. Consider, for example,

$$[\![((\lambda(a, x).\, a)\ z)]\!]\eta = \begin{cases} v_1 & \text{if } \eta z = (v_1, v_2) \\ \perp & \text{otherwise} \end{cases}$$

But, narrowing evaluates the expression to

$$[\![z = (a, x) \to a]\!]\eta = \begin{cases} v_1 & \text{if } \eta z = (v_1, v_2),\ \eta a = v_1,\ \eta x = v_2 \\ \perp & \text{otherwise} \end{cases}$$

While the original expression is defined in environments in which only $z$ is bound, the result expression is defined in environments in which $a$ and $x$ are also bound appropriately. In the next section, we shall introduce a construct for existential quantification using which this problem can be avoided. But, for now, observe that narrowing is complete provided we allow for the environments to be extended for new variables in the semantics of the result expressions.

In addition to soundness and completeness, narrowing also satisfies the property of *solution-completeness.* Let *basic environments* be

$$\mathit{Benv} = \mathit{Variable} \to B_{\perp\top}$$

Then the *basic parametric value* of a nonground expression is

$$\mathcal{B}[\![e]\!]\eta = \begin{cases} [\![e]\!]\eta & \text{if } \eta \in Benv \\ \bot & \text{otherwise} \end{cases}$$

For every nonground expression $e$, there is a set of terminal narrowings $\{e \Longrightarrow \sigma_i \to t_i\}_i$ where $t_i$ is a term or an abstraction, such that

$$\mathcal{B}[\![e]\!] = \bigsqcup_i [\![\sigma_i \to t_i]\!].$$

We call this ***basic solution-completeness*** of narrowing. It means that narrowing can solve for basic-valued instantiations of variables in nonground expressions. It cannot solve for function-valued instantiations because the APP rules do not allow variables as functions.

# 3 Existential quantification

Now that we have a basic narrowing functional language, we can extend it in a number of ways to add more expressive power. Note that the programs that can be written in the language so far are independent of whether reduction or narrowing is used as the operational semantics. Though narrowing gives us the capability to evaluate nonground expressions with free variables, this capability is exhibited only in the *use* of the programs, not in the programs themselves. So, the first extension we would like is to make this capability available in writing programs.

To introduce new "free" variables in expressions, we use a new binding construct in the syntax of expressions, *viz.*, existential quantification.

$$\begin{array}{rcl} e & ::= & (\exists x.\, e) \\ [\![(\exists x.\, e)]\!]\eta & = & \bigsqcup_{v \in Dcon} [\![e]\!](\eta; [v/x]) \end{array}$$

where $Dcon$ is the set $D - \{\top\}$. The construct $\exists x. e$ may be read as "for some value of $x$, the value of $e$". Intuitively, it means that every atomic piece information obtainable from $\exists x. e$ is obtainable from $e$ with some instantiation of $x$ from $Dcon$. If $e$ has no free occurrences of $x$ then $(\exists x.\, e)$ is trivially equivalent to $e$. If $e$ has free occurrences of $x$, then it may take different values for different instantiations of $x$. If all these values are consistent, then the information obtainable from $(\exists x. e)$ is the sum total of the information obtainable from the individual values. Otherwise, the expression denotes $\top$, and, operationally, it may evaluate to one or more of these values. Usually, we use the existential quantification in such a way that $e$ denotes a non-$\bot$ value for only one instantiation of $x$.

In logic programming parlance, variables introduced using the existential quantification are called *logical variables.* They provide a powerful programming paradigm discussed later in the section. Recall the semantic problems caused by the fact that the basic narrowing mechanism of section 2.4 introduces new free variables that are not present in the original expression. Using existential quantification, such new variables can be quantified so that they are not left free in the final expression. So, we modify the rules APP3′, EQ2′, and all the substitution propagation rules, including COMP, of section 2.4 as follows:

$$\begin{array}{llcl}
\text{APP3}': & ((\lambda(t_1,t_2)\, d)\, x) & \Longrightarrow & \exists x_1.\, \exists x_2.\, x = (x_1,x_2) \to ((\lambda t_1.\, \lambda t_2.\, d)\; x_1\; x_2)) \\
& & & \text{where } x_1 \text{ and } x_2 \text{ are new variables} \\
\text{EQ2}': & (x = (t_1,t_2)) & \Longrightarrow & \exists x_1. \exists x_2.\, x = (x_1,x_2) \to (x_1 = t_1 \wedge x_2 = t_2) \\
& & & \text{where } x_1 \text{ and } x_2 \text{ are new variables} \\
\text{PROP-PAIR1}: & ((\exists \overline{x}.\, \sigma \to e_1), e_2) & \Longrightarrow & (\exists \overline{x}.\, \sigma \to (e_1, \sigma[e_2])) \\
& & & \text{if the variables } \overline{x} \text{ are not free in } e_2, \\
& & \vdots & \\
\text{COMP}: & (\sigma_1 \to (\exists \overline{x}.\, \sigma_2 \to e)) & \Longrightarrow & (\exists \overline{x}.\, \sigma_1 \circ \sigma_2 \to e) \\
& & & \text{if the variables } \overline{x} \text{ are not free in } \sigma_1
\end{array}$$

The notation $\exists\overline{x}.$ stands for a series of *zero or more* variable bindings $\exists x_1 .\ldots\exists x_n$. Note that, in general, the terminal forms for successful narrowings are now of the form

$$\exists\overline{x}.\,\sigma \rightarrow t$$

and the propagation rules propagate the quantifiers as well as the substitutions outwards.

For the example mentioned in section 2.4, the new rules now produce

$$((\lambda(a,x).\,a)\ z) \Longrightarrow \exists a.\,\exists x.\,z = (a,x) \rightarrow a$$

The semantics of the result expression is

$$\begin{aligned}
&[\![\exists a.\,\exists x.\,z = (a,x) \rightarrow a]\!]\eta \\
&= \bigsqcup_{v_1,v_2\in Dcon} [\![z = (a,x) \rightarrow a]\!](\eta;[v_1/a, v_2/b]) \\
&= \bigsqcup_{v_1,v_2\in Dcon} \left\{ \begin{array}{ll} v_1 & \text{if } \eta z = (v_1,v_2) \\ \bot & \text{otherwise} \end{array} \right\} \\
&= \left\{ \begin{array}{ll} v_1 & \text{if } \eta z = (v_1,v_2) \\ \bot & \text{otherwise} \end{array} \right.
\end{aligned}$$

which is the same as the semantics of the original expression.

The $\exists$ quantification provides more justification for strict data structures. Consider

$$\exists x.\,(x = A \rightarrow 0,\ x)$$

If pairs were not strict, its denotation would be

$$\begin{aligned}
&\lambda\eta.\,\bigsqcup_{v\in Dcon} (\left\{ \begin{array}{ll} 0 & \text{if } v = A \\ \bot & \text{otherwise} \end{array} \right\}, v) \\
&= \lambda\eta.\,(0,\top)
\end{aligned}$$

Even though there is a unique consistent value for the second component whenever the first component is defined, the second component turns out to be $\top$ since no information can be exchanged between the two components. But, if pairs are strict, its denotation is

$$\begin{aligned}
&\lambda\eta.\,\bigsqcup_{v\in Dcon} \left\{ \begin{array}{ll} (0,A) & \text{if } v = A \\ \bot & \text{otherwise} \end{array} \right\} \\
&= \lambda\eta.\,(0,A)
\end{aligned}$$

Thus, while nonstrict data structures allow us to avoid $\bot$, strict data structures allow us to to avoid $\top$ when $\exists$ quantification is used.

The evaluation rules for the existential expressions are

| | | | |
|---|---|---|---|
| $\alpha$-EXIST: | $\exists x.\,t$ | $\Longrightarrow$ | $\exists x'.\,[x'/x][t]$<br>if $x'$ is not free in $t$ |
| EXIST1: | $(\exists x.\,t)$ | $\Longrightarrow$ | $t$ |
| EXIST2: | $(\exists x.\,\exists\overline{y}.\,x = t \rightarrow e)$ | $\Longrightarrow$ | $(\exists\overline{y}.\,e)$<br>if $x$ is not free in $t$ or $e$ |
| EXIST3: | $(\exists x.\,\exists\overline{y}.\,x = t \wedge \sigma \rightarrow e)$ | $\Longrightarrow$ | $(\exists\overline{y}.\,\sigma \rightarrow e)$<br>if $x$ is not free in $t$, $\sigma$ or $e$ |
| PROP-EXIST: | $(\exists x.\,\exists\overline{y}.\,\sigma \rightarrow e)$ | $\Longrightarrow$ | $(\exists\overline{y}.\,\sigma \rightarrow (\exists x.\,e))$<br>if $x$ is not free in $\sigma$ |

We now give a number of examples illustrating the paradigmatic use of logical variables.

**Example 1 (Member function)** We can define a predicate for testing membership in a list by

$$member(a, b.x) \equiv (a = b \rightarrow true;\ member(a, x))$$

Given ground arguments, it works the same way as a *member* function in a conventional functional language. For example,

$$\begin{aligned}&member(A, A.B.Nil)\\ &\Longrightarrow A = A \rightarrow true;\ member(A, B.Nil)\\ &\Longrightarrow true\end{aligned}$$

using the rule CHOICE1 to resolve the choice. If we use CHOICE2, we get

$$\begin{aligned}&member(A, B.nil)\\ &\Longrightarrow A = B \rightarrow true;\ member(A, Nil)\end{aligned}$$

and both the choices fail (denote $\bot$).

We can also invoke *member* with nonground arguments, just as we did with *append* in section 1. If we use a variable for the first argument as, for example,

$$member(e, A.B.Nil)$$

then we should expect to get all instantiations of $e$ for which $member(e, A.B.Nil)$ holds. This is indeed what is obtained by evaluating this expression. It evaluates nondeterministically to two terminal forms:

$$\begin{array}{ll}member(e, A.B.Nil) & \\ \Longrightarrow e = A \rightarrow true;\ member(e, B.Nil) & \text{APP}\\ \Longrightarrow e = A \rightarrow true \qquad (1) & \text{CHOICE1}\\ & \\ \Longrightarrow member(e, B.Nil) & \text{CHOICE2}\\ \Longrightarrow e = B \rightarrow true;\ member(e, Nil) & \text{APP}\\ \Longrightarrow e = B \rightarrow true \qquad (2) & \text{CHOICE1}\\ & \\ \Longrightarrow member(e, Nil) & \text{CHOICE2}\\ \quad \text{fails} & \end{array}$$

The use of a variable as the second argument to *member* is much more interesting, but it is also much more complex. The evaluation of

$$member(A, l)$$

should produce as an instantiation of $l$, a list that contains $A$. Since there are an infinite number of such lists, we get an infinite number of narrowings:

$$\begin{array}{ll}member(A, l) & \\ \Longrightarrow \exists b.\ \exists x.\ l = b.x & \text{APP}\\ \qquad\qquad \rightarrow (A = b \rightarrow true;\ member(A, x) & \\ \Longrightarrow \exists b.\ \exists x. l = b.x \rightarrow b = A \rightarrow true & \text{CHOICE1, EQ3}'\\ \Longrightarrow \exists b.\ \exists x. l = A.x \wedge b = A \rightarrow true & \text{COMP}\\ \Longrightarrow \exists x. l = A.x \rightarrow true & \text{EXIST3}\end{array}$$

Using CHOICE2 again produces an infinite number of further narrowings:

$$\begin{aligned}&\exists b.\ \exists x'. l = b.A.x' \rightarrow true\\ &\exists b.\ \exists b'.\ \exists x''. l = b.b'.A.x'' \rightarrow true\\ &\exists b.\ \exists b'.\ \exists b''.\ \exists x'''. l = b.b'.b''.A.x''' \rightarrow true\\ &\vdots\end{aligned}$$

Expressions with such infinite solutions are problematic since in most applications only one solution would do. In Prolog, the infinite solutions are inhibited using the nonlogical "cut" primitive. But, since we have no such primitive, they need to be inhibited through careful use of the $\exists$ quantification.

We can assert any number of values to be members of $l$. For example

$$member(A,l) \wedge member(B,l) \wedge member(C,l) \Longrightarrow \exists x.\, l = A.B.C.x \rightarrow true$$

However, there is a problem. If we evaluate

$$member(A,l) \wedge member(A,l)$$

we get not only the straightforward first narrowing

$$\exists x.\, l = A.x \rightarrow true$$

but also the one in which $l$ is bound to a list with two occurrences of $A$.

$$\exists x.\, l = A.A.x \rightarrow true$$

Avoiding such solutions requires the use of negation, discussed in section 4.3

**Example 2 (Retrieving attributes from association lists)** Let an association list be a list of pairs, whose first components are keys and second components are attributes, so that no key appears more then once in the association list. We can define a function to retrieve the attribute of a key as

$$retrieve(key, alist) \equiv \exists attr.\, member((key, attr), alist) \rightarrow attr$$

Operationally, its evaluation is done by first evaluating the body of the existential quantifier with *attr* as a free variable. When it narrows to a conditional term of the form

$$attr = A \rightarrow A$$

the condition is dropped by the rule EXIST2 and $A$ is produced as the value of the quantified expression. Since, by assumption, *alist* contains at most one pair for any key, the body of the quantifier has at most one successful narrowing. Hence, the definition of *retrieve* is consistent.

Like *member*, *retrieve* can also be invoked with a variable for either the first or the second argument. If a variable is used as the second argument, it gets instantiated to an association list that contains a pair for the given key. For example,

$$retrieve(A,l) \Longrightarrow \exists a.\, \exists x.\, l = (A,a).x \rightarrow a$$

But, if we try to retrieve the same key twice, it is possible to get different attributes. For example,

$$(retrieve(A,l),\ retrieve(A,l) \Longrightarrow \exists a.\, \exists a'.\, \exists x.\, l = (A,a).(A,a').x \rightarrow (a,a)$$

Even though we have assumed that an association list does not have multiple pairs for any value of the key, the definition of *retrieve* does not explicitly use this assumption. We can strengthen *retrieve* so that this assumption is built into it by using negation. We discuss this in section 4.3.

Note that, in a conventional functional language without logical variables, we would have to give an independent recursive definition for *retrieve* which compares only the keys in the association list with the given key. Logical variables allow us to reuse code in ways not possible in conventional functional languages.

**Example 3 (Address translation)** We want to define a function that translates a sequence of definitions and uses of symbolic addresses to one of numeric addresses. Suppose the input is a list containing two kinds of terms: $(Def, s)$ and $(Use, s)$, where $s$ is a symbol. For each symbol $s$ that appears in a $(Use, s)$ term, there is a unique term $(Def, s)$ in the list. The translation should translate each $(Def, s)$ to a term $(Assign, s, n)$ and each $(Use, s)$ to a term $(Use, n)$, such that

1. if $(Def, s)$ precedes $(Def, s')$ in the input sequence, then their translations $(Assign, s, n)$ and $(Assign, s', n')$ satisfy $n \leq n'$,
2. if $(Def, s)$ is translated to $(Assign, s, n)$ then every $(Use, s)$ is translated to $(Use, n)$, and
3. the numbers assigned to symbols are successive integers starting from 1.

What is interesting about the problem is that the use of a symbol can appear in the input list before its definition itself appears. So, two passes over the list are required in a conventional functional language: one to construct a table mapping symbols to numeric addresses, and a second to do the translation.

In the following program, we use a logical variable as a table. It gets constructed (bound) on the fly, as the translation is done.

$$
\begin{aligned}
& translate\ seq \equiv \\
& \quad letrec\ tran(Nil, table, n) \equiv Nil \\
& \qquad\quad tran((Def, s).seq, table, n) \equiv\ retrieve(s, table) = n \\
& \qquad\qquad\qquad\qquad\qquad\qquad\quad \rightarrow (Assign, s, n).tran(seq, table, n+1) \\
& \qquad\quad tran((Use, s).seq, table, n) \equiv (Use, retrieve(s, table)).tran(seq, table, n) \\
& \quad in\ \exists table.\ tran(seq, table, 1)
\end{aligned}
$$

This program is based on the fact that when *retrieve* is invoked with a variable as the second argument, it instantiates it to an association list containing a pair for the given key. Let us look at an example evaluation

$$
\begin{aligned}
& translate\ (Use, A).(Def, A).Nil \\
& \Longrightarrow \exists table.\ tran((Use, A).(Def, A).Nil), table, 1) \\
& \Longrightarrow \exists table.\ (Use, retrieve(A, table)) \,. \\
& \qquad\qquad\quad (retrieve(A, table) = 1 \rightarrow (Assign, A, 1)) \,. \\
& \qquad\qquad\quad Nil
\end{aligned}
$$

The subexpression $retrieve(A, table)$ has an infinite number of nondeterministic narrowings:

$$
\begin{aligned}
& \exists a.\ \exists t.\ table = (A, a).t \rightarrow a \\
& \exists k.\ \exists a.\ \exists a'.\ \exists t'.\ table = (k, a).(A, a').t' \rightarrow a \\
& \vdots
\end{aligned}
$$

Suppose we use the first narrowing. Then the original expression narrows to

$$
\begin{aligned}
\exists table.\ \exists a.\ \exists t.\ table = (A, a).t \rightarrow\ & (Use, a) \,. \\
& (retrieve(A, (A, a).t) = 1 \rightarrow (Assign, A, 1)) \,. \\
& Nil
\end{aligned}
$$

Again choosing the first narrowing for the remaining *retrieve* application, we obtain

$$
\begin{aligned}
& \exists table.\ \exists a.\ \exists t.\ table = (A, a).t \rightarrow\ (Use, a) \,. \\
& \qquad\qquad\qquad\qquad\qquad\qquad\quad (a = 1 \rightarrow (Assign, A, 1)) \,. \\
& \qquad\qquad\qquad\qquad\qquad\qquad\quad Nil \\
& \Longrightarrow \exists table.\ \exists a.\ \exists t.\ table = (A, 1).t \wedge a = 1 \rightarrow (Use, 1).(Assign, A, 1).Nil \\
& \Longrightarrow \exists t.\ (Use, 1).(Assign, A, 1).Nil \quad \text{by EXIST2, EXIST3} \\
& \Longrightarrow (Use, 1).(Assign, A, 1).Nil \qquad\ \text{by EXIST1}
\end{aligned}
$$

Thus, even though the attribute of $A$ is not known at the time $(Use, A)$ is translated, it is assumed to be a variable $a$ and is bound whenever $(Def, A)$ is translated.

But, there are other concerns about this program. Since each *retrieve* application has infinite narrowings, the body of the quantifier $tran(seq, table, 1)$ has infinite narrowings too. Is the definition of *translate* still consistent? Secondly, do we get infinite narrowings operationally or just one? To answer the first question, recall that each symbol has a $(Def, s)$ term in the input sequence. So, every variable temporarily assumed as the attribute of a symbol eventually gets bound when the $(Def, s)$ term is processed. So, the final term is variable-free. The various narrowings can bind table to contain other irrelevant pairs, e.g.,

$$\exists table.\, \exists k.\, \exists a.\, \exists t'.\, table = (k, a).(A, 1).t \rightarrow (Use, 1).(Assign, A, 1).Nil$$

But, the final result is the same, since the result term is variable-free. If the two retrieves produce different attributes for $A$ as discussed in example 2, then we can get a result term with variables, such as

$$\exists table.\, \exists a'.\, \exists t'.\, table = (A, 1).(A, a').t \rightarrow (Use, a').(Assign, A, 1).Nil$$

But, we shall give a stronger definition of *retrieve* in section 4.3 by which such narrowings are eliminated. So, the definition of *translate* is consistent. But, operationally, we do not yet have a framework by which the different nondeterministic branches can be collected together. So, we get an infinite number of narrowings all of which produce the same result. We will present a solution to this problem in section 4.2.

This example also illustrates our need for strict pairs. If pairs were not strict then the original *translate* application would be equivalent to

$$\begin{array}{l}(Use,\ \exists table.\, retrieve(A, table))\,.\\ (\exists table.\, retrieve(A, table) = 1 \rightarrow (Assign, A, 1))\,.\\ Nil\end{array}$$

Different components of the output sequence can use different instantiations of *table*, and it becomes impossible to enforce that we want a single *table* to be used for the entire translation.

**Example 4 (Permutation Inversion)** We are given a vector of $n$ elements each of which is an integer between 1 and $n$, so that no integer appears more than once. So, the vector represents a permutation of the integers $1 \ldots n$. We want to define a function that constructs a vector representing the inverse of the permutation.

Let us assume a primitive predicate $vector(v, n)$ which means "$v$ is a vector of $n$ elements". If $v$ is a variable, it binds it to a vector of $n$ existentially quantified variables. If $n$ is a variable, it binds it to the size of $v$. If both of them are variables, then the result is undefined.

Now, the function *inverse* can be defined as

$$\begin{array}{lll} inverse\ p & \equiv & \exists n.\, \exists v.\ (vector(p, n) \wedge \\ & & \qquad vector(v, n) \wedge \\ & & \qquad forall(ints\text{-}between(1, n),\ \lambda i.\, v(p(i)) = i)) \\ & & \qquad \rightarrow v \end{array}$$

$ints\text{-}between(m, n)$ generates a list of integers between $m$ and $n$. $forall(l, p)$ checks the predicate $p$ for all elements of the list $l$.

The body of the quantifier is evaluated with the logical variables $n$ and $v$. $vector(p, n)$ binds $n$ to the size of $p$. $vector(v, n)$ binds $v$ to a vector of $n$ variables. The *forall* expression binds every $p(i)$'th element of $v$ to $i$. At the end of this, there is a unique ground binding of $v$, since each $j \in 1 \ldots n$ appears exactly once in $p$.

Note that this it is quite hard to solve this problem efficiently in a conventional functional language without logical variables. The result vector $v$ would have to be defined by a function that maps an index $j$ to the corresponding element of $v$.

$$\textit{inverse } p \equiv \textit{make-vector}(\ \textit{size } p,\ \lambda j.\ \textit{such-that}(\textit{ints-between}(1,n),\ \lambda i.\ p(i) = j))$$

$\textit{make-vector}(n, f)$ generates a vector of size $n$, whose $j$'th element is $(f\ j)$. $\textit{such-that}(l,p)$ yields the first element of the list $l$ satisfying the predicate $p$. Since the element $v(j)$ can only be known by searching for $j$ in the input vector $p$, this solution takes $O(n^2)$ time. In contrast, the solution with logical variables takes only $O(n)$ time.

In conventional functional languages, a variable is always introduced together with its binding. Therefore, the variables do not play any useful role and can be eliminated. This is indeed demonstrated by the combinatory calculus. In a functional logic language, on the other hand, a variable can be introduced without a binding, and can be bound whenever information becomes available. For example, in the permutation inversion program, we introduce a vector of variables and then bind them independently in an arbitrary order. It is this decoupling of the introduction and the binding of variables, that allows us to solve the problem in $O(n)$ time. Similarly, in example 2, we introduce a variable for the attribute of a key, and bind the variable using *member*. This flexibility is reminiscent of imperative languages where we can allocate a location and use a pointer to it, before the contents of the location itself is defined. This is indeed how logical variables are implemented. But, they have several advantages over imperative variables. Firstly, the pointers are transparent. Secondly, variables are bound only once; they are not mutable. Therefore, the same operator can be used both for testing equality and for binding. When performing a unification, we do not need to know whether a variable is already bound or not.

However, there is a qualitative difference between the binding of logical variables and the binding done in function application. Consider the following two expressions that appear similar:

$$\textbf{let } x \equiv e_1 \textbf{ in } e_2$$

$$\exists x.\ x = e_1 \rightarrow e_2$$

The first is actually syntactic sugar for

$$((\lambda x.\ e_2)\ e_1)$$

The binding of $x$ in its evaluation is *strong* binding. It does not require the evaluation of $e_1$. But, the binding of $x$ in the evaluation of the second expression is *weak* binding. It is done only after $e_1$ is evaluated to a term. If, for example, $e_1$ is an infinite list and $e_2 = (\mathit{fst}\ x)$, then the first expression terminates under outermost evaluation, whereas the second one does not terminate even under outermost evaluation. This is because the $=$ construct denotes the application of a continuous equality function which cannot yield *true* if its arguments are infinite objects.

# 4 Failure and nondeterminism

## 4.1 Failure

By the evaluation rules we have seen so far, there are many terminal forms which simply denote $\perp$, e.g., $A = B$. They are the cases where our rules fail to reduce them to meaningful values. We can introduce an explicit failure expression which captures them all into a single terminal form. We use a special symbol $\omega$ for this purpose.

$$e ::= \omega$$

$$[\![\omega]\!]\eta = \perp$$

$\omega$ is also treated as a term.

The evaluation rules for failure are of two kinds: failure introduction rules and failure propagation rules. The failure introduction rules are

| | | | | |
|---|---|---|---|---|
| APP-FAIL1: | $(t\ e)$ | $\Longrightarrow$ | $\omega$ | |
| APP-FAIL2: | $((\lambda a.\ d)\ t)$ | $\Longrightarrow$ | $\omega$ | if $t \neq a$ |
| APP-FAIL3: | $((\lambda(t_1, t_2).\ d)\ a)$ | $\Longrightarrow$ | $\omega$ | |
| COND-FAIL: | $(t \rightarrow e)$ | $\Longrightarrow$ | $\omega$ | if $t \neq true$ |
| EQ-FAIL1: | $(a = b)$ | $\Longrightarrow$ | $\omega$ | if $a \neq b$ |
| EQ-FAIL2: | $(a = (e_1, e_2))$ | $\Longrightarrow$ | $\omega$ | |
| EQ-FAIL3: | $((e_1, e_2) = a)$ | $\Longrightarrow$ | $\omega$ | |
| EQ-FAIL4: | $(e_1, e_2)$ | $\Longrightarrow$ | $\omega$ | if $e_1$ or $e_2$ is an abstraction |
| CONJ-FAIL: | $(t_1 \wedge t_2)$ | $\Longrightarrow$ | $\omega$ | if $t_1 \neq true$ or $t_2 \neq true$ |

When $\omega$ appears in a strict position in a bigger expression, then the whole expression reduces to $\omega$. Many of these failure propagation cases are covered by the above rules. For example, $(\omega\ e) \Longrightarrow \omega$. The other cases are the following:

| | | | | |
|---|---|---|---|---|
| FAIL-APP2: | $((\lambda t.\ d)\ \omega)$ | $\Longrightarrow$ | $\omega$ | if $t$ is not a variable |
| FAIL-COND2: | $(\sigma \rightarrow \omega)$ | $\Longrightarrow$ | $\omega$ | |
| FAIL-PAIR1: | $(\omega, e)$ | $\Longrightarrow$ | $\omega$ | |
| FAIL-PAIR2: | $(e, \omega)$ | $\Longrightarrow$ | $\omega$ | |
| FAIL-EQ1: | $(\omega = (e_1, e_2))$ | $\Longrightarrow$ | $\omega$ | |
| FAIL-EQ2: | $((e_1, e_2) = \omega)$ | $\Longrightarrow$ | $\omega$ | |
| FAIL-EXIST: | $(\exists x.\ \omega)$ | $\Longrightarrow$ | $\omega$ | |

## 4.2 Nondeterminism

The rules for the choice operator ";" given in section 2.4 are nondeterministic. The evaluation can choose CHOICE1 or CHOICE2 along different branches. However, the nondeterminism does not commit to any one choice, but requires all the choices to be tried. This can be made explicit by carrying the choice operator until one of the choices evaluates to $\omega$. The basic choice rules are

| | | | |
|---|---|---|---|
| CHOICE-ELIM1: | $(\omega; e)$ | $\Longrightarrow$ | $e$ |
| CHOICE-ELIM2: | $(e; \omega)$ | $\Longrightarrow$ | $e$ |
| CHOICE-COM: | $(e_1; e_2)$ | $\Longrightarrow$ | $(e_2; e_1)$ |
| CHOICE-ASSOC: | $((e_1; e_2); e_3)$ | $\Longrightarrow$ | $(e_1; (e_2; e_3))$ |

We also need to propagate the choice operator from every strict evaluation position. This is done by the following rules:

| | | | |
|---|---|---|---|
| CHOICE-APP1: | $((e_1; e_2)\ e_3)$ | $\Longrightarrow$ | $(e_1\ e_3)\ ;\ (e_2\ e_3)$ |
| CHOICE-APP2: | $((\lambda t.\ d)\ (e_1; e_2))$ | $\Longrightarrow$ | $((\lambda t.\ d)\ e_1)\ ;\ ((\lambda t.\ d)\ e_2)$ if $t$ is not a variable |
| CHOICE-COND1: | $((e_1; e_2) \rightarrow e_3)$ | $\Longrightarrow$ | $(e_1 \rightarrow e_3\ ;\ e_2 \rightarrow e_3)$ |
| CHOICE-COND2: | $(\sigma \rightarrow (e_1; e_2))$ | $\Longrightarrow$ | $(\sigma \rightarrow e_1)\ ;\ (\sigma \rightarrow e_2)$ |
| CHOICE-EQ1: | $((e_1; e_2) = e_3)$ | $\Longrightarrow$ | $(e_1 = e_3)\ ;\ (e_2 = e_3)$ |
| CHOICE-EQ2: | $(e_1 = (e_2; e_3))$ | $\Longrightarrow$ | $(e_1 = e_2)\ ;\ (e_1 = e_3)$ |
| CHOICE-CONJ1: | $((e_1; e_2) \wedge e_3)$ | $\Longrightarrow$ | $(e_1 \wedge e_3)\ ;\ (e_2 \wedge e_3)$ |
| CHOICE-CONJ2: | $(e_1 \wedge (e_2; e_3))$ | $\Longrightarrow$ | $(e_1 \wedge e_2)\ ;\ (e_1 \wedge e_3)$ |
| CHOICE-PAIR1: | $((e_1; e_2), e_3)$ | $\Longrightarrow$ | $(e_1, e_3)\ ;\ (e_2, e_3)$ |
| CHOICE-PAIR2: | $(e_1, (e_2; e_3))$ | $\Longrightarrow$ | $(e_1, e_2)\ ;\ (e_1, e_3)$ |
| CHOICE-EXIST: | $(\exists x.\ (e_1; e_2)$ | $\Longrightarrow$ | $(\exists x.\ e_1)\ ;\ (\exists x.\ e_2)$ |

The rules CHOICE-PAIR1 and CHOICE-PAIR2 propagate the choice operator out of pairs, whereas the rule CHOICE-ELIM3 propagates it into pairs. Our intent is to use CHOICE-ELIM rules to eliminate as many choice operators as possible and then to propagate them to the outermost level. The terminal forms of the evaluation are given by

$$\begin{array}{lcl} \tau & ::= & \omega \mid \alpha\,\{;\alpha\}^* \\ \alpha & ::= & \exists\overline{x}.\,\sigma \to t \mid t \mid \exists\overline{x}.\,\sigma \to (\lambda t.e) \mid \lambda t.e \qquad \text{where } \overline{x} \notin DV(\sigma) \\ t & ::= & a \mid (t_1, t_2) \end{array}$$

In practice, the terminal form of an expression is produced "lazily" as a series of approximations, each of the form

$$\alpha_1; \cdots; \alpha_n; e$$

where $e$ is a continuation yet to be evaluated. The user is given the control to evaluate the continuation or to abort.

Note that, with our new choice rules, the evaluation is semantics-preserving. The only case where the earlier evaluation rules lost information was in CHOICE1 and CHOICE2. Since we have replaced them by semantics-preserving rules which carry the choice operators till the end, we avoid the loss of information. Though noble, this is not pragmatically desirable. Recall that in example 3 we have the situation of obtaining an infinite number of $\alpha$'s all of which are identical. At the cost of losing information only when the input expression is inconsistent, we can introduce a rule by which the infinite answers can be eliminated.

$$\text{CHOICE-ELIM3:} \quad t; e \implies t \qquad \text{if } t \text{ does not contain } \omega$$

The rationale is that since $t$ is completely defined, evaluating the continuation $e$ can at best produce an inconsistency. If the expression is consistent, $e$ would certainly evaluate to $t$ or $\perp$. So, if we do not care about finding inconsistencies we can throw $e$ away.

## 4.3 Alternative formulations

It may have appeared strange that we had only one truth value in our domain, and falsity was denoted by $\perp$. This makes it, in particular, impossible to use a negation operation because it would be nonmonotonic. Negation is, in fact, one of the big unsolved problems in logic programming. However, we have some preliminary ideas on how to treat it in functional logic languages. Our solution is to make the failure constant $\omega$ as a distinct value in the semantic domain.

$$D = \{\omega\}_{\perp\top} + B_{\perp\top} + [D \to D]$$

The choice operator is interpreted as the semantic operation $\vee$ defined by

$$\begin{array}{lcll} \omega \vee v & = & v & \\ v \vee \omega & = & v & \\ v_1 \vee v_2 & = & v_1 \sqcup v_2 & \text{if } v_1, v_2 \in B_{\perp\top} \\ (v_1 \vee v_2)x & = & v_1 x \vee v_2 x & \text{if } v_1 \text{ and } v_2 \text{ are functions} \end{array}$$

We can now define negation, using $\omega$ for falsity.

$$e \ ::= \ (\neg\, e)$$

$$[\![\neg\, e]\!]\eta \ = \ \begin{cases} \omega & \text{if } [\![e]\!]\eta = \mathit{true} \\ \mathit{true} & \text{if } [\![e]\!]\eta = \omega \\ \top & \text{if } [\![e]\!]\eta = \top \\ \perp & \text{otherwise} \end{cases}$$

Once we have negation, we can also introduce the classical if-then-else construct by syntactic sugaring:

$$(\textbf{if } e_1 \textbf{ then } e_2 \textbf{ else } e_3) \stackrel{\text{def}}{=} \textbf{let } p \equiv e_1 \textbf{ in } (p \rightarrow e_2; \neg\, p \rightarrow e_3)$$

Another possibility is to interpret ";" as a "sequential" choice operation, i.e.,

$$\begin{aligned} \omega \vee v &= v \\ v_1 \vee v_2 &= v_1 \qquad \text{if } v_1 \in B_\perp \end{aligned}$$

extended pointwise, as usual, for functions. With sequentiality, there is no inconsistency involved. If the first argument is not $\omega$, then the second argument is ignored. In particular, if the first argument is $\perp$, then the choice expression is also $\perp$. Since most implementations of logic languages implement ";" sequentially for the sake of efficiency, this would provide the appropriate semantics for such implementations.

The complete operational semantics of negation is outside the scope of this paper. But, let us briefly discuss how the problems mentioned in section 3 can be avoided by the use of negation. First of all, note that some negated equalities would be irreducible. So, they need to be treated as substitutions too.

$$\sigma ::= (x = t) \mid (\sigma_1 \wedge \sigma_2) \mid (\neg\, \sigma)$$

Now, we can rewrite the *member* function of example 1 using negation, as

$$member(a, b.x) \equiv (a = b \rightarrow true;\ \neg\, a = b \rightarrow member(a, x))$$

This definition makes explicit the assumption that an element appears only once in the list, by performing the recursive application of member only if $a \neq b$. First of all, this cuts down the computation when the arguments are ground. For example,

$$\begin{aligned} & member(A, A.B.Nil) \\ & \Longrightarrow A = A \rightarrow true;\ \neg\, A = A \rightarrow member(A, B.Nil) \\ & \Longrightarrow true; \omega \\ & \Longrightarrow true \end{aligned}$$

The recursive application of *member* is avoided, whereas in example 1 the search continues till the end of the list even though it is useless.

Secondly, if the second argument is a variable, it binds it to a list without duplicates. For example,

$$\begin{aligned} & member(A, l) \\ \Longrightarrow\ & (\exists x.\, l = A.x \rightarrow true); \\ & (\exists b.\, \exists x.\, l = b.x \wedge \neg\, b = A \rightarrow member(A, x)) \\ \Longrightarrow\ & (\exists x.\, l = A.x \rightarrow true); \\ & (\exists b.\, \exists x'.\, l = b.A.x' \wedge \neg\, b = A \rightarrow true); \\ & (\exists b.\, \exists b'.\, \exists x'.\, l = b.A.x' \wedge \neg\, b = A \wedge \neg\, b' = A \rightarrow member(A, x')) \\ \Longrightarrow\ & \quad \vdots \end{aligned}$$

We still get an infinite number of nondeterministic narrowings, but they all qualify the solutions by stating that the unbound elements of $l$ are not equal to $A$. This is useful in evaluating an expression such as

$$member(A, l) \wedge member(A, l)$$

Evaluating this by the member function of example 1 bound $l$ to lists with two occurrences of $A$. But, by our new definition of *member*, $l$ cannot get bound to a list with duplicates. Notice the second narrowing of this expression:

$$\begin{aligned}
&(\exists b.\, \exists x'.\, l = b.A.x' \wedge \neg\, b = A) \rightarrow member(A, b.A.x') \\
&\Longrightarrow \exists b.\, \exists x'.\, l = b.A.x' \wedge \neg\, b = A \rightarrow member(A, b.A.x') \\
&\Longrightarrow\ (\exists b.\, \exists x'.\, l = b.A.x' \wedge \neg\, b = A \rightarrow b = A \rightarrow true); \\
&\qquad (\exists b.\, \exists x'.\, l = b.A.x' \wedge \neg\, b = A \rightarrow \neg\, b = A \rightarrow member(A, A.x')) \\
&\Longrightarrow \omega;\ (\exists b.\, \exists x'.\, l = b.A.x' \wedge \neg\, b = A \rightarrow true) \\
&\Longrightarrow \exists b.\, \exists x'.\, l = b.A.x' \wedge \neg\, b = A \rightarrow true
\end{aligned}$$

The first choice in the above fails while taking the composition of substitutions $\neg\, b = A$ and $b = A$. Thus the first element of $l$ is prohibited from getting bound to $A$, no matter howmany times we test for the membership of $A$ in $l$.

Similarly, the *retrieve* function has to be defined making explicit use of the assumption that an association list contains at most one pair for each key.

$$retrieve(key, (k, a).x) \equiv (key = k \rightarrow a;\ \neg\ key = k \rightarrow retrieve(key, x))$$

Now, consider the evaluation of *translate*.

$$\begin{aligned}
&translate\ (Use, A).(Def, A).Nil \\
&\Longrightarrow \exists table.\ (Use, retrieve(A, table)).(retrieve(A, table) = 1 \rightarrow (Assign, A, 1)).Nil
\end{aligned}$$

If the first *retrieve* application binds *table* to $(k, a).(A.a').t'$, it also places the condition (negative substitution) that $(\neg\ k = A)$. So, the second application of *retrieve* yields $a'$ rather than $a$ as its result, which then gets bound to 1. Thus, independent of which narrowing of the *retrieve* application is used, the result is of the form

$$\exists \overline{x}.(Use, 1).(Assign, A, 1).Nil$$

and hence the definition of *translate* is consistent.

# 5 Set abstraction and nonground outputs

Even though our basic narrowing mechanism introduced new free variables in the evaluation of an expression, we saw that the new variables are not really free but are implicitly existentially quantified. We made this quantification explicit in section 3. Existential quantification also gives us a way to introduce an unbound variable and bind it within its scope. But, it does not allow us to export the variable outside its scope and bind it elsewhere. For example, we cannot say

$$\textbf{let}\ a \equiv \exists x.\, x\ \textbf{in}\ a = 1$$

because the expression $\exists x.\, x$ is inconsistent and denotes $\top$.

This example may appear somewhat esoteric, but exporting logical variables from their scope is quite useful. In the permutation inversion problem, we assumed a primitive $vector(v, n)$ which binds $v$ to a vector of $n$ variables if $v$ is a variable. Since, we know that $v$ is actually an output of this operation, we may consider using a primitive (*a-vector* $n$) which yields a vector of $n$ variables.

$$\begin{aligned}
&\exists v.\ v = (a\text{-}vector\ n) \wedge \\
&\quad forall(ints\text{-}between(1, n),\ \lambda i.\ v(p(i)) = i) \\
&\quad \rightarrow v
\end{aligned}$$

Id Nouveau [NPA86] has a primitive called *array* which behaves like *a-vector*. But, the primitive is inconsistent, just like $\exists x.\, x$, since it exports logical variables. Another example of the use of variable exportation is the following type inference program [Mil78]. The function $type(e, te)$ yields the type of $e$ in the type environment $te$ represented as an association list.

$$
\begin{array}{lcl}
type(x, te) & \equiv & (variable?\ x) \rightarrow retrieve(x, te) \\
type((Apply, f, e), te) & \equiv & \exists t.\ \exists u.\ type(f, te) = (Fun, t, u) \wedge type(e, te) = t \rightarrow u \\
type((Lambda, x, e), te) & \equiv & \exists t.(Fun, t, type(e, (x, t).te)
\end{array}
$$

The last equation can produce inconsistent results if $t$ does not get bound in the subexpression $type(e, (x, t).te)$. For example,

$$type((Lambda, X, X), Nil) \Longrightarrow \exists t.\ (Fun, t, t)$$

This is an inconsistent expression and denotes $\top$.

The solution to these problems, suggested in [Red86a,DFP86], is to treat such terms not as denoting values, but as denoting *sets of values*. We use the binding construct $(\bigcup x.\ \{e\})$ for this purpose. Whereas $(\exists x.\ x)$ denotes $\top$, $(\bigcup x.\ \{x\})$ denotes the set of all values. $(\bigcup t.\ \{(Fun, t, t)\})$ denotes the set of all terms of the form $(Fun, t, t)$. The expression $(Lambda, X, X)$ does not have a single type, but a whole set of types of the form $(Fun, t, t)$. Similarly, there is not a single vector of length $n$, but a whole set of vectors. The construct $(\bigcup x.\ \{e\})$ is called *absolute set abstraction* in [DFP86] and a *free set* in [Red86a].

Absolute set abstraction is not the only way to handle exportation of logical variables. We could treat all expressions as being indeterminate (implicitly denoting sets), and use many-to-many mappings instead of functions. Church-Rosser property would then be lost. This approach is taken in [Smo86].

We introduce the set concepts into our language by extending it with two new syntactic domains:

*Set*, ranged over by $s$

*Set-function*, ranged over by $\psi$

$$
\begin{array}{lcll}
s & ::= & \emptyset & \text{- empty set} \\
 & & \mid \{e\} & \text{- singleton} \\
 & & \mid (s_1 \cup s_2) & \text{- set union} \\
 & & \mid (e \parallel s) & \text{- apply to all} \\
 & & \mid (\psi\ e) & \text{- set function application} \\
 & & \mid (e \rightarrow s) & \text{- conditional set} \\
 & & \mid (\bigcup x.\ s) & \text{- absolute set abstraction} \\
\psi & ::= & (\lambda t.\ s) & \text{- abstraction} \\
 & & \mid (\psi_1 \cup \psi_2) & \text{- union of results}
\end{array}
$$

We also extend expressions by a new construct for testing set membership:

$$e\ ::=\ (e \in s)$$

Sets are not first-class values in the language. There are no functions on sets. There are also no higher type sets whose members are sets. However, the members of sets can be any values including functions.

We define the following constructs by syntactic sugar:

$$
\begin{array}{rcl}
\{e_1, e_2, \ldots, e_n\} & \stackrel{\text{def}}{=} & \{e_1\} \cup \{e_2\} \cup \cdots \cup \{e_n\} \\
\textbf{collect}\ t \leftarrow s\ \textbf{in}\ e & \stackrel{\text{def}}{=} & ((\lambda t.\ e) \parallel s) \\
(member\ s) & \stackrel{\text{def}}{=} & \exists x.\ x \in s \rightarrow x
\end{array}
$$

Th **collect** construct provides the qualified set abstraction used by Turner [Tur81]. Note that the binding of $x$ in this construct is *strong*, whereas the binding in the similar-looking expression

$$(\bigcup \overline{x}.\ t \in s \rightarrow e)$$

is *weak*. This is similar to the difference between let binding and = binding. If all the members of a set $s$ are consistent, we can get their lub using the construct (*member* $s$). Note that *member* cannot be defined as a function since functions cannot take sets as arguments.

The denotational semantics of the extended language uses new semantic domains:

$$\begin{aligned} [\![s]\!] &: Env \to P(D) \\ [\![\psi]\!] &: Env \to [D \to P(D)] \end{aligned}$$

where

$$Env = Variable \to D$$

We have not yet completely investigated the required structure of $P(D)$. But, since we do not have functions on sets, we believe it is adequate to use the Hoare powerdomain of $D$. Its elements are the nonempty, downward-closed and limit-closed subsets of $D$ and its partial order is the subset order. The semantics of the new constructs follows:

$$\begin{aligned}
[\![\emptyset]\!]\eta &= \{\bot\} \\
[\![\{e\}]\!]\eta &= \{[\![e]\!]\eta\} \\
[\![s_1 \cup s_2]\!]\eta &= [\![s_1]\!]\eta \cup [\![s_2]\!]\eta \\
[\![(e \parallel s)]\!]\eta &= \{([\![e]\!]\eta\ v) \mid v \in [\![s]\!]\eta\} \\
[\![(\psi\ e)]\!]\eta &= ([\![\psi]\!]\eta\ [\![e]\!]\eta) \\
[\![(e \to s)]\!]\eta &= \begin{cases} [\![s]\!]\eta & \text{if } [\![e]\!]\eta = true \\ D & \text{if } [\![e]\!]\eta = \top \\ \emptyset & \text{otherwise} \end{cases} \\
[\![(\bigcup x.\ s)]\!]\eta &= \bigcup_{v \in Dcon} [\![s]\!](\eta; [v/x]) \\
[\![\lambda t.\ s]\!]\eta &= \lambda v.\ \text{if } \exists \rho \in FV(t) \to D \text{ such that } [\![t]\!]\rho = v \text{ then } [\![s]\!](\eta;\rho) \text{ else } \{\bot\} \\
[\![\psi_1 \cup \psi_2]\!]\eta &= \lambda v.\ ([\![\psi_1]\!]\eta\ v) \cup ([\![\psi_2]\!]\eta\ v) \\
[\![e \in s]\!]\eta &= \begin{cases} true & \text{if } [\![e]\!]\eta \in B, [\![e]\!]\eta \in [\![s]\!]\eta \\ \top & \text{if } [\![e]\!]\eta = \top \\ \bot & \text{otherwise} \end{cases}
\end{aligned}$$

The operational semantics of the new constructs closely follows the rules we have given before. To see the correspondence, note that $\emptyset$ is similar to $\omega$, $\cup$ is similar to ";", and $\bigcup$ is similar to $\exists$. We again have several classes of rules: normal rules, failure rules, substitution propagation rules, failure (empty set) propagation rules and union propagation rules. The normal rules are given below. In addition, the rules APP1 through APP3′ hold for set function application, and COND and COND′ hold for set conditional.

| | | | |
|---|---|---|---|
| APP-ALL: | $(e \parallel \{e'\})$ | $\Longrightarrow$ | $\{(e\ e')\}$ |
| $\alpha$-ABS: | $(\bigcup x.\ s)$ | $\Longrightarrow$ | $(\bigcup x'.\ [x'/x][s])$ if $x$ is not free in $s$ |
| ABS1: | $(\bigcup x.\ \{t\})$ | $\Longrightarrow$ | $\{t\}$ if $x$ is not free in $t$ |
| ABS2: | $(\bigcup x.\ \bigcup \overline{y}.\ x = t \to s)$ | $\Longrightarrow$ | $(\bigcup \overline{y}.\ s)$ if $x$ is not free in $t$ or $s$ |
| ABS3: | $(\bigcup x.\ \bigcup \overline{y}.\ x = t \wedge \sigma \to s)$ | $\Longrightarrow$ | $(\bigcup \overline{y}.\ \sigma \to s)$ if $x$ is not free in $t$, $\sigma$ or $s$ |
| MEMBER: | $e \in \{e'\}$ | $\Longrightarrow$ | $e = e'$ |

Note the additional condition in ABS1 which was not present in EXIST1. If $x$ is free in $t$, then $(\bigcup x.\ \{t\})$ is a terminal form. The failure rules and failure propagation rules are:

$$\begin{array}{llcll}
\text{SET-APP-FAIL2:} & ((\lambda a.\, s)\ t) & \Longrightarrow & \emptyset & \text{if } t \neq a \\
\text{SET-APP-FAIL3:} & ((\lambda(t_1, t_2).\, d)\ a) & \Longrightarrow & \emptyset & \\
\text{SET-COND-FAIL:} & (t \rightarrow s) & \Longrightarrow & \emptyset & \text{if } t \neq \mathit{true} \\
\text{FAIL-SINGLE:} & \{\omega\} & \Longrightarrow & \emptyset & \\
\text{FAIL-APP-ALL:} & (e \parallel \emptyset) & \Longrightarrow & \{(e\ \omega)\} & \\
\text{FAIL-ABS:} & \bigcup x.\, \emptyset & \Longrightarrow & \emptyset & \\
\text{FAIL-MEMBER:} & e \in \emptyset & \Longrightarrow & \omega &
\end{array}$$

The substitution propagation rules are:

$$\begin{array}{llcl}
\text{PROP-APP-ALL2:} & (e \parallel (\bigcup \overline{x}.\, \sigma \rightarrow s)) & \Longrightarrow & \bigcup \overline{x}.\, \sigma \rightarrow (\sigma[e] \parallel s) \\
 & & & \text{if } \overline{x} \text{ are not free in } e \\
\text{PROP-SET-APP2:} & ((\lambda t.\, s)\ (\exists \overline{x}.\, \sigma \rightarrow e)) & \Longrightarrow & (\bigcup \overline{x}.\, (\sigma[\lambda t.\, s]\ e)) \\
 & & & \text{if } t \text{ is not a variable, and} \\
 & & & \overline{x} \text{ are not free in } \lambda t.\, s \\
\text{PROP-SET-COND:} & ((\exists \overline{x}.\, \sigma \rightarrow e) \rightarrow s) & \Longrightarrow & \bigcup \overline{x}.\, \sigma \rightarrow (e \rightarrow \sigma[s]) \\
 & & & \text{if } \overline{x} \text{ are not free in } s \\
\text{COMP-SET:} & (\sigma_1 \rightarrow (\bigcup \overline{x}.\, \sigma_2 \rightarrow s)) & \Longrightarrow & (\bigcup \overline{x}.\, \sigma_1 \circ \sigma_2 \rightarrow s) \\
 & & & \text{if the variables } \overline{x} \text{ are not free in } \sigma_1 \\
\text{PROP-ABS:} & (\bigcup x.\, \bigcup \overline{y}.\, \sigma \rightarrow s) & \Longrightarrow & (\bigcup \overline{y}.\, \sigma \rightarrow (\bigcup x.\, s)) \\
 & & & \text{if } x \text{ is not free in } \sigma \\
\text{PROP-MEMBER:} & e \in \bigcup \overline{x}.\, \sigma \rightarrow s & \Longrightarrow & \exists \overline{x}.\, \sigma \rightarrow (\sigma[e] \in s) \\
 & & & \text{if } \overline{x} \text{ are not free in } e
\end{array}$$

Finally, the union rules are

$$\begin{array}{llcl}
\text{UNION-ELIM1:} & \emptyset \cup s & \Longrightarrow & s \\
\text{UNION-ELIM2:} & s \cup \emptyset & \Longrightarrow & s \\
\text{UNION-COM:} & (s_1 \cup s_2) & \Longrightarrow & (s_2 \cup s_1) \\
\text{UNION-ASSOC:} & ((s_1 \cup s_2) \cup s_3) & \Longrightarrow & (s_1 \cup (s_2 \cup s_3)) \\
\text{UNION-APP-ALL2:} & (e \parallel (s_1 \cup s_2)) & \Longrightarrow & (e \parallel s_1) \cup (e \parallel s_2) \\
\text{CHOICE-APP-ALL2:} & (\lambda t.\, d \parallel \{e_1; e_2\}) & \Longrightarrow & (\lambda t.\, d \parallel \{e_1\}) \cup (\lambda t.\, d \parallel \{e_2\}) \\
 & & & \text{if } t \text{ is not a variable} \\
\text{UNION-SET-APP1:} & ((\psi_1 \cup \psi_2)\ e) & \Longrightarrow & (\psi_1\ e) \cup (\psi_2\ e) \\
\text{CHOICE-SET-APP2:} & ((\lambda t.\, s)\ (e_1; e_2)) & \Longrightarrow & ((\lambda t.\, s)\ e_1) \cup ((\lambda t.\, s)\ e_2) \\
 & & & \text{if } t \text{ is not a variable} \\
\text{CHOICE-SET-COND1:} & ((e_1; e_2) \rightarrow s) & \Longrightarrow & (e_1 \rightarrow s) \cup (e_2 \rightarrow s) \\
\text{UNION-SET-COND2:} & (\sigma \rightarrow (s_1 \cup s_2)) & \Longrightarrow & (\sigma \rightarrow s_1) \cup (\sigma \rightarrow s_2) \\
\text{UNION-ABS:} & (\bigcup x.\, s_1 \cup s_2) & \Longrightarrow & (\bigcup x.\, s_1) \cup (\bigcup x.\, s_2) \\
\text{UNION-MEMBER:} & e \in (s_1 \cup s_2) & \Longrightarrow & e \in s_1;\ e \in s_2
\end{array}$$

Conspicuously missing are the rules for propagating choice or quantifiers through the singleton construct. The problem is that a rule such as

$$\{(e_1; e_2)\} \Longrightarrow \{e_1\} \cup \{e_2\}$$

does not preserve the semantics. On the left, we have a singleton whose element is approximated by $e_1$ and $e_2$. On the right, we have a set of the two approximations, but the element itself is missing from it. But, we believe that a rule such as this would be practically useful. Further investigation is needed into the nature of this problem.

Let us now show some examples of the use of absolute set abstraction.

**Example 5 (Vectors using sets)** As promised at the beginning of the section, we can now write the permutation inversion function with exportation of logical variables. The primitive required for this is (*vectors n*) which yields the set of all vectors of length $n$ as an absolute set abstraction:

$$\bigcup x_1.\bigcup x_2.\ \ldots \bigcup x_n.\ \{[x_1, x_2, \ldots, x_n]\}$$

Now, *inverse* can be defined as

$$\begin{aligned} inverse\ p \equiv\ & \textbf{let}\ \ n \equiv (size\ p) \\ & \textbf{in}\ \ \exists v.\,(v \in (vectors\ n) \wedge forall(ints\text{-}between(1,n),\ \lambda i.v(p(i)) = i)) \rightarrow v \end{aligned}$$

This definition has the same effect as that in example 4 since, by rules PROP-MEMBER and MEMBER

$$(v \in \bigcup x_1.\ \ldots \bigcup x_n.\ \{[x_1, \ldots, x_n]\}) \Longrightarrow (\exists x_1.\ \ldots \exists x_n.\ v = [x_1, \ldots, x_n])$$

In Id Nouveau [NPA86], the array construct is found to be referentially opaque, since

$$\begin{aligned} & \textbf{let}\ \ x \equiv array(1..n) \\ & \textbf{in}\ \ (x, x) \end{aligned}$$

is not equivalent to

$$(array(1..n),\ array(1..n))$$

This can be avoided by using our *vectors* primitive:

$$\begin{aligned} & \textbf{collect}\ \ x \leftarrow (vectors\ n) \\ & \textbf{in}\ \ (x, x) \end{aligned}$$

is semantically different from

$$((vectors\ n),\ (vectors\ n))$$

The former is a set of pairs, and the latter is a pair of sets.

**Example 6 (Retrieving multiple attributes)** Suppose, unlike in example 2, we have association lists in which a key can appear in any number of pairs with different attributes. To obtain all the attributes of a key, we can use

$$retrieve\text{-}all(key, alist) \equiv \bigcup attr.\ member((key, attr), alist) \rightarrow \{attr\}$$

The body of the definition is our syntax for what we would write in set theory as

$$\{attr \mid member((key, attr), alist)\}$$

Here is a sample evaluation using it:

$$\begin{aligned} & retrieve\text{-}all(A, (A,1).(A,2).Nil) \\ & \Longrightarrow \bigcup attr.\ member((A, attr),\ (A,1).(A,2), Nil) \rightarrow \{attr\} \\ & \Longrightarrow \bigcup attr.\ (attr = 1 \rightarrow true;\ attr = 2 \rightarrow true) \rightarrow \{attr\} \\ & \Longrightarrow \bigcup attr.\ (attr = 1 \rightarrow \{1\}) \cup (attr = 2 \rightarrow \{2\}) \qquad \text{by CHOICE-SET-COND1} \\ & \Longrightarrow (\bigcup attr.\ attr = 1 \rightarrow \{1\}) \cup (\bigcup attr.\ attr = 2 \rightarrow \{2\}) \qquad \text{by UNION-ABS} \\ & \Longrightarrow \{1\} \cup \{2\} \qquad \text{by ABS2} \end{aligned}$$

**Example 7 (Type inference)** The type inference program discussed at the beginning of this section can be written using sets as

$$\begin{aligned} types(x, te) &\equiv (variable?\ x) \rightarrow \{retrieve(x, te)\} \\ types((Apply, f, e), te) &\equiv \bigcup t.\bigcup u.\ (Fun, t, u) \in types(f, te) \wedge t \in types(e, te) \rightarrow \{u\} \\ types((Lambda, x, e), te) &\equiv \bigcup t.\ \textbf{collect}\ u \leftarrow types(e, (x,t).te) \\ &\qquad\quad \textbf{in}\ (Fun, t, u) \end{aligned}$$

The multiple definitions for the function are combined using the *union* operator. So, these equations for *types* represent a definition of the form

$$\lambda(x, te). \ldots \cup \lambda((Apply, f, e), te). \ldots \cup \lambda((Lambda, x, e), te). \ldots$$

The type of the identity function is now inferred as:

$$\begin{aligned}
& types((Lambda, X, X), Nil) \\
& \Longrightarrow \bigcup t.\ \textbf{collect}\ u \leftarrow types(X, (X, t).Nil)\ \textbf{in}\ (Fun, t, u) \\
& \Longrightarrow \bigcup t.\ \textbf{collect}\ u \leftarrow \{t\}\ \textbf{in}\ (Fun, t, u) \\
& \Longrightarrow \bigcup t.\ \{(Fun, t, t)\}
\end{aligned}$$

and the type of a sample application is inferred as:

$$\begin{aligned}
& types((Apply, (Lambda, X, X), (Lambda, X, X)),\ Nil) \\
& \Longrightarrow \bigcup t. \bigcup u.\ ((Fun, t, u) \in \bigcup t'.\{(Fun, t', t')\}) \ \wedge\ (t \in \bigcup t''.\{(Fun, t'', t'')\}) \rightarrow \{u\} \\
& \Longrightarrow \bigcup t. \bigcup u.\ (\exists t'.\, t = t' \wedge u = t') \wedge (\exists t''.\, t = (Fun, t'', t'')) \rightarrow \{u\} \\
& \Longrightarrow \bigcup t. \bigcup u.\ (\exists t''.\, t = (Fun, t'', t'')) \wedge u = (Fun, t'', t'')) \rightarrow \{u\} \\
& \Longrightarrow \bigcup t. \bigcup u. \bigcup t''.\, t = (Fun, t'', t'') \wedge u = (Fun, t'', t'') \rightarrow \{(Fun, t'', t'')\} \\
& \Longrightarrow \bigcup t''.\ \{(Fun, t'', t'')\}
\end{aligned}$$

# 6 Conclusion

We first described our ideas on using narrowing as the operational semantics of functional languages in [Red85]. We then proceeded to build an implementation of this semantics, whose features are described in [Red86b]. This implementation effort showed us that our understanding of narrowing was still inadequate, in particular its relationship to lazy evaluation. We also found several problems regarding efficiency which we had no easy language to describe in or investigate. Hence we were led to a formalization of the basic semantics, presented here.

Representing substitutions as expressions and using explicit existential quantifiers are the novel features of this presentation. This formalization shows that logical variables do not go well with lazy data structures. This is clear from the fact that, for lazy pairs,

$$\exists x.\ (e_1, e_2) = (\exists x.e_1, \exists x.e_2)$$

So, each component of a lazy data structure can use its own instantiations of logical variables. This is quite contrary to our intuitive ideas of logic programming. Much further investigation is required into the interaction between lazy evaluation and logical variables.

Another aspect brought up by this formalization is the importance of negation. The examples 1, 2 and 3 show this clearly. Representing substitutions as expressions allows us to easily extend them to negative substitutions. But, several questions still need to be answered. For example, what is meant by applying a negative substitution on an expression? Similar questions also arise in higher-order narrowing, i.e. in solving for function variables. We shall present some solutions to these problems in a future paper.

# 7 Acknowledgements

Discussions with Arvind and Paul Hudak contributed to the ideas presented in this paper.

This research was supported by the University of Illinois at Urbana-Champaign.

# References


[BMS80] R. M. Burstall, D. B. MacQueen, and D. T. Sanella. Hope: an experimental applicative language. In *ACM LISP Conference*, pages 136–143, 1980.

[CKPR73] A. Colmerauer, H. Kanouri, R. Pasero, and P. Roussel. *Un Systeme de Communication Homme-machine en Francais.* Research Report, Groupe Intelligence Artificielle, Universite Aix-Marseille II, 1973.

[CM81] W. F. Clocksin and C. S. Mellish. *Programming in Prolog.* Springer-Verlag, 1981.

[DFP86] J. Darlington, A. J. Field, and H. Pull. The unification of functional and logic languages. In D. DeGroot and G. Lindstrom, editors, *Logic Programming: Functions, Relations and Equations*, pages 37–70, Prentice-Hall, 1986.

[Fay79] M. Fay. First-order unification in an equational theory. In *Fourth Workshop on Automated Deduction*, pages 161–167, Austin, Texas, 1979.

[Hul80] J-M. Hullot. Canonical forms and unification. In *Conference on Automated Deduction*, pages 318–334, 1980.

[Kow79] R. A. Kowalski. *Logic for Problem Solving.* North-Holland, 1979.

[Kow83] R. Kowalski. Logic programming. In R. E. A. Mason, editor, *Information Processing*, pages 133–145, North-Holland, 1983.

[Lin85] G. Lindstrom. Functional programming and the logical variable. In *ACM Symposium on Principles of Programming Languages*, 1985.

[Mil78] R. Milner. A theory of type polymorphism in programming. *Journal of Computer and System Sciences*, 17:348–375, 1978.

[Mil84] R Milner. A proposal for Standard ML. In *ACM Symposium on LISP and Functional Programming*, pages 184–197, 1984.

[NPA86] R.S. Nikhil, K. Pingali, and Arvind. *Id Nouveau.* Technical Report CSG 265, MIT, 1986.

[Red85] U. S. Reddy. Narrowing as the operational semantics of functional languages. In *Symposium on Logic Programming*, IEEE, Boston, 1985.

[Red86a] U. S. Reddy. On the relationship between logic and functional languages. In D. DeGroot and G. Lindstrom, editors, *Logic Programming: Functions, Relations and Equations*, pages 3–36, Prentice-Hall, 1986.

[Red86b] U.S. Reddy. *Logic Languages based on Functions: Semantics and Implementation.* Technical Report UIUCDCS-R-86-1305, University of Illinois at Urbana-Champaign, 1986. Ph.D. thesis done at University of Utah.

[Sla74] J. R. Slagle. Automated theorem-proving for theories with simplifiers, commutativity and associativity. *Journal of the ACM*, 21(4):622–642, 1974.

[Smo86] G. Smolka. Fresh: a higher-order language based on unification. In D. DeGroot and G. Lindstrom, editors, *Logic Programming: Functions, Relations and Equations*, pages 469–524, Prentice-Hall, 1986.

[SS84] M. Sato and T. Sakurai. Qute: a functional language based on unification. In *Intl. Conf. Fifth Generation Computer Systems*, pages 157–165, ICOT, 1984.

[Tur81] D.A Turner. The semantic elegance of applicative languages. In *Proceedings of the 1981 conference on Functional Programming Languages and Computer Architecture*, pages 85–92, ACM, 1981.

[Tur85] D.A. Turner. Miranda: a non-strict functional language with polymorphic types. In J-P. Jouannaud, editor, *Functional Programming Languages and Computer Architecture*, pages 1–16, Springer-Verlag, 1985.

# UNIFICATION OF QUANTIFIED TERMS

John Staples and Peter J. Robinson

Department of Computer Science, University of Queensland,
St. Lucia, Queensland 4067, Australia.

*Abstract.* Unification algorithms for quantified terms are needed for the implementation of extended functional and logic programming languages, and also for the implementation of other symbolic computation systems such as theorem provers and proof editors. This paper describes and proves correct such a unification algorithm. Although discussed here in a theoretically convenient way, the algorithm is suitable for enhancement of conventional unification algorithms for free variable terms, such as are found in, for example, Prolog interpreters. The algorithm has been demonstrated by modifying a conventional Prolog interpreter so as to interpret formulas which include previously declared quantifiers.

## 1. *Introduction*

There has been increasing interest recently in the development of interactive proof editors, for example to support proofs of program correctness. The essential difference between an interactive proof editor and an automatic theorem prover is that the design of proof editors recognizes that interesting theorems can be proved more efficiently when the user can control the proof strategy. To support the user in this task, proof editors need to support the high-level structure of human proofs. The concept of quantifier is one manifestation of this high-level structure.

The use of quantifiers illustrates the difference between human proof methods and those of classical automatic theorem provers. A wide range of quantifiers is used in human reasoning. As well as the usual logical quantifiers, quantifiers are used to create expressions whose values are of nonlogical types. Some examples are discussed below.

Because of the diversity of possible quantifiers we treat quantifiers as a syntactic category. We give an expanded definition of the syntax of terms, in which arbitrary quantifiers appear. We describe a unification algorithm which recognizes quantifiers at our syntactic level. Our algorithm does not attempt to recognize particular properties of particular, familiar quantifiers, though our approach is compatible with such refinements.

Within these limits there are various alternative forms of unification and matching which can be considered, and varying approaches to the syntactic characterization of quantifiers. We work in some detail a basic example in which quantifier syntax and unification appear in their most elementary forms.

1.1. EXAMPLES OF QUANTIFIERS. By far the best known examples of quantifiers are the logical quantifiers 'for all' and 'there exists'. In both these cases it is usual to write quantified terms in the form $Qxt$ where $Q$ is a quantifier, $x$ is the variable quantified and $t$ is the term to which quantification is applied. For example one writes *for-all* $xa(x)$ to express briefly the assertion 'for all $x$, $a(x)$'. This form of syntax for quantified terms will be used throughout this paper.

We take an abstract view of syntax in which no distinction is made between formulas, that is terms which are intended to denote boolean (logical) values, and terms intended to denote values of other types. Here they are all terms. Since the distinction between formulas and terms is widely made however, we use it to classify the following examples of quantifiers. In a quantified term $Qxt$, $t$ may have boolean type or not, and so may $Qxt$. Each of the four sorts provides familiar examples, as follows.

The work reported here was supported in part by the Australian Research Grants Scheme.

(a) $t$ and $Qxt$ both of boolean type. The familiar logical quantifiers *for-all* and *exists* are of this sort. So are various derived quantifiers, such as that used in 'for all $x$ greater than 0, $\cdots$'. Here we assume that the constraint on $x$ is part of the primitive concept of the quantifier. More flexible approaches to such constraints are possible, wherein each constraint is a parameter to the quantifier; see 1.1.3.

(b) $t$ of boolean type and $Qxt$ not. A familiar example of this sort is the comprehension quantifier in set theory. In this case $t$ is a suitable formula and $Qxt$, traditionally denoted $\{x: t\}$, denotes the set (or class) of values for $x$ which make $t$ true.

Another common example is 'the least $x$ such that $t$'.

(c) $Qxt$ of boolean type and $t$ not. In advanced work this sort of quantifier is often avoided with the aid of additional abstraction. Many assertions which appear in elementary work can however be formalised by such quantifiers. For example, taking $t$ to be $f(x)$, '$f(x)$ is continuous for all $x$' is often used instead of the more abstract '$f$ is continuous'. It can be viewed as quantifying $f(x)$ to produce a boolean-valued term.

(d) Neither $t$ nor $Qxt$ of boolean type. A basic example is when $t$ denotes the $x$-th term of an infinite sequence. Then, a notation such as $(t: x \geq 1)$, which denotes the whole sequence, is a quantification of $t$. A related example is where $t$ denotes the $x$-th term of an increasing sequence in a complete partial order, and is quantified in $\text{lub}_x t$ to denote the least upper bound of the sequence. Other sorts of limit notations are also quantifiers. For example $\Sigma_x t$ denotes the sum of the infinite arithmetical sequence whose $x$-th term is denoted $t$.

There are many extensions in use of the basic quantifier concept. This paper does not analyze any extensions, but here we briefly mention some and indicate the relationship with the basic case.

1.1.1. *Quantification of multiple variables.* It is natural to abbreviate $Qx_1 \cdots Qx_n t$ to $Qx_1, \cdots, x_n t$. It is trivial to extend the algorithm of this paper so that unification expands this abbreviation.

1.1.2. *Commutable quantifiers.* For some quantifiers such as *for-all* it is correct to commute adjacent occurrences. For example, *for-all x for-all y*$a(x, y)$ is logically equivalent to *for-all y for-all x*$a(x, y)$. For a purely syntactic algorithm such as ours, one must declare arbitrarily whether a quantifier is commutable in this way. Our unification algorithm can be extended so as to support commuting of quantifiers declared commutable, but that is beyond the scope of this paper. We plan to discuss it separately.

1.1.3. *Parameterized quantifiers.* Examples such as sums, integrals and restricted logical quantifiers show that additional terms are often associated with a quantifier, while being outside its scope. For example, $\int_u^v t\,dx$ is a quantifier which binds occurrences of $x$ in $t$, but not occurrences of $x$ in the terms $u$ or $v$. It is straightforward to extend our work to deal with this case, since from our purely syntactic point of view it does not differ significantly from a term of the form $a(u, v, Qxt)$.

1.2. SYNTACTIC CHARACTERIZATION OF QUANTIFIERS. Intuitively, the distinctive role of quantifiers has to do with the scope of variables. Quantifiers introduce block structure into terms, by allowing the declaration of uses for variables which are local to the body of the quantified subterms. Occurrences of a variable which are local to ('bound by') some quantifier have an intended meaning different from those ('free') occurrences of the variable which are not bound by any quantifier.

Operationally the significance of whether a variable occurrence $x$ in a term $u$ is free or bound is that it determines the way in which terms $t$ are substituted for that occurrence of $x$. In other words, syntactic recognition of quantifiers arises in the definition of the syntactic operation of substitution. As substitution is the basic syntactic operation involved in unification, the introduction of quantifiers has an immediate influence on the concept of unification, even when unification is not required to recognize particular properties of specific quantifiers. To motivate the discussion of substitution, we first discuss the syntactic role of bound variables.

1.3. CHANGES OF BOUND VARIABLE. The particular choice of a variable to be bound by a quantifier is insignificant, except to avoid ambiguity with other variables, because the intended meaning of the bound variable is local to the quantified subterm. Accordingly, changes of bound variable should be insignificant syntactically.

Two terms are called equal up to changes of bound variable if each can be converted to the other by changes of bound variable. In the theory of the lambda calculus this equivalence is called $\alpha$-equivalence. We use the same notation here.

The concept of $\alpha$-equivalence plays a fundamental role in the syntactic theory of quantified terms. For example, unification should not fail because of failure to notice that changing bound variables would achieve an identification of two terms. In this paper we turn to advantage the need to consider $\alpha$-equivalence. We extend it to incorporate also dereferencing of instantiated free variables, in a new syntax of terms which is theoretically convenient and close to practical implementations. The main novelty of this notation is that it supports the correct use of naive substitution.

Here are some examples, using conventional notation for quantified terms, which illustrate the influence of quantification on unification. A unifier $\sigma$ of two terms $t$ and $u$ is a substitution (if one exists) such that $\sigma(t)$ and $\sigma(u)$ are $\alpha$-equivalent. For example we require the following behaviours.

(a) *for-all* $xa(x)$ and *for-all* $ya(y)$ unify, for example by the identity substitution. This example illustrates that unification produces $\alpha$-equivalent terms, not necessarily identical terms. It also shows that a unification algorithm cannot treat bound variables in the same way as constants.

(b) *for-all* $xa(x)$ and *for-all* $xa(b)$, $b \neq x$, do not unify. This example shows that a unification algorithm cannot treat bound variables in the same way as free variables.

(c) *for-all* $x$ *for-all* $xa(x)$ and *for-all* $x$ *for-all* $ya(x)$ do not unify. This example illustrates that a unification algorithm cannot treat bound variables in any way which gives them an interpretation which is global to the whole term.

(d) *for-all* $xa(y)$ and *for-all* $ya(x)$ unify, but not to *for-all* $xa(x)$ or to *for-all* $ya(y)$. An example of a suitable unifier is the substitution which substitutes $x$ for free occurrences of $y$. The result of doing that to *for-all* $xa(y)$ may be denoted *for-all* $za(x)$. The new bound variable $z$ is introduced because correct substitution into conventional quantified terms must in some places change the names of bound variables, so as to avoid changes in the intended meanings of the terms: see 2.3.2.

1.4. SUBSTITUTION INTO QUANTIFIED TERMS. In the absence of quantifiers, substitution of a term $t$ for a variable $x$ in a term $u$ is mere textual replacement of all occurrences of $x$ in $u$ by copies of $t$. We call this approach naive substitution. It is also called instantiation of $x$ to $t$. It is naturally and efficiently implemented (in cases where $t$ includes no occurrence of $x$) by allowing sharing of common subexpressions, and by introducing an indirection at $x$ which points to (the original) $t$. This sort of implementation is by now common in such systems as Prolog interpreters.

It is highly desirable to preserve this naive approach to implementation of substitution when extending substitution to quantified terms, because of its simplicity, naturalness and widespread practical use. Unfortunately the conventional approach to the syntax of quantified terms, reviewed in 2.1.1, is inconsistent with naive substitution. That is not entirely surprising, since quantifiers create multiple roles for the same variable, but neither is it necessary. Substitution of $t$ for $x$ in a quantified term $u$ has only two sorts of error to avoid: the role of bound occurrences of $x$ should not be affected, and free variable occurrences within the substituted copies of $u$ should not become bound. This latter error is usually called capture of variables.

In this paper we define a concept of quantified term for which naive substitution is correct, basically by allowing each variable to retain its identity as a variable, even after instantiation. Specifically, each variable occurrence will be of the form $x(v)$, where $v$ is either *nil* (representing no substituted value) or else a term (a substituted value). In practice, sharing of common subexpressions can ensure that all occurrences of the variable have the same value, but for the sake of simplicity and generality that condition is not imposed in our theory. Our concept of substitution of $t$ for $x$ in $u$ is then; all occurrences of $x(nil)$ are naively (textually) replaced by $x(t)$.

For example, the result of substituting $t$ for $x$ in $x(nil)$ is $x(t)$. The result of substituting $t$ for $x$ in $y(nil)$, $y \neq x$, is $y(nil)$. The result of substituting $t$ for $x$ in $x(y(nil))$ is $x(y(nil))$. The result of substituting $t$ for $x$ in $z(x(nil))$ is $z(x(t))$, whether $z = x$ or not.

In practice, where all occurrences of $x$ are indirections to a single occurrence of $x$, if the single value of $x$ is $nil$, this value is updated to an indirection to $u$: a straightforward implementation of naive substitution (for cases where $u$ has no free occurrence of $x$). It is pleasing that this concept, which leads to a simple and natural theory and unification algorithm, also agrees with the actual practice of conventional Prolog interpreters.

The scheme just outlined need not cause problems for bound occurrences of $x$ because context can be used to decide whether a variable occurrence is free or bound, and accordingly whether its value is significant.

For example, the result of substituting $t$ for $x$ in *exists* $x(x(nil) = 0)$ is *exists* $x(x(t) = 0)$ . This change is however recognized by our definition of α-equivalence as insignificant. In our notation these terms are α-equivalent. They both represent the term conventionally denoted *exists* $x(x = 0)$. This convention may be summarized by the slogan

'quantification takes precedence over dereferencing'.

To avoid capture of variables our definition ensures, for all subterms $x(t)$ such that $t \neq nil$, that all variable occurrences in $t$ which are free in $t$ are globally free; that is, are outside the scope of any quantifiers in any surrounding term.

For example, for $y \neq x$ the result of substituting $y(nil)$ for $x$ in *exists* $y(x(nil) = 0)$ is *exists* $y(x(y(nil)) = 0)$. The latter corresponds to the conventional term *exists* $z(y = 0)$, not *exists* $y(y = 0)$, since dereferencing $x$ escapes the scope of the quantification of $y$. That can be summarized by the slogan

'dereferencing escapes the scope of all quantifiers'.

Note however that this principle is effective only when dereferencing actually occurs. It has no role when dereferencing is precluded by quantification.

Although the representation of terms considered here is adequate for our unification algorithm, it is too weak for some other purposes. Much of the difficulty of avoiding capture of variables in other contexts such as lambda calculus reductions comes from the fact that the free variables whose capture is to be avoided, though free within the subterm in which substitution is being performed, are bound at some more global level of a surrounding term. This phenomenon does not occur in elementary unification. As a familiar example where our simplified approach to representation of terms is too weak, consider evaluating the lambda expression

$$\lambda x((\lambda yy)x)$$

which in our notation might be written

$$\lambda x((\lambda yy(nil))x(nil))$$

Evaluation involves rewriting the subterm $(\lambda yy(nil))x(nil)$ to a subterm which represents the result of substituting $x(nil)$ for $nil$ in $y(nil)$. In our approach the resulting subterm is denoted $y(x(nil))$. That is correct considered as a term, but when used as a subterm in

$$\lambda x(y(x(nil)))$$

it results in an incorrect lambda expression, since dereferencing causes the occurrence of $x$ to be free instead of bound. A refinement of our approach to cover this point is beyond the scope of this paper, but we claim that our approach can be appropriately extended. We intend to discuss the extension separately.

1.5. COMPARISON WITH DE BRUIJN'S NOTATION. De Bruijn's notation [1] for bound variables is the best-known alternative to classical notation for quantified terms; see also [3, 4]. It does not use the same identifier for the several occurrences of a bound variable. Instead it represents each occurrence of the bound variable by a number representing the binding depth of that occurrence. In de Bruijn's notation the binding depth of an occurrence is the number of quantifiers between the bound variable occurrence and the binding quantifier. For example the term

$$\textit{for-all } x(a(x) \ \& \ \textit{exists } yb(x, y))$$

would be denoted

$$\textit{for-all } (a(0) \ \& \ \textit{exists } b(1, 0))$$

The notation to be introduced in this paper has little in common with de Bruijn's, since we rely heavily on using the same object to represent the several occurrences of a bound variable. In principle, our notation and de Bruijn's are equally successful in dealing with the elementary unification algorithm described in 5.2. In practice de Bruijn's notation has less space efficiency, since it allows fewer common subexpressions. The nonnaive substitution of de Bruijn's notation is also less time efficient.

We also claim that our notation supports extensions of our elementary unification algorithm more conveniently than does de Bruijn's. It is however beyond the scope of this paper to justify that claim.

1.6. AN INTRODUCTORY EXAMPLE. Here we sketch a simple application of the unification algorithm to be described in this paper. Consider the problem of unifying the two terms which are conventionally denoted

$$a(x, Qxb(x, z)) \ \text{ and } \ a(y, Qyb(y, x))$$

where $Q$ is a quantifier, $a$, $b$ are function symbols and $x$, $y$ and $z$ are variables.

In the notation introduced in 1.4, $x$, $y$ and $z$ are denoted $x(nil)$, $y(nil)$ and $z(nil)$ respectively, so that the above terms are denoted

$$a(x(nil), Qxb(x(nil), z(nil))) \ \text{ and } \ a(y(nil), Qyb(y(nil), x(nil))).$$

For readability in this example we abbreviate $x(nil)$, $y(nil)$ and $z(nil)$ to $x$, $y$ and $z$ respectively, so that initially our abbreviated notation is the same as the conventional notation.

The unification algorithm described in 5.2.1 unifies a sequence of pairs of terms, just as conventional free-variable unification does. Our algorithm also associates with each pair of terms in the sequence a pair of binding states. As the example will indicate, the left binding state records which variables in the left term are bound global to the term, and at which level; similarly for the right binding state. Binding levels will be positive integers. Binding states will be one-one partial functions from variables to binding levels and will be denoted as sets of equations. For example,

$$\{x = 1, y = 2\}$$

indicates a binding state in which the binding level of $x$ is 1 and the binding level of $y$ is 2.

For a typical application where it is desired to unify a single pair of terms, the algorithm starts from a sequence of terms comprising that one pair. The associated binding states are both empty. We write the pair of binding states as a subscript to the pair of terms. Thus the input to the unification algorithm in our example is

$$(a(x, Qxb(x, z)), \ a(y, Qyb(y, x)))_{\varnothing, \varnothing} \cdot []$$

Our algorithm begins to unify this sequence in the same way as the familiar free variable unification algorithm. It matches the outer function symbols of the first pair and generates two pairs corresponding to the two places of the function symbol. The result is to simplify the initial problem to

$$(x, y)_{\varnothing, \varnothing} \cdot (Qxb(x, z), Qyb(y, x))_{\varnothing, \varnothing} \cdot []$$

In accordance with the principle stated in 1.4, that binding takes precedence, the next step first checks whether $x$ has a binding level in $\varnothing$, and similarly for $y$. As neither does, the possibility of dereferencing $x$ or $y$ is considered, but neither can be dereferenced. Thus the situation is handled as in free variable unification. The instantiation of $x$ to $y$ is made part of the output substitution and (naive) substitution of $y$ for $x$ is applied to the remainder of the problem. The result is as follows. Note that our notation begins to depart from the conventional, as we record the instantiation of $x$ to $y$ by $x(y)$.

$$(Qxb(x(y), z), Qyb(y, x(y)))_{\varnothing, \varnothing} \cdot []$$

Next the quantifiers are matched and the bodies they quantify are taken as a pair of terms in a simplified problem. Also, the fact that $x$ is bound in the left side and $y$ is the corresponding bound variable in the right side is recorded by associating binding states

$$L = \{x = 1\} \quad \text{and} \quad R = \{y = 1\}.$$

Thus the simplified problem is

$$(b(x(y), z), b(y, x(y)))_{L, R} \cdot []$$

Since a function symbol is now at the top level, we again simplify as in free-variable unification, without changing the binding states. That results in the simplified problem

$$(x(y), y)_{L, R} \cdot (z, x(y))_{L, R} \cdot []$$

For each of $x$ and $y$ the algorithm first consults the associated binding state. As $x$ has the same value in $L$ as does $y$ in $R$, unification of this pair succeeds (without generating any increment to the output substitution). The remaining problem is then to unify

$$(z, x(y))_{L, R} \cdot []$$

As the terms are again both variables, the associated binding states are again consulted. On this occasion neither has a binding level. The next step is to see if either variable can be dereferenced. As $x$ has been instantiated to $y$, dereferencing of $x$ is the next step. Recall from 1.4 that dereferencing escapes the scope of all quantifiers. That is implemented by changing $R$ to $\varnothing$ for the simplified pair:

$$(z, y)_{L, \varnothing} \cdot []$$

As $z$ has no value in $L$ and $y$ has none in $\varnothing$, and as neither can be dereferenced, the instantiation of $z$ to $y$ is generated as the next and final part of the output substitution. Also this substitution is applied to the remainder of the sequence, though in this case there is nothing to do.

An empty sequence of term pairs has now been reached, signalling successful termination of the algorithm. The composition of the output substitutions is the overall unifying substitution generated by the algorithm. Note that this unifying substitution, when applied to the terms required to be unified, does not produce identical terms. In fact it produces

$$a(x(y), Qxb(x(y), z(y))) \quad \text{and} \quad a(y, Qyb(y, x(y)))$$

These terms are however equivalent modulo changes of bound variable and dereferencing. That is, they are $\alpha$-equivalent in the sense defined in 3.1.2.

## 2. *A New Representation of Terms*

Our approach departs from convention at a basic point: the representation of terms. Accordingly this section, after introducing the new concept of term, justifies it by describing a mapping from new terms to conventional terms, and by showing that naive substitution on new terms maps to conventional substitution. For simplicity this introductory discussion is given for substitutions for a single variable. There is no difficulty in extending it to parallel substitutions.

2.1. TERMS. Our definitions assume that the following primitive syntactic categories have already been fixed: an infinite set of variables; for each $n$ a set of $n$-place function symbols; a fixed set of quantifiers; and a final category comprising the single symbol *nil*. These categories are assumed to be disjoint, except that the sets of function symbols may overlap. We begin by reviewing the conventional definition of term.

More precisely, there appears to be no standard treatment of the syntax of arbitrary quantifiers. The concept we describe is a natural extrapolation from conventional treatments of familiar quantifiers. We restrict attention to the simplest case, of quantification over a single variable. Various extensions are possible but will be discussed separately.

2.1.1. *Definition.* Conventional terms are defined recursively as follows.

(a) Each variable is a term.

(b) For each $n$-place function symbol $f$ and each $n$-tuple $t_1, \cdots, t_n$ of terms, $f(t_1, \cdots, t_n)$ is a term. By convention we regard this definition as prescribing that each 0-place function symbol is a term.

(c) For each quantifier $Q$, each variable $x$ and each term $t$, $Qxt$ is a term.

For brevity we refer to conventional terms as old terms.

2.1.2. *Definition.* The terms introduced in this paper are defined in a similar recursive fashion. The only change is that 2.1.1 (a) is changed to:

(a) For each variable $x$ and each term $t$, $x(nil)$ and $x(t)$ are terms.

For brevity we call these terms new terms.

2.2. MAPPING FROM NEW TO OLD TERMS. To simplify our discussion of converting new terms to old we assume that an infinite sequence of variables is available which does not include any variables which occur free in the formula being transformed. It is insignificant how that sequence is obtained. For simplicity we shall assume that it is an arbitrary sequence, fixed throughout the paper. We also assume that our new representation of terms does not use those variables at all.

2.2.1. *Definition.* We use $s.1, s.2, \cdots$ to denote the variables in this special sequence.

Although our elementary unification algorithm unifies top level terms rather than subterms of other terms, we need to consider subterms when defining and reasoning about the algorithm. Indeed that requirement arises throughout the paper, starting with the mapping from new terms to old which we are about to define. The following definitions will be widely used.

2.2.2. *Definitions.* We define a binding state to be a one-one function from (finitely many) variables to positive integers. We shall often write, for a binding state $B$ and a variable $x$, $B(x) = \perp$ to denote that $x$ is not in the domain of $B$. We write $\varnothing$ for the binding state with empty domain.

It is convenient to say that a positive integer $n$ is above a binding state $B$ if for all $m$ in the range of $B$, $m < n$. Note that all positive integers are above $\varnothing$.

We write $\nu(B)$ for the least positive integer above $B$. We also write $\nu(L, R)$ for the least positive integer which is above both the binding states $L$ and $R$. We say that positive integers $n \geq \nu(B)$ are above $B$.

We shall make frequent use of an update operation $[n/x]$ on binding states, as follows. For each binding state $B$, positive integer $n \geq \nu(B)$ and variable $x$, $[n/x]B$ denotes the binding state such that

(a) $([n/x]B)(x) = n$;

(b) for $y \neq x$, $([n/x]B)(y) = B(y)$.

We write $[x]B$ for $[\nu(B)/x]B$. We also write, for a pair $L, R$ of binding states, $[n/x, y](L, R)$ for $([n/x]L, [n/y]R)$ and $[x, y](L, R)$ for $[\nu(L, R)/x, y](L, R)$. We abbreviate $[x, x](L, R)$ to $[x](L, R)$.

2.2.3. *Definition.* Our definition of a mapping $O$ from new terms to old is recursive. For the sake of the recursive definition we define more generally, for each binding state $B$, a mapping $O_B$. Then we finally define $O$ to be $O_\varnothing$.

The clauses of the recursive definition of $O_B$ are as follows.

(a) If $B(x) = \perp$ then $O_B(x(nil)) = x$.

(b) If $B(x) = \perp$ and $t \neq nil$ then $O_B(x(t)) = O_\varnothing(t)$.

(c) If $B(x) \neq \perp$ then for all terms $t$, $O_B(x(t)) = s.B(x)$, as defined in 2.2.1.

(d) For all $m \geq 0$, all $m$-place function symbols $f$ and all terms $t_1, \cdots, t_m$,

$$O_B(f(t_1, \cdots, t_m)) = f(O_B(t_1), ..., O_B(t_m)).$$

(e) For all quantifiers $Q$, variables $x$ and terms $t$,

$$O_B(Qxt) = Qs.\nu(B)O_{[x]B}(t)$$

It follows easily by induction following the definition of $O_B$ that

LEMMA 2.1. *$O_B(t)$ is defined for every new term $t$.*

2.3. SUBSTITUTIONS. We begin by reviewing the definition of substitution for old terms. The first step is to define the concept of free occurrences of variables in old terms, since that is used in the substitution definition. For completeness we define at the same time bound occurrences of variables, the quantifier which binds a bound occurrence, and the scope of quantifiers.

2.3.1. *Definitions.*

(a) In the term $x$ the displayed occurrence of $x$ is free. There are no bound occurrences of variables in $x$, no quantifiers and hence no scopes.

(b) In terms $t = f(t_1, \cdots, t_m)$ the free (respectively bound) occurrences of a variable $x$ in $t$ are the occurrences in each $t_i$ which are free (respectively bound) in $t_i$, $i = 1, \cdots, m$. The binding quantifier of a bound occurrence of $x$ in $t$ is its binding occurrence in $t_i$ and the scope of each quantifier occurrence in $t$ is its scope in the $t_i$ in which it occurs.

(c) In terms $t = Qxu$ there are no free occurrences of $x$. The occurrences of $x$ which are bound by the displayed quantifier are the occurrences which are free in $u$. The scope of the displayed quantifier is $u$. All other variable occurrences in $t$ are free (respectively bound) if and only if they are free (respectively bound) in $u$. The binding quantifiers of other bound occurrences are the same as in $u$, and the scope in $t$ of quantifiers occurring in $u$ is the same as in $u$.

Our discussion of the correctness of our new substitution makes use of the following observation.

LEMMA 2.2. *If $m$ is a positive integer and $B$ is a binding state such that for all variables $x$, $B(x) \neq m$, then $s.m$ does not occur free in $O_B(t)$.*

PROOF. The argument is a simple induction following the definition of terms $t$. The only interesting case is when $t = Qzu$, so that

$$O_B(t) = Qs.nO_{[z]B}(u), \qquad \text{where } n = \nu(B)$$

If $m = n$ then $s.m$ does not occur free, from clause (c) above. If $m \neq n$, then from the definition of update and the hypothesis about $B$, for all variables $x$, $([x]B)(x) \neq m$. The inductive hypothesis then shows that $m$ does not occur free in $O_{[z]B}(u)$. By definition of free therefore it does not occur free in $O_B(t)$; as required.

COROLLARY 2.3. *For all terms $t$ and positive integers $m$, $s.m$ does not occur free in $O(t)$.*

2.3.2. *Definition.* For all old terms $t$ and $u$, and variables $x$, we define in the conventional way the result $[t/x]u$ of substituting $t$ for the free occurrences of $x$ in $u$.

(a) $[t/x]x = t$.

(b) If $y$ is a variable and $y \neq x$ then $[t/x]y = y$.

(c) For terms $u = f(t_1, \cdots, t_n)$,

$$n \geq 0, [t/x]f(t_1, \cdots, t_n) = f([t/x]t_1, \cdots, [t/x]t_n).$$

(d) For terms $u = Qxv$, $[t/x]Qxv = Qxv$.

(e) If $y$ is a variable which is not $x$ and does not occur free in $t$ then $[t/x]Qyv = Qy[t/x]v$.

(f) Otherwise, writing $z$ for a variable which is not $x$ and does not occur free in $t$ or $v$, $[t/x]Qyv = [t/x]Qz[z/y]v$.

Strictly speaking the results of substitutions are not uniquely defined because the choice of $z$ in (f) is not uniquely defined. In fact the results of substitutions are unique up to changes of bound variables, but we do not discuss that here.

2.3.3. *Definition.* The definition of substitution for our new terms is simpler than for old terms. A preliminary definition of free occurrences is not required. Substitution of the new term $t$ for $x$ in the new term $u$ is just naive substitution of $t$ for *nil* in every occurrence of $x(nil)$ in $u$. To help avoid confusion we use the notation $<t/x>u$ to denote the result of this new substitution. It will be convenient for later discussion to state this definition recursively as follows.

(a) $<t/x>x(nil) = x(t)$.

(b) For every variable $y \neq x$, $<t/x>y(nil) = y(nil)$.

(c) For every variable $z$ and $u \neq nil$, $<t/x>z(u) = z(<t/x>u)$.

(d) $<t/x>f(t_1, \cdots, t_n) = f(<t/x>t_1, \cdots, <t/x>t_n)$.

(e) $<t/x>Qzu = Qz<t/x>u$, whether or not $x = z$, and regardless of occurrences of $z$ in $t$.

After we define parallel substitution in 4.1.1 we shall refer to this concept of substitution as single substitution.

The correctness of our approach to substitution can be supported as follows. First a generalized result is proved for the sake of the inductive argument.

PROPOSITION 2.4. *For all new terms $t$ and $u$ and variables $x$, $O_B(<t/x>u) = [O(t)/x]O_B(u)$.*

PROOF. We argue by induction following the definition of terms $u$. We distinguish several subcases when $u$ is a variable, as follows.

(aa) $u = z(v)$, where $z$ is a variable and $B(z) \neq \perp$. Then $<t/x>u = z(v')$, where the form of $v'$ depends on whether $v = nil$ and whether $z = x$. In any case, $O_B(<t/x>u) = s.B(z)$. Also, $O_B(u) = s.B(z)$, so that $[O(t)/x]O_B(u) = s.B(z)$ as required, since from Corollary 2.3, $s.B(z)$ does not occur free in $O(t)$.

(ab) $u = z(v)$, where $B(z) = \perp$ and $v \neq nil$. Then $<t/x>u = z(<t/x>v)$, so that by 2.2.3 (b), $O_B(<t/x>u) = O(<t/x>v)$. Similarly, $O_B(u) = O(v)$, and

$$[O(t)/x]O_B(u) = [O(t)/x]O(v) = O(<t/x>v)$$

by inductive hypothesis; as required.

(ac) If $u = x(nil)$, where $B(x) = \perp$, then $<t/x>u = x(t)$, so $O_B(<t/x>u) = O(t)$. Also, $O_B(u) = x$, so that $[O(t)/x]O_B(u) = O(t)$, as required.

(ad) If $u = y(nil)$, $y \neq x$, $B(y) = \perp$, then $<t/x>u = y(nil)$, so $O_B(<t/x>u) = y$. Also, $O_B(u) = y$, so that $[O(t)/x]O_B(u) = y$, as required.

(b) The case $u = f(t_1, \cdots, t_m)$ is straightforward and is omitted.

(c) Finally, if $u = Qzv$, for some variable $z$, then $<t/x>u = Qz<t/x>v$, so that

$$O_B(<t/x>u) = Qs.nO_{[z]B}<t/x>v, \text{ where } n = \nu(B).$$

Also $O_B(u) = Qs.nO_{[z]B}v$, and by Corollary 2.3, $s.n$ does not occur free in $O(t)$, so

$$\begin{aligned}[O(t)/x]O_B(u) &= [O(t)/x]Qs.nO_{[z]B}(v)\\ &= Qs.n[O(t)/x]O_{[z]B}(v)\\ &= Qs.nO_{[z]B}(<t/x>v) \text{ by inductive hypothesis,}\\ &= O_B(Qzv), \text{ as required.}\end{aligned}$$

COROLLARY 2.5 $O(<t/x>u) = [O(t)/x]O(u)$.

## 3. *Equivalence of New Terms*

We have already mentioned (1.3) that changing bound variables is insignificant for quantified terms. That is, unification should not fail because of differences in bound variable names. For the terms introduced in this paper dereferencing instantiated variables is also insignificant.

To express these requirements in our formalism we introduce an equivalence relation on terms. We follow convention in calling it α-equivalence. In this section we give its definition, a generalization which is used in our analysis, and some useful properties.

3.1. GENERALIZED α-EQUIVALENCE. Just as the mapping $O$ from new terms to old was generalized in 2.2 for the sake of the recursive definition, so we need to generalize the concept of α-equivalence. The generalization is also used later.

Since α-equivalence relates two terms, its generalization considers a pair of subterms together with a corresponding pair of binding states. In fact we define a relation $\sim_{L,R}$ for every pair $L$, $R$ of binding states and every positive integer $n$ which is above both $L$ and $R$.

For theoretical convenience the initial definition of $\sim_{L,R}$ is nonconstructive (in 3.1.1 (f)). An equivalent constructive definition is indicated in proposition 3.6 (f1).

3.1.1. *Definition.* The relation $\sim_{L,R}$ is defined recursively. For arbitrary terms $t$ and $u$, $t \sim_{L,R} u$ if and only if one of the following conditions is satisfied.

(a) $t = x(v)$, $u = y(w)$ and $L(x) = R(y) \neq \perp$.

(b) $t = x(v)$, $L(x) = \perp$, $v \neq nil$ and $v \sim_{\varnothing,R} u$.

(c) $u = y(w)$, $R(y) = \perp$, $w \neq nil$ and $t \sim_{L,\varnothing} w$.

(d) $t = x(nil)$, $L(x) = R(x) = \perp$ and $u = t$.

(e) $t = f(t_1, \cdots, t_m)$, $u = f(u_1, \cdots, u_m)$ and $t_i \sim_{L,R} u_i$, $i = 1, \cdots, m$.

(f) $t = Qxv$, $u = Qyw$ and there is a positive integer $k$ above both $L$ and $R$ such that for all $n \geq k$, $v \sim_{[n/x,y](L,R)} w$.

We shall abbreviate 'there is a positive integer $k$ above both $L$ and $R$ such that for all $n \geq k$' to 'for all $n$ sufficiently large'.

3.1.2. *Definition.* For arbitrary terms $t$ and $u$, $t \sim_\alpha u$ if and only if $t \sim_{\varnothing,\varnothing} u$.

3.2. BASIC PROPERTIES OF α-EQUIVALENCE. We shall often reason about α-equivalence by induction following the recursive definition. For this purpose it is sometimes inconvenient that clauses (b) and (c) of that definition overlap. The next two results show that one may choose whichever alternative is convenient.

LEMMA 3.1. *If* $t \sim_{L,R} u$, $t = x(v)$, $L(x) = \perp$ *and* $v \neq nil$, *then* $v \sim_{\varnothing,R} u$.

PROOF. We argue by induction following the definition of $\sim_{L,R}$. By definition of $\sim_{L,R}$, either (b) or (c) of that definition applies. The case (b) is trivially true, so consider case (c). $u = y(w)$, $R(y) = \perp$, $w \neq nil$ and $t \sim_{L,\varnothing} w$.

By inductive hypothesis $v \sim_{\varnothing,\varnothing} w$, so by 3.1.1 (c), $v \sim_{\varnothing,R} u$ as required.

COROLLARY 3.2. *If* $u = y(w)$, $R(y) = \perp$ *and* $w \neq nil$ *then* $t \sim_{L,\varnothing} w$.

It is now straightforward to show that $\sim_\alpha$ is an equivalence relation. The inductive argument uses the following properties of $\sim_{L,R}$.

LEMMA 3.3. *For all terms t, u and v, and all binding states B, L and R,*

(a) $t \sim_{B,B} t$;

(b) $t \sim_{L,R} u$ implies $u \sim_{R,L} t$;

(c) $t \sim_{L,R} v$ and $v \sim_{R,B} u$ implies $t \sim_{L,B} u$.

PROOF. Part (a) is proved straightforwardly by induction following the definition of $\sim_{B,B}$. Details are omitted. Similarly (b) is proved by induction following the definition of $\sim_{L,R}$. Again we omit details.

The proof of (c) is again similar, as follows. We argue by induction following the definition of $t \sim_{L,R} v$, and for each of those cases, by induction following the definition of $v \sim_{R,B} u$. The following cases correspond to the cases in the definition of $t \sim_{L,R} v$.

(a) $t = x(t')$, $v = z(v')$ and $L(x) = R(z) \neq \bot$. In view of the form of $v$, the definition of $v \sim_{R,B} u$ implies that $u = y(u')$ where $R(z) = B(y)$; which implies $t \sim_{L,B} u$.

(b) The inductive hypothesis applies to $t' \sim_{\varnothing,R} v$ and $v \sim_{R,B} u$ to show that $t' \sim_{\varnothing,B} u$. The required result follows immediately from the definition of $\sim_{L,R}$.

(c) $v = z(v')$, $v' \neq nil$, $R(z) = \bot$ and $t \sim_{L,\varnothing} v'$. By Lemma 3.1, $v' \sim_{\varnothing,B} u$, so the inductive hypothesis applies to give the required result.

(d) $t = x(nil) = v$ and $L(x) = R(z) = \bot$. From the definition of $v \sim_{R,B} u$, there are two possibilities. One case, (d), is that $R(x) = B(y)$ and $u = v$, when the result is immediate from the definition of $\sim_{L,R}$. The other case, (c), is that $u = y(u')$, $u' \neq nil$, $B(y) = \bot$ and $v \sim_{R,\varnothing} u'$. The inductive hypothesis then shows that $t \sim_{L,\varnothing} u'$, which satisifies the definition of $\sim_{L,B} u$.

(e) This case straightforward. Details are omitted.

(f) $t = Qxt'$, $v = Qzv'$ and $t' \sim_{[n/x,z](L,\ R)} v'$ for all $n$ sufficiently large. As $v \sim_{R,B} u$, then also $u = Qyu'$ and $v' \sim_{[n/z,y](R,\ B)} u'$ for all $n$ sufficiently large. Thus for all $n$ sufficiently large (a suitable lower bound is the larger of the lower bound of the two assumed cases) we have $t' \sim_{[n/x,y](L,\ B)} u'$ by inductive hypothesis. The result follows by definition of $\sim_{L,B}$.

COROLLARY 3.4. $\sim_\alpha$ is an equivalence relation.

It is also convenient to have alternative forms of clause (f) of the definition of $\sim_{L,R}$. For example a constructive alternative is needed for algorithmic implementation. First we note the following lemma.

LEMMA 3.5. *If q is a variable, if L, R are binding states such that $R(q) \neq \bot$ and $R(q)$ is not in the range of L, and if k is a positive integer not in the range of R, then for all terms t and u,*

$$t \sim_{L,R} u \text{ implies } t \sim_{L,\ [k/q]R} u.$$

PROOF. By induction following the definition of $\sim_{L,R}$. The following cases correspond to the cases in that definition.

(a) $t = x(v)$, $u = y(w)$ and $L(x) = R(y) \neq \bot$. Note that $q = y$ is excluded since $R(q)$ is not in the range of $L$, so $R(y) = ([k/q]R)(y)$, hence the result.

(b) $t = x(v)$, $v \neq \bot$, $L(x) = nil$ and $v \sim_{\varnothing,R} u$. The inductive hypothesis gives $v \sim_{\varnothing,\ [k/q]R} u$, from which the result is immediate.

(c) $u = y(w)$, $w \neq nil$, $R(y) = \bot$ and $t \sim_{L,\varnothing} w$. Now $y = q$ is excluded since $R(y) = \bot$, so $([k/q]R)(y) = R(y) = \bot$. The result follows.

(d) $t = x(nil)$, $L(x) = R(x) = \bot$ and $u = t$. Thus $x = q$ is excluded, so $([k/q]R)(x) = \bot$ and the result follows.

(e) The details are straightforward and are omitted.

(f) By definition of $\sim_{L,R}$ there is $m$ such that for all $n \geq m$, $v \sim_{[n/x,y](L,\ R)} w$. That is, $v \sim_{[n/x]L,\ [n/y]R} w$. We show that for all $n \geq \max\{m, k+1\}$,

$$v \sim_{[n/x]L,\ [n/y][k/q]R} w.$$

The result then follows.

In the case $q = y$, because of the definition of updates of binding states, $[n/y][k/q]R = [n/y]R$, so the result is immediate. In the case $y \neq q$, $[n/y][k/q]R = [k/q][n/y]R$, so the required result is provided by the inductive hypothesis. That completes the proof.

From the following result, any one of its three equivalent conditions may be used as as an alternative to clause (f) in the definition of $\sim_{L,R}$ (with $p$, $q$, $t$ and $u$ taken to be $x$, $y$, $v$ and $w$ respectively).

PROPOSITION 3.6. *For all terms t and u, all variables p and q and all binding states L and R, the following three assertions are equivalent.*

(f1) $t \sim_{[p,q](L,\ R)} u$.

(f2) For all $n$ not in the range of $L$ or $R$, $t \sim_{[n/p,q](L,\ R)} u$.

(f3) There exists $m$ not in the range of $L$ or $R$ such that $t \sim_{[m/p,q](L,\ R)} u$.

PROOF. Evidently (f2) implies (f1) and (f1) implies (f3). To complete a circle of implications we therefore show that (f3) implies (f2). We exclude $m = n$, since that case is given. The argument is by induction following the definition of $\sim_{[m/p,q](L,\ R)}$. The following cases correspond to the cases in that definition.

(a) $t = x(v)$, $u = y(w)$ and $([m/p]L)(x) = ([m/q]R)(y) \neq \bot$. It is enough to show that $([n/p]L)(x) = ([n/q]R)(y) \neq \bot$. Since binding states are one-one functions, $x = p$ if and only if $y = q$. In each of the cases $x = p$ and $x \neq p$ the result is straightforward. Details are omitted.

(b) $t = x(v)$, $v \neq nil$, $([m/p]L)(x) = \bot$ and $v \sim_{\varnothing,\ [m/p]R} u$. The conditions on $x$ and $p$ exclude $x = p$, so $([n/p]L)(x) = L(x) = ([m/p]L)(x) = \bot$. Lemma 3.5 ensures that $v \sim_{\varnothing,\ [n/p]R} u$, which gives the result.

(c) Symmetrical to (b).

(d) The argument is similar to that for (b). The conditions ensure that $([n/p]L)(x) = ([n/q]R)(x) = \bot$.

(e) The inductive hypothesis is applied straightforwardly. Details are omitted.

(f) $t = Qxv$, $u = Qyw$ and for all $k$ sufficiently large, (in particular $k > m$),

$$v \sim_{[k/x][m/p]L,\ [k/y][m/q]R} w$$

By further restricting $k$ to be also greater than $n$ we show that

$$v \sim_{[k/x][n/p]L,\ [k/y][n/q]R} w,$$

which is sufficient. We consider the following four cases.

(fa) $p = x$, $q = y$. Then $[k/x][m/p]L = [k/x]L = [k/x][n/p]L$ and similarly for $R$.

(fb) $p = x$, $q \neq y$. As in (fa), $[k/x][m/p]L = [k/x]L$. Also, $[k/y][m/q]R = [m/q][k/y]R$. Lemma 3.5 applies to

$$v \sim_{[k/x]L,\ [m/q][k/y]R} w$$

to show

$$v \sim_{[k/x]L,\ [n/q][m/q][k/y]R} w$$

which is

$$v \sim_{[k/x][n/q]L,\ [k/y][n/q]R} w$$

as required.

(fc) $p \neq x$, $q = y$. This case is symmetrical to (fb).

(fd) $p \neq x$, $q \neq y$. In this case, as for $R$ in (fb), $[k/x][m/p]L = [m/p][k/x]L$, and similarly for $R$. The inductive hypothesis then applies, after which the updates are again commuted to give the result.

3.2.1. *Definition.* For any pair $(L, R)$ of binding states the range intersection of $(L, R)$ is the pair

$(L', R')$ of binding states such that:

$$L' = \{(x, n) \in L: \text{there is } y \in R \text{ such that } n = R(y)\}.$$
$$R' = \{(y, n) \in R: \text{there is } x \in L \text{ such that } n = L(x)\}.$$

LEMMA 3.7. *If $(L, R)$ have range intersection $(L', R')$ and $L' \subseteq M \subseteq L$ and $R' \subseteq N \subseteq R$, then $t \sim_{L,R} u$ implies $t \sim_{M,N} u$.*

PROOF. By induction following the definition of $\sim_{L,R}$. The following cases are the same as the cases in the definition of $\sim_{L,R}$.

(a) $t = x(v)$, $u = y(w)$ and $L(x) = R(y) \neq \bot$. Then $M(x) = L(x) = R(y) = N(y)$, which gives the result.

(b) $t = x(v)$, $v \neq nil$, $L(x) = \bot$ and $v \sim_{\varnothing,R} u$. Note that the range intersection of $(\varnothing, R)$ is $(\varnothing, \varnothing)$, so the inductive hypothesis applied to $v \sim_{\varnothing,R} u$ certainly shows that $v \sim_{\varnothing,N} u$ for all $N$ such that $R' \subseteq N \subseteq R$. That gives the result.

(c) Symmetrical to (b).

(d) Straightforward; details omitted.

(e) Straightforward; details omitted.

(f) $t = Qxv$, $u = Qyw$ and $v \sim_{[n/x,y](L,R)} w$ for all $n$ sufficiently large. The range intersection of $[n/x]L$, $[n/y]R$ is $[n/x]L'$, $[n/y]R'$. Now $[n/x]L' \subseteq [n/x]M \subseteq [n/x]L$ and similarly for $R$, so the inductive hypothesis gives $v \sim_{[n/x,y](M,N)} w$, as required to complete the proof.

COROLLARY 3.8. *$t \sim_{\varnothing,R} u$ implies $t \sim_{\varnothing,\varnothing} u$.*

COROLLARY 3.9. *$t \sim_{L,\varnothing}$ implies $t \sim_{\varnothing,\varnothing} u$.*

LEMMA 3.10. *For all variables $x$ such that $L(x) = \bot$ and all terms $t$, $x(nil) \sim_{L,R} t$ if and only if $x(nil) \sim_{\varnothing,R} t$.*

PROOF. By definition of $\sim_{L,R}$, $t = y(u)$. If $R(y) \neq \bot$ then the assertion is vacuously true, so suppose that $R(y) = \bot$. If $u = nil$ the result is evident, since either equivalence implies $x = y$, which implies the other equivalence; so suppose $u \neq nil$. Then

$$\begin{aligned} x(nil) \sim_{L,R} t &\text{ if and only if } x(nil) \sim_{L,\varnothing} u \\ &\text{ if and only if } x(nil) \sim_{\varnothing,\varnothing} u, \text{ by inductive hypothesis,} \\ &\text{ if and only if } x(nil) \sim_{\varnothing,R} t, \text{ by definition of } \sim_{\varnothing,R}. \end{aligned}$$

COROLLARY 3.11. *If $R(x) = \bot$ then $t \sim_{L,R} x(nil)$ if and only if $t \sim_{L,\varnothing} x(nil)$.*

3.3. FREE AND BOUND VARIABLE OCCURRENCES IN TERMS. As noted in 2.3, in a conventional analysis of quantified terms every variable occurrence is classified as either free or bound, and this distinction is relied on heavily, for example in the definition of substitution.

The concepts of free and bound introduced here are somewhat different. First, they are relative to a binding state $B$. Then, not every variable occurrence will be either $B$-free or $B$-bound. We do not need these concepts for our definition of substitution. They are used however in our 'occurs check' in unification.

Intuitively, a variable occurrence is $B$-free if it is not bound, either by a quantifier within the term or by $B$. Likewise, $B$-bound means that the variable occurrence is not bound by a quantifier within the term, but is bound by $B$.

3.3.1. *Definition.* For every binding state $B$, the $B$-free and $B$-bound occurrences of variables in a term $t$ are defined recursively, following the construction of $t$, as follows.

(aa) $t = x(nil)$. The displayed occurrence of $x$ is $B$-bound if $B(x) \neq \bot$, and is $B$-free otherwise.

(ab) $t = x(v)$, $v \neq nil$. The displayed occurrence of $x$ is $B$-bound provided that $B(x) \neq \bot$ (but is not $B$-free otherwise). Provided $B(x) = \bot$, an occurrence of a variable in $v$ is $B$-free in $t$ if and only if it is $\varnothing$-free in $v$. No occurrences of variables in $v$ are $B$-bound in $t$.

(b) $t = f(t_1, \cdots, t_n)$. An occurrence of a variable in $t$ occurs in some $t_i$ and is $B$-free (respectively $B$-bound) in $t$ if and only if it is $B$-free (respectively $B$-bound) in $t_i$, $i = 1, \cdots, n$.

(c) $t = Qxv$. An occurrence of a variable in $t$ is $B$-free if and only if it occurs in $v$ and is $[x]B$-free in $v$. No occurrence of $x$ in $t$ is $B$-free. For variables $y \neq x$, an occurrence $y$ is $B$-bound in $t$ if and only if it is $[x]B$-bound in $v$.

LEMMA 3.12. *If $t \sim_{L,R} u$ then for every R-bound occurrence of a variable y in u there is an L-bound occurrence of a variable x in t such that $L(x) = R(y)$.*

PROOF. The argument is a straightforward induction following the definition of $\sim_{L,R}$. Details are omitted.

COROLLARY 3.13. *If $t \sim_{L,R} u$ then for every L-bound occurrence of a variable x in t there is an R-bound occurrence of a variable y in u such that $L(x) = R(y)$.*

COROLLARY 3.14. *If $t \sim_{\varnothing,R} u$ then u has no R-bound occurrences of variables.*

COROLLARY 3.15. *If $R(z) \neq \perp$, $R(z)$ is not a value of L and u has an R-bound occurrence of z then not $t \sim_{L,R} u$.*

LEMMA 3.16. *If m and n are not in the range of B then x occurs $[m/y]B$-free in t if and only if x occurs $[n/y]B$-free in t.*

PROOF. By symmetry it is enough to assume that $x$ occurs $[m/y]B$-free in $t$ and show it occurs $[n/y]B$-free. The argument is by induction following the construction of $t$.

(a) $t = z(v)$. Then $z \neq y$ so $([n/y]B)(z) = ([m/y]B)(z)$. Both equal $\perp$, since otherwise $x$ does not occur as assumed. If $z = x$ and $v = nil$ the result follows since $([m/y]B)(x) = B(x) = ([n/y]B)(x) = \perp$. Otherwise $x$ occurs $\varnothing$-free in $v$, which gives the result.

(b) $t = f(t_1, \cdots, t_n)$. The inductive step is straightforward and is omitted.

(c) $t = Qzv$. Then $x \neq z$ and $x$ occurs $[k/z][m/y]B$-free, say, in $v$. If $z = y$ then $[k/z][m/y]B = [k/z]B$. The inductive hypothesis then applies to adjust $k$ to $l = \nu([n/y]B)$, showing that $x$ occurs $[l/z]B$-free in $v$. That gives the result.

Finally, if $z \neq y$ it is enough to show that $x$ occurs $[k'/z][n/y]B$-free in $v$, where $k' = \max\{k, \nu([n/y]B)\}$, for then the inductive hypothesis shows that $x$ occurs $[\nu([n/y]B)/z][n/y]B$-free in $v$, as required. The inductive hypothesis also allows us to assume that $x$ occurs $[k'/z][m/y]B$-free in $v$. It follows, since $[k'/z][m/y] = [m/y][k'/z]$, that $x$ occurs $[m/y][k'/z]B$-free in $v$. The inductive hypothesis then applies once more to show that $x$ occurs $[n/y][k'/z]B$-free in $v$, and hence $[k'/z][n/y]B$-free in $v$, as required.

LEMMA 3.17. *If $t \sim_{L,R} u$ then z occurs L-free in t if and only if z occurs R-free in u.*

PROOF. By induction following the definition of $\sim_{L,R}$. The only interesting case is when $t = Qxv$, $u = Qyw$ and $v \sim_{[x,y](L,R)} w$. In this case, as $z$ occurs $L$-free in $t$, it occurs $[x]L$-free in $v$, and hence by Lemma 3.16, $z$ occurs $[\nu(L, R)/x]L$-free in $v$. The inductive hypothesis then shows $z$ occurs $[\nu(L, R)/y]R$-free in $w$, so Lemma 3.16 applies again to show that $z$ occurs $[y]R$-free in $w$, as required to complete the proof.

3.4. SIZE OF TERMS. The naive size of a term, measured by counting its symbols, is not particularly useful here since our use of dereferencing allows terms of various naive sizes to be α-equivalent. The following concept of size will be used to supplement the naive size concept.

3.4.1. *Definition.* $\text{size}_B(t)$ is defined for each binding state $B$ and term $t$, by recursion following the construction of $t$, as follows.

(aa) If $B(x) \neq \perp$ or $v = nil$ then $\text{size}_B(x(v)) = 1$.

(ab) If $B(x) = \perp$ and $v \neq nil$ then $\text{size}_B(x(v)) = \text{size}_\varnothing(v)$.

(b) $\text{size}_B(f(t_1, \cdots, t_n)) = 2 + \sum_{i=1}^{n} \text{size}_B(t_i)$.

(c) $\text{size}_B(Qxv) = 2+\text{size}_{[x]B}(v)$.

LEMMA 3.18. *$t \sim_{L,R} u$ implies $\text{size}_L(t) = \text{size}_R(u)$.*

The proof is an elementary induction following the definition of $\sim_{L,R}$, and is omitted.

LEMMA 3.19. *If m and n are positive integers not in the range of the binding state B then*

$$\text{size}_{[m/x]B}t = \text{size}_{[n/x]B}t$$

PROOF. By induction following the definition of $t$. The details are straightforward. In particular the case $t = Qzv$ is treated similarly to case (c) in the proof of Lemma 3.16.

3.5. A DEREFERENCING NOTATION. The following notation is sometimes convenient.

3.5.1. *Definition.* Terms $\downarrow_B t$ are defined for every binding state $B$ and every term $t$ recursively, following the construction of $t$, as follows.

(a) If $t = x(v)$, $v \neq nil$ and $B(x) = \bot$ then $\downarrow_B(t) = \downarrow_\varnothing(v)$.

(b) Otherwise $\downarrow_B(t) = t$.

Naive size of terms is useful for discussing dereferencing, since dereferencing reduces naive size. The following facts are elementary and so are not discussed in detail.

LEMMA 3.20. *For all terms u, the naive size of $\downarrow_B u$ is no greater than the naive size of u.*

LEMMA 3.21. *If the naive size of $\downarrow_B u$ is less than the naive size of u then $u = y(v)$ for some variable y such that $B(y) = \bot$ and some $v \neq nil$.*

COROLLARY 3.22. *$u = \downarrow_B u$ implies that u is not of the form $y(v)$ where $B(y) = \bot$ and $v \neq nil$.*

## 4. *Parallel Substitutions*

Unification constructs a parallel substitution which instantiates several variables independently. It is clumsy to define such substitutions in terms of substitutions for single variables. Therefore in this section we extend the definition of 2.3.3 to cover parallel substitution, and develop some properties. From here on, unqualified references to substitutions refer to parallel substitutions. The concept defined in 2.3.3 will henceforth be referred to as single substitution.

4.1. INTRODUCTION. As for single substitutions, our substitution concept is naive. For convenience in proofs we express it by a recursive definition.

4.1.1. *Definition.* For each natural number $m \geq 0$, each $m$-tuple $s_1, \cdots, s_m$ of terms and each $m$-tuple $x_1, \cdots, x_m$ of distinct variables, $<s_1, \cdots, s_m/x_1, \cdots, x_m>$ is a prefix operator notation for a function $\sigma$ which performs, for each $i = 1, \cdots, m$, naive substitution of $s_i$ for *nil* in all occurrences of $x_i(nil)$. A recursive definition is as follows.

(a) $\sigma(x_i(nil)) = x_i(s_i)$.

(b) For $x \neq x_i$, $i = 1, \cdots, m$, $\sigma(x(nil)) = x(nil)$.

(c) For all variables $x$ and all $t \neq nil$, $\sigma(x(t)) = x(\sigma(t))$.

(d) $\sigma(f(t_1, \cdots, t_n)) = f(\sigma(t_1), \cdots, \sigma(t_n))$.

(e) $\sigma(Qxt) = Qx\sigma(t)$.

For such a substitution $\sigma$ and $x = x_i$ we may write $\sigma_x$ to denote $s_i$. We also write $\sigma_x = \bot$ to denote that $x$ is none of $x_1, \cdots, x_m$.

The case $m = 0$ in the definition ensures that the identity function on terms is a substitution. We may denote it *id*.

LEMMA 4.1. *If $t_1, \cdots, t_m$ is a permutation of $s_1, \cdots, s_m$ and $y_1, \cdots, y_m$ is the same permutation of $x_1, \cdots, x_m$ then*

$$<s_1, \cdots, s_m/x_1, \cdots, x_m> = <t_1, \cdots, t_m/y_1, \cdots, y_m>.$$

The proof is a straightforward induction following the recursive definition of substitution, and is omitted.

On the other hand, this concept of substitution does not support identities such as $id = (x/x>$, as does substitution for old terms. The two substitutions *id* and $<x/x>$ are however equivalent in the sense to be defined in *AlEqOfSb*.

LEMMA 4.2. *If* $t \sim_{L,R} \sigma(t)$ *then* $L(y) = R(y)$ *for all variables y which occur R-bound in t.*

PROOF. By induction following the construction of $t$. The only interesting case is when $t = x(v)$, so that $\sigma(t) = x(v')$ say. If $R(x) \neq \bot$ then $t \sim_{L,R} \sigma(t)$ implies $L(x) = R(x)$, from the definition of $\sim_{L,R}$ and since Lemma 3.12 excludes case (b) in the definition 3.1.1 of $\sim_{L,R}$. From the definition of $R$-bound there are no other $R$-bound variable occurrences.

LEMMA 4.3. *If* $<u/x>t$ *has a B-free occurrence of x then u has an* $\emptyset$*-free occurrence of x.*

PROOF. By induction following the construction of $t$. The only interesting cases are the following three subcases of 2.1.2 (a).

(aa) $t = x(nil)$. Then $<u/x>t = x(u)$, so the result follows from the definition of $B$-free.

(ab) $t = y(nil)$, $y \neq x$. In this case the assertion is vacuously true.

(ac) $t = z(v)$, $v \neq nil$. The $<u/x>t = z(<u/x>v)$. If $B(z) \neq \bot$ then the assertion is vacuously true. If $B(z) = \bot$ then $<u/x>v$ has an $\emptyset$-free occurrence of $x$ and the inductive hypothesis applies.

4.2. α-EQUIVALENCE OF TERMS IS PRESERVED BY SUBSTITUTIONS. First we show the generalized result which is needed for the inductive hypothesis.

LEMMA 4.4. *For all terms t and u and all substitutions* $\sigma$, $t \sim_{L,R} u$ *implies* $\sigma(t) \sim_{L,R} \sigma(u)$.

PROOF. By induction following the definition of $t \sim_{L,R} u$. The following cases correspond to the cases in 3.1.1.

(a) $t = x(v)$, $u = y(w)$, $L(x) = R(y) \neq \bot$. Then $\sigma(t) = x(v')$ and $\sigma(u) = y(w')$, where $v'$ and $w'$ depend on whether $v$ and $w$ are *nil* or not. In any case, as $L(x) = R(x) \neq \bot$, $\sigma(t) \sim_{L,R} \sigma(u)$.

(b) $t = x(v)$, $L(x) = \bot$, $v \neq nil$ and $v \sim_{\emptyset,R} u$. Then $\sigma(t) = x(\sigma(v))$, so to show $\sigma(t) \sim_{L,R} \sigma(u)$ it is enough, by definition of $\sim_{L,R}$, to show that $\sigma(v) \sim_{\emptyset,R} \sigma(u)$. That follows from the inductive hypothesis.

(c) Symmetrical to (b).

Of the remaining cases, (d) is trivially true since $t = u$, and each of (e) and (f) is a straightforward application of the inductive hypothesis.

COROLLARY 4.5. $t \sim_\alpha u$ *implies* $\sigma(t) \sim_\alpha \sigma(u)$.

4.3. SUBSTITUTIONS, SIZES AND DEREFERENCING.

LEMMA 4.6. *For all terms t, binding states B and substitutions* $\sigma$, $\mathrm{size}_B(t) \leq \mathrm{size}_B(\sigma(t))$.

The proof is a straightforward induction following the construction of terms, and is omitted.

LEMMA 4.7. *If* $\sigma_x \neq \bot$ *and t has a B-free occurrence of x then* $\mathrm{size}_B(\sigma(t)) \geq \mathrm{size}_\emptyset(\sigma_x) + \mathrm{size}_B(t) - 1$.

PROOF. By induction following the construction of $t$. The only interesting case is when $t = y(v)$. If $v = nil$ then $x = y$ and $B(x) = \bot$ and $\mathrm{size}_B(t) = 1$, so $\mathrm{size}_B(\sigma(t)) = \mathrm{size}_B(x(\sigma_x)) = \mathrm{size}_\emptyset(\sigma_x)$ as required. If $B(y) \neq \bot$ then the result is vacuously true. If $B(y) = \bot$ and $v \neq nil$ then $\mathrm{size}_B(t) = \mathrm{size}_\emptyset(v)$, $\mathrm{size}_B(\sigma(t)) = \mathrm{size}_\emptyset(\sigma(v))$ and $t$ has a $\emptyset$-free occurrence of $x$. Thus the inductive hypothesis gives $\mathrm{size}_\emptyset(\sigma(v)) \geq \mathrm{size}_\emptyset(\sigma_x)) + \mathrm{size}_\emptyset(v) - 1$, from which the result is immediate.

4.4. COMPOSITION OF SUBSTITUTIONS. Since we have defined substitutions abstractly as functions, composition is just ordinary function composition.

4.4.1. *Definition.* For all functions $\sigma$ and $\tau$ on terms we define $\sigma \circ \tau$ to be the function on terms such that for all $t$, $(\sigma \circ \tau)(t) = \sigma(\tau(t))$.

Note that our definition of substitution does not make $s_i$ equal to $\sigma(x_i(nil))$, although the two are $\alpha$-equivalent.

As usual, composition is associative and the identity substitution is an identity for composition. We omit details. The basic property of composition of substitutions is that the result is a substitution:

LEMMA 4.8. *For all substitutions $\sigma$ and $\tau$, $\sigma \circ \tau$ is a substitution. In particular, writing $\sigma = \langle s_1, \cdots, s_m/x_1, \cdots, x_m\rangle$ and $\tau = \langle t_1, \cdots, t_n/y_1, \cdots, y_n\rangle$, where $x_i = y_i$, $i = 1, \cdots, k$, $k \geq 0$ and $x_{k+1}, \cdots, x_m$ is disjoint from $y_{k+1}, \cdots, y_n$,*

$$\sigma \circ \tau = \langle s_{k+1}, \cdots, s_m, \sigma(t_1), \cdots, \sigma(t_n)/x_{k+1}, \cdots, x_m, y_1, \cdots, y_n\rangle.$$

PROOF. It is straightforward to show by induction following the definition of terms $t$ that for all terms $t$,

$$(\sigma \circ \tau)(t) = \langle s_{k+1}, \cdots, s_m, \sigma(t_1), \cdots, \sigma(t_n)/x_{k+1}, \cdots, x_m, y_1, \cdots, y_n\rangle t.$$

4.5. $\alpha$-EQUIVALENCE OF SUBSTITUTIONS. The $\alpha$-equivalence relation on terms induces a corresponding relation on substitutions as follows.

4.5.1. *Definition.* For all substitutions $\sigma$ and $\tau$, $\sigma \sim_\alpha \tau$ if and only if for all terms $t$, $\sigma(t) \sim_\alpha \tau(t)$.

It is immediate from Corollary *AlEqGenEqv* that $\sim_\alpha$ for substitutions is an equivalence relation. Details are omitted.

Since we use generalized $\alpha$-equivalence relations on terms it is natural to make a similar generalization for substitutions.

4.5.2. *Definition.* For all substitutions $\sigma$ and $\tau$, $\sigma \sim_{L,R} \tau$ if and only if for all terms $t$, $\sigma(t) \sim_{L,R} \tau(t)$.

LEMMA 4.9. $\sigma \sim_{L,R} \tau$ *if and only if*

(i) $L = R$ and

(ii) for all $x$ such that $L(x) = \bot$, $\sigma(x(nil)) \sim_\alpha \tau(x(nil))$.

PROOF. First we assume that $\sigma \sim_{L,R} \tau$ and show (i) and (ii).

To show (i), we suppose that $L \neq R$ and derive a contradiction. Say $L(x) \neq R(x)$.

Choose $t = x(nil)$. If $L(x) \neq \bot$ then $R(x) = \bot$, since $R(x) \neq \bot$ implies, from the definition of $\sim_{L,R}$, that $L(x) = R(x)$, contradiction. But $L(x) \neq \bot$ and $R(x) = \bot$ together contradict $\sigma \sim_{L,R} \tau$, because they contradict $\sigma(x(nil)) \sim_{L,R} \tau(x(nil))$, by Corollary 3.13.

We conclude that $L(x) = \bot$. A symmetrical argument shows that $R(x) = \bot$. That contradicts our original supposition, as required to complete the proof.

Next we show (ii). By hypothesis and (i) we have, for each variable $x$ such that $L(x) = \bot$, that $\sigma(x(nil)) \sim_{L,L} \tau(x(nil))$. The following cases are exhaustive.

(a) $\sigma_x = \bot = \tau_x$. Then $\sigma(x(nil)) = x(nil) = \tau(x(nil))$.

(b) $\sigma_x = \bot$, $\tau_x \neq \bot$. By assumption $\sigma(x(nil)) = x(nil) \sim_{L,L} x(\tau_x) = \tau(x(nil))$. Thus by Corollary 3.2, $x(nil) \sim_{L,\varnothing} \tau_x$, so by Lemma 3.10, $x(nil) \sim_{\varnothing,\varnothing} \tau_x$, as required.

(c) $\sigma_x \neq \bot$, $\tau_x = \bot$. This case is symmetrical to the previous one.

(d) $\sigma_x \neq \bot$, $\tau_x \neq \bot$. As $L(x) = \bot$, then by Lemma 3.1 and Corollary 3.2 we have $\sigma_x \sim_{\varnothing,\varnothing} \tau_x$ as required.

Finally we show the converse, that (i) and (ii) imply for all terms $t$ that $\sigma(t) \sim_{L,L} \tau(t)$. Our argument is by induction on the construction of $t$. The only interesting cases are when $t = x(v)$. If $L(x) \neq \bot$ then the result is immediate from the definition of $\sim_{L,L}$, so we assume that $L(x) = \bot$.

The first case is when $v = nil$. We consider the following four subcases.

(a) $\sigma_x \neq \bot$, $\tau_x \neq \bot$. From the definition of substitution it is enough to show that $x(\sigma_x) \sim_{L,L} x(\tau_x)$. By definition of $\sim_{L,L}$ it is enough to show that $x(\sigma_x) \sim_{\varnothing,\varnothing} x(\tau_x)$. That is given by (ii).

(b) $\sigma_x = \bot$, $\tau_x \neq \bot$. To show $x(nil) \sim_{L,L} x(\tau_x)$ it is enough to show $x(nil) \sim_{L,\varnothing} \tau_x$. From Lemma 3.10 it is enough for that to show $x(nil) \sim_{\varnothing,\varnothing} \tau_x$, and that is given by (ii).

(c) $\sigma_x \neq \bot$, $\tau_x = \bot$. This subcase is treated symmetrically to the previous one.

(d) $\sigma_x = \bot = \tau_x$. This subcase is trivial.

The other interesting case is when $t = x(v)$ and $v \neq nil$. Here we have to show that $x(\sigma(v)) \sim_{L,L} x(\tau(v))$. By definition of $\sim_{L,L}$ it is enough to show that $\sigma(v) \sim_{\varnothing,\varnothing} \tau(v)$, and that is given by the inductive hypothesis.

4.5.3. *Definition.* In view of the result just proven, we introduce the notation $\sim_B$ to abbreviate $\sim_{B,B}$.

Next we show that $\alpha$-equivalence of substitutions is preserved by composition.

LEMMA 4.10. *For all substitutions $\sigma$, $\sigma'$ and $\tau$ such that $\sigma \sim_B \sigma'$, $\tau \circ \sigma \sim_B \tau \circ \sigma'$ and $\sigma \circ \tau \sim_B \sigma' \circ \tau$.*

PROOF. For the first result we show for all terms $t$ that $\tau(\sigma(t)) \sim_{B,B} \tau(\sigma'(t))$. By definition of $\sim_B$, $\sigma(t) \sim_{B,B} \sigma'(t)$, so the result follows from Lemma 4.4.

For the second result, $\sigma(\tau(t)) \sim_{B,B} \sigma'(\tau(t))$ immediately from the definition of $\sim_B$.

COROLLARY 4.11. *If $\sigma \sim_B \sigma'$ and $\tau \sim_B \tau'$ then $\sigma \circ \tau \sim_B \sigma' \circ \tau'$.*

PROOF. $\sigma \circ \tau \sim_B \sigma \circ \tau' \sim_B \sigma' \circ \tau'$, so the result follows by transitivity of $\sim_{L,R}$ for substitution.

The following results play a key role in our proof of correctness of our unification algorithm.

LEMMA 4.12. *If $\sigma = \langle s_1, \cdots, s_m/x_1, \cdots, x_m\rangle$ and $\tau = \langle s_2, \cdots, s_m/x_2, \cdots, x_m\rangle$ are substitutions and if $u$ is a term such that $\tau(u) \sim_\alpha s_1$, then for all binding states $B$, $\tau \circ \langle u/x_1\rangle \sim_B \sigma$.*

PROOF. We show for all terms $t$ that $\tau(\langle u/x_1\rangle t) \sim_{B,B} \sigma(t)$. The argument is by induction on $t$.

(a) $t = x_1(nil)$, $B(x_1) = \bot$. Then $\tau(\langle u/x_1\rangle t) = x_1(\tau(u))$. To show that is $\sim_B \sigma(x_1(nil)) = x_1(s_1)$ it is enough, by definition of $\sim_B$, to show that $\tau(u) \sim_\varnothing s_1$. That is given.

(b) $t = x_i(nil)$, $i = 2, \cdots, m$. Then $\tau(\langle u/x_1\rangle t) = x_i(s_i) = \sigma(t)$ as required.

(c) $t = y(nil)$, $y \neq x_i$, $i = 1, \cdots, m$. Then $\tau(\langle u/x_1\rangle t) = y(nil) = \sigma(t)$.

(d) $t = z(v)$, $v \neq nil$, $B(z) = \bot$. By definition of $\sim_{B,B}$, Lemma 3.1 and Corollary 3.2 it is enough to show that $\tau(\langle u/x_1\rangle v) \sim_\alpha \sigma(v)$. That is given by the inductive hypothesis.

The remaining cases are straightforward and are omitted.

LEMMA 4.13. *If $\sigma = \langle s_1, \cdots, s_m/x_1, \cdots, x_m\rangle$ and $\tau = \langle s_2, \cdots, s_m/x_2, \cdots, x_m\rangle$ are substitutions and $x = x_1$ does not occur $B$-free in $t$, then $\sigma(t) \sim_B \tau(t)$.*

PROOF. By induction following the construction of $t$. The only interesting case is when $t = y(v)$. The result is evident if $B(y) \neq \bot$, so we suppose $B(y) = \bot$. The result is evident if $v = nil$, whether or not $\sigma_y = \bot$, since in either case $\sigma_y = \tau_y$. Thus we suppose that $v \neq nil$, so that $\sigma(t) = y(\sigma(v))$ and $\tau(t) = y(\tau(v))$. Now by definition of $B$-free, $x$ does not occur $\varnothing$-free in $v$, so by inductive hypothesis $\sigma(v) \sim_\varnothing \tau(v)$, which gives the result.

4.6. ORDERING SUBSTITUTIONS. Our elementary unification algorithm is concerned with the calculation of most general unifiers. To define relative generality of substitutions we introduce the following orderings.

4.6.1. *Definition.* For all substitutions $\sigma$ and $\tau$ and all binding states $B$, $\sigma \subseteq_B \tau$ if and only if there is a substitution $\mu$ such that $\mu \circ \sigma \sim_B \tau$. We abbreviate $\subseteq_\varnothing$ to $\subseteq$.

We show that the ordering $\subseteq_B$ is a preorder, but it is not a partial order. It is not antisymmetric. Furthermore the equivalence relation it generates is coarser than $\sim_B$; but we do not consider that equivalence relation.

LEMMA 4.14. *For all substitutions* $\sigma$, $\tau$ *and* $\rho$,

(i) $\sigma \subseteq_B \sigma$;

(ii) $\sigma \subseteq_B \tau$ and $\tau \subseteq_B \rho$ implies $\sigma \subseteq_B \rho$.

PROOF.

(i) $id \circ \sigma = \sigma$.

(ii) By hypothesis there are $\mu$, $\mu'$ such that $\mu \circ \sigma \sim_B \tau$ and $\mu' \circ \tau \sim_B \rho$. Thus

$$\begin{aligned}(\mu' \circ \mu) \circ \sigma &= \mu' \circ (\mu \circ \sigma)\\ &\sim_B \mu' \circ \tau, \text{ by Lemma 4.10}\\ &\sim_B \rho, \text{ by hypothesis.}\end{aligned}$$

The following two results are straightforward consequences of the definitions.

LEMMA 4.15. *If* $\sigma \sim_B \tau$ *and* $\tau \subseteq_B \rho$ *then* $\sigma \subseteq_B \rho$.

LEMMA 4.16. *If* $\sigma \sim_B \tau$ *and* $\rho \subseteq_B \tau$ *then* $\rho \subseteq_B \sigma$.

LEMMA 4.17. *For all substitutions* $\rho$, $\sigma$ *and* $\tau$, $\sigma \subseteq \tau$ *implies* $\sigma \circ \rho \subseteq \tau \circ \rho$.

PROOF. Immediate from the definition of $\subseteq$ and Lemma 4.10.

4.7. SUBSTITUTIONS AND BOUND VARIABLES. In this section we collect some lemmas which are used in the analysis of the unification occurs check.

LEMMA. 4.18. *If* $t$ *has an* $R$*-bound occurrence of* $x$ *then so does* $\sigma(t)$ *for every substitution* $\sigma$. The proof is a straightforward induction on the construction of $t$.

LEMMA 4.19. *If* $L \subseteq R$ *and* $t$ *has no* $R$*-free occurrence of* $x$ *then for every variable* $y$ *such that* $L(y) = R(y)$ *and* $y$ *has an* $R$*-bound occurrence in* $<u/x>t$, $y$ *has an* $L$*-bound occurrence in* $t$.

PROOF. By induction on the construction of $t$. Note that since $L \subseteq R$, then $\nu(R) = \nu(L, R)$.

(aa) $t = z(nil)$. Then $z \neq x$ and so $<u/x>t = t$. Hence the result is evident.

(ab) $t = z(v)$, $v \neq nil$. Then $<u/x>t = z(<u/x>v)$. Assuming this term has an $R$-bound occurrence of $y$, then $z = y$, so the result is evident.

(b) $t = f(t_1, \cdots, t_n)$. The inductive hypothesis applies straightforwardly to this case.

(c) $t = Qzv$, so that $<u/x>t = Qz<u/x>v$. Now $z$ does not occur $R$-bound in $t$, so by the hypothesis on $y$, $y \neq z$ and $y$ has an $[z]R$-bound occurrence in $<u/x>v$. Also, whether or not $z = x$, $v$ has no $[z]R$-free occurrence of $x$. Hence the inductive hypothesis applies to $[z]R$ and $v$, to show that $y$ has an $[z]L$-bound occurrence in $v$, and hence an $L$-bound occurrence in $t$, as required.

LEMMA 4.20. *If* $t$ *has no* $R$*-free occurrence of* $x$, *and if* $L(y) = R(y)$ *for all* $y$ *which have an* $R$*-bound occurrence in* $<u/x>t$, *then* $t \sim_{L,R} <u/x>t$.

PROOF. By induction on the construction of $t$.

(aa) $t = y(nil)$ where $R(y) \neq \bot$. Then $L(y) \neq \bot$ so the result is evident, whether or not $y = x$.

(ab) $t = x(nil)$, $R(x) = \bot$ is excluded, since $t$ has no $R$-free occurrence of $x$.

(ac) $t = y(v)$, $v \neq nil$, $R(y) = \bot$. By definition of $R$-free, $v$ has no $\emptyset$-free occurrence of $x$. Thus by inductive hypothesis, $v \sim_{\emptyset,\emptyset} <u/x>v$. Hence by definition of $\sim_{L,R}$ and of substitution, $t \sim_{L,R} <u/x>t$.

(b) $t = f(t_1, \cdots, t_n)$. This case follows straightforwardly from the inductive hypothesis.

(c) $t = Qzv$. Write $(L', R')$ for $[z](L, R)$. By definition of $R$-free, $v$ has no $R'$-free occurrence of $x$. Also, by definition of $R$-bound, $L'(y) = R'(y)$ for all $y$ which occur $R'$-bound in $<u/x>v$, since such a $y$ is either $z$ or is $R$-bound in $<u/x>t$. Thus in each of the cases $z = x$ and $z \neq x$ the inductive hypothesis applies to give $v \sim_{[z](L, R)} <u/x>v$. The result follows by definition of $\sim_{L,R}$.

COROLLARY 4.21. *If t has no R-free occurrence of x and no R-bound occurrence of any variable then* $t \sim_{\emptyset,R} <u/x>t$.

PROOF. Take $L = \emptyset$ in Lemma 4.20. To see that there are no $y$'s with an $R$-bound occurrence in $t$, take $L = R$ in Lemma 4.19.

COROLLARY 4.22. *If t has no L-free occurrence of x and no L-bound occurrence of any variable then* $t \sim_{L,\emptyset} <u/x>t$.

LEMMA 4.23. *If* $\mathrm{size}_{\emptyset}(u) > 1$, *if t has an R-free occurrence of x and if* $L(y) = R(y)$ *for all variables y which occur R-bound in t, then* $\mathrm{size}_R(<u/x>t) > \mathrm{size}_L(t)$.

PROOF. By induction on the construction of $t$. If $t = y(nil)$ then $R(y) \neq \bot$ and $y \neq x$ are excluded by the assumption that $t$ has an $R$-free occurrence of $x$. Thus $y = x$ and $R(x) = \bot$, and so $\mathrm{size}_R(<u/x>t) = \mathrm{size}_{\emptyset}(u)$ which is by assumption greater than $1 = \mathrm{size}_L(t)$, whether or not $L(x) = \bot$.

If $t = y(v)$, $v \neq nil$, then $R(y) = \bot$, since $t$ has an $R$-free occurrence of $x$. For the same reason $v$ has a $\emptyset$-free occurrence of $x$. Trivially, $\emptyset(y) = \emptyset(y)$ for all variables $y$. Thus the inductive hypothesis applies to $v$ to show that $\mathrm{size}_{\emptyset}(<u/x>v) > \mathrm{size}_{\emptyset}(v)$. The result follows from the definitions of $\mathrm{size}_L$ and $\mathrm{size}_R$.

The case when $t = f(t_1, \cdots, t_n)$ is a straightforward application of the inductive hypothesis and is omitted.

The case $t = Qzv$ requires an application of Lemma 3.19 but is otherwise straightforward and is also omitted.

COROLLARY 4.24. *If* $t \sim_{L,R} <u/x>t$ *and* $\mathrm{size}_{\emptyset}(u) > 1$ *then t has no R-free occurrence of x.*

PROOF. If $t$ does have an $R$-free occurrence of $x$ then $\mathrm{size}_R(<u/x>t) > \mathrm{size}_L(t)$, from Lemma 4.2. But then Lemma 3.18 contradicts $t \sim_{L,R} <u/x>t$.

## 5. *Unifiers and a Unification Algorithm*

5.1. UNIFIERS. The concept of unification for quantified terms has to recognize $\alpha$-equivalence. We also need to define generalized concepts of unifier corresponding to the generalizations of $\alpha$-equivalence of terms given previously. Further, we need to consider unification of multiple pairs of terms, where each pair may be relative to a different pair of binding states. The following definitions are used to manage this situation.

5.1.1. *Definition.* A unification item is a subscripted pair $(t, u)_{L,R}$, where $L$ and $R$ are binding states and $t$, $u$ are terms. For brevity we may refer to unification items simply as items.

5.1.2. *Definition.* A substitution $\sigma$ is a unifier of an item $(t, u)_{L,R}$ if and only if $\sigma(t) \sim_{L,R} \sigma(u)$. An item which has a unifier is called unifiable.

5.1.3. *Definition.* A unifier $\sigma$ of an item $i$ is a most general unifier of $i$ if and only if for all unifiers $\rho$ of $i$, $\sigma \subseteq \rho$.

We also make use of the extensions of these concepts to sequences of items.

5.1.4. *Definition.* A substitution $\sigma$ is a unifier of a sequence $T$ of items if and only if $\sigma$ is a unifier of each item in the sequence. Similarly, $\sigma$ is a most general unifier of $T$ if and only if for all unifiers $\rho$ of $T$, $\sigma \subseteq \rho$.

The following lemmas are used in the proof of correctness of our unification algorithm.

LEMMA 5.1. *If* $t = x(v)$, $u = y(w)$, $L(x) = \bot$ *and* $R(y) \neq \bot$ *then* $(t, u)_{L,R}$ *is not unifiable.*

PROOF. By induction on the construction of $t$. At each stage we suppose $\sigma(t) \sim_{L,R} \sigma(u)$, and derive a contradiction. To do that, write $\sigma(t) = x(v')$ and $\sigma(u) = y(w')$. Then $v' \neq nil$, and

$v' \sim_{\varnothing,R} \sigma(u)$, from the definition of $\sim_{L,R}$. The definition of $\sim_{\varnothing,R}$ then requires that $v' = z(v^*)$, where $v^* \neq nil$. Trivially $\varnothing(z) = \bot$. The inductive hypothesis then gives the desired contradiction.

LEMMA 5.2. *If $L(x) = \bot$ and $u = \downarrow_R u$, then $(x(nil), u)_{L,R}$ is unifiable if and only if either*

(i) *$u = x'(nil)$ and $R(x') = \bot$, or*

(ii) *u has no R-free occurrence of x and no R-bound occurrence of any variable,*

*and in that case a unifier is $<u/x>$.*

PROOF. First we note that each of (i) and (ii) implies that $<u/x>$ is a unifier. That is evident for (i). For (ii) we have to show that $x(u) \sim_{L,R} <u/x>u$. From the definition of $\sim_{L,R}$ it is enough to show that $u \sim_{\varnothing,R} <u/x>u$. That follows from Corollary 4.21.

We show the converse. Say $\sigma(x(nil)) \sim_{L,R} \sigma(u)$. If $\sigma_x = \bot$ then (i) holds, so we suppose $\sigma_x \neq \bot$. Writing $\sigma(x(nil)) = x(v)$, from Lemma 3.1 we have $v \sim_{\varnothing,R} \sigma(u)$. Now by Corollary 3.14, $\sigma(u)$ has no $R$-bound variable occurrences, so by Lemma 4.18, $u$ has no $R$-bound variable occurrences. We consider $size_R(u)$.

If $size_R(u) = 1$ then $u = y(w)$ where from above $R(y) = \bot$. As $u = \downarrow_R u$, then from Corollary 3.22, $w = nil$. Thus (i) is satisfied.

If $size_R(u) > 1$, then (ii) is satisfied, because the assumption that $u$ has an $R$-free occurrence of $x$ implies by Lemma 4.7 that $size_R(\sigma(u)) > size_\varnothing(\sigma_x)$, which contradicts Lemma 3.18. Also, Lemma 3.12 implies that for all variables $y$, $R(y) = \bot$ as required, since the assumption $R(y) \neq \bot$ implies an $L$-bound occurrence of some variable in $\sigma(x(nil))$, contradiction.

The next result is a characterization of unifiability which can be regarded as a basis for the derivation of our unification algorithm.

PROPOSITION 5.3. *An item $(t, u)_{L,R}$ is unifiable if and only if one of the following conditions holds.*

(a) $t = x(v)$, $u = y(w)$ and $L(x) = R(y) \neq \bot$.

(b) $t = x(v)$, $v \neq nil$, $L(x) = \bot$ and $(v, u)_{\varnothing,R}$ is unifiable.

(c) $u = y(w)$, $w \neq nil$, $R(y) = \bot$ and $(t, w)_{L,\varnothing}$ is unifiable.

(d) $t = x(nil)$, $t' = x'(nil)$ and $L(x) = \bot = R(x)$.

(e) $t = x(nil)$, $L(x) = \bot$ and $u$ has no $R$-free occurrence of $x$, and no $R$-bound occurrence of any variable.

(f) $u = y(nil)$, $L(y) = \bot$ and $t$ has no $L$-free occurrence of $y$, and no $L$-bound occurrence of any variable.

(g) $t = f(t_1, \cdots, t_n)$, $u = f(u_1, \cdots, u_n)$ and the sequence

$$(t_1, u_1)_{L,R} \cdot \cdots \cdot (t_n, u_n)_{L,R}$$

is unifiable.

(h) $t = Qxv$, $u = Qyw$ and, $(v, w)_{[x,y](L,R)}$ is unifiable.

PROOF. First we observe that if any of the conditions (a) to (h) holds then $(t, u)_{L,R}$ is unifiable. Suitable unifiers are:

(a) *id*.

(b) Any unifier of $(v, u)_{\varnothing,R}$.

(c) Any unifier of $(t, w)_{L,\varnothing}$.

(d) If $x' = x$ then *id*, else $<x'/x>$.

(e) $<u/x>$.

(f) $<t/y>$.

(g) Any unifier of $(t_1, u_1)_{L,R} \cdot \cdots \cdot (t_n, u_n)_{L,R}$.

(h) Any unifier of $(v, w)_{[x,y](L,R)}$.

That these substitutions are unifiers as required is checked straightforwardly from the definition of substitution and the definition of $\sim_{L,R}$, in all cases except (e) and (f). In case (e) it is enough to show that $u \sim_{\varnothing,R} \langle u/x\rangle u$, which follows from Corollary 4.21. Case (f) is symmetrical.

Now we argue the converse. We suppose that there is $\sigma$ such that $\sigma(t) \sim_{L,R} \sigma(u)$, and show that one of the conditions holds. We consider the cases in the definition of $\sim_{L,R}$. The following cases correspond to the cases in that definition.

(a) $\sigma(t) = x(v)$, $\sigma(u) = y(w)$ and $L(x) = R(y) \neq \bot$. Then by definition of substitution $t = x(v')$ and $u = y(w')$, so (a) applies.

(b) $\sigma(t) = x(v)$, $v \neq nil$, $L(x) = \bot$ and $v \sim_{\varnothing,R} \sigma(u)$. If $t = x(v')$, $v' \neq nil$ and $v = \sigma(v')$ then $(v', u)_{\varnothing,R}$ is unifiable and (b) applies. Otherwise $t = x(nil)$ and $\sigma(x(nil)) \sim_{L,R} \sigma(u)$. If $u \neq \downarrow_R u$ then case (c) applies. Hence we assume that $u = \downarrow_R u$. Thus Lemma 5.2 shows that (d) or (e) applies.

(c) This case is symmetrical to (b).

The remaining cases (d), (e) and (f) are evident.

LEMMA 5.4. *For every substitution $\sigma$ which unifies $(x(nil), u)_{L,R}$, where $L(x) = \bot$ and $u = \downarrow_R u$, there is a substitution $\tau$ such that $\sigma \sim_\alpha \tau \circ \langle u/x\rangle$.*

PROOF. Throughout this proof we abbreviate $z(nil)$ to $z$, for all variables $z$. If $u = y$ for some variable $y$ then $R(y) = \bot$. If $y = x$ then $\sigma \sim_\alpha \sigma \circ id \sim_\alpha \sigma \circ \langle x/x\rangle$ as required. If $y \neq x$ then as $\sigma$ unifies, $\sigma$ instantiates at least one of $x$, $y$. If $\sigma$ instantiates both, then as $\sigma_x \sim_\alpha \sigma_y$, Lemma 4.12 gives $\sigma \sim_\alpha \sigma \circ \langle y/x\rangle$ as required. If $\sigma$ instantiates $y$ but not $x$ then $\sigma_y \sim_\alpha x$, so

$$\begin{aligned}\sigma &\sim_\alpha \sigma \circ \langle x/x\rangle, \text{ by Lemma 4.12},\\ &\sim_\alpha \sigma \circ (\langle x/y\rangle \circ \langle y/x\rangle)\\ &\sim_\alpha (\sigma \circ \langle x/y\rangle) \circ \langle y/x\rangle\end{aligned}$$

as required. If $\sigma$ instantiates $x$ but not $y$, then $\sigma_x \sim_\alpha y$ so $\sigma \sim_\alpha \tau \circ \langle y/x\rangle$ by Lemma 4.8.

Finally, if $u$ is not a variable then $\sigma$ instantiates $x$ and $\sigma_x \sim_\alpha \sigma(u)$, so by Lemma 4.12, $\sigma \sim_\alpha \tau \circ \langle u/x\rangle$ as required.

5.2. A UNIFICATION ALGORITHM. Our algorithm is similar in structure to the unification algorithms for free-variable formulas which are used in conventional Prolog interpreters. In general it calculates unifiers of sequences of items.

For convenience in our discussion of the cases when the algorithm detects that unification is impossible, our algorithm outputs a sequence $o_1, \cdots, o_k$ of objects, each of which is either a substitution or *fail*. The intention is that if any *fail* occurs in an output sequence, the input is not unifiable. Otherwise, the composition $o_k \circ \cdots \circ o_1$ of the output sequence of substitutions is intended to be a most general unifier of the input sequence. Our algorithm can be described as follows.

5.2.1. *Definition.* The function unify, which maps sequences of items to sequences $o_1, \cdots, o_k$ as described above, is defined recursively as follows. [] denotes the empty sequence.

(a) unify([]) = $[id]$.

(b) Otherwise, if $t = x(v)$ and $u = y(w)$ and $L(x) = R(y) \neq \bot$, then

$$\text{unify}((t, u)_{L,R}.T) = \text{unify}(T).$$

(c) Otherwise, if $t = x(v)$, $v \neq nil$ and $L(x) = \bot$ then

$$\text{unify}((t, u)_{L,R}.T) = \text{unify}((v, u)_{\varnothing,R}.T).$$

(d) Otherwise, if $u = y(w)$, $w \neq nil$ and $R(y) = \bot$ then

$$\text{unify}((t, u)_{L,R}.T) = \text{unify}((t, w)_{L,\varnothing}.T).$$

(e) Otherwise, if $t = u = x(nil)$ and $L(x) = R(x) = \bot$ then

$$\text{unify}((t, u)_{L, R}.T) = \text{unify}(T).$$

(f) Otherwise, if $t = x(nil)$ and $L(x) = \perp$ and $u$ has no $R$-free occurrence of $x$ and no $R$-bound occurrence of any variable then

$$\text{unify}((t, u)_{L, R}.T) = <u/x>.\text{unify}(<u/x>T).$$

(g) Otherwise, if $u = y(nil)$ and $L(y) = \perp$ and $t$ has no $L$-free occurrence of $y$ and no $L$-bound occurrence of any variable then

$$\text{unify}((t, u)_{L, R}.T) = <t/y>.\text{unify}(<t/y>T).$$

(h) Otherwise, if $T = f(t_1, \cdots, t_n)$ and $u = f(u_1, \cdots, u_n)$ then

$$\text{unify}((t, u)_{L, R}.T) = \text{unify}((t_1, u_1)_{L, R}. \cdots .(t_n, u_n)_{L, R}.T)$$

(i) Otherwise, if $t = Qxv$ and $u = Qyw$ then

$$\text{unify}((t, u)_{L, R}.T) = \text{unify}((v, w)_{[x, y](L, R)}.T)$$

(j) Otherwise $\text{unify}((t, u)_{L, R}.T) = [fail]$.

6. *Correctness of the Unification Algorithm*

There are two main aspects of the correctness of the algorithm. Termination is the requirement that unify should be defined for all input sequences. Also the output should have the properties indicated in 5.2.

6.1. PROOF OF TERMINATION. Trivially, cases (a) to (j) of 5.2.1 cover all sequences of items. For each sequence $T$ of items, unify($T$) is defined either absolutely (case (a)) or in terms of unify($U$), where $U$ is strictly smaller than $T$ in the following well-ordering.

(i) $T > U$ if the number of variables which have free occurrences in $T$ is greater than for $U$. Here a free occurrence of a variable in an item $(t,u)_{L, R}$ means an $L$-free occurrence in $t$ or an $R$-free occurrence in $u$.

(ii) $T > U$ if the number of variables which have free occurrences is the same for $T$ as for $U$, but $T$ has a greater naive size than $U$, where naive size of $T$ means the total number of characters in all terms of all items of $T$.

That $T$ is strictly greater than $U$ in this ordering is evident in all cases except (f) and the symmetrical case (g). In these cases the result is a consequence of Lemma 4.3.

6.2. SPECIFYING CORRECTNESS. To prove the algorithm correct we need a precise version of the informal requirements stated in 5.2.

6.2.1. *Definition.* For each sequence $T$ of items, unify($T$) is correct if and only if both

(a) if *fail* occurs in unify($T$) then $T$ has no unifier.

(b) if *fail* does not occur in unify($T$) $= o_1, \cdots, o_k$, then $o_k \circ \cdots \circ o_1$ is a most general unifier of $T$.

We call unify correct if it is correct for all sequences of items $T$.

6.3. UNIFICATION EQUIVALENCE OF ITEM SEQUENCES. As already noted, in most cases the recursive definition of unify($T$) defines it to be the same as unify($U$) for some related sequence $U$. The following concept allows these cases to be treated uniformly.

6.3.1. *Definition.* Two sequences $T$ and $U$ of items are $V$-unification equivalent, where $V$ is a third sequence of items, if and only if for all substitutions $\sigma$ which unify $V$, $\sigma$ unifies $T$ if and only if $\sigma$ unifies $U$.

We shall abbreviate $V$-unification equivalence to $V$-equivalence.

We also call $T$ and $U$ unification equivalent, or briefly equivalent, if $T$ and $U$ are []-equivalent.

We also call items $i$, $j$ $V$-unification equivalent, or $V$-equivalent for short, if the sequences $[i]$ and $[j]$ are $V$-equivalent.

LEMMA 6.1. *If item sequences $T$ and $U$ are equivalent and* unify($T$) = unify($U$) *then* unify($T$) *is correct if and only if* unify($U$) *is correct.*

PROOF. It is enough to assume that unify($T$) is correct and to show that unify($U$) is correct.

If unify($T$) includes an occurrence of *fail*, then by the assumed correctness $T$ has no unifier. Thus by hypothesis $U$ has no unifier; and *fail* occurs in unify($U$) as required.

If however unify($T$) has no occurrence of *fail*, then by the assumed correctness the composition $\sigma$ of unify($T$) is a most general unifier of $T$. By the equivalence hypothesis then, $\sigma$ is a unifier of $U$. Also $\sigma$ is a most general unifier of $U$ because if $\rho$ is another unifier of $U$ then by equivalence $\rho$ is also a unifier of $T$, so that $\sigma \subseteq \rho$.

In the next result $T.V$ denotes concatenation of the sequences $T$ and $V$.

LEMMA 6.2. *For all items $i$ and $j$ and all item sequences $T$, $i.T$ is $V$-equivalent to $j.T$ if and only if $i$ is $T.V$-equivalent to $j$.*

PROOF. A straightforward application of the definition of $V$-equivalence. Details are omitted.

### 6.4. PROOF OF CORRECTNESS.

THEOREM 6.3. unify *is correct.*

PROOF. We show by induction following the definition, 5.2.1, of unify($T$) that unify($T$) is correct. The following enumeration of cases matches that of 5.2.1.

(a) Clause 6.2.1 (b) applies with $k = 1$. Vacuously, all items in $T$ are unified by every substitution, and *id* is a most general substitution.

(b) By Lemma 6.1 it is enough to show that $(t, u)_{L, R}.T$ and $T$ are equivalent. That is evidently true in this case since every substitution unifies $(t, u)_{L, R}$.

(c) By Lemma 6.1 and Lemma 6.2 it is enough to show that $(t, u)_{L, R}$ and $(v, u)_{\varnothing, R}$ are $T$-equivalent. That means, show for all unifiers $\sigma$ of $T$ that $\sigma(t) \sim_{L,R} \sigma(u)$ if and only if $\sigma(v) \sim_{\varnothing,R} \sigma(u)$. That is immediate from the definitions of substitution and $\sim_{L,R}$.

(d) Symmetrical to *c*.

(e) Similar to (b).

(f) Write $i = (x(nil), u)_{L, R}$. If *fail* occurs in $\tau$ = unify($<u/x>T$) then $i.T$ is not unifiable. For, from Lemma 5.4 any unifier $\sigma$ of $i.T$ would satisfy $\sigma \sim_\alpha \tau' \circ <u/x>$. Thus $\tau' \circ <u/x>$ would unify $T$ and so $\tau'$ would unify $<u/x>T$, contradicting the inductive hypothesis. If however *fail* does not occur in $\tau$ then we show as follows that $\sigma = \tau \circ <u/x>$ is a most general unifier of $i.T$.

First we show $\sigma$ is a unifier. Now $\tau$ unifies $<u/x>T$ by inductive hypothesis, so $\tau \circ <u/x>$ unifies $T$. Also, $<u/x>$ unifies $i$, by Lemma 5.2, so $\sigma = \tau \circ <u/x>$ unifies $i$ also, and so is a unifier of $i.T$ as required.

To see that $\sigma$ is a most general unifier of $i.T$, say that $\rho$ is an arbitrary unifier of this sequence. Then $\rho$ is a unifier of $i$ and so from Lemma 5.4, $\rho \sim_\alpha \pi \circ <u/x>$, say. Then, as $\rho$ unifies $T$, $\pi$ unifies $<u/x>T$, so that by inductive hypothesis $\tau \subseteq \pi$. Thus from Lemma 4.17, $\sigma = \tau \circ <u/x> \subseteq \pi \circ <u/x>$, so from Lemma 4.16 $\sigma \subseteq \rho$ as required.

(g) Symmetrical to (f).

(h) Similar to (c).

(i) Similar to (c).

(j) In all other cases $T = i.U$ where by Proposition 5.3, $i$ has no unifier; so *fail* is correct.

## 7. *Discussion*

The unification algorithm whose theory is presented in this paper is adequate to support extensions of Prolog which incorporate arbitrary quantified terms. If such extensions are contemplated then it would be natural to provide interpretations of the usual logical quantifiers. A goal *?– exists XA* would be evaluated to *A*. A goal *?– for-all XA* could be evaluated to *?– ¬ exists X¬A*. Here, '¬' could denote some logically sound version such as Mu-Prolog's safe negation. That provides dynamic interpretation for logical quantifiers comparable to the static compilation of logical quantifiers by preprocessing proposed by Lloyd and Topor [2].

The elementary unification algorithm we have discussed is not however adequate to deal with the full range of situations which can arise in the implementation of proof editors, for various reasons. For example, it may be required to unify a formula with a subformula of another formula, where the subformula occurs within the scope of quantifiers. the scope of quantifiers. We plan to discuss separately an extension of the algorithm to deal with this situation.

As another example, consider the problem of implementing the Prolog extension which simplifies queries of the form *?– for-all XA* to *?– ¬ exists X¬A*. It is natural to propose that this be done by a rule

$$\textit{for-all } XA :\!- \neg \textit{ exists } X\neg A.$$

where *A* is a variable. That will not work satisfactorily using the syntax and algorithm of this paper. For, the variable *A* cannot be instantiated to terms which include *X*'s within the scope of the outer quantification of *X*. Thus an attempt to unify, for example, *for-all XA* and *for-all Xa*(*X*) will fail.

The reason for this failure is that in this situation *A* is being used as a meta variable, for which a different concept of substitution is appropriate. We plan to discuss meta variables separately. Note however that the problem is not simply that our unification algorithm excludes bound variables from instantiations of *A* in its extended occurs check. The use of meta variables also raises deeper issues for unification algorithms, but their discussion is beyond the scope of this paper.

The unification algorithm described in this paper has been implemented as a modification to the York Portable Prolog Interpreter. The issues arising in a practical implementation are beyond the scope of this paper.

*Acknowledgements*. Thanks to Roger Duke, Ian Hayes and Richard Hagen for criticism of drafts of this paper. Thanks to Richard Hagen also for carrying out the implementation of the algorithm.

REFERENCES

1. DE BRUIJN, N.G., Lambda calculus notation with nameless dummies. *Nederl. Akad. Wetensch. Series A 75* (1972) 381-392.

2. LLOYD, J.W. AND TOPOR, R., Making Prolog more expressive. *Univ. Melb. Dept Comp. Sci. Tech. Report 84/8* (1984).

3. O'DONNELL, M.J., Equational logic as a programming language. M.I.T. Press, Cambridge, Mass., 1985.

4. STAPLES, JOHN, A new technique for analysing parameter passing, applied to the lambda calculus. *Austral. Comp. Sci. Commun. 3* (1981) 201-210.